Unification of the Fundamental Particle Interactions II

ETTORE MAJORANA INTERNATIONAL SCIENCE SERIES

Series Editor:

Antonino Zichichi

European Physical Society
Geneva, Switzerland

(PHYSICAL SCIENCES)

Recent volumes in the series:

Volume 5 **PROBING HADRONS WITH LEPTONS**
Edited by Giuliano Preparata and Jean-Jacques Aubert

Volume 6 **ENERGY FOR THE YEAR 2000**
Edited by Richard Wilson

Volume 7 **UNIFICATION OF THE FUNDAMENTAL PARTICLE INTERACTIONS I**
Edited by Sergio Ferrara, John Ellis, and Peter van Nieuwenhuizen

Volume 8 **CURRENT ISSUES IN QUANTUM LOGIC**
Edited by Enrico G. Beltrametti and Bas C. van Fraassen

Volume 9 **ENERGY DEMAND AND EFFICIENT USE**
Edited by Fernando Amman and Richard Wilson

Volume 10 **INTERACTING BOSE–FERMI SYSTEMS IN NUCLEI**
Edited by F. Iachello

Volume 11 **THE SOLUTION OF THE INVERSE PROBLEM IN GEOPHYSICAL INTERPRETATION**
Edited by R. Cassinis

Volume 12 **NEUTRINO PHYSICS AND ASTROPHYSICS**
Edited by Ettore Fiorini

Volume 13 **UNCONVENTIONAL APPROACHES TO FUSION**
Edited by B. Brunelli and G. G. Leotta

Volume 14 **MINIATURIZATION OF HIGH-ENERGY PHYSICS DETECTORS**
Edited by A. Stefanini

Volume 15 **UNIFICATION OF THE FUNDAMENTAL PARTICLE INTERACTIONS II**
Edited by John Ellis and Sergio Ferrara

Unification of the Fundamental Particle Interactions II

Edited by

John Ellis

and

Sergio Ferrara
CERN
Geneva, Zwitzerland

Plenum Press • New York and London

Library of Congress Cataloging in Publication Data

Europhysics Study Conference on Unification of the Fundamental Particle Interactions (2nd: 1981: Erice, Italy)
Unification of the fundamental particle interactions II.

(Ettore Majorana international science series. Physical sciences; v. 15)
"Proceedings of the Europhysics Study Conference held October 6–14, 1981, in Erice, Sicily, Italy"—P.
Includes bibliographical references and index.
1. Grand unified theories (Nuclear physics)—Congresses. 2. Supersymmetry—Congresses. 3. Supergravity—Congresses. I. Ellis, John, 1946– . II. Ferrara, S. III. Title. IV. Series.
QC794.6.G7E94 1983 539.7′54 82-18900
ISBN 978-1-4615-9301-0 ISBN 978-1-4615-9299-0 (eBook)
DOI 10.1007/ 978-1-4615-9299-0

Proceedings of the Europhysics Study Conference held October 6–14, 1981, in Erice, Sicily, Italy

Softcover reprint of the hardcover 1st edition 1983
A Division of Plenum Publishing Corporation
233 Spring Street, New York, N.Y. 10013

PREFACE

Work on the unification of the fundamental particle interactions has continued vigorously since the first Europhysics Study Conference on this subject. At that time we emphasized the existence of two main approaches, one based on supersymmetry and possibly its local version, supergravity, and the other approach based on grand unified gauge theories. Discussion of the possible tests of these theoretical speculations included experiments on baryon decay and neutrino oscillations. In view of the uncertainties surrounding the observability of such phenomena, the early Universe was welcomed as a possible Laboratory for testing new theoretical ideas. At that time, we expressed the hope that the different gauge and supersymmetry approaches would cross-fertilize each other, and it is appropriate to ask now how much of that hope has been realized.

We believe there has recently been considerable theoretical rapprochement, which is amply reflected in these Proceedings. On the one hand it has been realized that many of the technical problems in grand unified gauge theories, such as arranging the hierarchy of different mass scales, may be alleviated using simple global supersymmetry. On the other hand there has been growing interest in the possibility that extended supergravity theories may furnish a suitable framework for the unification of all the fundamental particle interactions. Many physicists in fact now question actively whether the known "fundamental" particles are indeed elementary, or whether they are composite. The hopes and frustrations of such approaches, both within and without the supergravity framework, are also a common theme in these Proceedings.

Many theories of compositeness seek to populate the "desert" of unoccupied energy scales depressingly beloved of grand unified theories. Thus they offer our experimental colleagues some relief from the rather sparse diet of tests suggested by other unified theories. Progress in verifying or refuting these predictions has been slow, with tantalizing indications emerging that baryons may indeed decay and neutrinos may indeed have masses, but with no confirmation yet of these results. It seems that many new experiments on these subjects will soon be operational, and our theoretical speculations confronted with reality.

In the meantime, it is striking how cosmology and the very early Universe have become parts of a physicist's everyday life. Cosmological constraints on particle theories are treated very respectfully, and new attempts made to solve outstanding cosmological problems by appeals to microphysics. Here again we sense more cross-fertilization since the first Europhysics Study Conference in this series.

We hope that the material in this second set of Proceedings will prove useful for future developments in the unification of the fundamental particle interactions. We are optimistic that the different theoretical approaches will become more closely intertwined, and hopeful that experimental confirmation of some of these theoretical ideas may soon be provided.

The preparation of these Proceedings would have been impossible without the efficient, cheerful and industrious assistance of Monica O'Halloran and Anne-Marie Perrin. We thank them very sincerely for devoting so much of their own time and energy to this volume, and hope that is it worthy of their efforts.

John Ellis and Sergio Ferrara

CONTENTS

GUTs VERSUS SUSY GUTs

D.V. Nanopoulos

CERN, Geneva, Switzerland

GUTs

Grand unified theories (GUTs), theories that unify weak, electromagnetic and strong interactions, seem to be a major part of the physics culture nowadays, and very justifiably so. They do not simply contain the successful electroweak and strong interactions (QCD) theories, but they lead to numerous qualitative and quantitative results, most of them impossible to be derived in the absence of grand unification[1].

GUTs explain naturally the charge quantization, but, at the same time, entail that α, the electromagnetic fine structure constant, should be constrained to[2]

$$1/170 \leqslant \alpha \leqslant 1/120 \qquad (1)$$

rather severe bounds, consistent with $\alpha \simeq 1/137$.

GUTs demand that at some superhigh energy limit all three interactions have more or less the same strength, or inversely, that at low energies the three interactions should have different strengths, as is observed experimentally. In such theories one finds that the electroweak mixing angle, (θ_{e-w}), as measured at present energies, is given by[1]

$$\sin^2\theta_{e-w}(M_W) \simeq 0.214 \pm 0.002 \qquad (2)$$

which compares most favourably with the radiatively corrected experimental average.

$$\sin^2\theta_{e-w}(M_W) = 0.215 \pm 0.012 \quad (3)$$

GUTs not only explain naturally the similar behaviour under electroweak interactions of quarks and leptons and their disparity under strong interactions, but entail quark-lepton mass relations like[3,4]

$$m_b/m_\tau|_{\text{low energies}} \simeq 2.85 \quad (4)$$

in full accordance with what is observed experimentally. Incidentally, the spectacular agreement between Eq. (4) and its experimental value entails that[4] there are at most six flavours and that[4] the top quark mass has an upper bound

$$m_t < 155 \text{ GeV} \quad (5)$$

Being on the mass front, it is worth recalling that in GUTs neutrinos are either massless or acquire a very tiny mass (< 100 eV), in accordance with terrestrial and cosmological observations.

Certainly the most dramatic consequence of GUTs has to do with matter instability. As is well known, there are grand unified interactions that violate baryon (B) number and lepton (L) number conservation thus making the proton unstable. Fortunately, the characteristic scale of these interactions is superhigh, given by[5]

$$M_X \simeq (1 \text{ to } 4)10^{14} \text{ GeV} \quad (6)$$

with $\alpha_G \simeq 1/41$ so that protons are sufficiently stable[5]

$$\tau_p \simeq (0.6 \text{ to } 25) \cdot (M_X/5.10^{14} \text{ GeV})^4 \; 10^{30} \text{ years} \quad (7)$$

to satisfy the present lower bounds on the proton lifetime[6]

$$\tau_p \gtrsim 0 \; (10^{30} \text{ years}) \quad (8)$$

It is rather remarkable how the whole thing fits together.

GUTs also play a very important rôle in the evolution of the very early Universe. They not only contain all the highly desirable ingredients necessary for creating a baryon asymmetry, like B violation, CP violation and C violation, but in conjunction with an expanding Universe, one finds[7]

$$n_B/n_\gamma \sim 10^{-6} \text{ to } 10^{-12} \quad (9)$$

certainly including the observed number

$$n_B/n_\gamma \sim 10^{-8\pm 2} \quad (10)$$

It seems that one of the most worrisome problems of the standard Big Bang cosmology has found its natural resolution in the framework of GUTs.

Despite their success, GUTs have some problems too, which we now turn to.

GUT PROBLEMS

Here are some "hot" problems:

i) The gauge hierarchy problem

In GUTs one has at least two stages of spontaneous symmetry breaking (SSB): one that provides the grand unification scale (M_X) and one that gives the electroweak unification scale (M_W). What seems very bizarre is the gross difference between the two scales

$$M_W/M_X \lesssim 10^{-12} \tag{11}$$

because such a disparity in scales is very difficult first to be produced and then to be maintained in conventional field theories. The reason is simple. In conventional field theories containing scalar particles the SSB occurs thanks to the Higgs mechanism, but then the scale of SSB is more or less the same (give or take an order of magnitude) to the mass of the Higgs particle. But, the mass of the Higgs particle gets radiative corrections which are quadratically divergent from diagrams evolving fermion, gauge boson or scalar loops (Fig. 1)

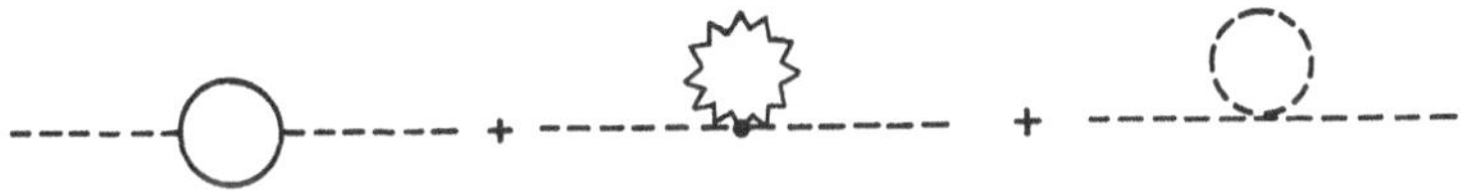

Figure 1

$$\delta m_H^2 \sim \Lambda^2 \sim M_X^2 \tag{12}$$

where, as indicated, the natural scale for the cut-off Λ is M_X. So, unless there is some magic reason, some kind of symmetry to make the sum of diagrams in Fig. 1 vanish, we are in trouble: $M_W/M_X \sim O(1)$ and not as given in Eq. (11). This is the heart of the gauge hierarchy problem. In conventional field theories there is no symmetry able to avoid the famous quadratic divergences of the scalar masses. Scale invariance is no good since it is broken in higher orders. To throw away altogether scalar particles in favour of techni- or hypercolour, looks like a hairy business[8] We should do better! Here it should be emphatically stressed that the gauge hierarchy problem does not belong only to GUTs, but also exists in the standard SU(3) × SU(2) × U(1) model. Indeed, even in the absence of grand unification, the mass of the Higgs boson that is responsible for the SU(2) × U(1) breaking is still quadratically divergent and the natural scale for the cut-off is now $M_{P\ell} \sim 10^{19}$ GeV, i.e.,

$$M_W/M_{P\ell} \sim 10^{-17} \tag{13}$$

even worse than before [cf. Eq. (11)].

ii) The family problem

This problem is self-explanatory. Why do we see this superfluous repetition of three families with identical SU(3) × SU(2) × U(1) quantum numbers but with such vastly different masses? How is it possible to explain the admittedly messy fermion mass spectrum and inter-family mixings?

iii) Gravity

How and where does gravity fit into the unification programme? After all, gravitational interactions seem to exist and play a rather important rôle in the physical world around us.

iv) Cosmological constant; monopoles

A common characteristic of the cosmological constant and magnetic monopoles is that both exist in GUTs, but in amounts catastrophically larger than what one sees experimentally. Namely, one finds

$$\Lambda_{\text{cosm.const.}}\Big|_{\text{GUTS}} \sim (M_X)^4 \tag{14}$$

to be compared with

$$\Lambda_{\text{cosm.const.}}\Big|_{\text{Experim.}} \sim (10^{-3}\ \text{eV})^4 \tag{15}$$

a difference of more than 100 (one hundred!) orders of magnitude and[9]

$$\frac{n_{\text{monop.}}}{n_\gamma}\Bigg|_{\text{GUTS}} \gtrsim 10^{-10} \tag{16}$$

to be compared with

$$\left.\frac{n_{monop.}}{n_{\gamma}}\right|_{Experim.} \leqq 10^{-25} \tag{17}$$

a difference of more than 15 orders of magnitude.

Certainly our GUTs need refinement and hopefully a supersolution to all our problems will emerge.

SUPERSYMMETRY (SUSY)

All the theories that we have considered up to now - electroweak, strong (QCD), GUTs - belong to the general class of renormalizable gauge field theories. A main characteristic of these theories is that they always transform fermions to fermions and bosons to bosons, acting, in general, on the internal quantum numbers carried by these particles. It is a legitimate question to ask whether there is some kind of theory which involves transformations changing fermions to bosons and vice versa. The answer is a big YES. These theories are called supersymmetric theories[10]. We thus have the following scheme:

$$\begin{array}{llll} \text{GAUGE SYMMETRY:} & \text{fermion} \leftrightarrow \text{fermion} & \Big\} & \text{change internal} \\ & \text{boson} \leftrightarrow \text{boson} & & \text{quantum numbers} \\ \text{SUPERSYMMETRY:} & \text{fermion} \leftrightarrow \text{boson} & & \text{change spin} \end{array} \tag{18}$$

The great hope is then that by building supersymmetric gauge theories, and exploiting the scheme of Eq. (18), we will finally have a realistic and possibly unique unification of all fundamental interactions of Nature. We will see that this hope is perhaps not that chimeric.

A unique and highly important feature of supersymmetric field theories is their mild and rather soft ultra-violet behaviour. These theories are indeed softening divergences. The root of this remarkable property lies, of course, in the existence of the high degree of fermion-boson symmetry. Indeed, in SUSY theories, there are the famous non-renormalization theorems[11], implying non-renormalization (finite or infinite) of certain parameters or of certain relations between parameters - something unimaginable in ordinary, conventional, standard field theories. I call this remarkable SUSY property the SET IT AND FORGET IT principle. Here lies the importance of SUSY theories. On the local front, supergravity[12] has much better ultra-violet behaviour than normal quantum gravity theories - it may be renormalizable or even finite. On the global front, N = 1 SUSY may solve[13] the notorious gauge hierarchy problem. What is happening is that in SUSY gauge theories the miracle occurs, and indeed the net result of adding up the diagrams of Fig. 1 is zero! Most remarkably, the diagrams add up to zero to any order in perturbation theory. The magic reason is called supersymmetry. Needless to stress that the above mechanism is correct as long as supersymmetry is an exact symmetry. In general, we would expect that if supersymmetry gets broken at some scale M_S, then Eq. (12) should be replaced by

$$\delta m_H^2 \sim M_S^2 \tag{19}$$

and since we would like to have $m_H^2 + \delta m_H^2 \lesssim O(M_W^2)$ in order to solve the gauge hierarchy problem it is rather obvious that we have to identify in general

$$M_S \sim O(M_W) \tag{20}$$

So, by looking for a solution to the gauge hierarchy problem, we not only make the physical significance of supersymmetry apparent, but we also determine the scale at which supersymmetry has to be broken. The answer to the question, "where are the SUSY partners (SO) of 'ordinary' particles (O)?" is then simple: in general

$$M_{SO}^2 - M_O^2 \simeq M_W^2 \tag{21}$$

i.e., they are heavy enough ($\sim M_W$) and so they have escaped objection, mainly for "energetic" reasons.

It seems that supersymmetry helps us on the superhigh energy front ($M_{P\ell} \sim 10^{19}$ GeV) by softening quantum gravity, as well as on the low energy front ($M_W \sim 100$ GeV) by solving the gauge hierarchy problem.

However, SUSY theories are not problem-free.

SUPER PROBLEMS

One of the basic problems of supersymmetry is its strong resistance against breaking. The reason being that in global supersymmetry the vacuum energy is positive or zero and that zero vacuum energy corresponds to a supersymmetric invariant vacuum[14]. In contrast with ordinary gauge theories, supersymmetric states are energetically favourable. So, if we want to spontaneously break supersymmetry, we had better exclude the supersymmetric states from our physically accessible ones. This turns out to be a formidable task.

It is not difficult to understand why, until today, we did not have[15] a successful low-energy supersymmetric model analogous to the standard SU(3) × SU(2) × U(1) model. There are many other problems that the SUSY model-builder has to face. By supersymmetrizing the low-energy world one necessarily buys charged and/or coloured scalar particles, which, if they do not behave properly, may blow up the whole theory. They can break charge or colour conservation by getting vacuum expectation values, or they may force protons to decay instantly ($\tau_p \sim$ few seconds!!!). Furthermore, they may severely damage all the successful low-energy phenomenology by upsetting g - 2, mediating flavour-changing neutral currents at intolerable rates[16], etc. The panacea seems to be the existence of an extra U(1)[15-17] broken at, or above, M_W. Again there are severe constraints on the nature of this new U(1)[15]. It should have mainly "axial" couplings to ordinary matter, it should be traceless, anomaly free (if it is broken around M_W) and naturally it should not spoil the successful neutral current phenomenology of the "standard" model. Again the task seems to be formidable, but not impossible. A lot of effort has been recently devoted to the above lines of thought[17,18].

Despite the present lack of a consistent low-energy SUSY model, some efforts have been made to construct supersymmetric grand unified models[19,13,18] which have some merit, and of which a discussion follows.

SUPERSYMMETRIC GUTS

The main reason for supersymmetrizing grand unified theories is of course the solution of the cumbersome gauge hierarchy problem. We have seen [see Eq. (20)] that a proliferation of the "low-energy" particle spectrum is then necessarily unavoidable. Every "known" particle, fermion, Higgs boson or gauge boson should have its corresponding superpartner with characteristic mass differences of order $O(M_W)$ [see Eq. (21)]. Additional problems to the ones discussed in the previous section appear. The new "low-energy" degrees of freedom will definitely modify the standard programme of grand unification and in general there is the danger that the whole programme will be mucked up. It is remarkable that in SUSY GUTs the standard success of ordinary GUTs remains more or less intact. So let us see how the unification programme changes. Our SUSY GUT should contain at least the supersymmetrized SU(3) × SU(2) × U(1) model. This piece of information is enough to give a kind of general analysis. It is clear from the beginning that the unification point is going to be raised. The new "light" degrees of freedom involve fermions and scalars, thus their contribution to the various β functions has the effect of delaying the change of the various coupling constants with energy. Notably, the strong coupling constant falls down with energy much smoother than before and so it will take "longer" for the different coupling constant to "meet". At the same time one expects a larger grand unification coupling constant. More precisely, in "minimal" type SUSY GUTs[19] one finds, for the coefficients of the SU(3), SU(2) and U(1) β-functions[20],

$$\beta_3 = 9 - f$$
$$\beta_2 = 6 - f - \frac{h}{2} \tag{22}$$
$$\beta_1 = -f - \frac{3h}{10}$$

where f represents the number of flavours (f $\gtrsim$ 6) and h stands for the number of "light" Higgs doublets (h $\gtrsim$ 2).

Concerning the coupling constants we get, using Eq. (22),

$$\frac{1}{\alpha_3(m)} = \frac{1}{\alpha_{SG}} - \frac{1}{2\pi}[9-f]\,\ell n\,[\frac{M_{SX}}{m}]$$
$$\frac{1}{\alpha_2(m)} = \frac{1}{\alpha_{SG}} - \frac{1}{2\pi}[6-f-\frac{h}{2}]\,\ell n\,[\frac{M_{SX}}{m}] \tag{23}$$
$$\frac{1}{\alpha_1(m)} = \frac{1}{\alpha_{SG}} - \frac{1}{2\pi}[-f-\frac{3h}{10}]\,\ell n\,[\frac{M_{SX}}{m}]$$

where as usual $\alpha_i \equiv \frac{g_i^2}{4\pi}(i=1,2,3)$, α_{SG} is the SUSY GUT unification fine structure constant and M_{SX} is the SUSY GUT unification mass and m a "low-energy" mass scale larger than or equal to $M_S \sim O(M_W)$ [see Eq. (20)]. We can recast Eqs. (23) in a more useful form[20]

$$\ell n(\frac{M_{SX}}{M_W}) = \frac{2\pi}{18+h}\left[\frac{1}{\alpha(M_W)} - \frac{8}{3}\frac{1}{\alpha_3(M_W)}\right] \tag{24}$$

$$\sin^2\theta_{e-w}(M_W) = \frac{(3+h/2)+(10-h/3)\alpha(M_W)/\alpha_3(M_W)}{18+h} \tag{25}$$

and

$$\frac{1}{\alpha_{SG}} = \frac{(9-f)1/\alpha(M_W)-(6-(8f/3)-h)1/\alpha_3(M_W)}{18+h} \tag{26}$$

where, for simplicity, we have identified the supersymmetry breaking scale (M_S) with M_W.

Using Eqs. (24) - (26), and taking into account higher order corrections, we get[21]

$$M_{SX} \simeq \begin{cases} 6.10^{16}\Lambda_{\overline{MS}} & \text{for } h = 2 \\ 3.10^{15}\Lambda_{\overline{MS}} & \text{for } h = 4 \end{cases} \tag{27}$$

where the present favourable value of $\Lambda_{\overline{MS}}$ (the QCD scale parameter evaluated in the modified minimal subtraction scheme with four flavours) is between 100 and 200 MeV. The electroweak angle is calculated to be[20,21]

$$\sin^2\theta_{e-w}(M_W) = \begin{matrix} 0.236\pm0.002 \text{ for } h = 2 \\ 0.259\pm0.002 \text{ for } h = 4 \end{matrix} \tag{28}$$

while $\alpha_{SG} \simeq 1/24$ to $1/25$ for six flavours and two light Higgs doublets.

We move next to the m_b/m_τ ratio in SUSY GUTs. Here we find[20,21]

$$\frac{(\frac{m_b}{m_\tau})_{SUSY}}{(\frac{m_b}{m_\tau})_{ORD}} = \left[\frac{\alpha_3(M_S)}{\alpha_{SG}}\right]^{8/9} \Big/ \left[\frac{\alpha_3(M_S)}{\alpha_G}\right]^{4/7} \tag{29}$$

and substituting $\alpha_{SG} \simeq 1/24$, $\alpha_G \simeq 1/41$ and $\alpha_3(M_S) \simeq 0.12$ we get[21]

$$\frac{(\frac{m_b}{m_\tau})_{SUSY}}{(\frac{m_b}{m_\tau})_{ORD}} = 1.0 \; ! \tag{30}$$

and thus by using Eq. (4), declaring that $(m_b/m_\tau)_{SUSY}$ is in full accordance with its experimental value. We find this "coincidence" remarkable. The situation is rather clear. As was expected the unification scale moves upward and the unification coupling constant increases as does the electroweak angle always compared to the ordinary GUTs results [cf. Eqs. (2) and (6)]. The m_b/m_τ remains unchanged, a surprise at least to me! Concerning the value of $\sin^2\theta_{e-w}$ it seems to be[21] a bit high for the case of two light Higgs doublets and certainly uncomfortably high for four light Higgs doublets compared with the experimental value [see Eq. (3)]. On the other hand, the increase of the unification scale by O(10) is very unfortunate for the proton decay experiments. Since conventionally [see Eq. (7)] $\tau_p \propto M_{SX}^4$, increasing M_{SX} by a factor of O(10) makes the proton lifetime longer by a factor of $O(10^4)$, not very pleasant news for experimentalists. It is rather unfortunate that the possibility of four light Higgs doublets, which make the unification scale smaller [see Eq. (27)], seems to be excluded experimentally [see Eq. (28)]. However, the show is not over! It has been remarked[17,22] that in a large class of SUSY grand unified theories, if there are no preventing symmetries, there are loop diagrams that may cause rapid proton decay. For example, by "dressing up" diagrams of the form

Figure 2

where s_f and $\tilde{H}_{SX}$ represent the SUSY partners of "light" fermions (f) and "superheavy" coloured Higgs triplets respectively, one may get "looping" proton decay

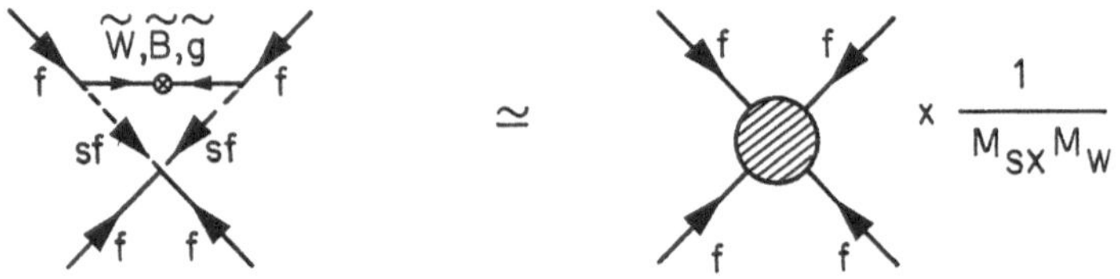

Figure 3

where again $\tilde{W}$, $\tilde{B}$, $\tilde{g}$ stand for the SUSY partners of the charged and neutral weak bosons and gluons respectively. The bizarre thing here is that $\tau_p \propto M_{SX}^2 M_S^2$ and not $\tau_p \propto M_{SX}^4$. One may then naively think that these kinds of SUSY theories are dead because they cause a too rapid proton decay[17,22]. A more careful analysis[21] showed though that things are different[18]. Indeed, we have found[21] that in such theories the proton lifetime can easily be 10^{31} years or a bit longer (not much longer though), and with the very "peculiar" characteristic decay mode[21] $\bar{\nu}_\tau K^+$. Thus, in the so-called softly broken SUSY GUTs[19] (without "preventing" symmetries) we find[21]

$$\tau_N \simeq O(10^{31\pm2}) \text{ years}$$

$$\Gamma(N \to \bar{\nu}_\tau K) >> \Gamma(N \to \bar{\nu}_\mu K) >> \Gamma(N \to \bar{\nu}\pi, \mu^+ K) >> \Gamma(N \to \mu^+\pi) >>$$
$$>> \Gamma(N \to e^+K) >> \Gamma(N \to e^+\pi) \tag{31}$$

But the surprises are not over. Very recently we have found[23] that SUSY GUT theories may solve naturally the monopole problem [see Eqs. (16),(17)]. In doing so though, we may upset the standard solution of the baryon asymmetry problem. One way to reconcile this puzzle and keep both solutions intact[23] is the existence of "light" superheavy Higgs triplets, i.e., $M_{H_3} \sim 10^{10}$ GeV *). But it is well known[24] that such Higgsons mediate proton decay with lifetime $\sim O(10^{31\pm2})$ years, and we found[23,25] that in SUSY GUTs the decay modes are given by[25]

$$\tau_N \simeq O(10^{31\pm2}) \text{ years}$$

$$\Gamma(\bar{\nu}_\mu K^+, \mu^+ K^o) : \Gamma(\bar{\nu}_e K^+, e^+ K^o, \mu^+ "\pi^o") : \Gamma(e^+\pi^o, \bar{\nu}_e \pi^+) \simeq$$

$$\simeq \quad 1 \quad : \quad \sin^2\theta_c \quad : \quad \sin^4\theta_c \tag{32}$$

$(\theta_c \cong$ Cabibbo angle)

All these predictions have to be contrasted with the ordinary GUT predictions[5] :

$$\tau_N \simeq 10^{31\pm2} \text{ years}$$

$$B(N \rightarrow e^+ \text{ non-strange}, \bar{\nu}_e \text{ non-strange}, \mu^+ \text{ or } \bar{\nu}_\mu \text{ strange}) :$$

$$B(N \rightarrow e^+ \text{ strange}, \bar{\nu}_e \text{ strange}, \mu^+ \text{ or } \bar{\nu}_\mu \text{ non-strange}) =$$

$$= 1 : \sin^2\theta_c \tag{33}$$

The contrast between Eqs. (31), (32) and (33) is rather dramatic. Apart from the case where the protons decay in the "conventional" way [Eq. (33)] but with $\tau_p \propto M_{SX}^4$ and M_{SX} as given by Eq. (27) (which will make life very, very difficult if not impossible), all other possibilities are very interesting and hopefully not impossible to test experimentally. We may know soon! I find it very remarkable that despite the proliferation of the "low energy" spectrum SUSY GUTs have succeeded in passing the tests of $\sin^2\theta_{e-w}$, m_b/m_τ, τ_p without much difficulty.

Still, SUSY GUTs have their own problems.

SUPERSYMMETRIC GUT PROBLEMS

Certainly the main advantage of SUSY GUTs is their capacity to provide a natural solution to the gauge hierarchy problem. It should be recalled that we would like to understand in a natural, satisfactory way:

---

*) Actually we find[25] in this case $\sin^2\theta_{e-w} \simeq 0.220$ much closer to the experimental value given by Eq. (2), than in other SUSY GUTs [see Eq. (28)].

1) why $(M_W/M_X) \lesssim 10^{-12}$?

2) how to separate, at the tree level, the masses of the Higgs doublet and its GUT partner, the coloured Higgs triplet ?

3) how to make the above mass-splitting between Higgs-doublet-triplet insensitive to radiative corrections ?

Clearly SUSY GUTs give a natural answer to problem 3) : why there are no radiative corrections in our mass splittings up to the level, of course, of SUSY breaking by the use of the SET IT AND FORGET IT principle. But it is fair to say that up to now we have no SUSY GUT that gives satisfactory answers to problems 1) and 2). It seems that problem 2) is not very difficult to solve but undoubtedly problem 1) seems quite hard. This problem is closely related to the problem of SUSY breaking which, as we saw, is hard to solve even prior to grand unification. There are some interesting ideas around involving non-perturbative effects as the agents of SUSY breaking[13], which seems to fit our purpose, since then one would naturally expect $M_W \sim M_{SX} \cdot e^{-(c/\alpha_{SG})}$. It remains to be seen if things will work out well. On the other hand, we stress again that one should watch out for the proliferation of "low-energy" degrees of freedom, since they have a tendency to raise the SUSY unification point. Hence our GUT programme will not make any sense, in its present form if $M_{SX} \sim M_{P\ell}$, because then gravitational interactions should not be neglected, on the contrary at energies around 10^{19} GeV they may be the dominant interactions. Anyway, what about gravitational interactions ? It seems more and more evident that the way to include gravitational interactions is through local supersymmetry, i.e., SUPERGRAVITY[12]. Supergravity seems not only able to offer a possibly consistent theory of quantum gravity BUT, in the context of N = 8 extended supergravity with a dynamically realized SU(8) [26], we may have for the first time a consistent unified theory of all interactions. Preliminary attempts[27] have shown that we are not that far from materializing the unification dream. A drawback to these earlier attempts[27] is the fact that supersymmetry is broken at the Planck scale (10^{19} GeV) thus making the solution of the gauge hierarchy problem with supersymmetry unbroken up to the Fermi scale (~100 GeV) impossible. Certainly the above mentioned attempts[27] need drastic modifications. Indeed we have suggested recently[28], always in the context of N = 8 extended supergravity with dynamically realized SU(8) gauge symmetry[26], a scenario where we obtain not only hierarchical gauge breaking but at the same time hierarchical supersymmetry breaking, thus having at our disposal an unbroken N = 1 global supersymmetry. The connection with SUSY GUTs then becomes apparent and as in the earlier attempts[27], the effective SUSY GUT Lagrangian is very much constrained in its particle content and the values of its parameters. An extra bonus of our approach[28] is a realistic materialization of an old idea[29] of how to avoid the cosmological constant in supergravity. It remains to be seen what other predictions (good or bad) such an approach may offer.

In conclusion, supersymmetry combined with gauge symmetry may provide a unified theory of all fundamental interactions which has the potential to solve problems like:

1) consistent theory of quantum gravity,

2) absence of cosmological constant,

3) gauge hierarchy,

4) family repetition,

5) ? ? ?

It seems clear that supersymmetry is here to stay. Experimentally a plethora of new particles are lurking around waiting to be discovered, at accessible (presently, or in the very near future) energies. What more can I say ?

JUST FIND THEM.

REFERENCES

1. For a review on GUTs see:
D.V. Nanopoulos, in Ecole d'Eté de Physique des Particules, Gif-sur-Yvette (1980) (IN2P3, Paris, 1980) p. 1;
P. Langacker, Phys. Rep. 72C:85 (1981).
2. J. Ellis and D.V. Nanopoulos, Nature 292:436 (1981).
3. A.J. Buras, J. Ellis, M.K. Gaillard and D.V. Nanopoulos, Nucl. Phys. B135:66 (1978).
4. D.V. Nanopoulos and D.A. Ross, Nucl. Phys. B157:273 (1979); Phys. Lett. 108B:351 (1982).
5. J. Ellis, M.K. Gaillard, D.V. Nanopoulos and S. Rudaz, Nucl. Phys. B176:61 (1980), and references therein.
6. J. Learned, F. Reines and A. Soni, Phys. Rev. Lett. 43:907 (1979);
M.L. Cherry et al., Phys. Rev. Lett. 47:1507 (1981);
M.R. Krishnaswamy et al., Phys. Lett. 106B:339 (1981).
7. For a review of the cosmological implications of GUTs, see:
D.V. Nanopoulos, in Progress in Particle and Nuclear Physics, Vol. 6:23 (1980).
8. See, for example:
E. Farhi and L. Susskind, Phys. Rep. 74C:277 (1981).
9. Ya.B. Zeldovich and M.Yu. Khlopov, Phys. Lett. 79B:239 (1978);
J.P. Preskill, Phys. Rev. Lett. 43:1365 (1979).
10. J. Wess and B. Zumino, Nucl. Phys. B70:39 (1974);
D. Volkov and V.P. Akulov, Phys. Lett. 46B:109 (1973);
Y.A. Gol'fand and E.P. Likhtman, Pisma Zh. Eksp. Teor. Fiz. 13:323 (1971).
11. J. Wess and B. Zumino, Phys. Lett. 49B:52 (1974);
J. Iliopoulos and B. Zumino, Nucl. Phys. B76:310 (1974);
S. Ferrara, J. Iliopoulos and B. Zumino, Nucl. Phys. B77:413 (1974);
S. Ferrara and O. Piguet, Nucl. Phys. B93:261 (1975);
M.T. Grisaru, W. Siegel and M. Roček, Nucl. Phys. B159:420 (1979).

12. S. Ferrara, P. van Nieuwenhuizen and D.Z. Freedman, Phys. Rev. D13:3214 (1976);
S. Deser and B. Zumino, Phys. Lett. 62B:335 (1976).
13. L. Maiani, in Proceedings of the Summer School of Gif-sur-Yvette (1979), p. 3;
E. Witten, Nucl. Phys. B188:513 (1981);
R.K. Kaul, Phys. Lett. 109B:19 (1982).
14. B. Zumino, Nucl. Phys. B89:535 (1975).
15. P. Fayet, in "Unification of the Fundamental Particle Interactions", Eds. S. Ferrara et al. (Plenum Press, N.Y. 1980), p. 587.
16. J. Ellis and D.V. Nanopoulos, Phys. Lett. 110B:44 (1982).
R. Barbieri and R. Gatto, Geneva University Preprint (1982).
17. S. Weinberg, Harvard Preprint HUTP-81/A047 (1981).
18. R. Barbieri, S. Ferrara and D.V. Nanopoulos, CERN Preprint TH. 3226 (1982) and unpublished.
19. S. Dimopoulos and H. Georgi, Nucl. Phys. B193:150 (1981);
N. Sakai, Z. für Phys. C11:153 (1981).
20. S. Dimopoulos, S. Raby and F. Wilczek, Phys. Rev. D24:1681 (1981);
L. Ibáñez and G.G. Ross, Phys. Lett. 105B:439 (1981);
M.B. Einhorn and D.R.T. Jones, University of Michigan Preprint UM-HE-81-55 (1981).
21. J. Ellis, D.V. Nanopoulos and S. Rudaz, CERN Preprint TH. 3199 (1981).
22. N. Sakai and T. Yanagida, Max Planck Institute Preprint MPI-PAE/PTh 55 (1981).
23. D.V. Nanopoulos and K. Tamvakis, CERN Preprint TH. 3227 (1982).
24. J. Ellis, M.K. Gaillard and D.V. Nanopoulos, Phys. Lett. 80B:360 (1979).
25. D.V. Nanopoulos and K. Tamvakis, CERN Preprint TH. 3247 (1982).
26. E. Cremmer and B. Julia, Nucl. Phys. B159:141 (1979).
27. J. Ellis, M.K. Gaillard and B. Zumino, Phys. Lett. 94B:343 (1980);
J.P. Derendinger, S. Ferrara and C.A. Savoy, Nucl. Phys. B188:77 (1981), and references therein.
28. R. Barbieri, S. Ferrara and D.V. Nanopoulos, Phys. Lett. 107B:275 (1981), unpublished, and work in progress.
29. S. Deser and B. Zumino, Phys. Lett. 62B:335 (1976).

RELATIONS BETWEEN GRAND UNIFIED AND MONOPOLE THEORIES

D.I. Olive

Blackett Laboratory
Imperial College
London SW7 2BZ. U.K.

INTRODUCTION

The theory of magnetic monopoles in spontaneously broken gauge theories is at an exciting but rather technical stage. Currently there is much progress in understanding static classical solutions describing one or more monopoles in arbitrary gauge groups(1). The outcome of these studies should shed light on duality conjectures which have been made concerning the possible quantum field theory of such soliton monopoles(2,3,4).

In this talk I propose not to talk about these developments, which are in a state of flux, but rather about two kinds of interrelationship between GUTs and monopole theories :

(a) one very speculative concerning how the duality conjectures could have a bearing on understanding GUTs and

(b) one relatively mundane as to how some of the mathematical technology used recently in monopole studies(4,5) can yield simple (Dynkin) diagrammatic rules for some of the common GUT group theory calculations. Thus I shall be interested in a cross-fertilization between the two subjects whose intimate relationship I explained in my previous Erice talk(6).

To an outsider the questions in GUTs are :-

i) which is the GUT group G and why?

ii) How is G broken and why?

iii) Into which representation of G is matter to be assigned and why?

First I explain the mathematical language needed to discuss these questions, then how monopole theory speculations may provide possible tentative answers which it should be the motivation of subsequent work to justify better, yet which compare reasonably with phenomenology. Finally I prove some theorems concerning special sorts of symmetry breaking and the decomposition of certain irreducible representations, showing how some calculations can be performed by simple graphical rules.

LIE ALGEBRA TECHNOLOGY

In order to discuss the above questions (i) - (iii) in a systematic way we need a compact notation for semisimple Lie algebras and their representations. This is supplied by the diagrams of Dynkin. The following brief summary is amplified in the famous lectures of Racah[7] and in several mathematical textbooks[8].

A basis of generators for the Lie algebra of G can be shown to consist of (r=rank G) mutually commuting (Cartan subalgebra) generators $H_1 \ldots H_r$, together with step operators $E_{\pm\alpha}$ associated with roots $\pm\alpha$

$$[\underline{H}, E_\alpha] = \underline{\alpha}\, E_\alpha \qquad (1)$$

Each root α can be expressed as a sum of r "simple roots" α_i, i=1....r.

$$\alpha = \sum_1^r n_i \alpha_i \qquad (2)$$

where the coefficients n_i are integers which are either all ≥ 0 or all ≤ 0. The system of simple roots can be reconstructed from the "Cartan matrix" of their scalar products in the sense:

$$K_{ij} = 2\alpha_i . \alpha_j / \alpha_j^2 \qquad i,j = 1....r \qquad (3)$$

This has integer entries with 2 on the diagonal and negative integers (≤ 0) off the diagonal. From K one can construct the Dynkin diagram D(G) which assigns points i to simple roots α_i with $K_{ij}K_{ji}$ lines joining points i and j. If $K_{ij}K_{ji}$ equals zero both factors vanish. If it is nonzero one factor at least equals -1. If it differs from 1 the roots α_i and α_j have different lengths and an arrow is added to signify which is shorter.

It is clear that given these properties, the Cartan matrix can be reconstructed from the Dynkin diagram, and from that the original Lie algebra (up to an isomorphism, by standard mathematical theorems). Thus all the information about the Lie algebra is encoded in the Dynkin diagram which, as we shall see, provides a

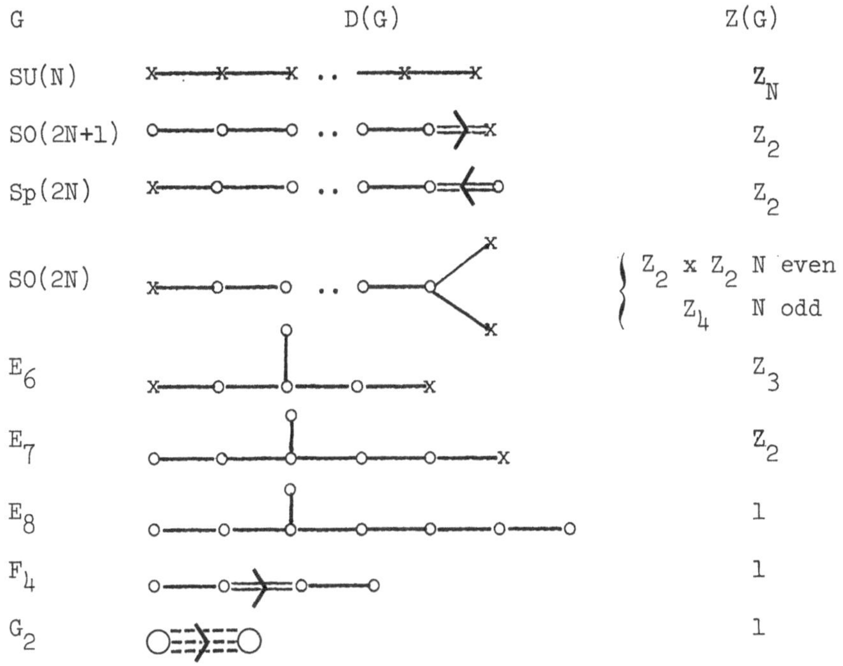

FIG. 1: The simple Lie algebras, their centres and their minimal weights.

very useful and convenient notation. The Dynkin diagrams for simple Lie algebras form connected tree diagrams and their catalogue is given below in Fig.1 together with the centre Z(G) of the universal covering group.

For physical applications we must understand the matrix representations. A state $|\lambda\rangle$ with "weight" $\underline{\lambda}$ in a representation satisfies

$$\underline{H}|\lambda\rangle = \underline{\lambda}\,|\lambda\rangle$$

The necessary and sufficient condition that λ be a weight of G is that

$$2\,\lambda.\alpha/\alpha^2 = \text{integer for each root } \alpha \text{ of } G \tag{4}$$

A linearly independent set of particular solutions to this condition is provided by the r "fundamental weights" $\lambda_1 \ldots \lambda_r$ defined by the orthogonality property with respect to the simple roots

$$2\lambda^i.\alpha^j/(\alpha_j)^2 = \delta^{ij} \tag{5}$$

Any weight can be written

$$\lambda = \sum_{i=1}^{r} m_i \lambda_i \qquad m_i = 0, \pm 1, \pm 2\ldots.. \tag{6}$$

and vice versa. It is a theorem that the irreducible representations of G correspond in a one-to-one way with the dominant weights, those for which $m_i \geq 0$. These positive integers can therefore be used to label the irreducible representations by annexing them to the Dynkin diagram. For example, if G = SU(2), $\underset{n}{x}$ means spin $n/_2$, while for G = SU(3):

$\underset{0}{x}$——$\underset{0}{x}$ = scalar, $\underset{1}{x}$——$\underset{0}{x}$ = triplet, $\underset{0}{x}$——$\underset{1}{x}$ = antitriplet, $\underset{1}{x}$——$\underset{1}{x}$ = octet

MINIMAL WEIGHTS AND REPRESENTATIONS

Because of a generalized Dirac quantization condition[8] it is known that magnetic monopoles can also be labelled by weights (actually of the dual group which has roots α/α^2 instead of α). Brandt and Neri[9] and Coleman [10] independently showed that the only stable monopoles corresponded to weights λ satisfying

$$2\lambda.\alpha/\alpha^2 = 0, \pm 1 \quad (\underline{\text{not}} \pm 2, \pm 3\ldots)\text{: } \alpha \text{ a root of G.} \tag{7}$$

Such weights are called "minimal". It is tempting to think of the weights corresponding to a monopole as defining a representation, and this is the content of one of the duality conjectures. The idea is that the quantum field operator of the monopole should transform with respect to the gauge group according to this representation[2]. These monopoles are massive and we are therefore lead to consider that matter transforms according to a minimal representation (one whose defining weight is minimal)[11].

It can be shown that the defining (dominant) weight of a minimal representation is a fundamental weight[4]. In the catalogue Fig. 1 of Dynkin diagrams the vertices corresponding to fundamental weights which are minimal are denoted by x's and the others by 0. In familiar terms the minimal representations of SU(N) correspond to antisymmetrical products of N's. For SO(2N+1) only the spinor representation is minimal. For SO(2N) both spinor representations and the defining representations are minimal. For the symplectic group Sp(2N) only the defining representation is minimal. Only E_7 and E_6 of the exceptional groups have minimal weights and will be discussed later.

As we shall see, it appears to be a phenomenological fact that matter does indeed favour such minimal representations. We wish to suggest that it is an important feature of monopole theory that it can potentially explain this fact for which we have seen no other argument.

To avoid confusion let us stress that we have used the words fundamental and minimal in a well-defined mathematical sense. Physicists have used the term "fundamental representation" in a very loose sense, roughly corresponding to minimal above.

Notice from Table 1 that if we count the scalar representation as minimal the number of minimal representations equals the number of elements of the centre of the universal covering group.

The weights of a given irreducible representation differ by sums of roots. The weights of G can be divided into equivalence classes, called cosets Λ/Λ_r, whose elements differ by sums of roots. It can be shown that the weights closest to the origin in each coset are minimal and that there are no other minimal weights[4].

Consider the reflection in a hyperplane perpendicular to a root α

$$x' = x - \left(\frac{2x.\alpha}{\alpha.\alpha}\right) \alpha \quad (8)$$

This is called a Weyl reflection. The combination of all these reflections yields a finite group, called the Weyl Group, which plays a fundamental role both in monopole theory and in representation theory. Because Weyl reflections correspond to gauge transformations, the weights of any representation split into orbits with respect to the Weyl group. Minimal representations are precisely those whose weights consist of just one such orbit

These facts will prove useful later.

SYMMETRY BREAKING

There seem to be two kinds of symmetry breaking in GUTs, at high and low mass scales. It seems that the validity of the monopole duality conjectures is most favoured when the Higgs field lies in the adjoint representation. Such a Higgs has several geometric advantages, having to do with extra dimensions and extended supersymmetry[12,13]. Apparently Nature chooses this mechanism for the high mass scale breaking whatever the group[11], and this is the mechanism we shall henceforth discuss. We shall say nothing about the low level breaking except to say that this Higgs appears to lie in a minimal representation.

Given G there are many possible exact symmetry groups H possible, depending on the G orbit of the adjoint Higgs ϕ . These are classified by gauge rotating the adjoint Higgs into the Cartan subalgebra, and indeed into the positive Weyl chamber (so that $\alpha_i\phi \geq 0$ for each simple root α_i). There is always an invariant U(1) factor in H generated by the charge Q in the direction of the asymptotic Higgs field[14]. Q is quantized if the remaining factor K is semisimple, and this seems desirable[5,6]. It can be proved[5] that the necessary and sufficient condition for K to be semisimple is that ϕ is gauge equivalent to the direction defined by a fundamental weight (eq. 5) which we shall call λ_ϕ. Then K is recognized from its Dynkin diagram D(K) which is obtained from D(G) by deleting the vertex corresponding to λ_ϕ and its links.

The U(1) charge (in the direction of the Higgs field) is

$$Q = e\hbar\phi.T/a \qquad \phi^2 = a^2 \tag{9}$$

Hence the gauge particles, corresponding to weights of G which are roots α, carry U(1) charges $q = e\,\hbar\lambda_\phi.\alpha/\sqrt{(\lambda_\phi)^2}$. By equations (2) and (5) q equals $n_\phi q_o$ where n_ϕ is the (integer) coefficient of the simple root α_ϕ in equation (2) and q_o is the U(1) charge of the α_ϕ gauge particle. By the Higgs-Kibble-Brout-Englert mechanism[15] the mass of the gauge particle is

$$m = a|q| = a|n_\phi q_o|$$

So gauge particles with $|n_\phi| > 1$ may dissociate into $|n_\phi|$ gauge particles with mass $a|q_o|$ while conserving Q[16]. This sort of instability may be undesirable and would be prevented if n_ϕ could only take the values 0, ±1. From equation (7) we see that $\lambda_\phi^V = \lambda_\phi/\alpha_\phi^2$ must then be a minimal weight[4] of the dual group, G^V, of G with roots α/α^2 instead of α. $D(G^V)$ and D(G) in Fig. 1 are related by reversing any arrows. Thus we see a second role for minimal fundamental weights : as well as defining minimal representations into which matter may be placed, they also define a special direction for the adjoint Higgs field yielding only one non-zero mass value for the gauge particles. Further limitations on K (and even G) result if we make similar demands on the soliton monopoles, conjectured to be gauge particles of G^V[3,4]. Then λ_ϕ itself must be minimal. That λ_ϕ and λ_ϕ^V be both minimal implies $D(G) = D(G^V)$ and so has only single links.

These arguments have been progressively weaker but do lead us to consider an interesting class of theories: SU(n+m) broken to SU(n)xSU(m)xU(1)/Z, SO(2N) broken to SU(N)xU(1)/Z or SO(2(N-1)xU(1)/Z, E_7 and E_6 broken to E_6xU(1)/Z and SO(10)xU(1)/Z respectively, with no other possibilities if G is simple. (Z is a cyclic subgroup of centre of K.) The suspicion is that these symmetry breakings are most favourable to the duality conjecture[4] (when extended supersymmetry is added).

In figure 2 appears the famous list of exceptional groups E_8, E_7, E_6, E_5 = SO(10), E_4 = SU(5), E_3 = SU(3)xSU(2), with minimal fundamental weights again denoted by a cross and now labelled by the dimensionality of the corresponding minimal representations. We have just seen that, for $3 \leq n \leq 6$ E_{n+1} can be broken down to E_nxU(1)/Z by the adjoint Higgs along a fundamental minimal weight of E_{n+1}.

This is because the right hand vertex of the Dynkin diagram D(En+1) is minimal for $3 \leq n \leq 6$ and its deletion leads to $D(E_n)$ and exposes a new minimal fundamental weight on the right.

The reader will notice that this list includes the most popular candidates for GUT groups[17] and that the minimal representations are the most popular assignments for matter. (The suffix n of E_n measures both the rank and the degree of implausibility.) What I claim to be new is the realization that there is a special role played by the minimal fundamental weights, both in defining matter multiplets and symmetry breaking. Monopole theory has focussed attention on the minimal weights and there are hints that further elucidation (possibly dependent on current developments) may fill out the picture.

Let us suppose E_n to be simply connected. Then its centre $Z(E_n)$ is given in the third column of Fig. 2 and always consists of a cyclic group, that with 9-n elements, denoted Z_{9-n}, so

$$|Z(E_n)| = 9-n \tag{10}$$

Let us comment on the global structure of the unbroken subgroup

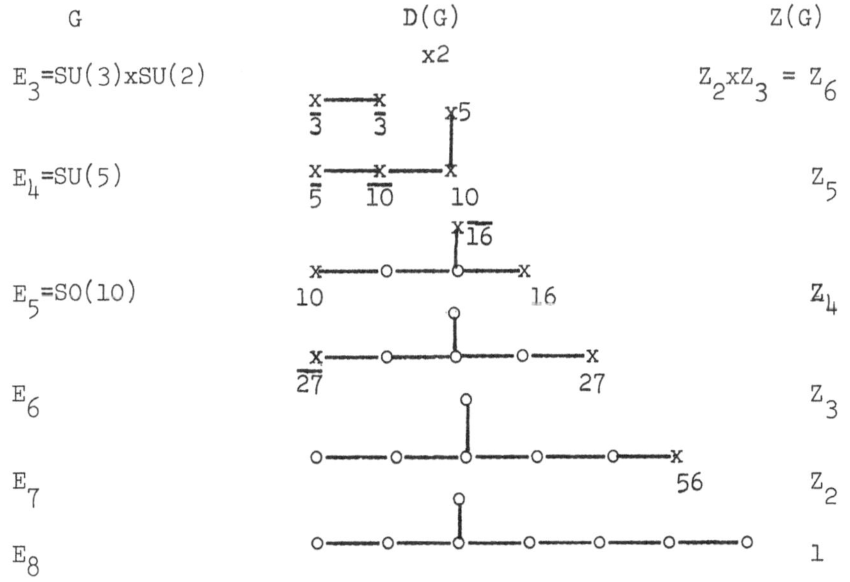

FIG. 2: The "exceptional" groups E_n with their Dynkin diagrams D and centre Z.

$E_n xU(1)/Z \subset E_{n+1}$ The general prescription for calculating this is known[5] and states that E_n is to be simply connected if E_{n+1} is. We said that the minimal weights correspond to points of the centre of E_{n+1}. In each case $3 \leq n \leq 6, \lambda_\phi$ actually corresponds to a generator of Z_{8-n}, the centre of E_{n+1}. By the general theory[5], Z the cyclic group of elements common to E_n and U(1) must then actually coincide with the centre of E_n, Z_{9-n}. Hence the exact symmetry subgroup of E_{n+1} is globally $E_n xU(1)/Z_{9-n}$. This was known for n=3 ($SU(5) \supset SU(3)xSU(2) \times U(1)/Z_6$)[18] and is now seen to apply in fact to all $7 \geq n \geq 3$.

It is intriguing that the same sequence of groups has occurred in the dimensional reduction of D = 11 supergravity. There the non-compact forms of these groups appear[19].

Progress[20] appears to be being made with some of the problems involved in schemes based on Fig.2: how to ballast unwanted representations, how to incorporate supersymmetry, and how to interpret the U(1) factors. (If Y_n is the U(1)) charge in $E_{n+1} \supset E_n xU(1)/Z_{9-n}$; Y_3 is weak hypercharge, Y_3, Y_4 and B-L are linearly related, while Y_5 appears to be a Fayet-Iliopoulos charge[21]).

RESULTS CONCERNING MINIMAL REPRESENTATIONS AND ADJOINT HIGGS BREAKING

Let us now put aside the speculative ideas which have led us to consider gauge groups G broken down to U(1)xK/Z (with K semi-simple) by an adjoint representation (AR) Higgs, which is therefore gauge equivalent to a fundamental weight, λ_ϕ. We shall now consider this system in its own right and ask how the representations of G will look when decomposed into irreducible representations of U(1)xK. We shall prove two theorems

1) Minimal representations (MR's) of G decompose into a sum of MR's of K with distinct U(1) charges.

2) If G has roots of equal length, the adjoint representation (AR) of G decomposes into the AR of U(1)xK (with zero U(1) charge) plus MR's of K with distinct, non-zero U(1) charges.

These theorems show that the AR, which is the most geometrical, automatically gives rise to MR's which themselves spawn MR's. Starting from AR's or MR's no other types of representation arise in this class of symmetry breaking. This could be an alternative explanation of their relevance in physics.

These theorems are well illustrated by the explicit results tabulated in Fig. 3 for $E_{n+1} \supset E_n xU(1)/Z_{9-n}$, n=7,....3.

Using ideas from recent papers on monopole theory[4,5] we shall prove the two theorems and then exhibit Dynkin diagrammatic

$E_8 \supset U(1) \times E_7/Z_2$

$$248 = (0, 1 \oplus 133) \oplus (\tfrac{1}{2}, 56) \oplus (-\tfrac{1}{2}, 56) \oplus (1,1) \oplus (-1,1)$$

$E_7 \supset U(1) \times E_6/Z_3$

$$133 = (0, 1 \oplus 78) \oplus (-\tfrac{2}{3}, 27) \oplus (\tfrac{2}{3}, \overline{27})$$
$$56 = (1,1) \oplus (\tfrac{1}{3}, 27) \oplus (-\tfrac{1}{3}, \overline{27}) \oplus (-1,1)$$

$E_6 \supset U(1) \times SO(10)/Z_4$

$$78 = (0, 1 \oplus 45) \oplus (-\tfrac{3}{4}, 16) \oplus (\tfrac{3}{4}, \overline{16})$$
$$27 = (1,1) \oplus (\tfrac{1}{4}, 16) \oplus (-\tfrac{1}{2}, 10)$$
$$\overline{27} = (-1,1) \oplus (-\tfrac{1}{4}, \overline{16}) \oplus (\tfrac{1}{2}, 10)$$

$SO(10) \supset U(1) \times SU(5)/Z_5$

$$45 = (0, 1 \oplus 24) \oplus (-\tfrac{4}{5}, 10) \oplus (\tfrac{4}{5}, \overline{10})$$
$$16 = (1,1) \oplus (\tfrac{1}{5}, 10) \oplus (-\tfrac{3}{5}, \bar{5})$$
$$\overline{16} = (-1,1) \oplus (-\tfrac{1}{5}, \overline{10}) \oplus (\tfrac{3}{5}, 5)$$
$$10 = (\tfrac{2}{5}, \bar{5}) \oplus (-\tfrac{2}{5}, 5)$$

$SU(5) \supset (1) \times SU(3) \times SU(2)/Z_6$

$$24 = (0,1,1) \oplus (0,8,1) \oplus (0,1,3) \oplus (\tfrac{5}{6}, \bar{3}, 2) \oplus (-\tfrac{5}{6}, 3, 2)$$
$$10 = (1,1,1) \oplus (\tfrac{1}{6}, 3, 2) \oplus (-\tfrac{2}{3}, \bar{3}, 1)$$
$$5 = (\tfrac{1}{2}, 1, 2) \oplus (-\tfrac{1}{3}, 3, 1)$$
$$\bar{5} = (\tfrac{1}{3}, \bar{3}, 1) \oplus (-\tfrac{1}{2}, 1, 2)$$
$$\overline{10} = (\tfrac{2}{3}, 3, 1) \oplus (-\tfrac{1}{6}, \bar{3}, 2) \oplus (-1,1,1)$$

Fig. 3. Decomposition of AR's and MR's

rules for calculating the decompositions in Fig.3 in a systematic way.

Let μ be a weight of the irreducible representation of G considered and decompose it into components parallel and orthogonal to λ_ϕ (the fundamental weight parallel to the Higgs field)

$$\mu = x(\mu)\lambda_\phi + \nu(\mu) \tag{11}$$

$x(\mu)$ is the U(1) charge of μ, normalized so that $x(\lambda_\phi)=1$. $\nu(\mu)$ is a weight of K, since if α is a root of K it is also a root of G and orthogonal to λ_ϕ, so that

$$2\nu(\mu).\alpha/\alpha^2 = 2\mu.\alpha/\alpha^2$$

Since the right hand side is integral, by equation(4), so is the left hand side and the result follows. If μ is minimal the right hand side equals 0, $\pm$ 1 and hence $\nu(\mu)$ is a minimal weight of K. Different μ's (in the same G irreducible representation) differ by sums of G roots, as we remarked earlier, which for given x are sums of K roots. So all $\nu(\mu)$ for a given x lie in the same coset Λ/Λ_r and are therefore weights of a single MR of K as claimed.

This proves the first theorem. The second follows similarly and is proven in(4).

Now we show how to use the idea of this proof to calculate the decompositions in Fig. 3 from the Dynkin diagrams in Fig.2.

To perform the decomposition we need expressions for the weights μ at each U(1) charge $x(\mu)$. These can be obtained from the defining (dominant) weight by judiciously chosen elements of the Weyl group (see equation(8)). Let λ_i denote a fundamental weight of D(G) and Λ_i the corresponding fundamental weight of D(K). Then, in the language of equation (11)

$$\nu(\lambda_i)=\Lambda_i \qquad \nu(\lambda_\phi) = 0 \tag{12}$$

The analysis is facilitated if λ_ϕ is the minimal weight defining the MR in question. Its Weyl reflection in α_ϕ is

$$\sigma_\phi(\lambda_\phi)=\lambda_\phi-\alpha_\phi = \lambda_\phi(1-x(\alpha_\phi)) + \sum\Lambda_{\phi+1} \tag{13}$$

by a result proven by Goddard and Olive(4,5). $\Sigma\,\Lambda_{\phi+1}$ denotes the sum of fundamental weights of D(K) corresponding to points linked directly to λ_ϕ in D(G). Goddard and Olive(5) also proved that

$$x(\alpha_\phi) = \frac{|\,Z(G)\,|}{|\,Z(K)\,|} \tag{14}$$

Hence when $G=E_{n+1}$, $K = E_n$, we have, by equation (10)

$$x(\alpha_\phi) = \frac{8-n}{9-n} \tag{15}$$

Another useful element of the Weyl group, used in monopole theory[22], is that unique one which turns each positive root into a negative one. For G we shall call this σ_0 and for K, Σ_0 . For all "real" simple Lie groups $\sigma_0 = -1$. The simple Lie groups which are not real are "complex" and are recognized from Table one by the fact that the centre, in the third column, possesses a complex element ($e^{2\pi i/k}, k \geq 3$). The complex groups are SU(N), ($N \geq 3$), SO(4N+2) and E_6 and it is seen from table 1 that their Dynkin diagrams admit a Z_2 symmetry interchanging simple roots α_i with α_j. For these groups

$$\sigma_0(\alpha_i) = -\alpha_j \quad \text{and} \quad \sigma_0(\lambda_i) = -\lambda_j \tag{16}$$

This means that the irreducible representations defined by λ_i and λ_j are conjugate.

These results provide enough information to perform all the calculations in Fig. 3. Let us illustrate by $E_6 \supset U(1) x E_5/Z_4 = U(1) x SO(10)/Z_4$. We have by equations (12), (13) and (16) and reference to $D(E_6)$ and $D(E_5)$ in Fig. 2 :

$$\nu(\lambda_{27}) = 0, \qquad \nu(\lambda_{27} - \alpha_{27}) = \Lambda_{16}$$

$$\nu(\sigma_0(\lambda_{27})) = \nu(-\lambda_{\overline{27}}) = -\Lambda_{10} = \Sigma_0(\Lambda_{10})$$

Hence the 27 of E_6 contains the 1, 16 and 10 of E_5. By dimension counting this is all. Hence

$$27 = (1,1) \oplus (\tfrac{1}{4},16) \oplus (-\tfrac{1}{2},10)$$

The U(1) charge starts at 1 as $x(\lambda_\phi) = 1$ and descends in units of $x(\alpha_\phi)$, which is $3/4$ by equation (15).

All the other decompositions in Fig. 3 for the MR's of G corresponding to λ_ϕ follow similarly, producing three MR's of K, one of which is scalar, one corresponding to the fundamental weight of K next to λ_ϕ in D(G) and finally one which is obtained from λ_ϕ by the application of σ_0 and Σ_0 via the automorphisms of D(G) and D(K), equation (16).

The decomposition of the remaining MR's follow similarly e.g. the 10 of G = SO(10) yields the 5 of K=SU(5) by equation (12). The $\bar{5}$ follows from $\nu(\sigma_0(\lambda_{10})) = \nu(-\lambda_{10}) = -\Lambda_5 = \Sigma_0(\Lambda_5)$. By dimension counting this is all. The U(1) charges of the $\bar{5}$ and 5 differ by $x(\alpha_\phi) = 4/5$ by equation (15). Their absolute value is deduced from the fact that the U(1) charge being a generator of G is traceless next to λ_ϕ in D(G) in any G representation, in particular the 10. Our method only yields 3 or 4 terms in the decomposition but that turns out to be sufficient.

For the decomposition of the AR's the zero U(1) charge x contribution is the AR of U(1) x K by theorem 2. The roots $\pm\alpha\phi$ furnish weights of the AR with U(1) charges given by equation (15). By equation (13) the MR of K corresponding to $-\alpha\phi$ is given by the fundamental weights of K corresponding to the vertices of D(G) linked to the λ_ϕ vertex. The MR of K conjugate to this is obtained by equation (16) and corresponds to $\alpha\phi$. When $\lambda\phi$ is minimal e.g. for $E_{n+1}\supset U(1) \times E_n/Z_{9-n}$, n=3...6, the x=0,$\pm$ x($\alpha\phi$) contributions are the only ones as we proved earlier. For $E_8\supset U(1)\times E_7/Z_2$, λ_ϕ cannot be minimal and there must be extra contributions. By dimension counting these are K scalars (which count as MR's) and must have $x = \pm 2x(\alpha_\phi) = \pm 1$ by equation (15).

Our point here is a pedagogical one: using concepts like that of the Weyl group (which is fundamental in monopole theory) it is possible to perform some practical calculations in a simple and unified way working directly with Dynkin diagrams, which encode the structure of the group.

I am grateful for discussions with P. Goddard and S. Rajpoot.

REFERENCES

1. R. Ward, Comm. Math. Phys. 79:317 (1981); Phys. Lett. 102B:136 (1981);
P. Forgács, Z. Horvath and L. Palla, Phys. Lett. 99B:232 (1981); Phys. Lett. 102B:131 (1981).
See also:
M.K. Prasad and P. Rossi, Phys. Rev. Lett. 46:806 (1981);
E. Corrigan and P. Goddard, Comm. Math. Phys. 80:575 (1981);
A.N. Leznov and M.V. Saveliev, Lett. Math. Phys. 3:489 (1979); Comm. Math. Phys. 74:111 (1980);
D. Olive, Imperial College preprint ICTP/80/81-1, to be published in Proceedings of the International Summer Institute on Theoretical Physics organized by Wuppertal University at Bad Honnef (Plenum Press, N.Y.);
N. Ganoulis, P. Goddard and D. Olive, Imperial College preprint ICTP/81/82-4 (1981);
W. Nahm, CERN preprint TH.3172 (1981);
N.J. Hitchin, Oxford preprint "Monopoles and Geodesics" (1981).
For background reviews see:
P. Goddard and D. Olive, Rep. Prog. Phys. 41:1357 (1978);
D. Olive, "Self-dual Monopoles", Bechyne lecture, Imperial College preprint ICTP/80/81-41 (1981);

2. P. Goddard, J. Nuyts and D. Olive, Nucl. Phys. B125:1 (1977).
3. C. Montonen and D. Olive, Phys. Lett. 72B:117 (1977);
F.A. Bais, Phys. Rev. D18:1206 (1978).
4. P. Goddard and D. Olive, Nucl. Phys. B191:528 (1981).
5. P. Goddard and D. Olive, Nucl. Phys. B191:511 (1981).
6. D. Olive, in "Unification of the fundamental particle interactions", eds. S. Ferrara, J. Ellis and P. van Nieuwenhuizen (Plenum Press, N.Y., 1980), p. 451.
7. G. Racah, "Lectures on Lie groups", CERN yellow report 61-8 and in "Group theoretical concepts and methods in elementary particle physics" (Gordon and Breach, N.Y., 1964), pp. 1 - 36.
8. J. Humphreys, "Introduction to Lie algebras and representation theory (Springer Verlag, Berlin , 1972).
9. F. Englert and P. Windey, Phys. Rev. D14:2728 (1976);
P. Goddard, J. Nuyts and D. Olive, Nucl. Phys. B125:1 (1975).
10. R. Brandt and F. Neri, Nucl. Phys. B161:253 (1979);
S. Coleman, Seminars given at CERN, unpublished (1979).
11. D. Olive, in "Mathematical Problems in Theorétical Physics", ed. K. Osterwalder, lecture notes in Physics 116 (Springer Verlag, Berlin, 1980), p. 249.
12. A. d'Adda, R. Horsley and P. Di Vecchia, Phys. Lett. 76B:298 (1978);
E. Witten and D. Olive, Phys. Lett. 78B:97 (1978).
13. D. Olive, Nucl. Phys. B153:1 (1979);
H. Osborn, Phys. Lett. 83B:321 (1979);
W. Nahm, Seminar at Imperial College, Spring 1979;
P. Rossi, Phys. Lett. 99B:229 (1981).
14. E. Corrigan, D. Olive, D.B. Fairlie and J. Nuyts, Nucl. Phys. B106:475 (1976).
15. P.W. Higgs, Phys. Rev. Lett. 12:132 (1964); Phys. Rev. Lett.13:508 (1964); Phys. Rev. 145:1156 (1966);
T.W.B. Kibble, Phys. Rev. 155:1557 (1967);
F. Englert and R. Brout, Phys. Rev. Lett. 13:321 (1964).
16. F.A. Bais, Phys. Rev. Lett. D18:1206 (1978);
E. Weinberg, Nucl. Phys. B167:500 (1980).
17. E_3: S. Weinberg, Phys. Rev. Lett. 19:1264 (1967);
A. Salam, Proc. 8th Nobel Symposium, ed. N. Svartholm (Wiley, N.Y., 1968).
E_4: H. Georgi and S.L. Glashow, Phys. Rev. Lett. 32:438 (1974).
E_5: H. Fritzsch and P. Minkowski, Ann. Phys. 93:193 (1975).
E_6: F. Gursey, P. Ramond and P. Sikivie, Phys. Lett. 60B:177 (1976);
Y. Achiman and B. Stech, Phys. Lett. 77B:389 (1978).
E_7: P. Sikivie and F. Gursey, Phys. Rev. D16:816 (1977).
E_8: N.S. Baaklini, Phys. Lett. 91B:376 (1980).
18. D.M. Scott, Nucl. Phys. B171:95 (1980);
M. Daniel, G. Lazarides and Q. Shafi, Nucl. Phys. B170 [FS1]:156 (1980);
C.P. Dokos and T.N. Tomaras, Phys. Rev. D21:2940 (1980).
19. R. Gilmore, "Lie groups, Lie algebras and some of their applications" (Wiley, N.Y., 1974), p. 314;
B. Julia, in "Superspace and Supergravity", eds. S. Hawking and M. Rocek (Cambridge University Press, 1981), p. 331 and references therein.

20. D.V. Nanopoulos, talk at this meeting.
21. P. Fayet and J. Iliopoulos, Phys. Lett. 31B:461 (1974).
22. N. Ganoulis, P. Goddard and D. Olive, Imperial College preprint ICTP/81/82-4 (1981).

KALUZA-KLEIN-KOUNTERTERMS

M.J. Duff *
CERN, Geneva, Switzerland

and

D.J. Toms
Blackett Laboratory, Imperial College, London

1. INTRODUCTION

The major obstacle in constructing a consistent quantum theory of gravitation is that Einstein's general theory of relativity seems, by conventional criteria, to be non-renormalizable and hence incapable of unambiguous prediction at the quantum level. Many ingenious suggestions have been made to circumvent this impasse but none with complete success[1]. In this paper, we wish to explore another possibility: to examine the problem within the framework of higher dimensional theories of the Kaluza-Klein[2] type.

It may seem paradoxical to attempt to improve the ultra-violet behaviour of a theory by increasing the dimension of space-time, but not if we are looking for finiteness as opposed to power-counting renormalizability. Moreover, the chances of finiteness seem to be better if the theory is supersymmetric and/or the dimension of space-time is odd. Indeed, the work presented here was largely inspired by Witten's[3] suggestion of obtaining an SU(3) × × SU(2) × U(1) gauge theory by applying Kaluza-Klein ideas to eleven-dimensional[4] supergravity.

*Permanent address: Blackett Laboratory, Imperial College, London.

Kaluza-Klein theories

In its modern form, a Kaluza-Klein theory works as follows. One starts with gravity, described by the Einstein Lagrangian, coupled to matter fields of lower spin in d space-time dimensions, where $d > 4$. Ideally, the resulting field equations should then admit of a stable ground-state solution for which the extra dimensions are compactified. The ground-state metric would then describe a product* manifold $M_1 \times M_2$, where M_1 is four-dimensional space-time with the usual signature (possibly Minkowski space, but more generally allowing a non-vanishing cosmological constant) and M_2 is a compact "internal" space of Euclidean signature. One then performs a Fourier expansion of all the fields, including the gravitational fluctuations about the ground state, with a suitable set of harmonics on M_2 and expansion coefficients depending only on the four co-ordinates of M_1. Integration over the extra co-ordinates then yields the field equations of an effective four-dimensional theory on M_1 describing a finite set of massless states, including the graviton, together with an infinite tower of massive states corresponding to the normal modes on M_2. The masses will be quantized in units of m, where m^{-1} is the typical "size" of M_2, and all states have spin $s \leq 2$. Thus these extra dimensions need not conflict with one's everyday sensation of inhabiting a four-dimensional world (with its inverse square law of gravitational attraction) provided the massive states are sufficiently heavy, i.e., provided the compactified dimensions are sufficiently small.

On the microscopic scale, we believe that the strong, weak and electromagnetic interactions are described by a Yang-Mills gauge theory. It is necessary therefore that M_2 be either a group space G or else a coset space G/H. In this case the spectrum of four-dimensional states will include, in addition to the massless graviton, massless Yang-Mills fields with gauge group G. Indeed, the whole beauty of Kaluza-Klein theories is that it is not necessary to postulate the separate existence of Yang-Mills fields; they are automatically begotten by gravity. Since G is presumed to contain $SU(3) \times SU(2) \times U(1)$, the number of extra dimensions must be enough to accommodate this**. Requiring that the corresponding gauge coupling constants be of order unity, or thereabouts, fixes

*We have in mind a global direct product, but it would be interesting to examine the consequences of a ground state which consists of a non-trivial fibre bundle. See Ref. 5. I am grateful to C.J. Isham for drawing my attention to this possibility.

**The most economical way to obtain the symmetry group G is to take $M_2 = G/H$ where H is a maximal subgroup of G. In any event, $d_2 = \dim M_2 = \dim G - \dim H$. Moreover, dim G cannot exceed that of a space of maximal symmetry, i.e., $\dim G \leq d_2(d_2+1)/2$. Hence $\dim H \leq d_2(d_2-1)/2$. I am grateful to D. Olive for discussions on this point.

the extra dimensions to be of the Planck size, i.e., $m^{-1} \sim 10^{-33}$cm. This nicely meets our requirement that they be small. In consequence, the massive states have masses in excess of 10^{19} GeV and cannot be identified with any known particles. In this approximation, all observed particles are massless. In a realistic theory, one's initial choice of spinor fields in d dimensions would be such as to yield massless spin-1/2 fermions in the right representations of SU(3) × SU(2) × U(1) to be identified with the known quarks and leptons. Their physical masses, and those of the W-bosons will then have to be provided by some other mechanism, perhaps by a Higgs-Kibble mechanism since in general scalar fields will also be fathered by gravity.

As far as low energy phenomenology is concerned, therefore, one can discard these massive Kaluza-Klein states and focus one's attention on the massless sector. <u>However, the inclusion of an infinite tower of massive states will drastically affect the ultraviolet behaviour of the theory</u>*. This is the subject of the present paper.

[To avoid confusion, one should contrast this picture of "spontaneous compactification"[6] with that of "dimensional reduction"[7]: another device which also appears in modern treatments of higher dimensional theories (like the bosonic string in d = 26 or the fermionic string in d = 10). In its original form, dimensional reduction simply consists of taking all fields to be independent of the extra co-ordinates. In the above Kaluza-Klein picture, this corresponds to a <u>flat</u> ground-state $M_1 \times M_2$, where M_1 is Minkowski space, and M_2 is <u>an</u> n-torus, (i.e., $S^1 \times S^1 \times \ldots S^1$) whose radii are then shrunk to zero to eliminate the massive states. An interesting modification, permissible when there is an internal symmetry, consists of taking the fields to be "twisted", i.e., not strictly periodic but only periodic up to a phase[8]. The special dependence on the extra co-ordinates then gives rise to symmetry breaking and mass generation. In neither form of dimensional reduction need one ascribe any physical significance to the extra dimensions, in sharp contrast with the spontaneous compactification described above.]

Supergravity

The Kaluza-Klein idea may, in principle, be applied to any theory of gravity plus matter in any dimension greater than four**.

---

*This infinite tower of massive states would presumably also have cosmological consequences.

**It has also been applied to theories with $s \leq 1$, but this seems the worst of all possible worlds. One acquires all the disadvantages of non-renormalizability without its only advantage: gravity.

Unfortunately, no-one yet knows whether there exists a choice of matter coupling and choice of dimension which fulfils all the requirements of a realistic theory listed above*. However, it is particularly compelling to apply the idea to supergravity, for the following reasons:

1) Supergravity severely restricts both the choice of matter couplings and the choice of dimension. Such a restriction must be a feature of any successful attempt at Kaluza-Klein unification. Otherwise, the mystery of the correct matter fields in four dimensions would simply be swapped for the same mystery in d dimensions, with the extra headache of which d to pick. In supergravity the maximum dimension permitting a consistent** ($s \leq 2$) theory is eleven[13]. In this $d = 11$ theory, all fields are gauge fields: the gravitational elfbein $e_\mu{}^a(x)$, the rank-three antisymmetric tensor $A_{\mu\nu\rho}(x)$, and a single Rarita-Schwinger Majorana spinor $\psi_\mu{}^\alpha(x)$ (hence $N = 1$ supersymmetry).

2) In Kaluza-Klein, the mass matrix for the fermion fields on M_1 is given by the fermion operator (Dirac or Rarita-Schwinger) on M_2. The number of massless fermions is therefore determined by the number of zero-eigenvalue modes of this fermion operator on the coset space G/H. With the exception of the n-torus (which featured in the dimensional reduction discussed earlier), all such spaces have positive scalar curvature (at least with their standard metric) and hence by a theorem due to Lichnerowicz[14] there are no zero-modes of the Dirac operator. One may circumvent this theorem by including Yang-Mills gauge fields in the higher dimension*** but this would be contrary to the spirit of Kaluza-Klein whereby the Yang-Mills fields emerge automatically. Fortunately, Lichnerowicz's theorem does not apply to the Rarita-Schwinger operator****. Thus to

---

* In particular, the stability of the Kaluza-Klein vacuum is an especially knotty problem. See Ref. 9 for a treatment of this problem in the case of M_1 = Minkowski space and $M_2 = S^1$.

** The problem of finding a consistent coupling of spin $\geq 5/2$ to gravity is qualitatively different from 3/2, and probably has no solution. See Refs. 10, 11 and 12.

*** After all, we know that the Dirac operator on S^4 has zero modes in the presence of Yang-Mills instantons.

**** Zero modes of the Rarita-Schwinger operator on compact spaces of Euclidean signature have been treated before in the context of index theorems and axial anomalies. See Refs. 15, 16, 17 and 10.

explain spin-1/2 quarks and leptons in four dimensions via Kaluza-Klein, spin-3/2 fields seem almost* compulsory. But the only known consistent spin-3/2 theory is supergravity !

3) Most higher dimensional theories do not exhibit spontaneous compactification**. In d = 11 supergravity, not only does spontaneous compactification work but the number of space-time dimensions is naturally singled out as four[20]! This is because a rank-three gauge field gives rise to a cosmological constant[21,22] and because its field strength $F_{\mu\nu\rho\sigma}$ has four indices. The field equation for F

$$d{*}F = F \wedge F \tag{1.1}$$

admits the solution (c = constant)

$$\begin{aligned} F_{\mu\nu\rho\sigma} &= c\ \varepsilon_{\mu\nu\rho\sigma}\ , \quad \mu,\nu,\rho,\sigma = 1,2,3,4 \\ &= 0 \quad\quad\ \ , \quad \text{otherwise} \end{aligned} \tag{1.2}$$

The resulting Einstein's equations then admit a ground state metric of the product form $M_1 \times M_2$, where M_1 is a four dimensional Einstein

* The absence of Dirac zero-modes on G/H has some fascinating exceptions involving non-standard metrics. It is not difficult to prove that on Einstein spaces ($R_{ij} = \Lambda\ \gamma_{ij}$) Killing vectors exist only when $R \geq 0$ and hence, by Lichnerowicz, there are no Dirac zero modes. However, if one is willing to relax the Einstein condition then spaces with negative R can admit both Killing vectors and Dirac zero modes. A classic example discussed by Hitchin in Ref. 18 is the three sphere S^3. Relative to the standard SU(2) × SU(2) invariant metric, it has positive R and no Dirac zero modes. However, there exists a one-parameter (λ) non-standard metric with SU(2) × U(1) symmetry and negative R which not only admits Dirac zero modes but for which the number of such modes depends on the value of λ. Counting them turns out to be a non-trivial exercise in number theory. (I am grateful for discussions on these points with R. Coquereaux.) Within a Kaluza-Klein framework, this could (?) yield a unified electroweak theory for which the number of quarks and leptons depended, via the prime numbers, on the value of θ_W.

** The introduction of Yang-Mills gauge fields sometimes helps but, as mentioned before, this is contrary to the spirit of Kaluza-Klein. Another possibility is to invoke higher derivatives, see Ref. 19.

space ($R_{\mu\nu} = \Lambda\, g_{\mu\nu}$) with Minkowski signature and $\Lambda = -4c^2 < 0$ *, while M_2 is a seven-dimensional Einstein space with Euclidean signature and $\Lambda = 2c^2 > 0$ and hence compact.

4) Seven is not merely the <u>maximum</u> dimension of $M_2 = G/H$ permitted by supersymmetry, but also the <u>minimum</u> dimension that will accommodate $G \supset SU(3) \times SU(2) \times U(1)$. This remarkable numerical "coincidence" was first observed by Witten[3], who showed that there are infinitely many seven-dimensional spaces with the right symmetry. A well-known example would be $CP^2 \times S^2 \times S^1$, though this example illustrates two potential problems. First, CP^2 does not admit of a spin structure, i.e., fermions cannot be globally defined. It does admit of a generalized ($spin^c$) structure[26], but it is unclear whether this is of any help in a Kaluza-Klein theory. Fortunately, there are other spaces with the right symmetry which do have a spin structure. Secondly, this way of obtaining the U(1) factor requires that one of the extra dimensions, the S^1 circle, be flat. It is notoriously difficult to find solutions of the field equations with some dimensions curved and some flat**. For example, the seven-dimensional Einstein space of positive curvature discussed in (3) above would allow S^7, $S^5 \times S^2$, $S^4 \times S^3$, $CP^2 \times S^3$ but not $CP^2 \times S^2 \times S^1$ or $S^6 \times S^1$.***. (The product of two Einstein spaces is also an

---

* The appearance at the classical level of an enormous cosmological constant for space-time M_1 is a general feature of Kaluza-Klein theories whenever M_2 is non-flat, i.e., whenever G is non-Abelian. Supergravity does not appear to solve this problem unless one can argue that the ground state need not satisfy the classical field equations. A related problem which supergravity might cure concerns the enormous Casimir vacuum energies induced by compactified dimensions of the Planck scale. See Ref. 23. One might anticipate that the fermion and boson contributions to these vacuum energies will cancel. See Refs. 24 and 25.

** For the same reason that a ground state M_1 = Minkowski space is difficult to achieve when M_2 is not flat.

*** Another possibility would be to start not with N = 1 supergravity in d = 11, but with N = 2 supergravity in d = 10, which already has a U(1) symmetry. (This theory has the distinction of being the zero-slope limit of the fermionic dual-string model.) One would then require compactification of the extra six dimensions, all of which may now be curved, to yield $SU(3) \times SU(2)$. This is possible since the $A_{\mu\nu\rho}$ field is also present in this theory. However, the fermions are neutral under this U(1). Moreover, in contrast to N = 1 in d = 11, this theory is not finite even at one loop. See Section 5.

Einstein space, provided they have the same Λ.) $S^5 \times S^2$ has the properties of admitting a spin structure, solving the field equations and accommodating the right symmetries. It would yield an SU(4) × SU(2) Yang-Mills theory. (If we are forced by the field equations to give S^5 and S^2 the same Λ, this Yang-Mills theory would have only one independent coupling constant.) On the other hand, Witten[3] has argued that the ground state need not necessarily be a solution of the field equations.

5) Finally, and this is the problem that concerns us most in this paper, any Kaluza-Klein theory must face up to the problem of non-renormalizability. The superficial degree of divergence of a Feynman graph in a d dimensional Einstein theory is given by[1]

$$D = (d-2)\,L + 2 \tag{1.3}$$

where L is the number of loops. Now although D is greater for $d = 11$ than $d = 4$, this is not so significant since all $d > 2$ theories are anyway power-counting non-renormalizable. An entirely different possibility is that the theory is finite owing to a mutual cancellation of ultra-violet divergences. This is known to happen to at least two loop order in the $N = 8$ supergravity in $d = 4$ obtained by dimensional reduction from $N = 1$ in $d = 11$ (and also in the $N < 8$ theories obtained by truncation from $N = 8$). However, if we are to take seriously the extra dimensions, the relevant question is not "Is $N = 8$ supergravity finite in $d = 4$?", but rather "Is $N = 1$ supergravity finite in $d = 11$?". Since the counterterm structure of simple and extended supergravity is still not well understood, finiteness in $d = 11$ seems a priori just as likely as in $d = 4$. The chances of a non-supersymmetric theory being finite, on the other hand, seem very remote whatever the dimension.

It is amusing to note from (1.3) that $d = 11$ supergravity is automatically finite at odd-loop order because there are no local invariants with an odd number of derivatives which could serve as counterterms*. Consequently, one expects that the effective four-dimensional theory will also be finite at odd loops provided one retains the infinite tower of massive states: something which would not necessarily be true if these states are discarded as in dimensional reduction.

* We are assuming a regularization scheme with a dimensionless regularizing parameter like dimensional or ζ function regularization, as opposed to Pauli-Villars for example. It should be added, however, that even in this case, matter fields could spoil the odd-loop finiteness of pure gravity. Fortunately, this does not happen in supergravity. See Section 5.

Whether or not a Kaluza-Klein theory is supersymmetric, it seems bizarre that an infinite number of states can yield a finite result, when a finite number of states leads to infinities. Equally strange is the situation where a non-vanishing counterterm in the higher dimensional theory survives as a counterterm in the effective four-dimensional theory. The puzzle is that it will have more derivatives than four-dimensional power-counting would allow. Similar puzzles occur for the axial and trace anomalies.

In this paper we attempt to clarify these issues. Although motivated by the possibility of a physically realistic Kaluza-Klein theory, we shall not invoke the full apparatus of eleven dimensional supergravity but content ourselves with some simple toy models. These examples, albeit unrealistic, will serve to illustrate some of the apparent paradoxes which arise when Kaluza-Klein theories are viewed first from the higher, and then from the lower dimensional standpoints.

Higher dimensional picture versus lower dimensional picture

In order to formulate some of these ultra-violet puzzles more precisely, consider the action for a single scalar field Φ on a five-dimensional manifold M(5) with metric $g_{\hat{\mu}\hat{\nu}}$:

$$S = \frac{1}{2}\int d^5x \sqrt{\det g_{\hat{\mu}\hat{\nu}}}\, g^{\hat{\mu}\hat{\nu}}\, \partial_{\hat{\mu}}\Phi\, \partial_{\hat{\nu}}\Phi \tag{1.4}$$

Suppose that one of the dimensions is compactified so that

$$M(5) = M(4) \times S^1 \tag{1.5}$$

where M(4) is unspecified. Let us write the co-ordinates

$$x^{\hat{\mu}} = (x^{\mu}, y) \tag{1.6}$$

where y is the S^1 co-ordinate

$$0 \leq y \leq \frac{2\pi}{m} \tag{1.7}$$

and m is the inverse radius of the circle. Since, in this simple example, the gravitational field acts as an external source with no dynamics of its own, we are free for illustrative purposes to choose the simple form

$$g_{\hat{\mu}\hat{\nu}}(x,y) = \begin{bmatrix} g_{\mu\nu}(x) & 0 \\ 0 & 1 \end{bmatrix} \tag{1.8}$$

Next, consider the Fourier expansion of the real scalar field

$$\Phi(x,y) = \sqrt{\frac{m}{2\pi}} \sum_{n=-\infty}^{\infty} \phi_n(x) \exp inmy \tag{1.9}$$

[Note that the expansion coefficients $\phi_n(x) = \phi^*_{-n}(x)$ have the correct canonical dimensions of a Bose field in four dimensions. In fact the canonical commutation relations for $\phi_n(x)$ follow from those of $\Phi(x,y)$ after integrating over y.] Substituting (1.8) and (1.9) into (1.4), and performing the y integration, one finds

$$S = \frac{1}{2} \sum_{n=-\infty}^{\infty} \int d^4x \sqrt{\det g_{\mu\nu}} \left[g^{\mu\nu} \partial_\mu \phi^*_n \partial_\nu \phi_n + n^2m^2 \phi^*_n \phi_n\right] \tag{1.10}$$

Thus, instead of regarding the theory as that of a single real massless scalar in five dimensions, we are at liberty to interpret it as the four-dimensional theory of a single real massless scalar (the n = 0 mode) plus an infinite tower of complex massive scalars with quantized masses

$$m_n^2 = n^2m^2 \tag{1.11}$$

Note that the masses of these complex scalars tend to infinity as we shrink the radius of the circle to zero.

The above example is a special case of a somewhat more general set-up where the scalar is defined on a product manifold

$$M = M_1 \times M_2 \tag{1.12}$$

with dimension

$$d = d_1 + d_2 \tag{1.13}$$

with co-ordinates

$$\hat{x}^{\hat{\mu}} = (x^\mu, y^i) \qquad \begin{aligned} \hat{\mu} &= 1,\ldots,d \\ \mu &= 1,\ldots,d_1 \\ i &= 1,\ldots,d_2 \end{aligned} \tag{1.14}$$

and product metric

$$g_{\hat{\mu}\hat{\nu}}(x,y) = \begin{bmatrix} g_{\mu\nu}(x) & 0 \\ 0 & \gamma_{ij}(y) \end{bmatrix} \tag{1.15}$$

From the higher, d dimensional point of view the action is

$$S = \int d^d x \, \frac{1}{2} \sqrt{\det g_{\hat{\mu}\hat{\nu}}} \, \Phi \, \Delta \, \Phi \tag{A.1}$$

where

$$\Delta = -\nabla^{\hat{\mu}} \nabla_{\hat{\mu}} \tag{1.16}$$

is the Laplacian on M which splits as the sum

$$\Delta = \Delta_1 + \Delta_2 \tag{1.17}$$

where Δ_1 and Δ_2 are the Laplacians on M_1 and M_2 respectively. Consequently, from the lower d_1 dimensional point of view the action is

$$S = \sum_n S_n = \sum_n \int d^{d_1}x \, \frac{1}{2} \sqrt{\det g_{\mu\nu}} \, [\phi_n \, \Delta_1 \, \phi_n + m_n^2 \, \phi_n \phi_n] \tag{B.1}$$

where

$$m_n^2 = \lambda_n \tag{1.18}$$

and λ_n are the eigenvalues of Δ_2 with eigenfunctions $Y_n(y)$

$$\Delta_2 Y_n(y) = \lambda_n Y_n(y) \tag{1.19}$$

The fields ϕ_n given by

$$\Phi(x,y) = \sum_n \phi_n(x) \, Y_n(y) \tag{1.20}$$

will belong to the appropriate representation of the symmetry group of M_2. For example, if M_1 = space-time with $d_1 = 4$ and M_2 = the two-sphere S^2 with $d_1 = 2$ and radius m^{-1}, then Y_n are just the spherical harmonics

$$\Phi(x,\theta,\phi) = \sum_{\ell m} \phi_{\ell m}(x) \, Y^{\ell}{}_{m}(\theta,\phi) \tag{1.21}$$

and $\phi_{\ell m}(x)$ belong to the $(2\ell+1)$ dimensional representation of SO(3) with masses

$$m_{\ell m}^2 = \ell(\ell+1)m^2 \tag{1.22}$$

Let us now turn to the quantization of this scalar field still keeping classical the gravitational background. We can choose to view the quantization from standpoint A, starting with the action (A.1) or from standpoint B, starting with the action (B.1).

In case A, the effective action W is given by the functional integral

$$\exp - W[g] = \int d\Phi \exp - S[g,\Phi] \qquad (A.2)$$

where, for convenience, we have Wick rotated to a Euclidean signature, and where S is given by (A.1). Performing the Gaussian integral yields

$$W = \frac{1}{2} \ln \det \Delta \qquad (A.3)$$

W is the generating functional for all one-loop graphs with internal Φ lines and with the external $g_{\hat{\mu}\hat{\nu}}(x,y)$ field acting at the vertices.

In case B, the effective action is given by

$$\exp - W[g] = \int \prod_n d\phi_n \exp - \sum_n S_n[g,\phi_n] \qquad (B.2)$$

where S_n are given by (B.1). Hence

$$W = \sum_n W_n = \frac{1}{2} \sum_n \ln \det (\Delta_1 + m_n^2) \qquad (B.3)$$

W_n is the generating functional for all one-loop graphs with internal ϕ_n lines and with the external $g_{\mu\nu}(x)$ field acting at the vertices.

From the properties of determinants, it is clear that the quantum effective action on the right-hand side of (A.3) is identical to the quantum effective action on the right-hand side of (B.3). At this formal level, therefore, the quantum theories as well as the classical theories are the same whether one adopts viewpoint A or viewpoint B. What is far from clear, however, is whether the ultraviolet divergences depend on which picture one adopts, A or B. The following questions spring to mind:

Q1) In picture A, the power-counting rules are clearly those appropriate to d dimensions. But consider again the example mentioned previously. A massive scalar in a four-dimensional theory ($d_1 = 4$) would normally give rise to one-loop counterterms of the form $m^4\sqrt{g}$, $m^2\sqrt{g}\,R$, and $\sqrt{g}(\text{Riemann})^2$. Yet a counterterm like $\sqrt{g}(\text{Riemann})^3$ should appear in picture A if we chose d = 6. In picture B, therefore, do we count powers in d_1 or in d dimensions?

Q2) Similarly, if we are to use dimensional regularization, does the location of the poles giving rise to the counterterms change in going from one picture to the other?

Q3) At the classical level, shrinking the radii of the extra dimensions sends the masses of all the massive modes to infinity: the only surviving propagating modes are the massless ones. After renormalization, however, does one also recover the quantum theory of the massless modes in this limit?

Q4) The effective action $W = \sum_n W_n$ of Eq. (B.3) will be divergent and will require the addition of a counterterm δW to render it finite. Similarly the individual effective actions W_n will require counterterms δW_n. Is δW equal to $\sum_n \delta W_n$?

Q5) Is δW influenced by the choice of extra dimensions? Consider again our d = 6 example; would it make any difference to choose $M_2 = S^2$ or $M_2 = S^1 \times S^1$?

Q6) Finally, if d is odd the theory is finite in picture A, i.e., $\delta W = 0$. On the other hand, we know that a single massive scalar in four dimensions certainly has divergences, i.e., $\delta W_n \neq 0$. If picture B is to agree with picture A, how can a theory with an infinite number of massive fields be finite when with a finite number of massive fields it is infinite?*

The most elegant techniques for resolving these puzzles are those of "generalized ζ functions" and the "heat kernel expansion" which we summarize in Section 2. In Section 3 we then apply these techniques to examine the problem of ultra-violet divergences in Kaluza-Klein theories from both the higher and lower dimensional points of view, establishing the equivalence of the two pictures and resolving the apparent paradoxes. The above puzzles also apply to the axial and trace anomalies. These are dealt with in Section 4, where a toy model with fermions is introduced and where we keep the spin 1 gauge fields in picture B in addition to the gravitational field. Of particular interest are the formulae describing the fermion zero modes on $M = M_1 \times M_2$ in terms of those on M_1 and M_2. We also comment on the universality of charge to mass ratios in Kaluza-Klein theories: "The spin-one equivalence principle". Section 5 brings the reader up to date with new developments since the Erice conference.

2. ZETA FUNCTIONS AND THE HEAT KERNEL

In this section we summarize the properties of generalized zeta functions and the heat kernel expansion which we employ in Section 3. See Ref. 27 for a review, and Ref. 28 for further references.

The heat kernel

In a compact manifold, M, a second order Laplace-type operator of the form $\Delta = -\Box + X$ will possess a discrete number of eigenfunctions ϕ_n and corresponding eigenvalues λ_n,

*Note incidentally that this Kaluza-Klein picture should be contrasted with the $N \to \infty$ limit of a 1/N expansion. There the infinitely many fields have the same mass and are in the same internal symmetry representations. Here, they have different masses and are in different representations.

$$\Delta\phi_n = \lambda_n \phi_n \tag{2.1}$$

Introduce a parameter t and define the traced "heat kernel"

$$\begin{aligned} Y(\Delta,t) &\equiv \mathrm{Tr}\, \exp - \Delta t \\ &= \sum_n \exp - \lambda_n t \end{aligned} \tag{2.2}$$

It can be shown, and we omit the proof, that Y(t) possesses an asymptotic expansion, valid as $t \to 0^+$, given by

$$Y(\Delta,t) \sim t^{-d/2} \sum_{k=0}^{\infty} A_k(\Delta) t^{k/2} \tag{2.3}$$

where d is the dimension of M, where

$$A_k(\Delta) = \frac{1}{(4\pi)^{d/2}} \int d^d x \sqrt{g}\, a_k(\Delta) \tag{2.4}$$

and where a_k are d independent invariants constructed from $g_{\mu\nu}$ and its derivatives of order k.

As an example, suppose Δ is the Laplacian on scalars of Section 1, then

$$a_0 = 1$$

$$a_2 = \frac{1}{6}R$$

$$\begin{aligned} a_4 &= \frac{1}{180}\left(R_{\mu\nu\rho\sigma}R^{\mu\nu\rho\sigma} - R_{\mu\nu}R^{\mu\nu} + \frac{5}{2}R^2\right) \\ a_6 &= \alpha R_{\mu\nu}{}^{\rho\sigma} R_{\rho\sigma}{}^{\alpha\beta} R_{\alpha\beta}{}^{\mu\nu} \\ &\quad + \beta\, R_{\mu\nu\rho\sigma} R^{\mu\alpha\rho\beta} R^{\nu}{}_{\alpha}{}^{\sigma}{}_{\beta} \\ &\quad + \gamma\, \nabla^{\alpha} R_{\mu\nu\rho\sigma} \nabla_{\alpha} R^{\mu\nu\rho\sigma} + \ldots . \end{aligned} \tag{2.5}$$

The precise form of a_6 is given in Ref. 29. The odd coefficients a_1, a_3, a_5,... vanish when M is compact without boundary*.

* In Kaluza-Klein M_2 will always be compact without boundary, whereas space-time M_1 would in general be non-compact. The techniques described here are valid if M_1 is non-compact with negligible boundary. The method also generalizes to the case of boundaries but will not be discussed here.

Zeta functions

In analogy with the Riemann ζ function[30]

$$\zeta_R(s) = \sum_{n=1}^{\infty} n^{-s} \qquad (2.6)$$

define a generalized ζ function formed from the eigenvalues of Δ

$$\zeta(\Delta,s) = \sum_n \lambda_n^{-s} \qquad (2.7)$$

This is related to $Y(\Delta,t)$ of (2.2) by a Mellin transform

$$\zeta(\Delta,s) = \frac{1}{\Gamma(s)} \int_0^\infty dt\, t^{s-1}\, Y(\Delta,t) \qquad (2.8)$$

$$\zeta(\Delta,s) = \frac{1}{\Gamma(s)} \int_0^1 dt\, t^{s-1}\, Y(\Delta,t) + \frac{1}{\Gamma(s)} \int_1^\infty dt\, t^{s-1}\, Y(\Delta,t) \qquad (2.9)$$

The second integral converges for all s, since $Y \sim \exp - \lambda_{min} t$ for large t, and the second term vanishes for $s = 0, -1, -2, \ldots$ In the first term we may use the asymptotic expansion (2.3). This shows that $\zeta(\Delta,s)$ is analytic for Re $s > d/2$. When d is <u>even</u>, $\zeta(\Delta,s)$ has only d/2 simple poles at $s = (d-k)/2$ for $k = 0, 2, 4, \ldots (d-2)$ with residue $A_{d-2s}/\Gamma(s)$. In particular, for $s = 0, -1, -2, \ldots$ $\zeta(s,\Delta)$ is <u>finite</u> and given by

$$\zeta(\Delta,s) = (-1)^s (-s)!\ A_{d-2s}(\Delta) \qquad (2.10)$$

When d is <u>odd</u>, $\zeta(\Delta,s)$ has an infinite number of simple poles at $s = (d-k)/2$ for $k = 0, 2, 4, \ldots$ with residue $A_{d-2s}/\Gamma(s)$, and <u>vanishes</u> for $s = 0, -1, -2$.

We note in particular that

$$\zeta(\Delta,0) = A_d(\Delta) \qquad (2.11)$$

gives a meaning, via (2.7), to the number of eigenmodes of the operator Δ, a quantity which is formally infinite. We have assumed so far that Δ is positive. In general, of course Δ will have zero modes, which must first be projected out before forming the generalized ζ functions. In this case (2.11) generalizes to

$$A_d(\Delta) = n(\Delta) + \zeta(\Delta,0) \qquad (2.12)$$

where $n(\Delta)$ = integer is the number of zero modes. Thus $A_d(\Delta)$ counts all the modes and $\zeta(\Delta,0)$, which is in general non-integer and sometimes even negative, "counts" the non-zero modes.

For example, suppose in our Kaluza-Klein theory of Section 1 Δ is the Laplacian on $M_2 = S^1$, i.e.,

$$\Delta = - \frac{\partial^2}{\partial y^2} \tag{2.13}$$

then

$$\lambda_n = m_n^2 = n^2 m^2, \quad n = 0, \pm 1, \pm 2, \ldots \tag{2.14}$$

and

$$\begin{aligned} \zeta(\Delta,s) &= 2 \sum_{n=1}^{\infty} (n^2 m^2)^{-s} \\ &= 2m^{-2s} \zeta_R(2s) \end{aligned} \tag{2.15}$$

where ζ_R is the Riemann ζ function (2.6). Thus the effective four-dimensional theory, given by the action (1.10) describes one massless state $[n(\Delta) = 1]$, while the "number" of massive states is

$$\zeta(\Delta,0) = 2 \zeta_R(0) = -1 \tag{2.16}$$

where we have used $\zeta_R(0) = -1/2$. See Ref. 30. Hence from (2.12)

$$A_1(\Delta) = 1 - 1 = 0 \tag{2.17}$$

in accordance with the vanishing of the $A_k(\Delta)$ coefficients when k is odd.

As another example, suppose Δ is the Laplacian on S^2 with radius m^{-1} and

$$\lambda_n = n(n+1)m^2 \tag{2.18}$$

with multiplicity $(2n+1)$. Then

$$\zeta(\Delta,s) = \frac{1}{m^{2s}} \sum_{n=1}^{\infty} \frac{(2n+1)}{[n(n+1)]^s} \tag{2.19}$$

Using the techniques described in Ref. 31, one can show that

$$\begin{aligned} \zeta(\Delta,0) &= \left[-\frac{3}{4} - \zeta_R(-1)\right] \\ \zeta(\Delta,-1) &= \frac{m^2}{4}\left[-\frac{1}{8} + \zeta_R(-1) - 7\,\zeta_R(-3)\right] \\ \zeta(\Delta,-2) &= \frac{m^4}{16}\left[\frac{1}{12} - \zeta_R(-1) + 14\,\zeta_R(-3) - 31\,\zeta_R(-5)\right] \end{aligned} \tag{2.20}$$

Since $\zeta_R(-1) = 1/6$, $\zeta_R(-3) = 1/120$ and $\zeta_R(-5) = -1/252$, one finds

$$
\begin{aligned}
A_0(\Delta) &&= m^{-2} \\
A_2(\Delta) &= 1 + \zeta(\Delta,0) &= \frac{1}{3} \\
A_4(\Delta) &= -\zeta(\Delta,-1) &= \frac{m^2}{15} \\
A_6(\Delta) &= \frac{1}{2}\,\zeta(\Delta,-2) &= \frac{4m^4}{315}
\end{aligned}
\tag{2.21}
$$

These results may be verified by inserting the curvature $R_{ijk\ell}$ for S^2 with metric γ_{ij}

$$R_{ijk\ell} = m^2(\gamma_{ik}\gamma_{j\ell} - \gamma_{i\ell}\gamma_{jk}) \tag{2.22}$$

into the formulae (2.4) and (2.5).

One-loop counterterms

Consider now a typical one-loop effective action given by

$$W = \frac{1}{2}\,\ell n \det \Delta = \frac{1}{2}\sum_n \ell n\, \lambda_n \tag{2.23}$$

W will be ultra-violet divergent. To regularize we may use dimensional regularization and replace the dimension d of M by d + ε. Thus (2.3) becomes

$$Y(\Delta,t) \sim t^{-(d+\varepsilon)/2} \sum_{k=0}^{\infty} A_k t^{k/2} \tag{2.24}$$

From (2.23) and (2.8),

$$
\begin{aligned}
W &= -\frac{d}{ds}\,\zeta(\Delta,s)\Big|_{s=0} \\
&= -\frac{1}{2}\int_0^\infty dt\; t^{-1}\, Y(\Delta,t)
\end{aligned}
\tag{2.25}
$$

Employing the same procedure as in (2.9) and using (2.24) we find, upon letting $\varepsilon \to 0$, the one-loop counterterm

$$\delta W \equiv -W_{DIVERGENT} = -\frac{1}{\varepsilon}\,A_d \tag{2.26}$$

Thus the calculation of the counterterm is trivial once we know the A_d coefficient. For example, massless scalars in d = 4 would give

$$
\begin{aligned}
\delta W &= -\frac{1}{\varepsilon}\,A_4 \\
&= -\frac{1}{\varepsilon}\,\frac{1}{(4\pi)^2}\int d^4x\,\sqrt{g}\,\frac{1}{180}\left[R_{\mu\nu\rho\sigma}R^{\mu\nu\rho\sigma} - R_{\mu\nu}R^{\mu\nu} + \frac{5}{2}R^2\right]
\end{aligned}
\tag{2.27}
$$

on using (2.5).

If, on the other hand, the scalars were massive with mass m, then

$$\Delta \to \Delta + m^2 \tag{2.28}$$

$$Y(\Delta,t) \to e^{-m^2t} \sum_n e^{-\lambda_n t} \tag{2.29}$$

$$A_d \to \sum_{k=0}^{d} \frac{(-m^2)^{(d-k)/2}}{[(d-k)/2]\,!} A_k \tag{2.30}$$

where to obtain (2.30) we have expanded the exponential factor in (2.29) as

$$e^{-m^2t} = 1 - m^2t + \frac{m^4}{2}t^2 - \ldots \tag{2.31}$$

In the above example, the counterterm would become

$$\delta W = -\frac{1}{\varepsilon}\left[\frac{m^4}{2}A_0 - m^2A_2 + A_4\right] \tag{2.32}$$

$$= -\frac{1}{\varepsilon}\frac{1}{(4\pi)^2} \int d^4x \sqrt{g}\left[\frac{m^4}{2} - m^2\frac{R}{6} + \frac{1}{180}(R_{\mu\nu\rho\sigma}R^{\mu\nu\rho\sigma} - R_{\mu\nu}R^{\mu\nu} + \frac{5}{2}R^2)\right] \tag{2.33}$$

on using (2.5).

3. DIVERGENCES IN KALUZA-KLEIN THEORIES

Let us now apply the techniques of Section 2 to our model scalar field theory of Section 1, first of all from the point of view of the higher dimension.

Higher dimensional picture (A)

Starting from Eqs. (A.1), (A.2) and (A.3) of Section 1, the counterterm will be given by

$$\delta W = -\frac{1}{\varepsilon} A_d(\Delta) \tag{3.1}$$

where Δ is the Laplacian on the d dimensional manifold M. When M is the product $M_1 \times M_2$, Δ splits into $\Delta_1 + \Delta_2$ as in (1.17). The beauty of this is that the heat kernel (2.2) factorizes

$$Y(\Delta,t) = Y(\Delta_1,t)Y(\Delta_2,t), \tag{3.2}$$

hence

$$t^{d/2} \sum_{k=0}^{\infty} A_k(\Delta)t^{k/2} = t^{d_1/2} \sum_{\ell=0}^{\infty} A_\ell(\Delta_1)t^{\ell/2} \cdot t^{d_2/2} \sum_{m=0}^{\infty} A_m(\Delta_2)t^{m/2} \tag{3.3}$$

Comparing powers of t,

$$A_k(\Delta) = \sum_{\ell=0}^{k} A_\ell(\Delta_1)A_{k-\ell}(\Delta_2) \tag{3.4}$$

and so (3.1) yields

$$\delta W = -\frac{1}{\varepsilon} \sum_{\ell=0}^{d} A_\ell(\Delta_1)A_{d-\ell}(\Delta_2) \tag{3.5}$$

(Note that the sum extends from 0 to d, the dimension of M.). Equations (3.1) and (3.5) have a simple interpretation. The $A_d(\Delta)$ in (3.1) will just provide the d dimensional counterterms, i.e.,

$$\frac{1}{(4\pi)^{d/2}} \int d^dx \sqrt{g}\ (\text{Riemann})^{d/2} \tag{3.6}$$

Whereas in (3.5), the $A_\ell(\Delta_1)$ will be four-dimensional counterterms, i.e.,

$$\frac{1}{(4\pi)^2} \int d^4x \sqrt{g}\ (\text{Riemann})^{\ell/2} \tag{3.7}$$

with numerical coefficients determined by the $A_{d-\ell}(\Delta_2)$. But this amounts to nothing more than computing the counterterm in d dimensions and then integrating out the extra y co-ordinates to obtain the counterterm for the effective four-dimensional theory - precisely the same procedure as adopted for the classical action.

Let us consider again the concrete example of d = 6:

$$\delta W = -\frac{1}{\varepsilon}\Big[A_0(\Delta_1)A_6(\Delta_2) + A_2(\Delta_1)A_4(\Delta_2) + A_4(\Delta_1)A_2(\Delta_2) + A_6(\Delta_1)A_0(\Delta_2)\Big] \tag{3.8}$$

and suppose $M_2 = S^2$. Then from (2.5) and (2.21)

$$\delta W = -\frac{1}{\varepsilon}\cdot\frac{1}{(4\pi)^2} \int d^4x \sqrt{g}\Big[\frac{4m^4}{315} + \frac{m^2}{90}R + \frac{1}{540}(R_{\mu\nu\rho\sigma}R^{\mu\nu\rho\sigma} - R_{\mu\nu}R^{\mu\nu} + \frac{5}{2}R^2) + \frac{1}{m^2}\ (\text{Riemann})^3\Big] \tag{3.9}$$

where

$$(\text{Riemann})^3 = \alpha\, R_{\mu\nu}{}^{\rho\sigma} R_{\rho\sigma}{}^{\alpha\beta} R_{\alpha\beta}{}^{\mu\nu} + \ldots \tag{3.10}$$

as in (2.5).

A few comments are now in order:

1) Comparison with (2.33) shows that the $\sqrt{g}$, $\sqrt{g}R$ and $\sqrt{g}R^2$ terms in (3.9) are only to be expected. The novel departure is the appearance of $\sqrt{g}(\text{Riemann})^3$ even though we are still at one loop.

2) Had we chosen $M_2 = S^1 \times S^1$ with inverse radii m and m', then $R_{ijk\ell} = 0$ and all the $A_k(\Delta_2)$ in (3.8) would vanish save for $A_0(\Delta_2)$. In this case <u>only</u> the $\sqrt{g}(\text{Riemann})^3$ survives with coefficient $(mm')^{-1}$.

3) Had we chosen d to be <u>odd</u>, then all the $A_k(\Delta_2)$ vanish irrespective of the choice of M_2 and $\delta W = 0$.

4) The above results indicate that we do not in general recover the corresponding four-dimensional massless theory as we shrink the extra dimensions. Compare (3.9) as $m \to \infty$ with (2.27).

Now let us repeat the above analysis from the point of view of the lower dimension.

<u>Lower dimensional picture (B)</u>

Starting from Eqs. (B.1), (B.2) and (B.3) of Section 1 the counterterms δW_n will be given by

$$\delta W_n = -\frac{1}{\varepsilon} A_{d_1}(\Delta_n) \tag{3.11}$$

where

$$\Delta_n = \Delta_1 + m_n^2 \tag{3.12}$$

and m_n^2 are the eigenvalues of Δ_2. From (2.30), (3.11) and (3.12),

$$\delta W_n = -\frac{1}{\varepsilon} \sum_{k=0}^{d_1} \frac{(-m_n^2)^{(d_1-k)/2}}{[(d_1-k)/2]\,!} A_k(\Delta_1) \tag{3.13}$$

<u>Naively</u>, one might suppose that, since $W = \Sigma_n W_n$, the total counterterms would be given by the sum of the individual counterterms, i.e.,

$$\delta W = \sum_n \delta W_n \tag{3.14}$$

Using (3.13) together with

$$\sum_n m_n^{-2s} = \zeta(\Delta_2, s) \tag{3.15}$$

and the relation (2.10) between ζ and A_k, one would then obtain

$$\delta W = -\frac{1}{\varepsilon} \sum_{\ell=0}^{d_1} A_\ell(\Delta_1) A_{d-\ell}(\Delta_2) \tag{3.16}$$

This is almost the same result as that obtained in the high dimensional picture given by (3.5), but there is one vital difference: the sum in (3.16) extends only as far as d_1.

In our d = 6 example (3.13) is

$$\delta W_n = -\frac{1}{\varepsilon} \int d^4x \sqrt{g} \Big[\frac{1}{2} m_n^4 - m_n^2 \frac{R}{6} + \frac{1}{180} (R_{\mu\nu\rho\sigma} R^{\mu\nu\rho\sigma} - R_{\mu\nu} R^{\mu\nu} + \frac{5}{2} R^2)\Big] \tag{3.17}$$

and (3.16) would become

$$\delta W = -\frac{1}{\varepsilon} \frac{1}{(4\pi)^2} \int d^4x \sqrt{g} \Big[\frac{4m^4}{315} + \frac{m^2}{90} R + \frac{1}{540} (R_{\mu\nu\rho\sigma} R^{\mu\nu\rho\sigma} - R_{\mu\nu} R^{\mu\nu} + \frac{5}{2} R^2)\Big] \tag{3.18}$$

In other words, because the sum in (3.16) extends only as far as four and not six, the $(\text{Riemann})^3$ term in the high dimension result (3.9) is missing.

Naively, therefore, it seems that the high dimension picture yields a result expected by power-counting in d dimensions, whereas the low dimension picture yields a result expected by power-counting in d_1 dimensions. Yet formally, the two pictures should describe the same theory. What is wrong ?

Reconciliation of pictures A and B

The answer lies in our seemingly innocuous assumption (3.14). This naive result would be perfectly legitimate if the summation extended over a finite number of fields. Because in our Kaluza-Klein theory the number of fields is infinite, however, then

$$\delta W \neq \sum_n \delta W_n \tag{3.19}$$

To understand the source of the error in the naive assumption (3.14), recall that in deriving (3.16) we employed (2.30), which in turn involved (2.31). In other words, we expanded

$$e^{-\lambda_n t} = 1 - \lambda_n t + \frac{\lambda_n^2}{2} t^2 - \ldots \tag{3.20}$$

So far, so good. It is then false, however, to claim

$$\begin{aligned}\sum_n e^{-\lambda_n t} &= \sum_n 1 - \sum_n \lambda_n t + \sum_n \frac{\lambda_n^2}{2} t^2 - \ldots \\ &= \zeta(\Delta_2,0) - \zeta(\Delta_2,-1)t + \zeta(\Delta_2,-2)\frac{t^2}{2} - \ldots \qquad (3.21) \\ &= \sum_{k=0}^{\infty} A_{k+d_2}(\Delta_2)t^{k/2}\end{aligned}$$

on using (2.10), because a comparison with the asymptotic expansion of the heat-kernel (2.3) shows that (3.21) has omitted all the negative powers of t.

The physical interpretation of this is that the individual effective actions W_n may have finite terms involving inverse powers of m_n which, when summed over n, yield infinities which therefore contribute to the infinite part of W. In our d = 6 example, W_n involves finite terms of the form $m_n^{-2}R^3$ which when summed over n yield $\zeta(\Delta_2,1)R^3$ which is divergent (see Section 2). To handle this divergence we note that in $d_2 + \varepsilon$ dimensions $\zeta(\Delta_2,s)$ has a pole at $\varepsilon = 0$ for $s = 1,2,\ldots,d_2/2$ with residue $A_{d_2-2s}(\Delta_2)$. In particular for $d_2 = 2$, $\zeta(\Delta_2,1)R^3$ is then replaced by $\varepsilon^{-1}\tilde{A}_0R^3$ which was just the missing term in (3.18).

With sufficient care, therefore, picture B gives the same result as picture A. We are now in a position to answer the questions posed at the end of Section 1.

Questions answered

A1) In both pictures, the power counting rules are those appropriate to the higher dimension, (as may also be understood by treating Σ_n like a momentum integration). Thus even the effective four-dimensional theory may have one-loop counterterms like R^3, etc.

A2) Similarly, with regard to the question of dimensional regularization, by the replacement

$$\int d^{d+\varepsilon}x \rightarrow \sum_n \int d^{d_1+\varepsilon}x \tag{3.21}$$

we see that the ε in the A counterterm (3.1) is the same ε as in the B counterterm (3.11). Dimensional regularization in $d = d_1 + d_2$ dimensions is seen to be equivalent to dimensional regularization in d_1 dimensions together with zeta-function regularization in d_2 dimensions.

A3) One does not recover the massless quantum theory by first performing the regularization and then shrinking the size of the extra dimensions. As the examples (2.27) and (3.9) illustrate, the divergences of the two theories are very different. Even the $(\text{Riemann})^2$ term, which is independent of m, has a different coefficient. Similar remarks apply to the finite contributions to the effective action, in particular the vacuum energies mentioned in Section 1. A single massless field in four dimensions induces no vacuum energy, whereas including all the massive modes and then sending $m \to \infty$ would yield an infinite vacuum energy (this is after the removal of ultra-violet divergences).

A4) Nor is it true that $\delta W = \Sigma_n \, \delta W_n$. As we have seen, it sometimes happens that

$$\sum_n (\text{finite})_n = \text{infinite.}$$

A5) The counterterms are influenced by the choice of extra dimensions e.g., $M_2 = S^2$ or $M_2 = S^1 \times S^1$. However, this is really an artefact of the toy models we have employed, in that the gravitational field was taken to be external with no dynamics of its own. In a realistic theory with all fields, including the gravitational fluctuations about the ground state, dynamical and depending on both x^μ and y^i this would not be the case. Different ground states would give different-looking counterterms in picture B, but all should be equivalent, in much the same way that the spontaneous breaking of a conventional gauge theory does not change its renormalizability properties. One would not expect finiteness in one phase and non-finiteness in another.

A6) Finally, although as mentioned in (A4) above, it sometimes happens that

$$\sum_n (\text{finite})_n = \text{infinite}$$

so it sometimes also happens that

$$\sum_n (\text{infinite})_n = \text{finite}$$

This explains how in picture B, the theory of an infinite number of massive states can be finite when the same theory of a single state is divergent. For example, taking the extra dimension to be a circle can yield poles in ε with residue

$$\sum_{n=-\infty}^{\infty} 1 \text{ and } \sum_{n=-\infty}^{\infty} n^2$$

But using ζ functions

$$\sum_{n=-\infty}^{\infty} 1 = 1 + 2\zeta_R(0) = 0 \tag{3.22}$$

and

$$\sum_{n=-\infty}^{\infty} n^2 = 2\zeta_R(-2) = 0 \qquad (3.23)$$

These results have an interesting application to axial anomalies when fermions are included in the Kaluza-Klein scheme.

4. FERMIONS AND THE AXIAL ANOMALY

In this section we describe a simple model with fermions, again unrealistic but nevertheless useful for illustration.

The fermion model

Consider the action for a single massless Dirac spinor in interaction with an external gravitational field in five dimensions

$$S = \int d^5x \det \hat{e}\ \bar{\Psi}\ i\ \gamma^{\hat{\mu}}\ \hat{D}_{\hat{\mu}}\ \Psi \qquad (4.1)$$

where $\hat{e}_{\hat{\mu}}{}^{\hat{a}}(x)$ is the fünfbein field (μ,a = 1,2,3,4,5), $\gamma^{\hat{a}}$ are the five-dimensional Dirac matrices

$$\{\gamma^{\hat{a}},\gamma^{\hat{b}}\} = 2\eta^{\hat{a}\hat{b}} \qquad (4.2)$$

and $\hat{D}_{\hat{\mu}}$ is the five-dimensional covariant derivative formed from the spin connection.

As in Section 1 we again consider the fifth dimension to be compactified to a circle with radius m^{-1}, hence

$$\Psi(x,y) = \sqrt{\frac{m}{2\pi}} \sum_{n=-\infty}^{\infty} \psi_n(x) \exp inmy \qquad (4.3)$$

This time, however, we generalize the ansatz for the five-dimensional gravity field to allow a non-vanishing gauge field in the effective four-dimensional theory:

$$\hat{e}_{\hat{\mu}}{}^{\hat{a}}(x,y) = \begin{bmatrix} e_\mu{}^a(x) & \kappa A_\mu(x) \\ 0 & 1 \end{bmatrix} \qquad (4.4)$$

where κ^2 is chosen to be Newton's constant times $16\pi^2$ in order that the five-dimensional Einstein action yield the correct Einstein-Maxwell four-dimensional action. Substituting (4.3) and (4.4) into (4.1) and performing the y integration yields

$$S = \sum_{n=-\infty}^{\infty} \int d^4x \det e\ \bar{\psi}_n\Big[i\ \gamma^\mu(D_\mu + iq_nA_\mu) + \\ + m_n\gamma^5 + \frac{i\kappa}{8}F_{\mu\nu}\gamma^\mu\gamma^\nu\gamma^5\Big]\psi_n \qquad (4.5)$$

where all quantities are now four-dimensional. Thus the effective theory describes one neutral massless fermion (the $n = 0$ mode) together with an infinite tower of charged massive fermions with quantized charges

$$q_n = n \kappa m \tag{4.6}$$

and quantized masses

$$m_n = nm \tag{4.7}$$

All states also couple non-minimally to electromagnetism via the Pauli term ($F_{\mu\nu} = \partial_\mu A_\nu - \partial_\nu A_\mu$).

Spin-one equivalence principle

The remarkable feature of the charged massive states is that from (4.6) and (4.7), the charge to mass ratio

$$\frac{q_n}{m_n} = \kappa \tag{4.8}$$

is a universal constant for all particles*. Thus even when considering purely electromagnetic effects, one cannot invoke any decoupling theorem to argue that a particle with mass a billion times the Planck mass, say, is any less important than a particle whose mass is "only" 10^{-5} grams, because this particle will couple a billion times more strongly. For example, the scattering of light by light is describable by a low-frequency Euler-Heisenberg effective Lagrangian of the form $q_n^4 F^4/m_n^4$ and each massive mode would contribute equally to this process. Of course the net effect is still small and so presumably negligible for phenomenology at current energies.

There is a simple physical interpretation of this universality effect. In the five-dimensional theory, the equivalence principle is at work ensuring that gravity couples to everything with equal strength, and in the effective four-dimensional theory the photon is "remembering" its five-dimensional ancestry. Similar effects would also occur for higher dimensions and non-Abelian gauge theories.

This "spin-one equivalence principle" will always be a feature of Kaluza-Klein theories.

---

* A similar phenomenon, though in the context of twisted fields and dimensional reduction rather than Kaluza-Klein, is responsible for Scherk's "antigravity". See Refs. 32 and 33.

The axial anomaly

Let us now consider the anomaly in the divergence of the axial vector current, following the approach of Refs. 34 and 35. There it was noted that in the functional integral approach to quantization, the origin of the anomaly may be understood by noting that the fermion measure is not invariant under chiral transformations. Let $d\mu[\bar{\psi}_n,\psi_n]$ represent the measure for the nth mode in the theory described by the action (4.5). Then under the transformation

$$\psi_n \to (\exp \gamma_5\ \alpha)\psi_n, \tag{4.9}$$

The measure transforms

$$d\mu[\bar{\psi}_n,\psi_n] \to d\mu[\bar{\psi}_n,\psi_n]\ e^{2i\alpha\nu_n} \tag{4.10}$$

where

$$\nu_n = \frac{q_n^2}{16\pi^2}\int d^4x\ \sqrt{g}\ \ {}^*F_{\mu\nu}F^{\mu\nu} - \frac{1}{384\pi^2}\int d^4x\ \sqrt{g}\ {}^*R_{\mu\nu\rho\sigma}R^{\mu\nu\rho\sigma} \tag{4.11}$$

is a topological invariant, and the stars denote the dual tensors. The complete functional measure in this theory is

$$\prod_{n=-\infty}^{\infty} d\mu[\bar{\psi}_n,\psi_n].$$

From (4.10) this transforms as

$$\prod_{n=-\infty}^{\infty} d\mu[\bar{\psi}_n,\psi_n] \to \prod_{n=-\infty}^{\infty} d\mu[\bar{\psi}_n,\psi_n]\cdot e^{2i\alpha\sum_{n=-\infty}^{\infty}\nu_n} \tag{4.12}$$

To interpret

$$\sum_{n=-\infty}^{\infty} \nu_n$$

we use the charge quantization (4.6) and ζ function regularization. Then from (4.11),

$$\sum_{n=-\infty}^{\infty} \nu_n = 2\zeta_R(-2)\cdot\frac{\kappa^2 m^2}{16\pi^2}\int d^4x\ \sqrt{g}\ {}^*F_{\mu\nu}F\mu\nu$$
$$- \left[1 + 2\zeta_R(0)\right]\frac{1}{384\pi^2}\int d^4x\ \sqrt{g}\ {}^*R_{\mu\nu\rho\sigma}R^{\mu\nu\rho\sigma} = 0 \tag{4.13}$$

on using (3.22) and (3.23). Thus the functional measure for this infinite tower of fermions is invariant under chiral transformations, even though that for a single fermion is not. Therefore the effective four-dimensional theory is anomaly-free which is entirely consistent with the five-dimensional viewpoint: there are no chiral anomalies in odd dimensions*.

The relation between the anomaly for a single mode (the lower, d_1 dimensional theory) and that for all the modes (the higher, d dimensional theory) for an arbitrary theory in arbitrary dimensions may be simply read off from the formulae for the tensor product of elliptic complexes. See Ref. 36. First we recall that in even dimensions the Atiyah-Singer theorem yields

$$N^+ - N^- = \nu \tag{4.14}$$

where N^+ and N^- are respectively the number of right and left handed zero modes of the Dirac operator and ν is topological invariant, of which (4.11) is a special case. Although there will in general be zero modes in odd dimensions also, the split into right and left is not defined. If the number of zero modes on $M = M_1 \times M_2$ is denoted by N and those on M_1 and M_2 by N_1 and N_2 respectively, then the relevant formulae are**

a) d_1 even, d_2 even

$$\begin{aligned} N^+ &= N_1^+ N_2^+ + N_1^- N_2^- \\ N^- &= N_1^+ N_2^- + N_1^- N_2^+ \end{aligned} \tag{4.15}$$

b) d_1 even, d_2 odd

$$N = (N_1^+ + N_1^-) N_2 \tag{4.16}$$

c) d_1 odd, d_2 odd

$$N^+ = N^- = N_1 N_2 \tag{4.17}$$

In case (a) we have from (4.14) and (4.15)

$$\nu = \nu_1 \nu_2 \tag{4.18}$$

---

*Here, as in Section 3, one can even turn the argument around: consistency of the higher and lower dimensional points of view demands the ζ function interpretation of divergent quantities.

**The formulae are in fact more complicated when M is not globally the direct product $M_1 \times M_2$, i.e., when the spin-one gauge fields in picture B have non-trivial topology. Incidentally, in conventional gauge theories some authors are happy to sum over different gauge field topologies, but baulk at the idea of summing over different space-time topologies. It is worth mentioning that, in Kaluza-Klein, the latter is necessary for the former.

In case (b), ν is not defined, and as illustrated in our five-dimensional example above there is no anomaly even when ν_1 is non-vanishing. In case (c), ν always vanishes.

The invariance of the measure in the effective four-dimensional theory when d is odd means that the θ parameter is zero. This has implications for the CP problem.

Equations (3.22) and (3.23), used to prove the vanishing of the axial anomaly, may also be used to prove the vanishing of the one-loop trace anomaly and β function, a result in any case expected since the theory is odd-loop finite*.

5. RECENT DEVELOPMENTS

Here we bring the reader up to date with some recent developments in Kaluza-Klein theories:

Mode expansions on G/H

The expansion of the fields in a Kaluza-Klein theory in terms of normal modes on the compact space M_2 is quite a non-trivial matter in the case that M_2 is a quotient space G/H, especially for fields with spin. This problem has recently been analyzed in some detail by Salam and Strathdee[38], who pay particular attention to the problem of embedding H in the tangent space group $SO(d_2)$. Of particular interest is their result that the massive particles belong to infinite dimensional representations of non-compact groups. For example, in the original 4 + 1 pure gravity theory the massive states are charged purely spin-two particles belonging to infinite dimensional representations of O(1,2). It would be interesting to see what kind of non-compact groups would emerge from eleven-dimensional supergravity.

Ultra-violet divergences

Further discussion of the kinds of problems arising with divergences and anomalies in higher dimensional theories, including the effects of taking the fields to be "twisted" in the higher dimensions, may be found in a recent paper by Duff and Toms[28].

Another development, due to Green, Schwarz and Brink[39] concerns the emergence of N = 1 supersymmetric Yang-Mills in d = 10 and N = 2 supergravity in d = 10 as the zero-slope limits of the fermionic string models. By considering the extra dimensions to

---

*The β function for a theory with an infinite number of fermions has also been studied in a different context by Nicolai[37].

be compactified as circles and then letting the radii of the circles tend to zero, they are able to analyze the ultra-violet divergences of the theories for different values of d. Interestingly, they find that at one-loop both theories are ultra-violet finite for $d < 8$ and infra-red finite for $d > 4$ *, the infra-red divergence being milder in the gravitational case.

Note however that this analysis does not include N = 1 in d = 11 which differs from N = 2 in d = 10 by an infinite tower of massive states. Indeed the eleven-dimensional theory is finite at all odd loops, in particular at one-loop. [Actually, it is not quite so obvious as it is for pure gravity that there are no invariants with an odd number of derivatives. For example, one could imagine a counterterm containing the $F_{\mu\nu\rho\sigma}$ of Eq. (1.1) raised to an odd power. Fortunately there is a discrete symmetry, (time reversal together with $A_{\mu\nu\rho} \to -A_{\mu\nu\rho}$) which rules these out.] Possible on-shell counterterms do exist at even loop order, however. They are of the symbolic form

$$(\kappa^2)^{L-1} \int d^{11}x \; d^{32}\theta [D^2]^\beta [W]^\gamma \tag{5.1}$$

where $W_{[\mu\nu\rho\sigma]}(x,\theta)$ is the super field strength[40], D stands for its supercovariant derivative, and the θ's are the fermionic co-ordinates of superspace. The requirement that (5.1) be dimensionless and compatible with the above discrete symmetry yields

$$\beta + \gamma = 9L - 14 = \text{positive even integer} \tag{5.2}$$

Thus $\beta + \gamma = 4 \bmod 18$ and candidate counterterms exist at L = 2,4,6 etc. [I am very grateful to P. Howe, R. Kallosh and M. Sohnius for discussions on these points. An analysis of supergravity counterterms in both d = 4 and d > 4 supergravity has also been given recently by Grisaru and Siegel[41]. These authors make plausible the assumption that only integrals over all superspace, as in (5.1), ever appear as counterterms.] One should still not be too pessimistic about the finiteness of d = 11 supergravity, however, since time and again in supersymmetry and supergravity one finds that a priori possible counterterms nevertheless appear with zero coefficient[1].

More about Kaluza-Klein supergravity

In Section 1 we described some of the things supergravity can do for Kaluza-Klein. We finish this paper by considering what Kaluza-Klein can do for supergravity. After all, whether or not a realistic SU(3) × SU(2) × U(1) theory emerges in the way described, the Kaluza-Klein idea still provides a deeper understanding of the

* Should one invoke the avoidance of infra-red divergences as another reason for Kaluza-Klein theories ?

structure of supergravity in four dimensions.

The Cremmer-Julia[4] version of N = 8 supergravity in d = 4 may be obtained as follows. First, one adopts the ground state solution of Minkowski space × 7-torus. The gauge group is then simply 7 U(1)'s. Secondly, one discards the massive modes. The remaining massless states belong to representations of a rigid SO(7). In particular, some of the spin-0 fields are realized as antisymmetric tensors. Finally, only after making (topologically non-trivial) duality transformations, which transform antisymmetric tensors to scalars, does one obtain the version with the E_7 (rigid) × SU(8) (local) invariance. The geometrical origin of these hidden symmetries is thus far from clear.

Although in this pioneering work Cremmer and Julia were aware of the possibility of choosing the extra seven dimensions to be something other than a 7-torus, they confined their explicit calculations to this case. So leaving aside the question of a realistic gauge theory for the moment, it is still of interest for a supergravity enthusiast to ask what kind of d = 4 theories would emerge with different ground states. After all, there are potentially infinitely many such theories and it would be nice to have a concrete example of just one.

The task of carrying out the complete Kaluza-Klein reduction is very difficult but the most important piece of information is the number of massless states, their spins, and internal quantum numbers. Rules for counting these massless states will be given in a forthcoming publication by Duff and Pope[42]. Though straightforward in principle these rules can be complicated to apply in all but the simplest cases, and next to the 7-torus the simplest ground state solution, as discussed in Section 1, is the 7-sphere with its standard metric*, S^7 = SO(8)/SO(7). This yields an effective d = 4 theory with local SO(8) invariance and negative cosmological constant. Applying the counting rules for the massless modes, which fall naturally into SO(8) representations, we find 1 graviton, 8 gravitini, 28 vectors (from the Killing vectors on S^7), and 56 spin 1/2 fermions. A direct count of the massless spin-0 fields is rather more difficult, but indirectly one can argue that it must be 70 owing to the 8-fold supersymmetry, which in turn is ensured by the presence of 8 massless spin-3/2 particles. [The difference in Betti numbers of the 7-sphere from the 7-torus turns out to guarantee that these 70 states will be genuine scalars all coming from the elfbein $e_\mu{}^a(x)$, rather than antisymmetric tensors coming from the 3-index field $A_{\mu\nu\rho}(x)$.]

*Seven-spheres hold great fascination for the mathematicians. See, for example, "On manifolds homeomorphic to the 7-sphere" by Milnor in Ref. 43. Will all this differential geometry eventually be important for supergravity ?

But an N = 8 supergravity theory in d = 4 with all the above attributes already exists ! It is the recently constructed gauged theory of De Wit and Nicolai[44] which has, in addition to the obvious local SO(8) invariance, a hidden local SU(8). One recovers the E_7 (rigid) × SU(8) (local) version as one switches off the gauge coupling. Thus Kaluza-Klein puts into perspective the different N = 8 supergravities known to exist in d = 4 (and which have a different ultra-violet behaviour[1].) Taking a 7-sphere and expanding its radius to ∞ leads directly to the E_7 × SU(8) theory, which was obtained only indirectly by starting from the 7-torus. Indeed, one might now hope that S^7 will yield a more transparent geometrical origin of these hidden symmetries*.

The 7-torus and the 7-sphere are, however, very special cases. A generic ground state would have fewer than 8 gravitini, (i.e., fewer than 8 supersymmetries) and perhaps none at all. Indeed, this may be the most welcome feature of Kaluza-Klein ideas applied to eleven-dimensional supergravity: it would give a supersymmetric explanation for the absence of any observed supersymmetry in particle physics at current energies.

ACKNOWLEDGEMENTS

We are very grateful for conversations with A. Chockalingham, R. Coquereaux, S. Ferrara, G. Gibbons, P. Howe, C. Isham, R. Kallosh, W. Mecklenburg, H. Nicolai, D. Olive, D. Pollard, C. Pope, A. Salam, M. Sohnius, K. Stelle, P. Stephenson, J. Strathdee, P. Townsend, A. Trautman, N. Warner, P. West, C. Wetterich and E. Witten.

REFERENCES

1. For a recent review, see M.J. Duff, Ultraviolet Divergences in Extended supergravity, CERN preprint TH.3232 (1982), to appear in "Supergravity 81", Eds. S. Ferrara and J.G. Taylor, Cambridge University Press.
2. Th. Kaluza, Sitzungsber. Preus. Akad. Wiss. Berlin, Math. Phys. Kl:966 (1921); O. Klein, Z. Phys. 37:895 (1926).
3. E. Witten, Nucl. Phys. B186:412 (1981); this paper gives a comprehensive list of references to earlier work. See also Ref. 28.
4. E. Cremmer and B. Julia, Nucl. Phys. B159:141 (1979).
5. A. Chockalingham, Ph.D. Thesis, Imperial College (1981).
6. J. Scherk and J.H. Schwarz, Phys. Lett. B57:463 (1975).
7. For a review see E. Cremmer, preprint LPTENS 81/18 (1981).
8. J. Scherk and J.H. Schwarz, Phys. Lett. B82:60 (1979).
9. E. Witten, Instability of the Kaluza-Klein vacuum, Princeton preprint (1981).

*I am very grateful to Sergio Ferrara and Hermann Nicolai for discussions on this point.

10. S.M. Christensen and M.J. Duff, Nucl. Phys. B154:301 (1979).
11. F.A. Berends, J.W. van Holten, B. de Wit and P. van Nieuwenhuizen Nucl. Phys. B154:261 (1979).
12. C. Aragone and S. Deser, Phys. Lett. 85B:161 (1979).
13. W. Nahm, Nucl. Phys. B135:149 (1978).
14. A. Lichnerowicz, C.R. Acad. Sci. Paris Sér. A-B257:7 (1963).
15. S.M. Christensen and M.J. Duff, Phys. Lett. 76B:571 (1978)
16. S.W. Hawking and C.N. Pope, Nucl. Phys. B146:381 (1978).
17. N.K. Nielsen, M. Grisaru, H. Römer and P. van Nieuwenhuizen, Nucl. Phys. B140:477 (1978).
18. N. Hitchin, Adv. Math. 14:1 (1974).
19. C. Wetterich, Spontaneous compactification in higher dimensional gravity, CERN preprint TH.3239 (1982).
20. P.G.O. Freund and M.A. Rubin, Phys. Lett. B97:233 (1980).
21. M.J. Duff and P. van Nieuwenhuizen, Phys. Lett. B94:179 (1980).
22. A. Aurelia, H. Nicolai and P.K. Townsend, Nucl. Phys. B176:509 (1980).
23. S. Unwin, Phys. Lett. B103:18 (1981).
24. B. Zumino, Nucl. Phys. B89:535 (1975).
25. P.C. West, Nucl. Phys. B106 (1976) 219.
26. S.W. Hawking and C.N. Pope, Phys. Lett. 73B:42 (1978).
27. S.W. Hawking, Commun. Math. Phys. 55:133 (1977).
28. M.J. Duff and D.J. Toms, Divergences and anomalies in Kaluza-Klein theories, CERN preprint TH.3248 (1982). Talk given by M.J. Duff at the Second Quantum Gravity Seminar, Moscow, October 1981.
29. P.B. Gilkey, J. Diff. Geom. 10:601 (1975).
30. E.T. Whittaker and G.N. Watson, A course of modern analysis, Cambridge University Press, (1927).
31. S.M. Christensen and M.J. Duff, Nucl. Phys. B170 [FS1]:480 (1980).
32. J. Scherk, From supergravity to antigravity, in Supergravity, Eds. D.Z. Freedman and P. van Nieuwenhuizen, North Holland (1980).
33. D. Pollard, Antigravity and classical solutions of five-dimensional Kaluza-Klein theory, Imperial College preprint ICTP/81/82-16 (1982).
34. K. Fujikawa, Phys. Rev. Lett. 42:1195 (1979)
35. S. Deser, M.J. Duff and C.J. Isham, Phys. Lett. 93B:419 (1980).
36. M.F. Atiyah and R. Bott, Ann. of Math. 86:374 (1967).
37. H. Nicolai, Phys. Lett. B84:219 (1979).
38. A. Salam and J. Strathdee, On Kaluza-Klein theories, ICTP Trieste preprint IC/81/211 (1981).
39. M.B. Green, J.H. Schwarz and L. Brink, N = 4 Yang-Mills and N = 8 supergravity as limits of string theories, Caltech preprint CALT-68-880 (1981).

40. E. Cremmer and S. Ferrara, Phys. Lett. 91B:61 (1980);
L. Brink and P. Howe, Phys. Lett. 91B:384 (1980);
S. Ferrara, Eleven-dimensional supergravity in superspace, in Unification of the Fundamental Particle Interactions, Eds. S. Ferrara, J. Ellis and P. van Nieuwenhuizen, Plenum Press (1980).
41. M.T. Grisaru and W. Siegel, Supergraphity II, Caltech preprint CALT-68-882 (1982).
42. M.J. Duff and C.N. Pope, to be published.
43. J. Milnor, Ann. of Math. 65:399 (1956).
44. B. de Wit and H. Nicolai, Extended supergravity with local SO(8) × SU(8) invariance, Phys. Lett. 108B:285 (1982).

LOCAL SO(8) × SU(8) INVARIANCE IN N = 8 SUPERGRAVITY AND ITS IMPLICATION FOR SUPERUNIFICATION*

B. de Wit

NIKHEF-H, Amsterdam

H. Nicolai

CERN, Geneva, Switzerland

INTRODUCTION

The unification of elementary particles and their fundamental forces is probably one of the most daring endeavours in theoretical physics. It necessarily involves the extrapolation in energy over many orders of magnitude where no direct experimental information is, or is expected to become, available. Thus one has to rely mainly on theoretical prejudices. Supergravity offers a unique and fascinating starting point for such a unification, which includes the gravitational forces ab initio, and which treats fermions and bosons in a symmetric fashion. The fact that gravity is an intrinsic part of these theories may seem to have a deep theoretical significance, but in practice it is one of the major obstacles for their application since the mass scale at which supergravity is defined is thus set by the Planck mass: 10^{19} proton masses. Nevertheless, there have been courageous attempts to investigate the implications of supergravity for phenomenologically relevant energies which have met with some success[1].

Extended supergravity is characterized by the number of independent supersymmetries under which the theory is invariant. These N supersymmetries are naturally fused with a rigid SO(N) symmetry group. The largest supergravity theory is defined for N = 8 [2-4]. In their pioneering paper[4], Cremmer and Julia demonstrated that extended supergravity theories have more symmetries than the aforementioned rigid SO(N), which are hidden in the conventional approach. First, there is an (off-shell) local group H without a kinetic term for the H gauge connections. Secondly there is a rigid (on-shell) non-compact symmetry group G which is an invariance

* Talk given by H. Nicolai

of the equations of motion. H is then isomorphic to the maximal compact subgroup of G, and the scalar fields of the theory are described as the coset space G/H. For N = 8, G = $E_{7(+7)}$ and H = = SU(8), and it is this case which we will consider here. Lower N supergravities may be obtained by consistent truncation. It is possible to impose a gauge condition for the local H symmetry, which corresponds to choosing a specific parametrization of the coset space. In that case, the equations of motion are invariant under a rigid H group; its SO(N) subgroup is a manifest symmetry of the Lagrangian in this formulation, which corresponds precisely to the SO(N) group mentioned above.

Since N extended supergravity contains $\binom{N}{2}$ Abelian gauge fields in the adjoint representation of SO(N) one may attempt to promote the rigid SO(N) to a local gauge group. This programme has been carried out for $N \leq 4$ [5] and, more recently, for N = 5 [6]. The latter paper also contains lowest order results for N = 8. Thus these theories contain a gauge coupling constant g in addition to the gravitational coupling constant κ (which we put equal to one in what follows).

In this contribution we wish to report on a successful attempt to formulate N = 8 supergravity with local SO(8) gauge invariance[7]. An important aspect of this work is that the local SU(8) invariance group can be preserved in the process of gauging SO(8). To clarify the significance of this fact we recall that previously (i.e., for $N \leq 5$) the gauging was achieved in the manifestly SO(N) invariant formulation, which is obtained in the local H gauge. The SO(N) group that was gauged was thus embedded in the rigid SU(N) × SU(N) of G × H. Here we proceed differently by leaving the local SU(8) intact and by gauging the SO(8) subgroup of E_7 instead. The conventional form with local SO(8) and without E_7 × SU(8) may be recovered by fixing the local SU(8) gauge afterwards. At present it is not known whether this theory follows from a non-trivial reduction from 11 dimensional supergravity[8].

There are essentially two reasons for keeping the local SU(8) throughout the calculation, the first one being technical: the E_7/SU(8) coset structure turns out to be just as indispensable for the construction and the consistency proof of gauged N = 8 supergravity as in the ungauged case, even though E_7 is no longer a symmetry of the theory since it is broken by the SO(8) gauging. The second reason is even more important. It has been known for quite some time[9] that an SO(8) gauge group is too small to comprise the observed particle states, and on the basis of this observation it has been argued that an SO(8) gauging is undesirable. Our results show that such an objection is not valid since both SU(8) and SO(8) may be relevant for the particle spectrum. In fact, we believe that the existence of N = 8 supergravity with local SO(8) × SU(8) has important implications for possible superunification scenarios;

some of these will be discussed in Section 4. We should add that some interesting features of this theory have been known for some time. For instance, the β function of the SO(8) gauge coupling constant g vanishes in the one-loop approximation[10)]. Furthermore, the gauging generates a potential for the scalar fields which is not bounded from below. Although this poses serious difficulties, it is conceivable that such a potential may ultimately trigger spontaneous breaking of supersymmetry, SU(8), or both. We will comment further on this in Section 4. Whether such a mechanism is compatible with previously proposed ones to obtain spontaneously broken N = 8 supergravity without gauging[11] remains an interesting problem.

RESULTS

In our conventions we will closely follow Refs. 3 and 6. The field multiplet of the N = 8 theory contains one graviton e_μ^a which is a singlet with respect to SO(8) × SU(8), eight gravitinos ψ_μ^i and 56 spin 1/2 fields χ^{ijk} which are assigned to representations of chiral SU(8) and singlets under SO(8), and 28 vector fields A_μ^{IJ} which transform as a 28 under SO(8). The scalar fields are represented by a 56-bein ("Sechsundfünfzigbein")

$$\mathcal{V} = \begin{bmatrix} u_{ij}{}^{IJ} & v_{ijKL} \\ v^{k\ell IJ} & u^{k\ell}{}_{KL} \end{bmatrix} \tag{2.1}$$

which is an element of E_7 in the fundamental representation. $\mathcal{V}$ transforms under local SU(8) from the left, and under E_7 from the right; its inverse can be written in terms of the same submatrices u and v

$$\mathcal{V}^{-1} = \begin{bmatrix} u^{ij}{}_{IJ} & -v_{k\ell IJ} \\ -v^{ijKL} & u_{k\ell}{}^{KL} \end{bmatrix} \tag{2.2}$$

All assignments have been collected in the Table below. Note that these are off-shell assignments in contradistinction to the ungauged

	e_μ^a	ψ_μ^i	A_μ^{IJ}	χ^{ijk}	$u_{ij}{}^{IJ}$	v^{ijIJ}
SO(8)	1	1	28	1	28	28
SU(8)	1	8	1	56	$\overline{28}$	28

case with E_7(rigid) × SU(8)(local) where only the field strengths and their duals, but not the vector fields themselves, could be fitted into on-shell representations of E_7. We write the N = 8 Lagrangian as follows:

$$\begin{aligned}\mathcal{L} = &-\frac{1}{2} e\, R(e,\omega) - \frac{1}{2} \varepsilon^{\mu\nu\rho\sigma} \bar{\psi}^i_\mu \gamma_\nu \overleftrightarrow{D}_\rho \psi_{\sigma i} - \\ &-\frac{1}{8} e \left[F^+_{\mu\nu IJ} F^+_{\mu\nu KL} (2S^{IJ,KL} - \delta^{IK}\delta^{JL}) + \text{h.c.}\right] - \\ &-\frac{1}{2} e \left[F^+_{\mu\nu IJ} O^{+\,KL}_{\mu\nu} S^{IJ,KL} + \text{h.c.}\right] - \\ &-\frac{1}{2} e\, \bar{\chi}^{ijk} \overleftrightarrow{\not{D}} \chi_{ijk} - \frac{1}{96} e\, |\mathcal{A}^{ijk\ell}_\mu|^2 - \\ &- \frac{1}{12} e \left[\bar{\chi}_{ijk} \gamma^\nu \gamma^\mu \psi_{\nu\ell} \mathcal{A}^{ijk\ell}_\mu + \text{h.c.}\right]\end{aligned} \tag{2.3}$$

where $F^+_{\mu\nu IJ}$ ($F^{-\,IJ}_{\mu\nu}$) denotes the self-dual (antiself-dual) Abelian fields strengths of the 28 vectors A^{IJ}_μ

$$F_{\mu\nu}{}^{IJ} = F^+_{\mu\nu IJ} + F^-_{\mu\nu}{}^{IJ} = 2\partial_{[\mu} A^{IJ}_{\nu]} \tag{2.4}$$

$S^{IJ,KL}$ is defined in terms of the submatrices of the 56-bein by the condition

$$(u_{ij}{}^{IJ} + v_{ijIJ})\, S_{IJ,KL} = u_{ij}{}^{KL} \tag{2.5}$$

and $O^{+\,IJ}_{\mu\nu}$ is defined by

$$\begin{aligned} u^{ij}{}_{IJ} O^{+\,IJ}_{\mu\nu} = &\frac{\sqrt{2}}{144} \eta\, \varepsilon^{ijk\ell mnpq} \bar{\chi}_{k\ell m} \sigma_{\mu\nu} \chi_{npq} - \\ &- \frac{1}{2} \bar{\psi}_{\lambda k} \sigma_{\mu\nu} \gamma^\lambda \chi^{ijk} + \\ &+ \frac{\sqrt{2}}{2} \bar{\psi}^i_\rho \gamma^{[\rho} \sigma_{\mu\nu} \gamma^{\sigma]} \psi^j_\sigma \end{aligned} \tag{2.6}$$

The derivatives in (2.3) are covariant with respect to local Lorentz and local SU(8) transformations; hence, besides the standard spin connection ω^{ab}_μ, we have SU(8) gauge fields $\mathcal{B}_\mu{}^i{}_j$ which satisfy

$$(\mathcal{B}_\mu{}^i{}_j)^* = -\mathcal{B}_\mu{}^j{}_i \; ; \; \mathcal{B}_\mu{}^i{}_i = 0 \tag{2.7}$$

These gauge fields occur in D_μ according to

$$D_\mu \phi^i = \partial_\mu \phi^i + \frac{1}{2} \mathcal{B}_\mu{}^i{}_j \phi^j \tag{2.8}$$

where ϕ^i is an SU(8) vector in the fundamental representation.

The SU(8) gauge fields $\mathcal{B}_{\mu j}^{i}$ do not correspond to dynamic degrees of freedom (at least classically); they can be expressed in terms of the physical fields of N = 8 supergravity. This dependence can be viewed as the result of an algebraic equation of motion (first order form) or of a conventional constraint (second order form). For our purposes it is most convenient to choose the second option; this brings our results in direct correspondence with those of Ref. 3. Also, the quantity $\mathcal{A}_{\mu}^{ijk\ell}$, which characterizes the scalar kinetic terms in (2.3), is dependent. The dependence of $\mathcal{A}_{\mu}$ and $\mathcal{B}_{\mu}$ on the 56-bein is determined by the requirement of E_7 invariance; the only quantity of that kind which contains one derivative is given by $D_{\mu}\mathcal{V}\cdot\mathcal{V}^{-1}$. This 56 × 56 matrix transforms covariantly under local SU(8) and takes its values in the Lie algebra of E_7. The dependence of $\mathcal{A}_{\mu}$ and $\mathcal{B}_{\mu}$ is now defined by

$$D_{\mu}\mathcal{V}\cdot\mathcal{V}^{-1} = \begin{bmatrix} 0 & -\frac{\sqrt{2}}{4}\mathcal{A}_{\mu ijk\ell} \\ -\frac{\sqrt{2}}{4}\mathcal{A}_{\mu}^{mnpq} & 0 \end{bmatrix} \tag{2.9}$$

The diagonal blocks of (2.9) characterize the SU(8) subalgebra and are used to express $\mathcal{B}_{\mu}$ in terms of the submatrices of $\mathcal{V}$ and their covariant derivatives. The part of the algebra orthogonal to SU(8) defines $\mathcal{A}_{\mu}$ in a similar fashion. Note that $\mathcal{A}_{\mu}$ does not explicitly depend on $\mathcal{B}_{\mu}$ in this way. Its classification as a component of the E_7 Lie algebra implies

$$\mathcal{A}_{\mu}^{ijk\ell} = \frac{1}{24}\eta\,\varepsilon^{ijk\ell mnpq}\mathcal{A}_{\mu mnpq} \tag{2.10}$$

This is a typical example of the kind of argument that is of crucial importance throughout this paper. It is not possible to show the validity of (2.10) directly from the explicit dependence of $\mathcal{A}_{\mu}$ on the 56-bein; instead, we have to rely on group-theoretic arguments based on E_7.

It is possible to view the matrix $\begin{bmatrix}\mathcal{B} & \mathcal{A}^* \\ \mathcal{A} & \mathcal{B}^*\end{bmatrix}$ as a connection of a local E_7 group. In that context, (2.9) specifies that the connection is pure gauge and has vanishing E_7 field strengths. Indeed, application of a second SU(8) covariant derivative D_{ν} on (2.9) and antisymmetrization in μ and ν leads to

$$([D_{\mu},D_{\nu}]\mathcal{V})\mathcal{V}^{-1} = -\frac{1}{8}\begin{bmatrix}0 & \mathcal{A}^{*}_{\mu} \\ \mathcal{A}_{\mu} & 0\end{bmatrix}\begin{bmatrix}0 & \mathcal{A}^{*}_{\nu} \\ \mathcal{A}_{\nu} & 0\end{bmatrix} - \frac{\sqrt{2}}{4}\begin{bmatrix}0 & D_{\mu}\mathcal{A}^{*}_{\nu} \\ D_{\mu}\mathcal{A}_{\nu} & 0\end{bmatrix} - (\mu \leftrightarrow \nu) \tag{2.11}$$

On the other hand, the commutator of two covariant derivatives is equal to the field strength; as SU(8) acts on $\mathcal{V}$ from the left, the left-hand side of (2.11) takes the simple form

$$([D_\mu, D_\nu]\mathcal{V}) \cdot \mathcal{V}^{-1} = \begin{bmatrix} \delta^{[i}_{[m} \, \mathcal{F}(\mathcal{B})_{\mu\nu n]}{}^{j]} & 0 \\ 0 & \delta^{[k}_{[p} \, \mathcal{F}(\mathcal{B})_{\mu\nu}{}^{\ell]}{}_{q]} \end{bmatrix} \tag{2.12}$$

where $\mathcal{F}(\mathcal{B})_{\mu\nu}{}^{i}{}_{j}$ denotes the SU(8) field strengths. Equations (2.11) and (2.12) play a crucial role in establishing the invariance of the supergravity action; this requires a number of partial integrations which lead to SU(8) field strengths by means of the Ricci identity (2.12) or to derivatives on $\mathcal{A}_\mu$. By using (2.11) and (2.12) we find that the antisymmetric derivative of $\mathcal{A}_\mu$ vanishes and that the SU(8) field strength can be expressed in terms which cancel against other variations. Indeed, (2.11) and (2.12) were previously found by requiring supersymmetry invariance of the action.

The gauging of SO(8) is now affected by further extending the covariant derivative with respect to local SO(8) embedded in the E_7 group. For instance, we define

$$D_\mu u_{ij}{}^{IJ} \equiv \partial_\mu u_{ij}{}^{IJ} + \mathcal{B}_\mu{}^{k}{}_{[i} u_{j]k}{}^{IJ} - 2g\, A_\mu^{K[I} u_{ij}{}^{J]K} \tag{2.13}$$

where A_μ^{IJ} are the 28 vectors of N = 8 supergravity. At the same time we replace the field strengths $F_{\mu\nu}^{IJ}$ by their fully SO(8) covariant counterparts

$$F_{\mu\nu}{}^{IJ} \equiv 2\partial_{[\mu} A_{\nu]}{}^{IJ} - 2g\, A_{[\mu}{}^{IK} A_{\nu]}{}^{KJ} \tag{2.14}$$

The presence of the order g terms in the Lagrangian and transformation rules violates the supersymmetry invariance of the original action. To re-establish the invariance one has to introduce new terms in the Lagrangian and transformations. These can be parametrized in terms of three tensorial functions A_{1-3} which depend on the scalars contained in the 56-bein. The parametrization takes the following form[6]

$$\begin{aligned} \delta_g \bar{\psi}_\mu^i &= -\sqrt{2}\, g\, \bar{\varepsilon}_j \gamma_\mu A_1^{ji} \\ \delta_g \bar{\chi}^{ijk} &= -2g\, \bar{\varepsilon}^\ell A_{2\ell}{}^{ijk} \end{aligned} \tag{2.15}$$

$$
\begin{aligned}
\mathcal{L}_g &= \sqrt{2} g\, e\, A_{1ij}\, \bar{\psi}^i_\mu \sigma^{\mu\nu} \psi^i_\nu + \text{h.c.} \\
&+ \frac{1}{6} g\, e\, A^i_{2jk\ell}\, \bar{\psi}^\mu_i \gamma_\mu \chi^{jk\ell} + \text{h.c.} \\
&+ g\, e\, A_3^{ijk,\ell mn}\, \bar{\chi}_{ijk} \chi_{\ell mn} + \text{h.c.}
\end{aligned}
\tag{2.15}
$$

$$
L_{g^2} = g^2\, e\, \left(\frac{3}{4}|A_1^{ij}|^2 - \frac{1}{24}|A^i_{2jk\ell}|^2\right)
$$

Note that the SU(8) tensors A_{1-3} must satisfy certain symmetry properties as a consequence of the way in which they appear in (2.15). The solution for A_{1-3} can be found by requiring that all $gF\bar{\varepsilon}\psi$ and $gF\bar{\varepsilon}\chi$ variations of the Lagrangian cancel:

$$
\begin{aligned}
A_1^{ij} &= -\frac{4}{21} T_m{}^{ijm} \\
A_{2m}{}^{ijk} &= -\frac{4}{3} T_m{}^{[ijk]} \\
A_3^{ijk,\ell mn} &= \frac{\sqrt{2}}{108} \eta\, \varepsilon^{ijkpqr[\ell m} T^{n]}{}_{pqr}
\end{aligned}
\tag{2.16}
$$

where the SU(8) tensor T is defined by

$$
T_\ell{}^{kij} \equiv (u^{ij}{}_{IJ} + v^{ijIJ})(u_{\ell m}{}^{JK} u^{km}{}_{KI} - v_{\ell mJK} v^{kmKI}) \tag{2.17}
$$

which is manifestly antisymmetric in i and j. Note that T is not invariant with respect to the full E_7 group, but it is invariant under local SO(8) and covariant under SU(8).

In order for the solution (2.16) to be consistent, the tensor T has to obey a number of non-trivial identities; these follow from the E_7 structure of the 56-bein. For instance, one can prove that T admits the following decomposition

$$
T_\ell{}^{kij} = T_\ell{}^{[kij]} + \frac{2}{7} \delta_\ell^{[i} T_m{}^{j]mk} \tag{2.18}
$$

where $T_m{}^{jmk}$ is symmetric in j,k and $T_\ell{}^{[kij]}$ is completely antisymmetric and traceless, i.e.,

$$
T_k{}^{[kij]} = 0 \tag{2.19}
$$

For the proof of invariance of the Lagrangian in order g and g^2, one also needs the following identities in terms of A_1 and A_2

$$D_\mu A_1^{ij} = -\frac{1}{24}\sqrt{2}\,\mathcal{A}_\mu^{ik\ell m}\,A^j_{2k\ell m} - (i \leftrightarrow j)$$

$$D_\mu A_{2i}^{jk\ell} = -\frac{\sqrt{2}}{2} A_{1im}\,\mathcal{A}_\mu^{mjk\ell} - \frac{3}{4}\sqrt{2}\,\mathcal{A}_\mu^{mn[jk}A^{\ell]}_{2imn} - \frac{\sqrt{2}}{4}\,\mathcal{A}_\mu^{mnp[j}\delta_i^k A^{\ell]}_{2mnp}$$

$$A_{2i}^{k\ell m}\,A^j_{2klm} - 18\,A_{1ik}A_1^{kj} = \tag{2.20}$$

$$= \frac{1}{8}\,\delta_i^{\,j}\,(A_{2p}^{k\ell m}A^p_{2klm} - 18\,A_{1k\ell}A_1^{k\ell})$$

$$- A_{1it}A^t_{2jk\ell} + \frac{\eta}{24}\,\varepsilon_{abcpqr[jk}A_{2\ell]}^{abc}A_{2i}^{pqr} =$$

$$= \text{antisymmetric and self-dual in } [ijk\ell]$$

where the defining equation (2.9) for $\mathcal{A}_\mu$ now includes the covariantization with respect to local SO(8). The proof of (2.20) is again based on the E_7 properties of $\mathcal{V}$.

SPECIAL GAUGE CHOICE

It is instructive to impose an SU(8) gauge choice which corresponds to an explicit parametrization of the E_7/SU(8) coset space. In the symmetric gauge[4], the 56-bein becomes

$$\mathcal{V} = \exp\begin{bmatrix} 0 & -\frac{\sqrt{2}}{4}\phi_{ijk\ell} \\ -\frac{\sqrt{2}}{4}\bar{\phi}^{mnpq} & 0 \end{bmatrix} \tag{3.1}$$

and introducing the variable y [4]

$$y_{ij,k\ell} \equiv \phi_{ijmn}\left(\frac{\tanh\sqrt{(1/8)\bar{\phi}\phi}}{\sqrt{\bar{\phi}\phi}}\right)^{mn}{}_{k\ell} \tag{3.2}$$

(Matrix-multiplication always over pairs of indices.) We obtain an explicit representation for $\mathcal{V}$ [4]

$$\mathcal{V} = \begin{bmatrix} P^{-\frac{1}{2}} & -P^{-\frac{1}{2}}\, y \\ -\bar{P}^{-\frac{1}{2}}\, \bar{y} & \bar{P}^{-\frac{1}{2}} \end{bmatrix} \tag{3.3}$$

with

$$P_{ij}{}^{k\ell} \equiv \delta_{ij}{}^{k\ell} - y_{ijmn}\, \bar{y}^{mnk\ell} \tag{3.4}$$

[the notation is the same as in Ref. 3]. The inverse 56-bein is given by formula (2.2). Inserting this parametrization into $T_i{}^{jkl}$, we obtain the functions A_{1-3} as infinite series of the self-dual scalar fields $\phi^{ijk\ell}$. For A_1 and A_2 these expressions read as follows up to cubic order

$$A_1{}^{ij} = (1 - \tfrac{1}{96}|\phi|^2)^{-\frac{1}{2}}\, \delta^{ij} + + \frac{\sqrt{2}}{96}\phi^{ikmn}\phi_{mnpq}\phi^{pqkj} + O(\phi^4)$$

$$A_{2\ell}{}^{ijk} = -\frac{\sqrt{2}}{2}(1 - \tfrac{1}{144}|\phi|^2)\phi^{ijkl} - \tfrac{3}{8}\,\phi_{mnl[i}\phi^{jk]mn} + \frac{\sqrt{2}}{16}\,\phi_{1pqr}\phi^{pqs[i}\phi^{jk]\,rs} + O(\phi^4) \tag{3.5}$$

$$|\phi|^2 \equiv \phi^{ijkl}\,\phi_{ijkl}$$

A_3 is related to A_2 by (2.16). In the reduction to N = 5, these results can be compared with those of Ref. 6. Both SO(8) and SU(8) are affected by the gauge choice (3.1); however, the non-trivial SO(8) subgroup acting on both SO(8) and SU(8) indices is preserved. Determining $\mathcal{B}_{\mu}{}^{i}{}_{j}$ from the SO(8) covariantized version of (2.9), we find

$$\mathcal{B}_{\mu}{}^{i}{}_{j} = \frac{2}{3}\Big\{ (\partial_\mu P^{-\frac{1}{2}} \cdot P^{-\frac{1}{2}})_{jm}{}^{im} - (\partial_\mu(P^{-\frac{1}{2}}y) \cdot \bar{P}^{-\frac{1}{2}}\,\bar{y})_{jm}{}^{im} - - 2g\, A_\mu^{mp}(P^{-\frac{1}{2}}{}_{jr}{}^{qm}P^{-\frac{1}{2}}{}_{pq}{}^{ir} - (P^{-\frac{1}{2}}y)_{jrqm}(\bar{P}^{-\frac{1}{2}}\bar{y})^{pqir}\Big\} \tag{3.6}$$

In order to rewrite (3.6) in a more suggestive form we introduce the fully SO(8) covariant derivative $\mathcal{D}_\mu$

$$\mathcal{D}_\mu P^{-\frac{1}{2}}{}_{jr}{}^{pq} \equiv \partial_\mu P^{-\frac{1}{2}}{}_{jr}{}^{pq} - 2g\, A_{\mu t[j}P^{-\frac{1}{2}}{}_{r]\,t}{}^{pq} - - 2g\, A_{\mu t[p}P^{-\frac{1}{2}}{}_{jr}{}^{q]\,t} \tag{3.7}$$

which now acts on all indices. In terms of $\mathcal{D}_\mu$, (3.6) becomes

$$\mathcal{B}_{\mu\ j}^{\ i} = \frac{2}{3}\left\{\mathcal{D}_\mu P^{-\frac{1}{2}} \cdot P^{-\frac{1}{2}})_{jm}^{\ \ im} - (\mathcal{D}_\mu(P^{-\frac{1}{2}}y) \cdot \bar{P}^{-\frac{1}{2}}\bar{y})_{jm}^{\ \ im}\right\} + \tag{3.8}$$
$$+ \frac{4}{3}g\ A_{\mu t[j}\delta^{ir}_{r]}\ t$$

where we have also used the fact that $P^{-1} - (P^{-\frac{1}{2}}y)(\bar{P}^{-\frac{1}{2}}\bar{y}) = \mathbb{1}$. Hence,

$$\mathcal{B}_{\mu\ j}^{\ i} = -2g\ A_\mu^{ij} + \text{non-linear SO(8) covariant terms} \tag{3.9}$$

In the special gauge (3.1) there is no longer a distinction between SO(8) and SU(8) indices, and Eq. (3.9) shows that A_μ^{IJ} will now couple minimally to all fields that carry SO(8) and/or SU(8) indices. This corresponds to the standard formulation of supergravity with local SO(N) that has been obtained for $N \leq 5$.

We now turn to the discussion of the cosmological term induced by the gauging. It is important that for $N \geq 4$ this is not a cosmological constant but rather a scalar field potential which is furthermore rather badly behaved being unbounded from below. From (2.15) it is evident that the potential

$$g^{-2}\ \mathcal{P}(\mathcal{V}) = \frac{1}{24}\,|A_{2i}^{\ \ jk\ell}|^2 - \frac{3}{4}\,|A_1^{\ ij}|^2 \tag{3.10}$$

is the difference between two positive definite terms and as such can take values between $+\infty$ and $-\infty$. However, one should keep in mind that A_1 and A_2 are not independent but depend on the 56-bein in the way specified in the preceding section. To analyze this aspect a little further it is advantageous to go to the special gauge, and we consider here the cases N = 4 and N = 5.

i) For N = 4 [5], there is one complex scalar ϕ and the potential

$$\mathcal{P}(\phi) = -2 - \frac{4}{1-|\phi|^2}\ ,\quad |\phi|^2 \leq 1 \tag{3.11}$$

is unbounded from below but bounded from above. The singularity at finite $|\phi|^2$ is an artifact of the coset parametrization since the variable ϕ corresponds to the variable y introduced in (3.2).

ii) For N = 5 [6] there are five complex scalars ϕ^i and the potential is $(\phi^i \equiv \phi_i^*)$

$$\mathcal{P}(\phi_i) = -2 - \frac{4}{1-\phi_i\phi^i} + \tag{3.12}$$
$$+ \frac{1}{2}\frac{1}{(1-\phi_i\phi^i)^2}\ \left[(\phi_i\phi^i)^2 - |\phi_i\phi_i|^2\right],\quad \phi_i\phi^i \leq 1$$

which, of course, is still unbounded from below. The Cauchy-Schwarz inequality tells us that

$$|\phi_i\phi^i|^2 - |\phi_i\phi_i|^2 \geq 0 \tag{3.13}$$

and is saturated if and only if

$$\phi_i = e^{i\alpha}\,\phi^i \tag{3.14}$$

for some $\alpha \in [0, 2\pi)$. In all those directions the potential tends to $-\infty$ as $|\phi|^2 \to 1$; this set in particular includes all possible N = 4 truncations. In any other direction, (3.13) is strictly positive and as its coefficient in (3.12) diverges faster than the second term in (3.12) for $|\phi|^2 \to 1$, the potential tends to $+\infty$ there. It is amusing that the set (3.14) with "bad behaviour" constitutes a set of measure zero on the sphere $|\phi|^2 = 1$ in $\mathbb{C}^5$.

For N = 8 one may also insert the special gauge choice into (3.1) thereby obtaining the N = 8 potential explicitly in terms of the self-dual scalar fields. Unfortunately, the resulting expression is rather unwieldy since it cannot be reduced to simply contracted structures such as $\phi^{ijk\ell}\phi_{ijk\ell}$, and therefore we shall not discuss it here. The simplicity of (3.10) in terms of A_1 and A_2 is entirely lost in any special gauge.

CONCLUDING REMARKS

The result that N = 8 supergravity exists with local SO(8) × × SU(8) widens the range of possible scenarios for superunification. One possibility is that SO(8) × SU(8) is actually relevant for the particle spectrum at the Planck mass scale. The problem of embedding the SU(3) × SU(2) × U(1) invariance of strong, weak and electromagnetic interactions into SO(8) × SU(8) may then be resolved in a variety of ways. At any rate, the SO(8) gauge interaction breaks the E_7 invariance of the theory. As is well known, unitary representations of the non-compact E_7 group are infinite dimensional and therefore superunification based on E_7 × SU(8) would lead to a proliferation of particle states at the Planck mass[12]. One has argued that an infinite number of particle states is a welcome feature which leads to a necessary re-arrangement of helicity states. However, the fact that another version of N = 8 supergravity is now available may require a fundamental re-orientation of the superunification strategy.

Since the local SO(8) and SU(8) are realized in a different fashion in the Lagrangian they will presumably play different roles. One possibility which we find rather intriguing is the following "preconfinement" hypothesis: one may assume that the SO(8) Yang-Mills interactions provide the force which confines the preons at the Planck mass scale. In that case, all "observable" states would have to be SO(8) singlets. Since the only supergravity fields that carry SO(8) indices are the spin 0 and spin 1 fields, we are led to the conclusion that only the graviton, the gravitinos and the spin 1/2 fields correspond to observable particle states. An important aspect of this mechanism is that it gives us a well-defined confinement criterion [namely "SO(8)-neutralness"] which would explain why the graviton is not confined. Such a criterion appears to be lacking in the conventional approach to superunification. The SO(8) preconfinement hypothesis also leads to supersymmetry breaking: because only part of the basic supermultiplet is confined, N = 8 supersymmetry can no longer be realized. Some dynamical restoration of symmetry may take place because the bound states may fill the gap and lead to a supersymmetric spectrum again. However, this would presumably lead to a lower N version of rigid supersymmetry which may be relevant at much lower mass scales than local N = 8 supersymmetry. If the latter is broken at the Planck scale one is no longer forced to put the bound states into N = 8 supermultiplets which could possibly circumvent some problems of the standard scheme. In view of the fact that the β function is known to vanish at one loop and is conjectured to vanish to all orders, the SO(8) preconfinement would have to be non-perturbative. Further studies of the dynamical aspects are badly needed and we wish to emphasize that these should now be undertaken within a broader context than before.

We have already mentioned the fact that the potential induced by the gauging has the annoying feature of being unbounded from below. At present one can envisage (at least) two possible resolutions to this problem. The first is that the potential may not be all that significant any more if it lies in the confined sector, and we remind the reader that unbounded potentials have been encountered and dealt with in physics before. Another possibility which may be related to the first is that additional terms in the potential are generated dynamically by the same mechanism that produces the kinetic terms for the SU(8) connections. Such additional terms could remedy the defect of unboundedness and would correspond to a quantum mechanical stabilization of the potential. The effective potential would then be a sum of two terms such that

$$\mathcal{P}_{\text{eff}}(\mathcal{V}) = \mathcal{P}(\mathcal{V}) + \mathcal{P}_{\text{dyn.}}(\mathcal{V}) \geq \text{const.} \tag{4.1}$$

Indeed, the form of $\mathcal{P}(\mathcal{V})$ in (3.10) bears a striking resemblance to a Higgs-like potential with two mass terms for the SU(8) composite tensors A_1 and A_2 but with higher order (ϕ^4-like) interactions missing. If those were generated dynamically with the correct sign one would obtain a spontaneous breaking of SU(8) via a non-vanishing vacuum expectation value

$$\langle A_1^{ij} \rangle \neq 0 \tag{4.2}$$

for example. We stress that in the ungauged theory there is no such possibility for breaking SU(8) due to the absence of a scalar field potential: if one insists on E_7 invariance, there is no way of constructing non-trivial SU(8) objects from the 56-bein $\mathcal{V}$. Non-vanishing vacuum expectation values for A_1 and A_2 would give masses to the spinors via (2.15) and this might entail some interesting phenomenological consequences. Thus, the gauging of SO(8) might well turn out to be indispensable for bridging the gap between N = 8 supergravity and present-day phenomenology.

ACKNOWLEDGMENTS

We thank the participants at this meeting and C.B. Lang for stimulating discussions.

REFERENCES

1. J. Ellis, M.K. Gaillard, L. Maiani and B. Zumino in Unification of the Fundamental Particle Interactions, Plenum Press (1980);
J. Ellis, M.K. Gaillard and B. Zumino, Phys. Lett. 94B:343 (1980);
J.P. Derendinger, S. Ferrara and C.A. Savoy, Nucl. Phys. B188:77 (1981);
J. Ellis, M.K. Gaillard and B. Zumino, preprint LAPP-TH-44 (1981).
2. B. de Wit and D. Freedman, Nucl. Phys. B130:105 (1977).
3. B. de Wit, Nucl. Phys. B158:189 (1979).
4. E. Cremmer and B. Julia, Phys. Lett. 80B:48 (1978); Nucl. Phys. B159:141 (1979).
5. D.Z. Freedman and A. Das, Nucl. Phys. B120:221 (1977);
E. Fradkin and M.A. Vasiliev, Lebedev Inst. preprint (1976);
D.Z. Freedman and J. Schwarz, Nucl. Phys. B137:333 (1978);
E. Cremmer, S. Ferrara and J. Scherk, unpublished.
6. B. de Wit and H. Nicolai, Nucl. Phys. B188:98 (1981).
7. B. de Wit and H. Nicolai, Phys. Lett. 108B:285 (1982).
8. E. Cremmer, B. Julia and J. Scherk, Phys. Lett. 76B:409 (1978).
9. M. Gell-Mann, unpublished.
10. S.M. Christensen, M.J. Duff, G.W. Gibbons and M. Rocek, Phys. Rev. Lett. 45:161 (1980);
T. Curtright, Phys. Lett. 102B:17 (1981).
11. E. Cremmer, J. Scherk and J. Schwarz, Phys. Lett. 84B:83 (1979);
A. Aurilia, H. Nicolai and P.K. Townsend, Nucl. Phys. B176:509 (1980).
12. M. Günaydin, these proceedings.

RECENT DEVELOPMENTS IN THE GROUP-MANIFOLD APPROACH

R. D'Auria, P. Fré and T. Regge

Istituto di Fisica Teorica, Università di Torino, Italy
and
Istituto Nazionale di Fisica Nucleare, Sezione di Torino

The "group-manifold" scheme for the formulation of supersymmetric theories, its goals and advantages have already been reviewed in other conferences: in particular we refer the reader to Ref. 1 and to all the papers quoted therein.

Here we plan to describe the most recent results and developments within this approach. There are three of them and, correspondingly, this contribution is subdivided into three chapters covering three different but related topics. The first chapter is a report on an investigation by R. D'Auria and T. Regge on Gravitational Instantons. It is shown how the use of first order formalism, typical of the group manifold approach, allows for the existence of non-trivial topological solutions, lost in the second order formalism, which might be of relevance in the quantum theory.

Chapter II is an introduction to a powerful group theoretical technique for the analysis of Bianchi identities and group-manifold equations of motion, essential in deriving geometric Lagrangians and Auxiliary Fields. Described in short this technique is a systematics of Fierz identities, which are reinterpreted as Clebsh-Gordan decompositions of representation products.

Chapter III is a summary of the geometric derivation of D = 11 Supergravity, recently obtained by R. D'Auria and P. Fré, and briefly illustrates the hidden supergroup underlying the theory, whose identification is yielded by the geometric formulation.

REFERENCE

1. For a general review see: R. D'Auria, P. Fré and T. Regge Trieste preprint IC/81/54 (1981), Lectures given at the Spring School on Supergravity, Trieste April 1981.

CHAPTER I - A GRAVITATIONAL INSTANTON

Presented by: T. Regge
Reference: R. D'Auria and T. Regge, Torino preprint IFTT 409 (1981), Nucl. Phys. B in press.

INTRODUCTION

The problem of characterizing vacuum solutions in conventional gravity has attracted much attention in recent years. In particular excellent reviews on Euclidean vacuum solutions and the gauge structure of gravity have appeared in Refs 1 and 2. These field configurations have much in common with conventional Yang-Mills instantons discovered by Belavin et al.[3]. In particular they possess a non-trivial topology and a self-dual or antiself-dual curvature tensor.

But just for these reasons none of these solutions is asymptotically flat in the global sense: loosely speaking the set of points lying at a fixed large distance from the instanton do not form a manifold S^3 but rather a quotient S^3/Γ where Γ is a discrete subgroup $S^3(SU(2))$. In general relativity the vanishing of torsion implies that the SO(4) bundle is a sub-bundle of the tangent bundle.

This forces the topological invariants of SO(4) to be identified with those of the tangent bundle; the base manifold cannot be globally R^4 or S^4, it must have handles if the instanton is nontrivial.

And in turn the vanishing of torsion is a consequence of the first order formulation of gravity as introduced by Palatini. In Ref. 4 it was conjectured that the vanishing of torsion could be a dynamical effect in many ways similar to the Meissner effect in superconductivity. This leaves room for "torsion vortices", analogs of the Abrikosov tubes in superconductors, in which torsion does not necessarily vanish and creates subtle topological effects. In Ref. 4 some torsion vortices appeared as special configurations of affine geometry in the formulation of Cartan. These configurations did not correspond to any solution of any known field theory generalizing conventional gravity.

In other words they did not follow from an analogue of Landau-Ginzburg equations for superconductors. They exhibited however many of the interesting features of the Landau-Ginzburg equations. In particular the vanishing of the gap inside a flux tube had a perfect analogue in the vanishing of the vierbein inside the instanton.

As was already discussed in Ref. 4 the vanishing of the vierbein at some point is not a disastrous feature of theory. In ordinary gravity it is indeed an utterly undesirable condition because one insists on keeping the torsion zero at all times. This in turn implies that the curvature tensor becomes infinite at such a point, the action itself diverges and one speaks of a true singularity. But here one simply deals with fields which, including curvature, are everywhere regular (in fact C^∞); the action is finite and the configuration acceptable. We do not know yet if a theory of this kind has any chance to be realistic. Quite possibly some of the relevant features could reappear in more structured attempts. It would also be interesting to see whether there is possibility of phase transitions into the non-superconducting phase. Here instantons form a condensate and the vierbein, analogue of the gap, should vanish everywhere along with the metric tensor. Concepts of this sort could be useful near singularities like the big bang or a black hole.

1. - THE GRAVITATIONAL MEISSNER EFFECT

We start from the vierbein one form V^a related to the metric tensor through:

$$g_{\mu\nu} = \eta_{ab} V^a_\mu V^b_\nu \tag{1.1}$$

Next we introduce the Lorentz connection one-form $\omega^{ab} \equiv \omega^{ab}_\mu dx^\mu$. The torsion R^a is given by:

$$R^a = dV^a - \omega^{ab} \wedge V_b \equiv \mathcal{D} V^a \tag{1.2}$$

and the curvature R^{ab} by

$$R^{ab} = d\omega^{ab} - \omega^{ac} \wedge \omega_c^{\cdot\, b} \tag{1.3}$$

The R^a and R^{ab} are two-forms. Ordinary Einstein theory is based on the vanishing of torsion and on the metricity condition:

$$\omega^{ab} = -\omega^{ba} \tag{1.4}$$

Is the vanishing of torsion really that fundamental ? In order to discuss this point we draw an analogy with superconductors. The Landau-Ginzburg theory contains a Maxwell field A_μ coupled to a scalar field f. The gap f obeys the equation of motion:

$$D_\mu D^\mu f = g f\left(|f|^2 - \lambda^2\right) \qquad \left(D_\mu \equiv \partial_\mu + ieA_\mu\right) \tag{1.5}$$

Near the symmetry-breaking vacuum the magnitude of f is constant:

$$|f| = \lambda \tag{1.6}$$

Suppose now we work in the static case: Eq. (1.5) is elliptical and far from walls and impurities the general solution drifts into an asymptotic regime in which it is covariantly constant:

$$D_\mu f \equiv \left(\partial_\mu + ieA_\mu\right) f = 0 \tag{1.7}$$

By differentiating again one obtains:

$$[D_\mu, D_\nu] f = ie F_{\mu\nu} f = 0 \tag{1.8}$$

so that either f or $F_{\mu\nu}$ must vanish. This is the Meissner effect. In a type II superconductor we have $F_{\mu\nu} = 0$ almost everywhere with the exception of the inside of the Abrikosov tubes. Outside we have:

$$A_\mu = -\frac{i}{e}\partial_\mu \theta \tag{1.9}$$

$$f = \lambda e^{i\theta} \tag{1.10}$$

where θ is a phase. But consider a path enclosing the tube: along the path θ varies from 0 to 2π. If f is differentiable it must vanish somewhere along a line inside the flux tube. The map $S^1 \to U(1)$ defined by f has a definite homotopy type which depends on the flux of the magnetic field.

A gravitational analogue of this structure may arise by considering objects which are covariantly constant in the "superconducting region" where Einstein equations are valid. An obvious candidate is the vierbein since the vanishing of torsion indeed implies that the covariant derivative, according to the group SO(4) and the connection ω^{ab} vanishes everywhere. From the vanishing of the torsion, taking derivatives once more we find:

$$\mathcal{D}\mathcal{D}V^a = -R^{ab} \wedge V_b = 0 \tag{1.11}$$

Therefore ordinary Riemannina space-time is the analogue of the flux-free superconductor. The condition $R^a = 0$ locks together the SO(4) gauge group and the GL(4,r) group of co-ordinate transformations. The topological invariants of SO(4) are tied up with those of the underlying differential manifold and are identifiable with the Euler characteristics and Hirzebruch signature of the base. If on the other hand the vierbein vanishes somewhere this is no longer true and also it cannot be true that the torsion vanishes everywhere. A torsion vortex is a finite region of space-time combining both features of flux tubes and gravitational instantons. The following table will be useful in comparing gravity with superconductors:

	Gravity	Superconductivity
Bundle Group	$SO(4) \subset GL(4)$	$U(1) \subset \mathbb{C}$
Connection one-form	ω^{ab}	A
Field Strength	R^{ab}	$F = dA$
Covariantly Constant Object	$\begin{cases} V^a_\mu \in GL(4) \\ R^a = \mathcal{D}V^a = 0 \end{cases}$	$\begin{cases} f \in C \\ D_\mu f = 0 \end{cases}$
Meissner Effect	$R^{ab} \wedge V_b = 0$	$F_{\mu\nu} f = 0$
Topological Object	Torsion Vortex (Poinlike)	Magnetic Vortex (Line)
Enclosing Manifold	S^3 = three-sphere	S^1 = circle
Insidc Vortex	$R^a \neq 0$; $R^{ab} \wedge V_b \neq 0$	$F_{\mu\nu} \neq 0$
Vortex Map	Iwasawa Decomposition of V^a_μ	$U(1)$: $f = \lvert f \rvert e^{i\theta}$
Topological charge	Euler and Pontryagin numbers	Quantized Flux of F
Fundamental Equation	See action (2.24)	Landau - Ginzburg

We obtain a very simple space-time with torsion vortices by "deforming" the Levi-Civita connection of an ordinary flat Euclidean metric. We introduce co-ordinates $\hat{x}^\mu$ on R^4. Let $r = \rho^2 = y_\mu y^\mu$ and let a system of flat space vierbeins be defined in a non-standard way as follows:

$$\begin{aligned} e^0 &= 2y_\mu dy^\mu = dr \\ e^1 &= y_0 dy_1 - y_1 dy_0 + y_2 dy_3 - y_3 dy_2 = r\sigma_1 \\ e^2 &= y_1 dy_2 - y_2 dy_1 + y_3 dy_0 - y_0 dy_3 = r\sigma_2 \\ e^3 &= y_2 dy_3 - y_3 dy_2 + y_0 dy_1 - y_1 dy_0 = r\sigma_3 \end{aligned} \tag{1.12}$$

where σ_1, σ_2, σ_3 are the SU(2) left-invariant forms. The one-forms e^a obey the differential equations:

$$\begin{aligned} de^0 &= 0 \\ de^i &= \frac{1}{r}\left(e^0 \wedge e^i + \epsilon^{ijk} e^j \wedge e^k\right) \end{aligned} \tag{1.13}$$

We see that the e^a are C^∞ one-forms. The Levi-Civita connection forms derived from Cartan's structural equation turn out to be self-dual and equal to:

$$\omega^{i0} = \omega^{jk} = -\frac{1}{r} e^i \quad (i, jk = \text{cyclic}) \tag{1.14}$$

The ω^{ab} are singular at the origin as expected since the vierbein vanishes there. If we calculate the curvature tensor we find that it vanishes identically. This space-time is indeed flat. We consider now a regularized connection:

$$\omega^{i0} = \omega^{jk} = -\varphi(t) e^i \quad (i, jk = \text{cyclic}) \tag{1.15}$$

where φ is a regular and differentiable function which vanishes like $1/r$ at infinity. The torsion is now given as:

$$\begin{cases} R^{0} = 0 \\ R^{i} = \left(-\varphi + \frac{1}{r}\right)\left(e^{0}\wedge e^{i} + \epsilon^{ijk} e^{j} \wedge e^{k}\right) \end{cases} \tag{1.16}$$

The curvature becomes:

$$R^{10} = R^{23} = -\left(\varphi' + \frac{\varphi}{r}\right) e^{0} \wedge e^{1} - 2\left(\frac{\varphi}{r} - \varphi^{2}\right) e^{2} \wedge e^{3} \quad \text{(cyclic)} \tag{1.17}$$

Quite obviously both torsion and curvature are differentiable. The choice:

$$\varphi = -\frac{1}{r}\left(1 - e^{-r/\lambda}\right) \tag{1.18}$$

yields a finite action:

$$\frac{1}{4}\int R^{ab} \wedge e^{c} \wedge e^{d} \epsilon_{abcd} = \int R^{\mu\nu}{}_{,\mu\nu} \sqrt{-g}\, d^{4}x = \frac{\lambda^{2}}{16} \tag{1.19}$$

With this choice the Euler characteristic χ and Pontryagin number P are changed. P is now given by:

$$P = -\frac{1}{8\pi^{2}}\int R^{ab} \wedge R_{ab} = 2 \tag{1.20}$$

instead of P = 0 in the flat case. Moreover (see Ref. 4):

$$\chi = 0 \tag{1.21}$$

while in flat space $\chi = 1$. The SO(4) bundle is not equivalent to the bundle of orthonormal tangent frames of Euclidean space. The proposed structure has the effect of detaching the geometry of the underlying differentiable base manifold from that of the principal bundle. This means heuristically that gravity is no longer a geometrical theory at least globally.

2. - FIRST ORDER GEOMETRICAL COUPLING OF GRAVITY TO PSEUDOSCALAR MATTER

In order to find a configuration with the above properties satisfying a dynamical equation we consider the following action

$$\mathcal{A} = \text{const} \times \int \Big[R^{ab} \wedge V^c \wedge V^d \epsilon_{abcd} + 2 \mathcal{F} R^{ab} \wedge V_a \wedge V_b + 2\gamma R^a \wedge V_a \wedge d\mathcal{F} - \frac{1}{8} U(\mathcal{F}) V^a \wedge V^b \wedge V^c \wedge V^d \epsilon_{abcd} \Big] \tag{2.1}$$

The first term is the usual Einstein-Cartan first order Lagrangian of gravity. The second term defines a particular coupling of a pseudoscalar field $\mathcal{F}(x)$ with gravity. The third term does not change the second order version of the theory but contains a free parameter γ which later will be fixed to a particular value in order to have a regular connection. Finally we add a potential or "mass term" $U(\mathcal{F})$; its explicit form will be fixed in the following in order to obtain a suitable instanton configuration. Upon variation of ω^{ab}, V^a and $\mathcal{F}$ one finds the following set of first order equations:

$$\epsilon_{abcd} R^c \wedge V^d + \mathcal{F}(R_a \wedge V_b - R_b \wedge V_a) + (1+\gamma) d\mathcal{F} \wedge V_a \wedge V_b = 0 \tag{2.2}$$

$$R^{ab} \wedge V^c \epsilon_{abcd} + \mathcal{F} R^{ad} \wedge V_a - 2(1+\gamma) R^d \wedge d\mathcal{F} = \frac{1}{4} U(\mathcal{F}) \epsilon_{abcd} V^a \wedge V^b \wedge V^c \wedge V^d \tag{2.3}$$

$$(1+\gamma) R^{ab} \wedge V_a \wedge V_b + \gamma R^a \wedge R_a = \frac{1}{16} \frac{\delta U}{\delta \mathcal{F}} \epsilon_{abcd} V^a \wedge V^b \wedge V^c \wedge V^d \tag{2.4}$$

Equation (2.2) has the solution

$$R^a = (1+\gamma) \Big[-\frac{1}{2} \frac{\mathcal{F}}{\mathcal{F}^2 - 1} d\mathcal{F} \wedge V^a + \frac{1}{2} \frac{\partial_d \mathcal{F}}{\mathcal{F}^2 - 1} V_b \wedge V_c \epsilon^{abcd} \Big] \tag{2.5}$$

from which we get

$$\omega^{ab} = \overset{\circ}{\omega}{}^{ab} + \frac{\gamma+1}{\mathcal{F}^2 - 1} \left(\partial^{[a} \mathcal{F} V^{b]} + \epsilon^{abcd} V_c \partial_d \mathcal{F} \right) \equiv \overset{\circ}{\omega}{}^{ab} + h^{ab} \tag{2.6}$$

where $\overset{0}{\omega}{}^{ab}$ is the torsionless part of the spinor connection satisfying $R^a = 0$. Inserting Eq. (2.6) into (2.3) and (2.4) we get the second order formulation of the theory. After lengthy but straightforward calculations one finds

$$R_{\mu\nu} - \frac{1}{2} g_{\mu\nu} R = \frac{3}{4}\left(\partial_\mu\phi\partial_\nu\phi - \frac{1}{2} g_{\mu\nu}\partial^\alpha\varphi\partial_\alpha\varphi\right) - \frac{3}{8} U(\varphi) g_{\mu\nu} \tag{2.7}$$

$$\nabla_\mu \partial^\mu \varphi = \frac{1}{2}\frac{\delta U}{\delta\varphi} \tag{2.8}$$

where we have set

$$\mathcal{F} = \cos\left(\frac{1}{1+\gamma}\varphi\right) \tag{2.9}$$

Here $\nabla_\mu u^\nu$ denotes the co-ordinate-covariant derivative related to the Lorentz covariant derivative $\mathcal{D}_\mu u^r$ through

$$\nabla_\mu u^\nu = (V^{-1})^\nu_r \overset{0}{\mathcal{D}}_\mu(\overset{0}{\omega}) u^r ; \quad \mathcal{D}u^r = \overset{0}{\mathcal{D}}u^r - h^{rs} u_s \tag{2.10}$$

Let us note explicitly that the unorthodox coupling of $\mathcal{F}(x)$ as given in (2.1) reproduces for any value of γ (in particular $\gamma = 0$) the usual minimal coupling of (pseudo)-scalar matter to gravity. The instanton solution will be now constructed as a particular isotropic solution of Eqs. (2.7) and (2.8) in Euclidean space.

3. - THE INSTANTON SOLUTION

According to the plan stated in the previous sections we seek solutions of the form:

$$V^a = e^{-\lambda(x)} e^a \Rightarrow g_{\mu\nu} = e^{-2\lambda(x)} \delta_{\mu\nu} \tag{3.1}$$

which is conformally scaled with respect to the standard flat Euclidean space-time vierbeins e^a introduced in Section 4. We also suppose that both $\mathcal{F}$ and λ depend only on the radial variable r. Under these conditions the equations of motion become:

$$\lambda'' + (\lambda')^2 - \frac{\lambda'}{r} + \frac{3}{4}(\varphi')^2 = 0 \tag{3.2a}$$

$$\lambda'' - \frac{1}{2}(\lambda')^2 + 2\frac{\lambda'}{r} + \frac{3}{8}\left(\varphi'^2 + e^{-2\lambda}U(\varphi)\right) = 0 \qquad (3.2b)$$

$$\varphi'' + \left(\frac{3}{r} - 2\lambda'\right)\varphi' - \frac{1}{2}e^{-2\lambda}\frac{\delta U}{\delta\varphi} = 0 \qquad (3.3)$$

where the last equation is a consequence of the Einstein equations (3.2a), (3.2b) via the Bianchi identities. Next we impose boundary conditions on $\mathcal{F}$ and λ in such a way as to reproduce the same topological configuration described in the preceding section.

In fact let us suppose that $\mathcal{F}$ and λ are differentiable functions at r = 0 such that:

$$\lim_{r\to 0} \mathcal{F} = 1 \qquad (3.4)$$

$$\lim_{r\to 0} \lambda = 1 \qquad (3.5)$$

Since

$$d\mathcal{F} = \dot{\mathcal{F}}\,dr = \dot{\mathcal{F}}\,e^0 \qquad (3.6)$$

$$\partial_a \mathcal{F} = \dot{\mathcal{F}}\,\delta^0_a \qquad (3.7)$$

asymptotically as $r \to 0$ we find:

$$R^a \underset{\sim}{\ } -\frac{1}{4}(\gamma+1)\frac{1}{r}\left(e^0\wedge e^1 - \epsilon^{abc0}\, e^b \wedge e^c\right) \qquad (3.8)$$

and this form turns out to be regular everywhere by virtue of Eq. (1.13). Moreover, conditions (3.4) ensure that $R^a \to 0$ as $r \to \infty$. Moreover one may easily check that the choice $\gamma = -5$ removes the singularity of the flat part of the spinor connection [Eq. (1.14)] giving an ω^{ab} regular at the origin. Since we have got the same topological configuration as that described in Section 1, the Pontryagin and Euler numbers are the same, namely P = 2, X = 0. The problem is now that of finding solutions of the field equations which correspond to the prescribed boundary conditions (3.4), (3.5) and give a finite value to the action (2.1). Using the equations of motion (2.2), (2.3) and (2.4) the action takes the following value:

$$\mathcal{A} = \text{const} \int e^{-4\lambda}\, U(\varphi)\, d^4x \tag{3.9}$$

Setting

$$u = e^{\lambda(z)} \quad ; \quad z = r^2 \tag{3.10}$$

eq. (3.2) and (3.3) become:

$$u'' + \frac{3}{4}\varphi'^2 u = 0 \tag{3.11a}$$

$$u'^2 - \frac{uu'}{z} + \frac{1}{4}\varphi'^2 u^2 = \frac{1}{16}\frac{U(\varphi)}{z} \tag{3.11b}$$

$$\varphi'' + 2\varphi'\left(\frac{1}{z} - \frac{u'}{u}\right) = \frac{1}{8u^2}\,\frac{1}{z}\,\frac{\delta U}{\delta \varphi} \tag{3.11c}$$

where differentiation is now with respect to z. An exact solution of these equations may be found when $U \equiv 0$. In this case one finds:

$$u = A\cos\left(\frac{\varphi - \varphi_0}{2}\right) \tag{3.12a}$$

$$\varphi' = \frac{2}{B}\sin^2\left(\frac{\varphi - \varphi_0}{2}\right) \tag{3.12b}$$

$$z = -B\,\mathrm{cotg}\left(\frac{\varphi - \varphi_0}{2}\right) \tag{3.12c}$$

when A, B and φ_0 are integration constants. One verifies immediately that $u \equiv e^{-\lambda}$ as given by (3.12a) does not satisfy the boundary condition (3.4); at the origin the vierbein V^a diverges. In order to stabilize this configuration a "mass" term $U(\phi) \neq 0$ is therefore needed.

By developing $u(z)$ and $\varphi(z)$ in power series in the neighbourhood of $z = 0$ we see that the singularity can be cancelled if

$$\frac{U(\varphi)}{16} \underset{z \to 0}{\sim} -1 + \text{const}\,\varphi + \mathcal{O}(\varphi^2) \tag{3.13}$$

Actually the mass term does not contribute at infinity if

$$U(\varphi_0) = U'(\varphi_0) = 0 \qquad \left(U' = \frac{\delta U}{\delta \varphi} \right) \tag{3.14}$$

In this case the numerical analysis shows that the asymptotic regime for large z described by Eqs. (3.12) is reached. In general, however, conditions (3.14) still give a logarithmically divergent action (3.9), unless higher derivatives of M vanish as well:

$$U''(\varphi_0) = 0 \tag{3.15}$$

The simplest choice satisfying (3.13), (3.14) and (3.15) is the following

$$U(\varphi) = -\frac{8}{3} \sin^4\left(\frac{\varphi - \varphi_0}{2}\right) \equiv -\frac{8 \cdot 32}{3}\left(1 - y^2\right) y^4 \tag{3.16}$$

The Eq. (3.11) have some obvious symmetries which allow us to restrict φ_0 to the interval $0 < \varphi_0 < \pi$. The solution for $\varphi_0 = 0$ yields $u = \varphi \equiv 0$; there is no regular solution for $\varphi = \pi$. Numerically a mass function with $U(\varphi_0) = 0$, $U'(\varphi_0) = 0$ does not lead to an acceptable solution, since in this case either u increases indefinitely or it vanishes, both cases being unphysical.

REFERENCES

1. T. Eguchi and A.J. Hanson, Ann. Phys. (N.Y.) 120:82:106 and references therein.
2. T. Eguchi, P.B. Gilkey and A.J. Hanson, Phys. Rep. 66:213 (1980).
3. A. Belavin, A. Polyakov, A. Schwaz and Y. Tyupkin, Phys. Lett. 59B:85 (1975).
4. A.J. Hanson and R. Regge, "Torsion and Quantum Gravity", Proceedings of the Joint Conference on Group Theory and Mathematical Physics, the University of Texas at Austin (1978).
5. K.S. Stelle and P.C. West, Phys. Rev. D21:1466 (1980).

CHAPTER II - FIERZ IDENTITIES AND GROUP THEORY

Presented by: T. Regge
Reference: R. D'Auria, P. Fré, E. Maina, T. Regge, Torino preprint IFTT 410 (1981), to be published in Annals of Physics.

The equations one encounters in the group-manifold approach, both in the analysis of Bianchi Identities while looking for auxiliary fields, and in the analysis of equations of motion while constructing a Lagrangian, are exterior form equations. They state that certain p-forms, constructed out of the fundamental supergroup potential μ^A only by use of the exterior algebra operations d and Λ, must be zero. In order to extract the information they contain it is necessary to project them on a basis of independent p-forms, namely of monomials $\mu^{A_1} \wedge \mu^{A_2} \wedge \ldots \wedge \mu^{A_p}$. In supergravity theories there is usually a pair (G,H) of a supergroup G, on whose Lie algebra G the index A of μ^A is running, and a subgroup $H \subset G$ under those action the Lagrangian is exactly gauge invariant. As a result dynamics takes place in the coset space G/H and the phenomenon of factorization shows up. Namely all equations contain only bare μ^K forms (the index K runs on G/H) and all the group curvatures $R^A = d\mu^A + 1/2\, C^A_{\cdot BC}\mu^B \wedge \mu^C$ have non-vanishing components only the G/H directions. For instance in N-extended D = 4 supergravity we have G = Osp(4/N) (or its contracted version $\overline{\text{Osp}(4/N)}$) and H = SO(1,3) $\otimes$ O(N). All the p-forms one deals with are, therefore, to be expanded in the basis of the exterior algebra generated by the vierbein one-form V^a and the gravitino one-form ψ_A spanning the cotangent space to G/H.

The basic idea of our method is the following. V^a and ψ^α_A are irreducible representations of the exact gauge group H: indeed V^a is an SO(1,3) vector and ψ^α_A is both an SO(1,3) spinor and an SO(N) vector (the index A of ψ_A runs on N-values). Correspondingly the exterior products

$$\Omega^{a_1\ldots a_n \alpha_1\ldots\alpha_m}_{A_1\ldots A_m} = V^{a_1}\wedge\ldots\wedge V^{a_n}\wedge \psi^{\alpha_1}_{A_1}\wedge\ldots\wedge\psi^{\alpha_m}_{A_m} \qquad (1)$$

are tensor products of irreducible representations and they can be decomposed into irreducible representations via a Clebsch-Gordan series. The relevant point is that

$$\Omega^{a_1\ldots a_n\alpha_1\ldots\alpha_m}_{A_1\ldots A_m}$$

is antisymmetric in

$$\{a_1 \leftrightarrow a_2 \ldots \leftrightarrow a_n\}$$

and symmetric in

$$\left\{ \begin{matrix}\alpha_1 \\ A_1\end{matrix} \leftrightarrow \begin{matrix}\alpha_2 \\ A_2\end{matrix} \leftrightarrow \cdots \leftrightarrow \begin{matrix}\alpha_m \\ A_m\end{matrix} \right\}.$$

This implies that only certain irreducible representations appear in its decomposition while the others are ruled out. This decomposition, with explicitly calculated coefficients, contains the same information as the list of all possible Fierz identities and provides a systematic way to perform calculations.

Indeed what one has to do is to decompose into irreducible representations every object which appears in the equation, collect the terms multiplying every basis irreducible representation and set them separetely to zero.

The method is best illustrated by an explicit example. We choose N = 1 supergravity. Given the Bianchi identities:

$$\nabla R^a = \mathcal{D} R^a + R^{ab} \wedge V_b - i \bar{\Psi}_A \wedge \gamma^a \rho_A = 0 \tag{2a}$$

$$\nabla R^{ab} = \mathcal{D} R^{ab} = 0 \tag{2b}$$

$$\nabla \rho = \mathcal{D}\rho - \frac{i}{2} \Sigma_{ab} \psi \wedge R^{ab} = 0 \tag{2c}$$

the problem of auxiliary fields is that of finding a parametrization of the curvatures R^{ab}, R^a, ρ consistent with Eq. (2), containing a minimal number of objects and not implying the propagation equations of the space-time components $R^{ab}{}_{mn}$ and ρ_{mn}, namely:

$$R^{am}{}_{\cdot bm} - \frac{1}{2}\delta^a_b R^{mn}{}_{\cdot mn} = 0 \; ; \quad \gamma^m \rho_{mn} = 0 \tag{3}$$

To solve this problem we observe that the left-hand side of the Bianchi identity is a three-form and that the most general three-form can be written as follows

$$\Omega = \Omega_{abc} V^a \wedge V^b \wedge V^c + \Omega^{(1)}_{ab|\alpha} V^a \wedge V^b \wedge \psi^\alpha + \\ + \Omega^{(2)}_{a|i} V^a \wedge X^i + \Omega^3_i \, \Xi^i \tag{4}$$

where the X^i are the two-form irreducible representations appearing in the decomposition of $\psi^\alpha \wedge \psi^\beta$

$$\psi^\alpha \wedge \psi^\beta = f_i^{\alpha\beta} X^i \tag{5}$$

and the Ξ^i are the three-form irreducible representations appearing in the decomposition of $\psi^\alpha \wedge \psi^\beta \wedge \psi^\gamma$

$$\psi^\alpha \wedge \psi^\beta \wedge \psi^\gamma = f_i^{\alpha\beta\gamma} \, \Xi^i \tag{6}$$

Therefore we begin by classifying all the SO(1,3) representations occuring in the tensor product of two or three spin 1/2 representations.

In table 1 the numbers of the extreme left are the eigenvalues of the Casimir operators of SO(1,3) whose rank is two and the plus and minus superscripts refer to the parity eigenvalues. The dimensions can be calculated from Ref. 1.

They can be also obtained by more elementary means. For instance Ξ^{ab} has dimension eight because it is a spinor tensor $(4 \times 6 = 24)$ satisfying $4 \times 4 = 16$ conditions $\gamma^b \Xi_{ab} = 0$.

Table 1 exhausts the list of relevant representations because of the following decomposition rules:

$$[\tfrac{1}{2},\tfrac{1}{2}] \otimes [\tfrac{1}{2},\tfrac{1}{2}] = \underbrace{[1,1] \oplus [1,0]^+}_{\text{symm}} \oplus \underbrace{[1,0]^- \oplus [0,0]^+ \oplus [0,0]^-}_{\text{antisym}} \tag{7a}$$

$$[\tfrac{1}{2},\tfrac{1}{2}] \otimes [1,1] = [\tfrac{3}{2},\tfrac{3}{2}] \oplus [\tfrac{3}{2},\tfrac{1}{2}] \oplus [\tfrac{1}{2},\tfrac{1}{2}] \tag{7b}$$

$$[\tfrac{1}{2},\tfrac{1}{2}] \otimes [1,0] = [\tfrac{3}{2},\tfrac{1}{2}] \oplus [\tfrac{1}{2},\tfrac{1}{2}] \tag{7c}$$

$$[\tfrac{1}{2},\tfrac{1}{2}]\otimes[0,0]=[\tfrac{1}{2},\tfrac{1}{2}] \qquad (7d)$$

Equation (2.7a) says that any 4 × 4 matrix can be expanded in a complete Dirac basis.

Table 1: Representations of SO(1,3)

Representation Type	Dimension	Corresponding Tensor, Spinor or Spinor Tensor
$[1,1]$	6	$X_{ab} = -X_{ba}$ tensor
$[1,0]^{(+)}$	4	$\overset{(+)}{X}{}^{a}$ vector
$[1,0]^{(-)}$	4	$\overset{(-)}{X}{}^{a}$ axial vector
$[0,0]^{(+)}$	1	$\overset{(+)}{X}$ scalar
$[0,0]^{(-)}$	1	$\overset{(-)}{X}$ pseudoscalar
$[\tfrac{3}{2},\tfrac{3}{2}]$	8	$\Xi_{ab} = -\Xi_{ba}$; $\gamma^{b}\Xi_{ab} = 0$ irred. spinor tensor
$[\tfrac{3}{2},\tfrac{1}{2}]^{(+)}$	12	$\Xi^{(+)}_{a}$; $\gamma^{a}\Xi^{(+)}_{a} = 0$ irred. spinor vector
$[\tfrac{3}{2},\tfrac{1}{2}]^{(-)}$	12	$\Xi^{(-)}_{a}$; $\gamma^{a}\Xi^{(-)}_{a} = 0$ irred. spinor axial vector
$[\tfrac{1}{2},\tfrac{1}{2}]^{(+)}$	4	$\Xi^{(+)}$ Majorana spinor
$[\tfrac{1}{2},\tfrac{1}{2}]^{(-)}$	4	$\Xi^{(-)}$ Majorana pseudospinor

Equations (7b) and (7c) correspond to the familiar decomposition of a spinor-vector (or a spinor-tensor) into a traceless part plus a trace. Let us for instance consider a spinor vector ξ_a. We can write:

$$\xi_a = \xi_a^{(12)} + \frac{1}{4}\gamma_a \xi^{(4)} \tag{8}$$

where

$$\xi_a^{(12)} = \xi_a - \frac{1}{4}\gamma_a \gamma^m \xi_m \tag{9}$$

is the irreducible [3/2,1/2] part satisfying:

$$\gamma^a \xi_a^{(12)} = 0 \tag{10}$$

and

$$\xi^{(4)} = \gamma \cdot \xi \tag{11}$$

is the [1/2,1/2] part.

Similarly given a spinor tensor ξ_{ab} we can write:

$$\xi_{ab} = \xi_{ab}^{(8)} - \gamma_{[a} \xi_{b]}^{(12)} + \frac{i}{6} \Sigma_{ab} \xi^{(4)} \tag{12}$$

where

$$\xi_{ab}^{(8)} = \xi_{ab} + \gamma_{[a} \gamma^m \xi_{b]m} - \frac{i}{3} \Sigma_{ab} \gamma^m \gamma^n \xi_{mn} \tag{13a}$$

$$\xi_a^{(12)} = \gamma^m \xi_{am} - \frac{1}{4}\gamma^a \gamma^m \gamma^n \xi_{mn} \tag{13b}$$

$$\xi^{(4)} = \gamma^m \gamma^n \xi_{mn} \tag{13c}$$

are the irreducible representations [3/2,3/2], [3/2,1/2] and [1/2,1/2] respectively. From this point on we shall define the irreducible representations assuming Eqs. (8) and (12) as the standard expansion of any spinor-vector or spinor-tensor.

The quantity $\psi^\alpha \wedge \psi^\beta$ has 10 components. Looking at Table 1, on the other hand, we realize that

$$10 = \dim [1,1] + \dim [1,0]^{(+)} = 6+4 \tag{14}$$

which are precisely the representations occurring in the symmetric product decomposition. In the N = 1 case no contribution from the antisymmetric $[1,0]^{(-)}$, $[0,0]^{(+)}$ and $[0,0]^{(-)}$ representations is allowed. Equation (2.5) therefore reduces to

$$\psi \wedge \bar{\psi} = \frac{1}{4} \gamma^a \overset{(+)}{X}_a + \frac{1}{2} \Sigma_{ab} X^{ab} \tag{15}$$

Going now to the $\psi^\alpha \wedge \psi^\beta \wedge \psi^\gamma$ sector we see that it has 20 components; moreover, from Table 1 it is evident that the only way to obtain 20 is by setting

$$20 = 8 + 12 \tag{16}$$

This means that the only representations which are completely symmetric in α, β, γ are [3/2,3/2] and [3/2,1/2]. This is the origin of all Fierz identities. The explicit construction is the following. If we have the wedge product of three $\psi \wedge \psi \wedge \psi$ we can start by decomposing two of them according to (15).

In this way we end up with the following spinor-vector and spinor-tensor:

$$\Theta_a = \psi \wedge \bar{\psi} \wedge \gamma_a \psi \tag{17a}$$

$$\Theta_{ab} = \psi \wedge \bar{\psi} \wedge \Sigma_{ab} \psi \tag{17b}$$

Because of the previous discussion we can set

$$\Theta_a = \alpha\, \Xi^{(12)}_a \tag{18a}$$

$$\Theta_{ab} = \Xi^{(8)}_{ab} - \gamma_{[a} \Xi^{(12)}_{b]} \tag{18b}$$

where $\Xi^{(12)}_a$ is an irreducible [3/2,1/2] representation (satisfying $\gamma^a \Xi^{(12)}_a = 0$ and $\Xi^{(8)}_{ab}$ is an irreducible [3/2,3/2] representation (satisfying $\gamma^b \Xi^{(12)}_{ab} = 0$). We compute the coefficient by applying Eq. (15) once again in the definition (17a) of θ_a. We get

$$\psi \wedge \bar{\psi} \wedge \gamma^a \psi = \alpha \, \Xi^{(12)}_a = \frac{1}{2} \psi \wedge \bar{\psi} \wedge \gamma_a \psi - i \gamma^b \psi \wedge \bar{\psi} \wedge \Xi_{ab} \psi \tag{19}$$

and hence:

$$\psi \wedge \bar{\psi} \wedge \gamma_a \psi = \alpha \, \Xi^{(12)}_a = -2i \gamma^b \psi \wedge \bar{\psi} \wedge \Xi_{ab} \psi = -2i \, \Xi^{(12)}_a \tag{20}$$

With Eq. (20) we have completed the construction of an irreducible basis for N = 1 d = 4 superspace. The result is summarized in Table 2

Table 2: Irreducible Basis of N = 1 d = 4 Superspace

$\psi \wedge \bar{\psi} = \frac{1}{4} \gamma_a \overset{(+)}{X}{}^a + \frac{1}{2} \Sigma_{ab} X^{ab}$
$\psi \wedge \bar{\psi} \wedge \gamma^a \psi = \psi \wedge \overset{(+)}{X}{}^a = -2i \Xi^{(12)}_a$
$\psi \wedge \bar{\psi} \wedge \Sigma^{ab} \psi = \psi \wedge X_{ab} = \Xi^{(8)}_{ab} - \gamma_{[a} \Xi^{(12)}_{b]}$

The auxiliary fields of N = 1 supergravity are now easily obtained. The group-manifold equations of motion imply that the translational torsion is zero:

$$R^a = 0 \tag{21}$$

Equation (21) is obtained through variation of the non-propagating field ω^{ab}. Equation (21) is therefore a good candidate for a kinematical constraint. This assumption can be tested as follows. We write the most general decompositions of R^{ab} and ρ:

$$R^{ab} = R^{ab}{}_{\cdot mn} V^m \wedge V^n + \bar{\Theta}^{ab|m} \psi \wedge V_m + \bar{\psi} \wedge K^{ab} \psi \tag{22a}$$

$$\rho = \rho_{mn} V^m \wedge V^n + H_m \psi \wedge V^m + \Omega_{\alpha\beta} \psi^\alpha \wedge \psi^\beta \tag{22b}$$

and we insert them in Eq. (2a), together with the condition (21). The constraints imposed by Bianchi identities can be easily solved and the spinor $\Theta^{ab|m}$, the matrices K^{ab} and H_m, and the rank three spinor $\Omega_{\alpha\beta}$ determined. First we consider the ψVV sector of (2a) which reads:

$$\bar{\Theta}^{ab|m} \psi \wedge V_m \wedge V_b - i \bar{\psi} \wedge \gamma^a \rho_{mn} V^m \wedge V^n = 0 \tag{23}$$

The solution is:

$$\bar{\Theta}^{ab|m} = 2i \bar{\rho}^{m[a} \gamma^{b]} + i \bar{\rho}^{ab} \gamma^m \tag{24}$$

Then we look at the $\psi\psi V$ sector. We get

$$\bar{\psi} \wedge K^{ab} \psi \wedge V_b - i \bar{\psi} \wedge \gamma^a H^b \psi \wedge V_b = 0 \tag{25}$$

This equation splits in two:

$$\bar{\psi} \wedge \gamma^{\{a} H^{b\}} \psi = 0 \tag{26a}$$

$$\bar{\psi} \wedge \gamma^{[a} H^{b]} \psi = - i \bar{\psi} \wedge K^{ab} \psi \tag{26b}$$

Equation (26a) is easily solved. Decomposing H^b in a complete Dirac basis:

$$H_b = H_b^{(+)} \mathbb{1} + i H_b^{(-)} \gamma^5 + \overset{(-)}{H}{}_b^a \gamma^5 \gamma_a + i \overset{(+)}{H}{}_b^a \gamma_a + i H_b^{mn} \Sigma_{mn} \tag{27}$$

where all the coefficients are real and the i factors have been inserted in such a way as to make $H^b\psi$ a Majorana spinor. We ask then that $\gamma^{\{a}H^{b\}}$ is a linear combination of $\mathbb{1}$, γ_5, and $\gamma_5\gamma_a$ so that Eq. (26) is satisfied. This requirement gives two linear equations (corresponding to the forbidden matrices γ_m and Σ_{mn}) whose general solution is:

$$H_m = i\mathcal{A}_m\gamma_5 + \mathcal{P}\gamma_5\gamma_m + i\mathcal{S}\gamma_m + 2\gamma_5\Sigma_{mn}\mathcal{A}'^n \tag{28}$$

where $\mathcal{A}_m$ and $\mathcal{A}'_m$ are two axial vectors, $\mathcal{S}$ and $\mathcal{P}$ a scalar and a pseudoscalar respectively. Substitution of Eq. (28) into (26) then gives:

$$K^{ab} = -2\mathcal{P}\gamma_5\Sigma^{ab} + 2i\mathcal{S}\Sigma^{ab} - i\epsilon^{abcd}\mathcal{A}'_c\gamma_d \tag{29}$$

Finally we consider the $\psi\wedge\psi\wedge\psi$ sector of Eq. (2a). Setting

$$\Omega = \Omega_{\alpha\beta}\,\psi^\alpha\wedge\psi^\beta \tag{30}$$

we get:

$$-i\bar{\psi}\wedge\gamma^a\Omega = 0 \Longleftrightarrow \bar{\Omega}\wedge\gamma^a\psi = 0 \tag{31}$$

This constraint is solved as follows. In full generality we can write:

$$\Omega = i\chi_a \overset{(+)}{X}{}^a + i\xi_{ab}X^{ab} \tag{32}$$

where χ_a and ξ_{ab} are a spinor-vector and a spinor-tensor respectively. They can be decomposed according to Eqs. (8) and (12)

$$\chi_a = \chi_a^{(12)} + \frac{1}{4}\gamma_a\chi^{(4)} \tag{33a}$$

$$\xi_{ab} = \xi_{ab}^{(8)} - \gamma_{[a}\xi_{b]}^{(12)} + \frac{i}{6}\Sigma_{ab}\xi^{(4)} \tag{33b}$$

Inserting Eq. (32) into (31) and using again Table 2 we obtain:

$$\begin{aligned}\bar{\Omega}\wedge\gamma^m\psi &= i\bar{\chi}^a\gamma^m\psi\wedge\overset{(+)}{X}_a + i\bar{\xi}_{ab}\gamma^m\psi\wedge X^{ab} \\ &= i\bar{\chi}^a\gamma^m\left(-2i\,\boxplus_a^{(12)}\right) + i\bar{\xi}^{ab}\gamma_m\left(-\gamma_{[a}\boxplus_{b]}^{(12)}\right) \\ &\quad + i\bar{\xi}^{ab}\gamma_m\boxplus_{ab}^{(8)} = 0\end{aligned} \tag{34}$$

Since $\Xi_a^{(12)}$ and $\Xi_{ab}^{(8)}$ are independent, their coefficients must vanish separately. In this way we obtain a set of equations for each of the irreducible components introduced in Eq. (33a) and (33b). More precisely, we find:

$$\begin{cases} \xi_{ab}^{(8)} = 0 \\ \xi_a^{(12)} = \chi_a^{(12)} = 0 \\ \xi^{(4)} = -3\chi^{(4)} \end{cases} \tag{35}$$

This means that the most general solution of Eq. (31) is

$$\Omega = -\frac{1}{4}\gamma^a \chi\, \overline{\Psi} \wedge \gamma_a \psi + \frac{1}{2} \Sigma_{ab} \chi\, \overline{\Psi} \wedge \Sigma^{ab} \psi \tag{36}$$

where χ is a Majorana spinor. An equivalent convenient way of writing the solution (36) is the following one

$$\Omega = \gamma_5 \psi \wedge \overline{\Psi} \wedge \gamma_5 \chi \tag{37}$$

In this way we have completely solved the torsion Bianchi identity (2a) and we have found a parametrization of the curvatures in terms of the set of auxiliary fields:

$$\mathcal{A}_a,\ \mathcal{A}_a',\ \mathcal{S},\ \mathcal{P},\ \chi$$

This is the so-called non-minimal formulation of $N = 1$ supergravity first studied by Brown and Gates[2], shown to be necessary in the geometrical coupling of supergravity to the Wess-Zumino multiplet[3]. The possibility of obtaining a minimal formulation with a smaller number of auxiliary fields is suggested by two considerations:

i) numerology: counting the bosons and fermions we have

Bosons			Fermions		
V_μ^a	= graviton	= 6	ψ_μ	= gravitino	= 12
$\mathcal{A}_a$		= 4	χ		= 4
$\mathcal{S}$		= 1			16
$\mathcal{P}$		= 1			
$\mathcal{A}'_a$		= 4			
		16			

Therefore taking out an axial vector from the bosons and the χ from the fermions we still have an equal number of bosonic and fermionic states.

ii) The condition $\Omega = 0 \Rightarrow \chi = 0$ is indeed yielded by the equations of motion.

Now we check that $\chi = 0$ is also a kinematical constraint by analyzing the remaining Bianchi identities, (2b) and (2c). Considering first the $\psi\psi\psi$ sector of Eq. (2c) and assuming $\chi = 0$ we find:

$$\begin{aligned} & -\frac{i^2}{2} A_m \gamma^5 \psi \wedge \bar{\psi} \wedge \gamma^m \psi - \frac{i}{2} P \gamma_5 \gamma_m \psi \wedge \bar{\psi} \wedge \gamma^m \psi \\ & -\frac{i^2}{2} S \gamma_m \psi \wedge \bar{\psi} \wedge \gamma^m \psi - i \gamma_5 \Xi_{ab} \psi A^{a\prime} \wedge \bar{\psi} \wedge \gamma^b \psi \\ & -\frac{i}{2} \Xi_{ab} \psi \wedge \left(-2P \bar{\psi} \wedge \gamma^5 \Xi^{ab} \psi + 2i S \bar{\psi} \wedge \Xi^{ab} \psi \right. \\ & \qquad \left. - i \epsilon^{abcd} A_c' \bar{\psi} \wedge \gamma_d \psi \right) = 0 \end{aligned} \tag{38}$$

Substituting the $\psi \wedge \psi \wedge \psi$ decomposition of Table 2, Eq. (38) immediately reduces to:

$$\frac{1}{2} \left(A_m - A_m' \right) \gamma_5 (-2i) \Xi_m^{(12)} = 0 \tag{39}$$

which implies:

$$A_m' = \frac{1}{2} A_m \tag{40}$$

Therefore the conjecture that an axial vector and the χ field can be eliminated together is indeed consistent with the Bianchi identities.

Summarizing we get the following parametrization of the OSP(4/1) curvatures:

$$R^a = 0 \tag{41a}$$

$$\begin{aligned} \rho = {} & \rho_{ab} V^a \wedge V^b + i A_a \gamma_5 \psi \wedge V^a + \gamma_5 \Xi^{ab} \psi \wedge A_a V_b \\ & + i S \gamma_a \psi \wedge V^a + P \gamma_5 \gamma_a \psi \wedge V^a \end{aligned} \tag{41b}$$

$$R^{ab} = R^{ab}{}_{\cdot mn} V^m \wedge V^n - 2i\,\bar\psi \gamma^{[a} \rho^{b]}{}_{m} \wedge V^m - i\,\bar\psi \wedge \gamma^m \rho_{ab} \wedge V_m$$
$$+ 2i\,\mathcal{S}\,\bar\psi \wedge \Xi^{ab} \psi - 2\mathcal{P}\,\bar\psi \wedge \gamma_5 \Xi^{ab} \psi \qquad (41c)$$
$$+ \frac{i}{2} \epsilon^{abcd}\,\bar\psi \wedge \gamma_c \psi\, \mathcal{A}_d$$

There are still two projections of the Bianchi identity (2c) to be checked (ψVV and $\psi\psi V$). The first is uninteresting because it determines the spinor derivative $\mathcal{D}_\alpha \rho_{ab}$, i.e., the action of supersymmetry on ρ_{ab} which is just a derivative of the field ψ_μ. The second projection ($\psi\psi V$) however gives the spinor derivatives $\mathcal{D}_\alpha \mathcal{A}_a$, $\mathcal{D}_\alpha \mathcal{S}$ and $\mathcal{D}_\alpha \mathcal{P}$, i.e., the supersymmetry transformations of the auxiliary fields. Let us set:

$$\mathcal{D}\mathcal{A}_m = \mathcal{D}_n \mathcal{A}_m V^n + \bar\psi \alpha_m \qquad (42a)$$

$$d\mathcal{S} = \mathcal{D}_m \mathcal{S} V^m + \bar\psi \sigma \qquad (42b)$$

$$d\mathcal{P} = \mathcal{D}_m \mathcal{P} V^m + \bar\psi \pi \qquad (42c)$$

The ($\psi\psi V$) sector of Eq. (2c) now reads

$$i\rho_{ab}\,\bar\psi \wedge \gamma^a \psi \wedge V^b - i\gamma_5 \psi \wedge \bar\psi \alpha_b \wedge V^b - i\gamma_b \psi \wedge \bar\psi \sigma \wedge V^b$$
$$- \gamma_5 \Xi^{ab} \psi \wedge \bar\psi \alpha_a \wedge V_b - \gamma_5 \gamma_b \psi \wedge \bar\psi \pi \wedge V^b \qquad (43)$$
$$- \frac{1}{2} \Xi^{rs} \psi \wedge \bar\psi \left(2\gamma_{[r} \rho_{s]b} + \gamma_b \rho_{rs}\right) \wedge V^b = 0$$

Using the decomposition of Table 2 for $\psi \wedge \bar\psi$ we can separate Eq. (43) in two, one being the coefficient of $\overset{(\pm)}{X}{}^m$ the other being the coefficient of X_{mn}. Explicitly we find

$$i\rho_{mb} - \frac{i}{4}\gamma_5 \gamma_b \gamma_m \pi - \frac{i}{4}\gamma_5 \gamma_m \alpha_b - \frac{1}{4}\gamma_5 \Xi^{ab} \gamma_m \alpha_a$$
$$- \frac{i}{4}\gamma_b \gamma_m \sigma - \frac{1}{4}\Xi^{rs} \gamma_m \gamma_r \rho_{sb} - \frac{1}{8}\Xi^{rs} \gamma_m \gamma_b \rho_{rs} = 0 \qquad (44)$$

$$\Big(-\frac{i}{2}\gamma_5 \vec{\Xi}_{mn}\alpha_b - \frac{i}{2}\gamma_5 \vec{\Xi}^{ab}\vec{\Xi}_{mn}\alpha_a - \frac{i}{2}\gamma_b \vec{\Xi}_{mn}\sigma - \frac{1}{2}\gamma_5\gamma_b \vec{\Xi}_{mn}\pi - \vec{\Xi}^{rs}\vec{\Xi}_{mn}\gamma_r\rho_{sb} - \frac{1}{2}\vec{\Xi}^{rs}\vec{\Xi}_{mn}\gamma_b\rho_{rs}\Big) = 0 \tag{45}$$

Decomposing ρ_{ab}, α_a into irreducible representations

$$\rho_{ab} = \rho^{(8)}_{ab} - \gamma_{[a}\rho^{(12)}_{b]} + \frac{i}{6}\vec{\Xi}_{ab}\rho^{(4)} \tag{46a}$$

$$\alpha_a = \alpha^{(12)}_a + \frac{1}{4}\gamma_a\alpha^{(4)} \tag{46b}$$

Eqs. (44) become relations of the type:

$$\alpha^{(12)}_b = \mathrm{const} \times \gamma_5\rho^{(12)}_b \qquad \alpha^{(4)} = \mathrm{const}\,\rho^{(4)} \tag{47a}$$

$$\sigma = \mathrm{const} \times \rho^{(4)} \tag{47b}$$

$$\pi = \mathrm{const} \times \gamma_5\rho^{(4)} \tag{47c}$$

yielding the transformation rules of $\mathcal{S}$, $\mathcal{P}$ and $\mathcal{A}_a$. Since $\mathcal{A}_a$, $\mathcal{S}$, $\mathcal{P}$ transform into the [3/2,1/2] and [1/2,1/2] parts of ρ_{ab} they are indispensable auxiliary fields. Indeed the Rarita-Schwinger equation

$$\gamma^m\rho_{am} = 0 \iff \epsilon^{abcd}\gamma_5\gamma_b\rho_{cd} = 0 \tag{48}$$

states that ρ_{ab} is a pure [3/2,3/2]. Hence $\rho^{(12)}_b$ and $\rho^{(4)}$ are the off-shell parts of ρ_{ab}. In this way we have recovered the minimal set of N = 1 auxiliary fields[4]. The technique we have illustrated works with obvious modifications in any extended supergravity in every dimension. What has to be done is to look for the irreducible representations of the suitable H group and at their product decompositions. In the main reference the method was applied to N = 2 D = 5 [5] supergravity for which theory we obtained the so far unknown multiplet of auxiliary fields listed in the following table:

Table 3: $48 \oplus 48$ Multiplet of $d = 5$ Supergravity

BOSONS			FERMIONS		
V^a_μ	= graviton	10	ψ^A_μ	= gravitino	32
B_μ	= photon	4	χ^A	= spin 1/2 O(2) doublet	8
$T_{\mu\nu}$	= antisymmetric tensor	10	$\chi^{A'}$	= spin 1/2 O(2) doublet	8
$\mathcal{A}_\mu$	= vector	5			
$\mathcal{A}^{\boxed{A}\boxed{B}}$	= vector, symmetric and traceless in O(2) space	10			
$\mathcal{S}$	= scalar	1			
$\mathcal{S}^{\boxed{A}\boxed{B}}$	= scalar, symmetric and traceless in O(2) space	2			
$\mathcal{A}'_\mu$	= vector	5			
$\mathcal{S}'$	= scalar	1			
TOTAL NUMBER OF OFF-SHELL DEGREES OF FREEDOM		48			48

Of interest for the topics of the following chapter is the systematics of Fierz identities in D = 11 supergravity. We start by giving the dimensionality of the SO(1,10) representations appearing in the symmetric product of two, three and four gravitino one-forms ψ. (ψ is a spin 1/2 Majorana one-form.)

Table 4: Dimensions of SO(1,10) irreducible representations appearing in the symmetric products of 2, 3, 4 irreducible representations $(1/2)^5$

BOSE IRREDUCIBLE REPRESENTATIONS		FERMI IRREDUCIBLE REPRESENTATIONS	
TYPE	DIMENSION	TYPE	DIMENSION
$(0)^5$	1	$(1/2)^5$	32
$(1)(0)$	11	$(3/2)(1/2)^4$	320
$(1)^2(0)^3$	55	$(3/2)^2(1/2)^3$	1408
$(1)^3(0)^2$	165		
$(1)^4(0)$	330		
$(1)^5$	462	$(3/2)^5$	4224
$(2)(0)^4$	65		
$(2)(1)(0)^3$	429		
$(2)^2(0)^3$	1144		
$(2)(1)^4$	4290		
$(2)^2(1)^3$	17160		
$(2)^5$	32604		

The notations are easily explained.

The eleven-dimensional Lorentz group SO(1,10) has rank five and therefore its irreducible representations are labeled by five integer or half integer numbers. In the integer case we are dealing with a Bosonic representation and the five numbers $\lambda_1 \geq \lambda_2 \geq \lambda_3 \geq \geq \lambda_4 \geq \lambda_5$ labeling it can be identified with the number of boxes in each row of a Young tableau. In this way the representation $(1)^2(0)^3$ corresponds, for instance, to the tableau $\begin{array}{|c|}\hline \; \\ \hline \; \\ \hline\end{array}$ namely to an antisymmetric tensor $T_{a_1a_2}$. Analogously $(2)^2(0)^3$ corresponds to the tableau

$$\begin{array}{|c|c|}\hline a_1 & a_3 \\ \hline a_2 & a_4 \\ \hline\end{array}$$

that is to the tensor

$$T_{\substack{a_1 a_2 \\ a_3 a_4}}$$

while $(1)^5$ is a skew symmetric five index tensor

$$\begin{array}{|c|}\hline \cdot \\ \hline \cdot \\ \hline \cdot \\ \hline \; \\ \hline \; \\ \hline\end{array} \sim T_{a_1 \ldots a_5}$$

In the half-integer case the representation is of the Fermi type. The corresponding object is a spinor tensor having in its vectorial indices the symmetry of the Young tableau $\lambda_1 - 1/2, \lambda_2 - 1/2, \ldots, \lambda_5 - 1/2$. Moreover it is irreducible in the sense that whatever trace can be obtained contracting it with Γ matrices is zero.

For instance the irreducible representation $(3/2)(1/2)^4$ is a spinor tensor with the symmetry $(1)(0)^4$ in its Bose indices, namely Ξ_a. The irreducibility means $\Gamma^a\Xi_a = 0$. Analogously $(3/2)^2(1/2)^3$ is a spinor tensor with Bose indices of the type $(1)^2(0)^3$ namely $\Xi_{a_1a_2}$ (skew symmetric). The irreducibility condition is $\Gamma^{a_2}\Xi_{a_1a_2} = 0$.

The use of numerology provides an easy tool to work out the representations appearing in each symmetric product. We find:

$$\left\{\left(\tfrac{1}{2}\right)^5 \otimes \left(\tfrac{1}{2}\right)^5\right\}_{sym} = (1)(0)^4 \oplus (1)^2(0)^3 \oplus (1)^5 \tag{49}$$

$$\frac{32 \times 33}{2} = 528 \qquad = \quad 11 + 55 + 462$$

$$\left\{\left(\tfrac{1}{2}\right)^5 \otimes \left(\tfrac{1}{2}\right)^5 \otimes \left(\tfrac{1}{2}\right)^5\right\}_{sym} = \left(\tfrac{1}{2}\right)^5 \oplus \left(\tfrac{3}{2}\right)\left(\tfrac{1}{2}\right)^4 \oplus \left(\tfrac{3}{2}\right)^2\left(\tfrac{1}{2}\right)^3 \oplus \left(\tfrac{3}{2}\right)^5$$

$$\frac{32 \times 33 \times 34}{2 \cdot 3} = 5984 = 32 + 320 + 1408 + 4224 \tag{50}$$

$$\left\{\left(\tfrac{1}{2}\right)^5 \otimes \left(\tfrac{1}{2}\right)^5 \otimes \left(\tfrac{1}{2}\right)^5 \otimes \left(\tfrac{1}{2}\right)^5\right\}_{sym} = (0)^5 \oplus (1)^3(0)^2 \oplus (1)^5 \oplus (2)(0)^4$$

$$\oplus (2)(1)(0)^3 \oplus (2)^2(0)^3 \oplus (2)^2(1)^3 \oplus (2)^5 \tag{51}$$

$$\frac{32 \times 33 \times 34 \times 35}{2 \cdot 3 \cdot 4} = 52360 = 1 + 165 + 330 + 462 + 65 + 429 + 1144 + 17160 + 32604$$

These decompositions are made explicit in the following way. Let ψ be the Majorana gravitino one-form and $\bar{\psi} = \psi\,\Gamma_0$

Table 5

EXPLICIT FIERZ DECOMPOSITION IN D = 11

$$\psi\wedge\bar{\psi} = \frac{1}{32}\left(\Gamma_a\bar{\psi}\wedge\Gamma^a\psi - \frac{1}{2}\Gamma_{ab}\bar{\psi}\wedge\Gamma^{ab}\psi + \frac{1}{5!}\Gamma_{a_1\ldots a_5}\bar{\psi}\wedge\Gamma^{a_1\ldots a_5}\psi\right)$$

$$\psi\wedge\bar{\psi}\wedge\Gamma_a\psi = \Xi_a^{(320)} + \frac{1}{11}\Gamma_a\,\Xi^{(32)}$$

$$\psi\wedge\bar{\psi}\wedge\Gamma_{a_1a_2}\psi = \Xi^{(1408)}_{a_1a_2} - \frac{2}{9}\Gamma_{[a_1}\Xi^{(320)}_{a_2]} + \frac{1}{11}\Gamma_{a_1a_2}\Xi^{(32)}$$

$$\begin{aligned}\psi\wedge\bar{\psi}\wedge\Gamma_{a_1\ldots a_5}\psi &= \Xi^{(4224)}_{a_1\ldots a_5} + 2\Gamma_{[a_1\ldots a_3}\Xi^{(1408)}_{a_4a_5]}\\ &\quad + \frac{5}{9}\Gamma_{[a_1\ldots a_4}\Xi^{(320)}_{a_5]} - \frac{1}{77}\Gamma_{a_1\ldots a_5}\Xi^{(32)}\end{aligned}$$

$$\bar{\psi}\wedge\Gamma_{a_1}\psi\wedge\bar{\psi}\wedge\Gamma_{a_2}\psi = X^{(65)}_{\substack{a_1\\ a_2}} + \frac{1}{11}\delta_{a_1a_2}X^{(1)}$$

$$\bar{\psi}\wedge\Gamma_{a_1a_2}\psi\wedge\bar{\psi}\wedge\Gamma_{a_3}\psi = X^{(429)}_{\substack{a_1a_2\\ a_3}} + X^{(165)}_{a_1a_2a_3}$$

$$\begin{aligned}\bar{\psi}\wedge\Gamma_{a_1a_2}\psi\wedge\bar{\psi}\wedge\Gamma_{a_3a_4}\psi &= X^{(1144)}_{\substack{a_1a_2\\ a_3a_4}} + X^{(330)}_{a_1a_2a_3a_4}\\ &\quad + \frac{44}{81}\delta^{[a_1}_{[a_3}X^{(65)\,a_2]}_{\;\;a_4]} - \frac{2}{11}\delta^{a_1a_2}_{a_3a_4}X^{(1)}\end{aligned}$$

$$\begin{aligned}\bar{\psi}\wedge\Gamma_{a_1\ldots a_5}\psi\wedge\bar{\psi}\wedge\Gamma_{a_6}\psi &= \varepsilon_{a_1\ldots a_6b_1\ldots b_5}X^{(462)}_{b_1\ldots b_5}\\ &\quad + X^{(4290)}_{\substack{a_1\ldots a_5\\ a_6}} + \frac{15}{7}\,{}_{a_6[a_1}X^{(330)}_{a_2\ldots a_5]}\end{aligned}$$

$$\begin{aligned}\bar{\psi}\wedge\Gamma_{a_1\ldots a_5}\psi\wedge\bar{\psi}\wedge\Gamma_{a_6a_7}\psi &= \frac{i}{56}\varepsilon_{a_1\ldots a_7b_1\ldots b_4}X^{(330)}_{b_1\ldots b_4}\\ &+ X^{(17160)}_{\substack{a_1\ldots a_5\\ a_6a_7}} - \frac{i}{300}\varepsilon_{b_1\ldots b_5a_1\ldots a_5[a_6}X^{(4290)}_{\substack{b_1\ldots b_5\\ a_7]}}\\ &- \frac{180}{21}\delta^{[a_6a_7]}_{[a_1a_2}X^{(165)}_{a_3\ldots a_5]} - 1200i\;{}^{[a_6}\tilde{X}^{(462)}_{[a_1\;a_2\ldots a_5]a_7]}\end{aligned}$$

where

$$\Xi^{(32)}, \quad \Xi^{(320)}_{a} \left(\Gamma^{a}\,\Xi^{(320)}_{a} = 0\right), \quad \Xi^{(1408)}_{a_1 a_2} \left(\Gamma^{a_2}\,\Xi_{a_1 a_2} = 0\right),$$
$$\Xi^{(4224)}_{a_1 - a_5} \left(\Gamma^{a_5}\,\Xi^{(4224)}_{a_1 - a_5} = 0\right)$$

are, respectively the irreducible representations $(1/2)^5$, $(3/2)(1/2)^4$, $(3/2)^2(1/2)^3$, $(3/2)^5$ listed in Table 4. Similarly

$$X^{(1)}, \quad X^{(65)}_{\substack{a\\ b}}, \quad X^{(165)}_{a_1 \cdots a_3}, \quad X^{(330)}_{a_1 - a_4}, \quad X^{(462)}_{a_1 - a_5}, \quad X^{(429)}_{\substack{a_1 a_2\\ a_3}},$$
$$X^{(1144)}_{\substack{a_1 a_2\\ a_3 a_4}}, \quad X^{(4290)}_{\substack{b_1 - b_5\\ a_1}}, \quad X^{(17160)}_{\substack{a_1 - a_5\\ b_1 b_2}}$$

are respectively the bosonic reducible representations $(0)^5$, $(2)(0)^4$, $(1)^3(0)^2$, $(1)^4(0)$, $(1)^5$, $(2)(1)(0)^3$, $(2)^2(0)^3$, $(2)(1)^4$, $(2)^2(1)^3$ also listed in Table 4. Moreover we have

$$\tilde{X}^{(462)}_{a_1 - a_6} = \epsilon_{a_1 - a_6 b_1 - b_5}\, X^{(462)}_{b_1 - b_5} \tag{52}$$

As we have explained the decomposition of Table 4 is a substitute for all Fierz identities which correspond to the appearance of the same irreducible representations in several different products of fermionic currents. The irreducible representations Ξ and X form a complete and orthonormal basis for the decomposition of, respectively, 3 - ψ and 4 - ψ terms.

REFERENCES

1. F. Murnaghan, The Theory of Group Representations, (Johns Hopkins Press, Baltimore, 1938).
2. M. Brown and S.J. Gates Jr., Nucl. Phys. B165:445 (1980).
3. R. D'Auria, P. Fré and T. Regge, Nucl. Phys. B 188:342 (1981).
4. S. Ferrara and P. van Nieuwenhuizen, Phys. Lett. 74B:333 (1978); K. Stelle and P.C. West, Phys. Lett. 74B:330 (1978).
5. R. D'Auria, P. Fré, E. Maina and T. Regge, Annals of Physics 135:237 (1981).

CHAPTER III - D = 11 SUPERGRAVITY

Presented by: R. D'Auria and P. Fré
Reference: "Geometric Supergravity in D = 11 and its hidden Supergroup", Torino preprint IFTT 415 (1981), submitted to Nucl. Phys. B.

1. INTRODUCTION

Simple supergravity in D = 11 was introduced by Cremmer, Julia and Scherk in Ref. 1 and later formulated by Cremmer and Ferrara in superspace[2]. It is the maximally extended supertheory containing at most spin two particles; by dimensional reduction[2] it yields N = 8 supergravity in four dimensions which is considered, with increasing interest, a possibly viable theory for the unification of all interactions.

An up to now unsolved problem was the identification of the supergroup underlying this theory.

This no academic question, rather a fundamental one. Indeed, supergravity claims to be the local theory of a suitable supergroup allowing the unification of all truly elementary particles in a single supermultiplet; therefore, a supergravity theory whose supergroup is unknown is somehow incomplete. The need for a supergroup was already felt by the inventors of the theory who, in their original paper[1], proposed Osp(32/1) as the most likely candidate. This proposal is based on two facts:

i) Osp(32/1) is the minimal grading of Sp(32) which on the other hand, is the maximal bosonic group preserving the Majorana property of a Majorana spinor.

ii) The generators of Osp(32/1) are, with respect to the Lorentz subgroup $SO(1,10) \subset Osp(32/1)$, the following tensors (or spinors):

$$P_a \, , \; J_{ab} \, , \; Z_{a_1 \ldots a_5} \, , \; Q_\alpha \tag{1.1}$$

where J_{ab} and $Z_{a_1 \ldots a_5}$ are skew symmetric. The quantities J_{ab}, P_a, Q_α can be interpreted respectively as the Lorentz, translation and supersymmetry generators. The five-index skew symmetric generator $Z_{a_1 \ldots a_5}$ on the other hand, can be seen as associated to the physical $A_{\mu\nu\rho}$ field appearing in D = 11 supergravity in the following indirect way. The potential associated to $Z_{a_1 \ldots a_5}$ is a one-form $B_{a_1 \ldots a_5}$: multiplying $B_{a_1 \ldots a_5}$ by five elfbeins $V^{a_1} \; V^{a_2} \; \ldots \; V^{a_5}$ (the gauge fields of the generator P_a) we obtain a six-form B:

$$\mathcal{B} = B^{a_1 \ldots a_5} \wedge V_{a_1} \wedge \ldots \wedge V_{a_5} \tag{1.2}$$

Calling $B_{\mu_1 \ldots \mu_6}$ its space-time components and $\mathcal{F}_{\mu_1 \ldots \mu_7}$ their curl:

$$\mathcal{F}_{\mu_1 \ldots \mu_7} = \partial_{[\mu_1} \mathcal{B}_{\mu_2 \ldots \mu_7]} \tag{1.3}$$

it is attractive to assume that $\mathcal{F}_{\mu_1 \ldots \mu_7}$ is related to the curl of $\mathcal{A}_{\mu\nu\rho}$ by a duality relation:

$$\mathcal{F}_{\mu_1 \ldots \mu_7} = \text{const} \times \epsilon_{\mu_1 \ldots \mu_7 \nu_1 \ldots \nu_4} \partial^{\nu_1} A^{\nu_2 \nu_3 \nu_4} \tag{1.4}$$

If this is the case, then there should be a formulation of D = 11 supergravity which utilizes $B_{\mu_1 \ldots \mu_6}$ as a fundamental field instead of $A_{\mu_1 \mu_2 \mu_3}$. H. Nicolai and P. van Nieuwenhuizen tried to find it[3]. In this respect it must be noted that in the graded Lie algebra of Osp(32/1) the generators $Z_{a_1 \ldots a_5}$ are not Abelian and mix, in a non-trivial way, with the space-time symmetries P_a, J_{ab}. Indeed Osp(32/1) is described by the following curvatures:

$$R^{ab} = \mathcal{R}^{ab}(\omega) + \alpha_2 V^a \wedge V^b + \alpha_3 \bar{\psi} \wedge \Gamma^{ab} \psi + \alpha_4 B^{ac_1 \ldots c_4} \wedge B^b{}_{c_1 \ldots c_4} \tag{1.5a}$$

$$R^a = \mathcal{D} V^a - \frac{i}{2} \bar{\psi} \wedge \Gamma^a \psi + \alpha_1 \epsilon^{a b_1 \ldots b_5 c_1 \ldots c_5} \mathcal{B}_{b_1 \ldots b_5} \wedge B_{c_1 \ldots c_5} \tag{1.5b}$$

$$R^{a_1 \ldots a_5} = \mathcal{D} B^{a_1 \ldots a_5} - \frac{i}{2} \bar{\psi} \wedge \Gamma^{a_1 \ldots a_5} \psi + \alpha_5 \epsilon^{a_1 \ldots a_5 b_1 b_2 b_3 c_1 c_2 c_3} B_{b_1 b_2 b_3 \ell_1 \ell_2} \wedge B_{c_1 c_2 c_3}{}^{\ell_1 \ell_2} \tag{1.5c}$$

$$\rho = \mathcal{D}\psi + i\alpha_6 \Gamma_a \psi \wedge V^a + \alpha_7 \Gamma_{a_1 \ldots a_5} \psi \wedge B^{a_1 \ldots a_5} \tag{1.5d}$$

where $\mathfrak{D}$ denotes the Lorentz-covariant derivative and Q^{ab} is defined as

$$\mathcal{R}^{ab} = d\omega^{ab} - \omega^{ac} \wedge \omega_c{}^b$$

where $\alpha_1, \alpha_2, \ldots, \alpha_7$ are numerical constants, fixed by Jacobi identities (that is integrability conditions (dd = 0) of Eqs. (1.5) at zero curvature). Because of this property of the algebra a theory based on Osp (32/1) is bound to violate the Coleman-Mandula theorem[4] since it will provide a non-trivial unification of internal and external symmetries at the bosonic level[5]. Therefore, before looking into a $B_{\mu_1 \ldots \mu_6}$ formulation of D = 11 supergravity it is advisable to perform an Inönü-Wigner contraction of Osp(32/1) by setting:

$$\omega^{ab} \to \omega^{ab} \qquad R^{ab} \to R^{ab} \tag{1.6a}$$

$$V^a \to eV^a \qquad R^a \to eR^a \tag{1.6b}$$

$$B^{a_1 \ldots a_5} \to eB^{a_1 \ldots a_5} \qquad R^{a_1 \ldots a_5} \to eR^{a_1 \ldots a_5} \tag{1.6c}$$

$$\psi \to \sqrt{e}\,\psi \qquad \rho \to \sqrt{e}\,\rho \tag{1.6d}$$

where e is a scaling parameter. In the contraction limit $e \to 0$ one obtains the contracted Osp (32/1) supergroup:

$$R^{ab} = \mathcal{R}^{ab} \tag{1.7a}$$

$$R^a = \mathfrak{D}V^a - \frac{i}{2}\bar{\psi} \wedge \Gamma^a \psi \tag{1.7b}$$

$$R^{a_1 \ldots a_5} = \mathfrak{D}B^{a_1 \ldots a_5} - \frac{i}{2}\bar{\psi} \wedge \Gamma^{a_1 \ldots a_5} \psi \tag{1.7c}$$

$$\rho = \mathfrak{D}\psi \tag{1.7d}$$

which is free from the Coleman-Mandula disease since now $Z_{a_1 \ldots a_5}$ is Abelian. Even with these precautions, however, the result of Nicolai, Townsend and van Nieuwenhuizen was negative. The six-form formulation of D = 11 supergravity does not seem to exist[3]. As the reader will see we reach the same conclusion in a totally different way.

This being the state of the art, the situation we had to face was the following:

i) D = 4 and D = 5 simple supergravities are interpretable as local theories of a suitable supergroup. Their Lagrangians can be retrieved in a systematic way using the group manifold approach[6] which utilizes the one-form potential of the supergroup as the only fundamental field and the geometric operations d (= exterior derivative), Λ (= wedge product) as the only allowed manipulations in the construction of the action.

ii) The supergroup interpretation of D = 11 supergravity and, hence, its geometric formulation within the group manifold approach is not straightforward, essentially because of the following fact: the field $A_{\mu\nu\rho}$ of the Cremmer-Julia-Scherk theory is a three-form rather than a one-form and therefore it cannot be interpreted as the potential of a generator in a supergroup. The solution of the dilemma shows up almost naturally when the problem is formulated in these terms. Since the Cremmer-Julia-Scherck theory contains forms of higher degree, the physical fields are not one-form potentials of a super Lie algebra, rather they are p-form potentials of a generalized Cartan integrable system. The notion of Cartan integrable system (CIS in the following), discussed in Section 2, is a natural generalization to the case of p-forms of the Maurer-Cartan equations defining a (super) Lie algebra. All the concepts advocated by the group manifold framework, namely curvature, covariant exterior derivative, cosmococycle condition for the existence of the vacuum solution and rheonomy can be almost trivially extended to the case of a CIS-manifold.

In this paper we first introduce the notion of Cartan integrable system and then, after showing the existence of a specific CIS in D = 11 we construct supergravity as a geometric theory on this CIS-manifold. Later, once the theory has been obtained, we address the question whether our CIS is equivalent to an ordinary supergroup, namely whether our three-form A can be viewed as a polynomial in a set of ordinary one-forms in such a way that, giving the exterior derivatives of these latter we recover the exterior derivative of the former (A).

The answer is yes and we actually get a double solution: there are two different supergroups whose one-form potentials can be used interchangeably to parametrize the three-form A. Both in establishing the integrability of our CIS and in solving the cosmo-cocycle condition for the linear part of the Lagrangian a central role is played by Fierz identities. An account of the systematics of D = 11 Fierz identities, following the group theoretical technique fully explained in Ref. 7, has been given in Chapter II. In this respect we want to point out that Fierz identities in D = 11 and also the specific CIS we use were already derived by A. D'Adda and T. Regge in some unpublished notes[8] which were very inspiring for us.

2. CARTAN INTEGRABLE SYSTEM

It is very well known that a (super) Lie algebra can be described in two equivalent ways. The first is provided by the familiar commutation relations among the generators. One starts with a set of operators $\vec{T}_A$ forming the basis of the tangent space T(M) to a manifold M. If we can write a set of commutation relations

$$[\vec{T}_A, \vec{T}_B\} = C^L{}_{AB}\, \vec{T}_L \tag{2.1}$$

where C^L_{AB} are structure constants satisfying the Jacobi identities:

$$[\vec{T}_A,[\vec{T}_B,\vec{T}_C\}\} + (-)^{A(B+C)}[\vec{T}_B,[\vec{T}_C,\vec{T}_A\}] + (-)^{B(C+A)}[\vec{T}_C,[\vec{T}_A,\vec{T}_B\}\} = 0 \tag{2.2}$$

then the manifold M is a (super) Lie group and (2.1) is its (super) Lie algebra. The Jacobi identities (2.2) are all we have to check in order to be sure that (2.1) defines a viable (super) Lie algebra.

The second description of a (super) group, equally well-known but, only for historical reasons, less used in the physics literature, consists of the Maurer-Cartan equations.

In this set up one considers a manifold M and its cotangent space CT(M): CT(M) is the vector space of one-forms on the manifold M. Given a basis σ^A of CT(M) the exterior derivative $d\sigma^A$ is a two-form and can be decomposed in the basis provided by $\sigma^B \wedge \sigma^C$

$$d\sigma^A = F^A{}_{BC}\, \sigma^B \wedge \sigma^C \tag{2.3}$$

If we can find a set $\{\sigma^A\}$ such that the F^A_{BC} are constants:

$$F^A{}_{BC} = -\frac{1}{2} C^A{}_{BC} \tag{2.4}$$

consistent with the integrability condition dd = 0, namely if we can set

$$d\sigma^A + \frac{1}{2} C^A{}_{BC}\, \sigma^B \wedge \sigma^C = 0 \tag{2.5}$$

then using (2.5) we automatically obtain:

$$dd\sigma^A = - C^A{}_{BC}\, d\sigma^B \wedge \sigma^C = -\frac{1}{2} C^A{}_{BC}\, C^B{}_{RS}\, \sigma^R \wedge \sigma^S \wedge \sigma^C = 0 \tag{2.6}$$

and M is a (super) Lie group and (2.5) are its Maurer-Cartan equations. The (super) Lie algebra of M is obtained via the introduction of a dual basis in the tangent space T(M): indeed if $\{\vec{T}_A\}$ is a set of tangent vectors such that

$$\sigma^A(\vec{T}_B) = \delta^A_B \tag{2.7}$$

Eq. (2.5) implies Eq. (2.1) and vice versa. In the same way Eq. (2.6) implies Jacobi identities (2.2) and vice versa. Therefore all we have to do in order to be sure that Eq. (2.5) defines a true (super) Lie group is to check whether Eq. (2.6) holds. Equation (2.6) is the integrability condition of the Maurer-Cartan equations (2.5).

As we have already pointed out, the two ways of describing a Lie algebra are totally equivalent, yet the first is more customary in physics. Dealing with gravity and supergravity theories, however, the second approach is more appropriate for the following reason. Since the ultimate goal is the construction of an action integral for the (super) group potentials, if we start with the Maurer-Cartan equations (2.5) the transition to the potentials is simply performed via the replacement of the one-forms σ^A satisfying (2.5) (left invariant one-forms) with a set of one-forms μ^A which do not satisfy (2.5) (soft forms or supergroup potentials). The two-forms:

$$R^A = R^A[\mu] = d\mu^A + \frac{1}{2} C^A{}_{BC}\, \mu^B \wedge \mu^C \tag{2.8}$$

expressing the deviation from the Maurer-Cartan equations are called the curvatures of μ^A. The physical action is the integral of a polynomial (in the exterior algebra sense) in μ^A and R^A with the eventual addition of some 0-forms. The rules of this game, which goes under the name of group manifold approach, are discussed for example in Ref. 6 or with more details in Ref. 9: all supergravity theories so far examined fit nicely into this framework.

The notion of Cartan integrable system appears to be a most natural generalization of the concept of (super) Lie group if we adopt the language of the Maurer-Cartan equations as the primary description of the group structure.

Suppose that we have a manifold M whose dimension, however, is not at this point fixed. (In the case of the proper super Lie group instead the dimension of M is just equal to the number of generators $\vec{T}_A$ or, equivalently, of left-invariant one-forms σ^A.) Suppose that on M we define a set of p-forms of various degree $\{\Theta^{A(p)}\}$ whose exterior derivative $d\Theta^{A(p)}$ can still be expressed as a polynomial in $\Theta^{A(p)}$ with constant coefficients:

$$d\Theta^{A(p)} + \sum_{n=1}^{N} \frac{1}{n} C^{A(p)}_{B_1(p_1)\cdots B_n(p_n)} \Theta^{B_1(p_1)} \wedge \cdots \wedge \Theta^{B_n(p_n)} = 0 \tag{2.9}$$

The number N is equal to $p_{max} + 1$ where p_{max} is the highest degree in the set $\{\Theta^{A(P)}\}$.

Obviously, since all the terms in Eq. (2.9) have to be (p+1)-forms, the constants $C^{A(p)}_{B_1(p_1)\dots B_n(p_n)}$ are different from zero only if

$$p_1 + \cdots + p_n = p+1 \tag{2.10}$$

Moreover they have the proper symmetry in the exchange of any two neighbouring indices:

$$C^{A(p)}_{B_1(p_1)\dots B_i(p_i) B_{i+1}(p_{i+1})\dots B_n(p_n)} = (-1)^{B_i B_{i+1} + p_i p_{i+1}} C^{A(p)}_{B_1(p_1) B_2(p_2)\dots B_{i+1}(p_{i+1}) B_i(p_i)\dots B_n(p_n)} \tag{2.11}$$

We say that Eq. (2.9) is a generalized Maurer-Cartan equation (GMCE) and that it describes a Cartan integrable system (CIS) if and only if the integrability condition $dd\,\Theta^{A(P)} = 0$ follows automatically from (2.9). Explicitly the condition for (2.9) to be a CIS is the following one:

$$dd\Theta^{A(p)} = -\sum_{n=1}^{N}\sum_{m=1}^{N} C^{A(p)}_{B_1(p_1)\cdots B_n(p_n)} C^{B_1(p_1)}_{D_1(q_1)\cdots D_m(q_m)} \Theta^{D_1(q_1)} \wedge\cdots\wedge \Theta^{D_m(q_m)} \wedge \Theta^{B_2(p_2)} \wedge\cdots\wedge \Theta^{B_n(p_n)} = 0 \tag{2.12}$$

Equation (2.12) is the analogue of Eq. (2.6) and therefore it is just the analogue of the Jacobi identies (2.2) of on ordinary Lie algebra.

Given a CIS all concepts advocated by the group manifold approach can be naturally extended. Let us go through their list.

i) Soft-forms or CIS-potentials

A set $\{\theta^{A(p)}\}$ satisfying the GMCE (2.9) is named a left-invariant set.

A new set $\{\Pi^{A(p)}\}$ which does not satisfy (2.9) will instead be a soft-set. The $\Pi^{A}(p)$ may be viewed as the Yang-Mills potentials of the CIS, the same way as μ^A are the Yang-Mills potentials of the ordinary super group described by the ordinary Maurer-Cartan Eq. (2.5).

ii) CIS-curvatures, CIS-Bianchi identities and CIS-covariant derivatives

Given a soft-set $\Pi^{A(p)}$ its deviation from the GMCE (2.9) is named the curvature set of $\{\Pi^{A}(p)\}$

$$R^{A(p+1)} = d\Pi^{A(p)} + \sum \frac{1}{\alpha} C^{A(p)}_{B_1(p_1)\cdots B_\alpha(p_\alpha)} \Pi^{B_1(p_1)} \wedge \cdots \wedge \Pi^{B_\alpha(p_\alpha)} \tag{2.13}$$

The integrability of the CIS, that is condition (2.12), yields a differential identity on the curvatures $R^{A}(p+1)$ which is worthy of the name of a Bianchi identity:

$$\nabla R^{A(p+1)} = dR^{A(p)} + \sum \frac{1}{\alpha} C^{A(p)}_{B_1(p_1)\cdots B_\alpha(p_\alpha)} R^{B_1(p_1)} \wedge \cdots \wedge \Pi^{B_\alpha(p_\alpha)} \tag{2.14}$$

In complete analogy with what one does in Chevalley cohomology theory (see Ref. 9) we say that the left-hand side of Eq. (2.14) defines the covariant derivative of an adjoint set.

Suppose $H^{A}(p+1)$ is a set of (p+1)-forms: the combination

$$\nabla H^{A(p+1)} = dH^{A(p+1)} + \sum_{\alpha=1}^{N} C^{A(p)}_{B_1(p_1)\cdots B_\alpha(p_\alpha)} H^{B_1(p_1+1)} \wedge \Pi^{B_2(p_2)} \wedge \cdots \wedge \Pi^{B_\alpha(p_\alpha)} \tag{2.15}$$

will be named the covariant adjoint derivative of $H^{A(p+1)}$. With this definition the Bianchi identity (2.14) just states that the covariant adjoint derivative of the curvature is zero as happens with ordinary supergroups. Let us now assume that we have a multiplet $\nu_{A(d-p-1)}$ of forms whose degree is the complement of the degree of $H^{A(p+1)}$ with respect to some fixed number d. We say that $\{\nu_{A(d-p-1)}\}$ is a <u>coadjoint set</u> of forms if $I^{(d)}$, obtained multiplying $H^{A(p+1)}$ with $\nu_{A(d-p-1)}$ is an invariant:

$$I^{(d)} = H^{A(p+1)} \wedge \nu_{A(d-p-1)} \tag{2.16}$$

Invariant just means the following: the covariant derivative of $I^{(d)}$ coincides with its ordinary exterior derivative:

$$\begin{aligned} \nabla I^{(d)} &= \nabla H^{A(p+1)} \wedge \nu_{A(d-p-1)} + (-)^{p+1} H^{A(p+1)} \wedge \nabla \nu_{A(d-p-1)} = \\ &= dI^{(d)} = dH^{A(p+1)} \wedge \nu_{A(d-p-1)} + (-1)^{p+1} H^{A(p+1)} \wedge \nabla \nu_{A(d-p-1)} \end{aligned} \tag{2.17}$$

Equation (2.17) provides the definition of <u>coadjoint covariant derivative</u>. Indeed in order for (2.17) to be true we must have:

$$\nabla \nu_{A(d-p-1)} = d\nu_{A(d-p-1)} - (-)^{p+1} \sum_{n=1}^{N} C^{B_1(p_1)}_{A(p) B_2(p_2) \cdots B_n(p_n)} \Pi^{B_2(p_2)} \wedge \ldots \wedge \Pi^{B_n(p_n)} \wedge \nu_{B_1(d-p_1-1)} \tag{2.18}$$

where

$$p_1 + 1 = p + p_2 + p_3 + \cdots + p_n$$

iii) <u>Contraction</u>

The notation of contraction of a generic polynomial Ω in the soft forms $\Pi^{A(p)}$ coincides with the concept of functional variation. Therefore we set:

$$\underline{A}\rfloor \Omega = \frac{\delta}{\delta \Pi^{A(p)}} \Omega \tag{2.19}$$

3. CARTAN INTEGRABLE SYSTEM FOR D = 11 SUPERGRAVITY

We first narrow down our hunting ground by taking into account the following remarks.

i) Since supergravity contains ordinary gravity plus the Rarita-Schwinger field, our CIS must be an extension of the following ordinary Maurer-Cartan equations:

$$d\omega^{ab} - \omega^{ac} \wedge \omega_c{}^b = 0 \tag{3.1a}$$

$$dV^a - \omega^{ab} \wedge V_b - \frac{i}{2} \bar{\psi} \wedge \Gamma^a \psi = 0 \tag{3.1b}$$

$$d\psi - \frac{1}{4} \omega^{ab} \Gamma_{ab} \psi = 0 \tag{3.1c}$$

which correspond to the super Lie algebra of the graded Poincaré group in eleven dimensions. The indices a, b, c run from 0 to 10 and the standard Minkowskian metric

$$\eta_{ab} = \begin{pmatrix} 1 & \dots & & 0 \\ 0 & -1 & \dots & 0 \\ \vdots & & \ddots & \vdots \\ 0 & \dots & & -1 \end{pmatrix} \tag{3.2}$$

is used in the raising and lowering operations.

The skew-symmetric $\omega^{ab} = -\omega^{ba}$ is the Lorentz connection one-form, V^a is the elfbein one-form and ψ is the Majorana gravitino one-form.

ii) Since in D = 11 there is no internal symmetry group whose indices can be used and since we admit only massless particles of spin smaller than two the only other Bose fields which might enter the Lagrangian are skew-symmetric tensors of the type $A_{\mu_1 \dots \mu_p}$. These latter are nothing else than p-forms.

iii) If we assume that supersymmetry is linearly realized, the transformation rule of $A_{\mu_1 \dots \mu_p}$ must be of the following type:

$$\delta A_{\mu_1 \dots \mu_p} = \text{const} \times \bar{\varepsilon}\, \Gamma_{[\mu_1 \dots \mu_{p-1}} \psi_{\mu_p]} \tag{3.3}$$

Equation (3.3) means that, in the vacuum which is what matters for the derivation of generalized Maurer-Cartan equations, the exterior derivative of $A^{(p)}$ has to be the following one:

$$dA^{(p)} = \alpha_p \bar{\psi} \wedge \Gamma^{a_1 \dots a_{p-1}} \psi \wedge V_{a_1} \wedge \dots \wedge V_{a_{p-1}} \tag{3.4}$$

where α_p is some non-zero constant.

Since the only non-vanishing currents are those corresponding to symmetric Γ-matrices, namely

$$\bar{\psi} \wedge \Gamma^{a} \psi \quad ; \quad \bar{\psi} \wedge \Gamma^{a_1 a_2} \psi \quad ; \quad \bar{\psi} \wedge \Gamma^{a_1 \dots a_5} \psi \tag{3.5}$$

and their duals:

$$\bar{\psi} \wedge \Gamma^{a_1 \dots a_{10}} \psi \quad ; \quad \bar{\psi} \wedge \Gamma^{a_1 \dots a_9} \psi \quad ; \quad \bar{\psi} \wedge \Gamma^{a_1 \dots a_6} \psi \tag{3.6}$$

we conclude that the only a priori viable forms are $A^{(2)}$, $A^{(3)}$, $A^{(6)}$, $A^{(7)}$, $A^{(10)}$ and $A^{(11)}$. The Cartan system obtained by the addition of Eq. (3.4) to Eqs. (3.1) must however be integrable, namely we must have:

$$\begin{aligned} ddA^{(p)} &= \alpha_p \mathcal{D} (\bar{\psi} \wedge \Gamma^{a_1 \dots a_{p-1}} \psi \wedge V_{a_1} \wedge V_{a_2} \wedge \dots \wedge V_{a_{p-1}}) \\ &= p \alpha_p \frac{1}{2} \bar{\psi} \wedge \Gamma^{a_1 \dots a_{p-1}} \psi \wedge \bar{\psi} \wedge \Gamma_{a_1} \psi \wedge V_{a_2} \wedge \dots \wedge V_{a_{p-1}} = 0 \end{aligned} \tag{3.7}$$

Whether Eq. (3.7) holds depends on the structure of the quadrilinear Fierz identities. Indeed in order for (3.7) to be true we must have

$$\bar{\psi} \wedge \Gamma^{m a_1 \dots a_{p-2}} \psi \wedge \bar{\psi} \wedge \Gamma_m \psi = 0 \tag{3.8}$$

which happens only if

$$\begin{array}{ll} p-2 = 1 & p-2 = 10 \\ p-2 = 2 & p-2 = 9 \end{array} \tag{3.9}$$

Conditions (3.9) are easily understood recalling Table 5 of Chapter II which states that the only antisymmetric tensors absent in the decomposition of $\{(1/2)^5 \otimes (1/2)^5 \otimes (1/2)^5 \otimes (1/2)^5 \otimes (1/2)^5\}$ are $(1)(0)^4$, $(1^2)(0^3)$ and obviously their duals $(1^{10})(0)$, $(1^9)(0)$.

Therefore the viable p-forms which can be embedded together with ω^{ab}, V^a, ψ in an integrable Cartan system are those among $p = 2,3,6,7,10,11$ which also satisfy Eq. (3.9) namely

$$p = 3 \quad , \quad p = 11 \tag{3.10}$$

Now since $A^{(11)}$ is a form of maximum degree its curl (= exterior derivative) cannot enter the Lagrangian of $D = 11$. Hence it is to be dismissed.

Therefore we conclude that the Cartan integrable system corresponding to a linear representation of supersymmetry in eleven dimensions, later to be recalled with the name of C_{11}, is described by the following generalized curvatures:

Cartan Integrable System C_{11}:

$$R^{ab} = d\omega^{ab} - \omega^{ac} \wedge \omega_c{}^b \tag{3.11a}$$

$$R^a = \mathcal{D} V^a - \frac{i}{2} \bar{\psi} \wedge \Gamma^a \psi \tag{3.11b}$$

$$\rho = \mathcal{D} \psi \tag{3.11c}$$

$$R^{\square} = dA - \frac{1}{2} \bar{\psi} \wedge \Gamma^{ab} \psi \wedge V_a \wedge V_b \tag{3.11d}$$

The GMCE obtains when ω^{ab}, V^a, ψ, A are left-invariant and the curvatures are set to zero. In the soft-case, when the curvatures are different from zero the integrability of the system shows up as Bianchi identities.

CIS-Bianchi of C_{11}:

$$\nabla R^{ab} = \mathcal{D} R^{ab} = 0 \tag{3.12a}$$

$$\nabla R^a = \mathcal{D} R^a + R^{ab} \wedge V_b - i \bar{\psi} \wedge \Gamma^a \rho = 0 \tag{3.12b}$$

$$\nabla \rho = \mathcal{D}\rho + \frac{1}{4}\Gamma_{ab}\,\psi \wedge R^{ab} = 0 \tag{3.12c}$$

$$\nabla R^{\square} = dR^{\square} - \bar{\psi}\wedge\Gamma^{a_1a_2}\rho\wedge V_{a_1}\wedge V_{a_2} + \bar{\psi}\wedge\Gamma^{a_1a_2}\psi\wedge R_{a_1}\wedge V_{a_2} = 0 \tag{3.12d}$$

If $\{\nu_{ab},\nu_a,n,\nu_\square\}$ is a coadjoint set where ν_{ab}, ν_a, n are of degree (d-2) and $\nu_\square$ is of degree d-4, and we write the invariant:

$$I = R^{ab}\wedge\nu_{ab} + R^{a}\wedge\nu_{a} + R^{\square}\wedge\nu_{\square} + \bar{\rho}\wedge n \tag{3.13}$$

the procedure outlined in Section 2 [Eq. (2.16) and following ones] yields the definition of the coadjoint covariant derivative:

<u>Coadjoint covariant derivative of</u> C_{11}:

$$\nabla\nu_{ab} = \mathcal{D}\nu_{ab} + V_{[a}\wedge\nu_{b]} + \frac{1}{4}\bar{\psi}\wedge\Gamma_{ab}n \tag{3.14a}$$

$$\nabla\nu_{a} = \mathcal{D}\nu_{a} - \bar{\psi}\wedge\Gamma_{a}\psi\wedge V^{b}\wedge\nu_{\square} \tag{3.14b}$$

$$\nabla\nu_{\square} = d\nu_{\square} \tag{3.14c}$$

$$\nabla n = \mathcal{D}n - \Gamma_{a_1a_2}\psi\wedge V^{a_1}\wedge V^{a_2}\wedge\nu_{\square} - i\,\Gamma_{a}\psi\wedge\nu_{a} \tag{3.14d}$$

Being through with these preliminaries, we can now start turning the crank and constructing our geometric Lagrangian based on C_{11}.

4. THE GEOMETRICAL ACTION

According to the prescriptions of the group manifold approach the action $\mathcal{A}$ of d = 11 supergravity will be written as the sum of two pieces:

$$\mathcal{A} = \mathcal{A}_0 + \mathcal{A}_1 \tag{4.1}$$

$\mathcal{A}_0$ is the integral of a polynomial (in the exterior calculus sense) quadratic in the curvatures*) $R^A \equiv (R^{ab}, R^a, R^{\square}, \rho)$

*) The restriction to quadratic polynomials avoids the possibility of propagation equations of order higher than two (and vertices with more than 4ψ fields).

$$\mathcal{A}_0 = \int_{M_{11}} \left(\Lambda + R^A \wedge \nu_A + R^A \wedge R^B \wedge \nu_{AB} \right) \tag{4.1a}$$

M_{11} is an arbitrary eleven dimensional submanifold (to be identified later with the physical space-time) floating inside the CIS manifold, and all the terms in the integrand are 11-forms.

are polynomials in the CIS potentials $\mu^A \equiv (\omega^{ab}, V^a, A, \psi)$ of suitable degree.

Besides $\mathcal{A}_0$ which is built entirely in terms of the CIS potentials μ^A a geometrical action may or may not contain a further piece $\mathcal{A}$ with new 0-form independent fields which are identified by their equations of motions with the intrinsic components of the spin-1 curvatures. ($\mathcal{A}_1$ simply generalizes the Maxwell Lagrangian in its first order form.) While $\mathcal{A}_1$ is absent in the so-called "pure theories" (like d = 4 N = 1 and d = 5 N = 2 supergravities), its presence is mandatory in other cases (d = 4 N = 2,3 supergravities) to avoid the collapse to a trivial theory and to allow for the propagation of the spin 1-fields. It will turn out that its presence is necessary in our case and its form will be explicitly written.

We begin to determine the explicit form of $\mathcal{A}_0$, that is of the polynomials Λ, ν_{ab}, ν_a, n, $\nu_{\square}$, ν_{AB} using the following building principles:

a) <u>the action is locally Lorentz invariant</u>. This means that Λ, ν_{ab}, ν_a, n, $\nu_{\square}$, ν_{AB} are polynomials in V^a, ψ, A, the spin connection ω^{ab} being excluded. Moreover everything is a good SO(1,9) tensor.

b) <u>The vacuum ($R^{ab} = R^a = R \;= \rho = 0$) is a solution</u>. This condition is fulfilled if the following cosmococycle conditions are satisfied by the multiplet:

$$\nabla \nu_{ab} = 0 \tag{4.2a}$$

$$\underline{a}\rfloor \Lambda + \nabla \nu_a = 0 \tag{4.2b}$$

$$\underline{\alpha}\rfloor \Lambda + \nabla n^{\alpha} = 0 \tag{4.2c}$$

$$\underline{\square}\rfloor \Lambda + \nabla \nu_{\square} = 0 \tag{4.2d}$$

$$\text{at } R^{ab} = R^a = R^{\square} = \rho = 0$$

c) The equations of motion are invariant under the scale transformation which leaves the generalized Maurer-Cartan equations invariant; this last requirement needs further explanation. Let us first note that the definitions (3.11) of the CIS-curvatures are invariant under the following scale transformation

$$\begin{aligned} \omega^{ab} &\to \omega^{ab} & R^{ab} &\to R^{ab} \\ V^a &\to e V^a & R^a &\to e R^a \\ \psi &\to \sqrt{e}\,\psi & \rho &\to \sqrt{e}\,\rho \\ A &\to e^3 A & R^{\square} &\to e^3 R^{\square} \end{aligned} \tag{4.3}$$

where e is a real parameter. Since the equation of motions of the theory are relations among the curvatures and the potentials, in order to be consistent, they must not depend on the specific choice of e.

Indeed every value of e singles out an element in an equivalence class of isomorphic Cartan integrable systems. The equations of motion of the dynamical theory should depend only on the equivalence class and not on the specific element in the class. Otherwise it is almost evident that the theory will be trivial admitting, at most, the vacuum solution. In fact if the equations of motion depend on e they will provide relations among the curvature components also depending on e.

This scale criterion is very powerful and easily implemented: it is just sufficient that, under the transformation (4.3) all terms in the action (4.1) scale with the same power of e. Since the Einstein term

$$R^{ab} \wedge V^{c_1} \wedge \dots \wedge V^{c_9}\, \epsilon_{abc_1\dots c_9} \tag{4.4}$$

has to be there and it has scale dimension e^9, this fixes the scale of all other terms.

The scale criterion was not clearly stated in previous work on the group-manifold approach but it can be checked that in existing theories like D = 4 and D = 5 supergravity it just kills those terms which have to be suppressed in order for the theory to be non-rigid (for example in D = 4 the requirement of Lorentz and parity invariance plus the vacuum condition yields the action

$$\mathcal{A} = \int_{M^4} \{ R^{ab} \wedge V^c \wedge V^d \epsilon_{abcd} + 4\bar{\psi} \wedge \gamma_5 \gamma_a \rho \wedge V^a + a R^{ab} \wedge \bar{\psi} \wedge \gamma_5 \Sigma_{ab} \psi \} \tag{4.5}$$

which is trivial unless a = 0 (for a discussion of this point (see p. 26 of Ref. 9). Now it happens that the last term in (4.5) has scale dimension e while all the others have scale dimension e^2: hence it must be suppressed.) The most general form of the polynomials Λ, ν_{ab}, ν_a, $\nu_\square$, n, ν_{AB} which fulfills criteria a) and c) is the following:

$$\begin{aligned}
\Lambda &= a\bar{\psi} \wedge \Gamma^{a_1 a_2} \psi \wedge \bar{\psi} \wedge \Gamma^{a_3 a_4} \psi \wedge V^{a_5} \wedge \ldots \wedge V^{a_{11}} \epsilon_{a_1 \ldots a_{11}} \\
&\quad + b\bar{\psi} \wedge \Gamma^{a_1 a_2} \psi \wedge \bar{\psi} \wedge \Gamma^{a_3 a_4} \psi \wedge V_{a_1} \wedge \ldots \wedge V_{a_4} \wedge A \\
\nu_{ab} &= -\frac{1}{9} \epsilon_{abc_1 \ldots c_9} V^{c_1} \wedge \ldots \wedge V^{c_9} \\
\nu_a &= i\beta_1 V_a \wedge \bar{\psi} \wedge \Gamma^{c_1 \ldots c_5} \psi \wedge V^{c_6} \wedge \ldots \wedge V^{c_{11}} \epsilon_{c_1 \ldots c_{11}} \\
&\quad + i\beta_2 \bar{\psi} \wedge \Gamma_{ac_1 \ldots c_4} \psi \wedge V^{c_1} \wedge \ldots \wedge V^{c_4} \wedge A \\
\nu_\square &= i k_1 \bar{\psi} \wedge \Gamma_{a_1 \ldots a_5} \psi \wedge V^{a_1} \wedge \ldots \wedge V^{a_5} + k_2 \bar{\psi} \wedge \Gamma^{ab} \psi \wedge V_a \wedge V_b \wedge A \\
n &= h_1 \Gamma_{c_1 \ldots c_8} \psi \wedge V^{c_1} \wedge \ldots \wedge V^{c_8} + i h_2 \Gamma_{c_1 \ldots c_5} \psi \wedge V^{c_1} \wedge \ldots \wedge V^{c_5} \wedge A \\
R^A \wedge R^B \wedge \nu_{AB} &= \gamma R^\square \wedge R^\square \wedge A
\end{aligned} \tag{4.6}$$

where a, b, β_1, β_2, h_1, h_2, k_1, k_2, γ are the numerical constants. All these with the exception of γ are determined by the vacuum conditions (4.2). Actually one may easily see that adding to the Lagrangian

$$\mathcal{L}_0 = \Lambda + R^{ab} \wedge \nu_{ab} + R^a \wedge \nu_a + R^\square \wedge \nu_\square + \bar{\rho} \wedge n$$

the total divergence:

$$\alpha\, d\left(i\,\bar{\psi}\wedge\Gamma^{a_1\ldots a_5}\psi\wedge V_{a_1}\wedge\ldots\wedge V_{a_5}\wedge A\right)$$

we may put $\beta_2 = 0$. Implementing Eqs. (4.2), after extremely long but straightforward manipulations which make essential use of the Fierz decomposition of table 5 we arrive at the following solution:

$$a = \frac{1}{4}\;;\quad b = -210\;;\quad k_1 = -84\;;\quad k_2 = -840$$

$$\beta_1 = \frac{7}{30}\;;\quad h_1 = 2\;;\quad \beta_2 = h_2 = 0.$$

Finally the coefficient γ of the quadratic term is fixed by the requirement of gauge invariance of the action (4.19) under the transformation

$$A \rightarrow A + \delta A \quad ; \quad \delta A = d\varphi$$

where φ is an arbitrary two-form. The motivations of this requirement are the following ones:

a) analogy with D = 5 supergravity where the gauge invariance under

$$\delta B = d\varphi \tag{4.8}$$

fixes the coefficients of the quadratic terms in such a way as to guarantee non-triviality of the theory[12].

b) analogy with the Cremmer-Julia-Scherk formulation where

$$\delta A_{\mu\nu\rho} = \partial_{[\mu}\,\epsilon_{\nu\rho]} \tag{4.9}$$

is indeed an invariance of the action.

c) actual inspection of the equations of motion which reveals the following: if the terms with a bare A do not cancel identically in all equations the only possible solution is the vacuum ($R^{ab} = R^a = R = \rho = 0$).

Performing the explicit variation of $\mathcal{L}_0$ we obtain

$$\delta A = d\varphi \Rightarrow \delta\mathcal{L}_0 = -840\, R^{\square}\wedge\bar{\psi}\wedge\Gamma_{ab}\psi\wedge V^a\wedge V^b\wedge d\varphi$$
$$+\gamma\, R^{\square}\wedge R^{\square}\wedge d\varphi - 210\,\bar{\psi}\wedge\Gamma_{a_1a_2}\psi\wedge\bar{\psi}\wedge\Gamma_{a_3a_4}\psi\wedge V^{a_1}\wedge\ldots\wedge V^{a_4}\wedge d\varphi. \tag{4.10}$$

An integration by parts shows that $\delta\mathcal{L}_0$ is a total divergence only if

$$\gamma = -840 \tag{4.11}$$

Substituting the values of the parameters in (4.6) we obtain the final form of $\mathcal{A}_0$:

$$\begin{aligned}\mathcal{A}_0 = \int \Big\{ & -\frac{1}{9} R^{a_1a_2} \wedge V^{a_3} \wedge \dots \wedge V^{a_{11}} \epsilon_{a_1 \dots a_{11}} \\ & + \frac{7i}{30} R^{a} \wedge V_a \wedge \bar{\psi} \wedge \Gamma^{b_1 \dots b_5} \psi \wedge V^{b_6} \wedge \dots \wedge V^{b_{11}} \epsilon_{b_1 \dots b_{11}} \\ & - i\,84\, R^{\square} \wedge \bar{\psi} \wedge \Gamma_{a_1 \dots a_5} \psi \wedge V^{a_1} \wedge \dots \wedge V^{a_5} \\ & - 840\, R^{\square} \wedge \bar{\psi} \wedge \Gamma_{ab} \psi \wedge V^{a} \wedge V^{b} \wedge A \\ & + 2 \bar{\rho} \wedge \Gamma_{c_1 \dots c_8} \psi \wedge V^{c_1} \wedge \dots \wedge V^{c_8} \\ & + \frac{1}{4} \bar{\psi} \wedge \Gamma^{a_1a_2} \psi \wedge \bar{\psi} \wedge \Gamma^{a_3a_4} \psi \wedge V^{a_5} \wedge \dots \wedge V^{a_{11}} \epsilon_{a_1 \dots a_{11}} \\ & - 210\, \bar{\psi} \wedge \Gamma^{a_1a_2} \psi \wedge \bar{\psi} \wedge \Gamma^{a_3a_4} \psi \wedge V_{a_1} \wedge \dots \wedge V_{a_4} \wedge A \\ & - 840\, R^{\square} \wedge R^{\square} \wedge A \Big\} \end{aligned} \tag{4.12}$$

At this point however we may see very clearly the necessity of adding a further piece $\mathcal{A}_1$ to (4.12); in fact the propagation of the $A_{\mu\nu\rho}$ field on the physical space-time demands a kinetic term of the type

$$^{*}R^{\square} \wedge R^{\square} \tag{4.13}$$

involving the notion of Hodge duality on the space-time sub-manifold. As is well known the Hodge dualization is a meaningless operation in the geometric group manifold approach and terms like (4.13) have to emerge in the second order Lagrangian after the elimination of some non-propagating fields appearing in the first order one. So far only two mechanisms are known to get this result. One was found in D = 5 supergravity[11,12] and also in the coupling of a scalar field to gravity[15]. In D = 5 supergravity it works in the following way.

The torsion equation, obtained through the ω^{ab} variation, yields:

$$\epsilon_{abc_1c_2c_3} R^{c_1} \wedge V^{c_2} \wedge V^{c_3} + \eta V_a \wedge V_b \wedge R^{\otimes} = 0 \qquad (4.14)$$

where $\eta = \pm 1$, R^c is the supertorsion and

$$R^{\otimes} = dB^{\otimes} - \frac{i}{2} \bar{\xi} \wedge \xi \qquad (4.15)$$

is the curvature associated to the Maxwell one-form $B^{\otimes} = B^{\otimes}_{\mu}\, dx^{\mu}$. Equation (3.20) implies that the supertorsion $R^a_{\otimes}$ has space-time components proportional to the curl of $B^{\otimes}_{\mu}$. Indeed the solution of (3.20) is:

$$R^{\otimes} = F_{ab} V^a \wedge V^b \qquad (4.16)$$

$$R^a = -\frac{\eta}{4} \epsilon^{abcdf} F_{bc} V_d \wedge V_f \qquad (4.17)$$

Inserting (3.22) back into the first order Lagrangian one realizes that the geometric, Hodge dual-free, term:

$$R^{\otimes} \wedge R^a \wedge V_a \qquad (4.18)$$

becomes the kinetic term

$$F^{rs} F_{rs} V^{a_1} \wedge V^{a_2} \wedge \ldots \wedge V^{a_5} \epsilon_{a_1 \ldots a_5} \qquad (4.19)$$

of the $B^{\otimes}_{\mu}$ field. Unfortunately this beautiful mechanism is not accessible to the $A_{\mu\nu\rho}$ field of D = 11 supergravity simply because varying $\mathcal{A}_0$ with respect to ω^{ab} one obtains $R^a = 0$ [see Eqs. (5.1a), and (5.2)]. Moreover, varying $\mathcal{A}_0$ with respect to A one gets:

$$15 R^{\square} \wedge R^{\square} + 15 R^{\square} \wedge \bar{\psi} \wedge \Gamma^{ab} \psi \wedge V_a \wedge V_b + i \bar{\psi} \wedge \Gamma_{a_1 \ldots a_5} \rho \wedge V^{a_1} \wedge \ldots \wedge V^{a_5} = 0 \qquad (4.20)$$

Projecting out eight-elfbeins Eq. (4.20) yields

$$R^{\square}_{a_1 \ldots a_4} R^{\square}_{b_1 \ldots b_4} \epsilon^{a_1 \ldots a_4 b_1 \ldots b_4 c_1 c_2 c_3} = 0 \qquad (4.21)$$

which instead of being the Maxwell equation for the space-time components $R_{a_1 \dots a_4}$ of R is an algebraic constraint on the latter implying $R_{a_1 \dots a_4} = 0$. Inserted into the other equations of motion this result would imply $R^{ab} = R^a = R = \rho = 0$ and the theory would collapse to a trivial one.

The second mechanism for the geometric generation of the dual was introduced in N = 2 and N = 3 supergravity by one of us[13]. It consists of the addition of the 0-form F_{ab} as an independent dynamical field and it corresponds to a first-order formulation of the Maxwell Lagrangian.

The analogue of this mechanism in D = 11 supergravity would be the addition of a 0-form $F_{a_1 \dots a_4}$. We thus introduce the following extra term:

$$\mathcal{A}_1 = \int \Big\{ m\, F^{a_1 \dots a_4} R^{\square} \wedge V^{a_5} \wedge \dots \wedge V^{a_{11}} \epsilon_{a_1 \dots a_{11}} + n\, F_{a_1 \dots a_4} F^{a_1 \dots a_4} V^{c_1} \wedge \dots \wedge V^{c_{11}} \epsilon_{c_1 \dots c_{11}} \Big\} \tag{4.22}$$

where $F_{a_1 \dots a_4}$ is a four-index skew-symmetric 0-form and m,n two numerical parameters. Equation (4.22) corresponds to the first order formulation of the Maxwell Lagrangian.

In the next section we show that, provided m and n take specific values, the action $\mathcal{A}_0 + \mathcal{A}_1$ is non-trivial and describes a rheonomic theory.

5. EQUATIONS OF MOTION - NON-TRIVIALITY AND RHEONOMY

The equations of motion of the theory (4.29) are the following ones.

Torsion Equation (variation in ω^{ab}):

$$\epsilon_{ab\, c_1 \dots c_9} R^{c_1} \wedge V^{c_2} \wedge \dots \wedge V^{c_9} = 0 \tag{5.1a}$$

First Maxwell Equation ($F_{a_1 \dots a_4}$ variations)

$$m\, R^{\square} \wedge V^{a_5} \wedge \dots \wedge V^{a_{11}} \epsilon_{a_1 \dots a_4 a_5 \dots a_{11}} + 2n\, F_{a_1 \dots a_4} V^{c_1} \wedge \dots \wedge V^{c_{11}} \epsilon_{c_1 \dots c_{11}} = 0 \tag{5.1b}$$

Second Maxwell Equation (variation in A)

$$168i\,\bar\psi\wedge\Gamma_{a_1\ldots a_5}\rho\wedge V^{a_1}_{\wedge\ldots\wedge}V^{a_5}-2520\,\bar\psi\wedge\Gamma_{ab}\psi\wedge V^{a}_{\wedge}V^{b}\wedge R^{\square}$$
$$-2520\,R^{\square}\wedge R^{\square}+m\,\mathcal{D}F_{a_1\ldots a_4}\wedge V_{a_5}\wedge\ldots\wedge V_{a_{11}}\,\epsilon^{a_1\ldots a_{11}} \tag{5.1c}$$
$$+i\,\frac{7}{2}m\,F_{a_1\ldots a_4}\bar\psi\wedge\Gamma_{a_5}\psi\wedge V_{a_6}\wedge\ldots\wedge V_{a_{11}}\,\epsilon^{a_1\ldots a_{11}}=0$$

Gravitino Equation (variation in $\bar\psi$)

$$4\,\Gamma_{a_1\ldots a_8}\rho\wedge V^{a_1}_{\wedge\ldots\wedge}V^{a_8}-168i\,\Gamma_{a_1\ldots a_5}\psi\wedge V^{a_1}_{\wedge\ldots\wedge}V^{a_5}_{\wedge}R^{\square}$$
$$-m\,\Gamma_{ab}\psi\wedge V^{a}_{\wedge}V^{b}_{\wedge}F_{c_1\ldots c_4}V_{c_5}\wedge\ldots\wedge V_{c_{11}}\,\epsilon^{c_1\ldots c_{11}}=0 \tag{5.1d}$$

Einstein Equation (variation in V^n)

$$-R^{a_1a_2}_{\wedge}V^{a_3}_{\wedge\ldots\wedge}V^{a_{10}}\epsilon_{a_1\ldots a_{10}n}$$
$$+i\,\frac{7}{15}R_n\wedge\bar\psi\wedge\Gamma_{b_1\ldots b_5}\psi\wedge V_{b_6}\wedge\ldots\wedge V_{b_{11}}\,\epsilon^{b_1\ldots b_{11}}$$
$$+i\,\frac{7}{5}R^{a}_{\wedge}V_a\wedge\bar\psi\wedge\Gamma_{b_1\ldots b_5}\psi\wedge V_{b_6}\wedge\ldots\wedge V_{b_{10}}\,\epsilon^{b_1\ldots b_{10}n}$$
$$+\frac{7i}{15}V_n\wedge\bar\psi\wedge\Gamma_{b_1\ldots b_5}\rho\wedge V_{b_6}\wedge\ldots\wedge V_{b_{11}}\,\epsilon^{b_1\ldots b_{11}} \tag{5.1e}$$
$$-i\,\frac{7}{5}\bar\psi\wedge\Gamma_{b_1\ldots b_5}\psi\wedge V_n\wedge R_{b_6}\wedge V_{b_7}\wedge\ldots\wedge V_{b_{11}}\,\epsilon^{b_1\ldots b_{11}}$$
$$-420i\,R^{\square}\wedge\bar\psi\wedge\Gamma^{a_1\ldots a_4n}\psi\wedge V^{a_1}_{\wedge\ldots\wedge}V^{a_4}+16\,\bar\rho\wedge\Gamma_{c_1\ldots c_7n}\psi\wedge V^{c_1}_{\wedge\ldots\wedge}V^{c_7}$$
$$+11m\,F_{a_1\ldots a_4}F^{a_1\ldots a_4}{}_{\wedge}V^{c_1}_{\wedge\ldots\wedge}V^{c_{10}}\epsilon_{c_1\ldots c_{10}n}$$
$$+7m\,F_{a_1\ldots a_4}V_{a_5}\wedge\ldots\wedge V_{a_{10}}\wedge R^{\square}\,\epsilon^{a_1\ldots a_{10}n}$$
$$-m\,F_{a_1\ldots a_4}\bar\psi\wedge\Gamma_{an}\psi\wedge V^{n}_{\wedge}V_{a_5}\wedge\ldots\wedge V_{a_{11}}\,\epsilon^{a_1\ldots a_{11}}=0$$

Considering the first Eq. (5.1a) we immediately obtain

$$R^a = 0 \tag{5.2}$$

Therefore the supertorsion vanishes on-shell just as in N = 1 and N = 2 four-dimensional supergravities. Equation (5.2) can be solved for the connection ω^{ab}_{μ} as a functional of V^a_μ and ψ_μ. Explicitly:

$$\omega^{ab}{}_{\mu} = \overset{0}{\omega}{}^{ab}{}_{\mu} - \frac{i}{4}\left(\bar{\Psi}_\mu \Gamma_\lambda \psi_\nu + \bar{\Psi}_\lambda \Gamma_\nu \psi_\mu + \bar{\Psi}_\lambda \Gamma_\mu \psi_\nu - [\lambda \leftrightarrow \nu]\right) V^{\lambda|a} V^{\nu|b} \tag{5.3}$$

where $\overset{0}{\omega}{}^{ab}$ is the usual connection satisfying the space-time torsionless condition

$$\partial_{[\mu} V^a_{\nu]} - \omega^{ab}_{[\mu} V_{\nu].b} = 0 \tag{5.4}$$

Considering next Eq. (5.1b) we obtain

$$R^{\square} = -\frac{11!\,2n}{7!\,4!\,m} F_{a_1 \ldots a_4} V^{a_1}_{\wedge \ldots \wedge} V^{a_4} \tag{5.5}$$

Therefore if we set

$$n = -\frac{m\,7!\,4!}{2.11!} = -\frac{m}{660} \tag{5.6}$$

$F_{a_1 \ldots a_4}$ can be identified with the space-time components of the curvature $R^{\square}$:

$$F_{a_1 \ldots a_4} = R^{\square}{}_{a_1 \ldots a_4} \tag{5.7}$$

which, because of Eq. (5.5), has no outer spinorial components:

$$R^{\square} = F_{a_1 \ldots a_4} V^{a_1}_{\wedge \ldots \wedge} V^{a_4} \tag{5.8}$$

The choice (5.6) amounts to nothing else but a field redefinition of $F_{a_1 \ldots a_4}$.

Using now Eqs. (5.2) and (5.8) into Eqs. (5.1c) and (5.1d) we obtain the following result:
If the parameter m takes the value:

$$m = 2 \tag{5.9}$$

then the gravitino equation is consistent with the second Maxwell equation and we have the solution:

$$\rho = \rho_{ab} V^a_\wedge V^b - \frac{1}{3}\left(i\,\Gamma^{a_1a_2a_3}\psi_\wedge V^{a_4} + \frac{1}{8}\Gamma^{a_1\ldots a_5}\psi_\wedge V_{a_5}\right) F_{a_1\ldots a_4} \quad (5.10)$$

where $_{ab}$ and $F_{a_1\ldots a_4}$ satisfy the following propagation equations:

$$\Gamma^{abc}\rho_{bc} = 0 \quad (5.11a)$$

$$\mathcal{D}_m F^{mc_1c_2c_3} - \frac{1}{2\cdot 4!\,7!} F_{a_1\ldots a_4} F_{a_5\ldots a_8}\, \epsilon^{a_1\ldots a_8 c_1c_2c_3} = 0 \quad (5.11b)$$

On the other hand if $m \neq 2$ we obtain $F_{a_1a_2a_3a_4} = 0$ and the only solution is $R^{ab} = R^a = F_{a_1\ldots a_4} = \rho = 0$. The various projections of the Einstein Eq. (5.1e) do not pose any further threat and besides yielding the graviton propagation equation:

$$R^{am}{}_{\cdot bm} - \frac{1}{2}\delta^a_b R^{mn}{}_{\cdot mn} = 3\left(F^{ac_1c_2c_3} F_{bc_1c_2c_3} + \frac{1}{8}\delta^a_b F^{c_1\ldots c_4} F_{c_1\ldots c_4}\right) \quad (5.12)$$

they give rheonomic conditions which express the outer components $R^{ab}_{\alpha m}$ and $R^{ab}_{\alpha\beta}$ in terms of the inner ones ρ_{ab} and $F_{a_1\ldots a_4}$ (see Table 1).

Therefore when $m = 2$ and $n = -1/330$ the theory described by action (4.29) becomes non-trivial and rheonomic: it goes without saying that upon transition to space-time second order formalism it coincides with the Cremmer-Julia-Scherk theory[1].

We think it proper to conclude this section with a summary of the final result. It is given in Table 1.

Table 1

SUMMARY OF D = 11 - SUPERGRAVITY

CARTAN INTEGRABLE SYSTEM :

$$R^{ab} = d\omega^{ab} - \omega^{ac} \wedge \omega_{c}{}^{b}$$

$$R^{a} = \mathcal{D}V^{a} - \frac{i}{2}\bar{\psi} \wedge \Gamma^{a}\psi$$

$$\rho = \mathcal{D}\psi$$

$$R^{\square} = dA - \frac{1}{2}\bar{\psi} \wedge \Gamma^{ab}\psi \wedge V_{a} \wedge V_{b}$$

$$\begin{aligned}
\mathcal{A} = \int_{M_{11}} \Big\{ & -\frac{1}{9}R^{a_1 a_2} \wedge V^{a_3} \wedge \ldots \wedge V^{a_{11}} \varepsilon_{a_1 \ldots a_{11}} \\
& + \frac{7i}{30}R^{a} \wedge V_{a} \wedge \bar{\psi} \wedge \Gamma^{b_1 \ldots b_5}\psi \wedge V^{b_6} \wedge \ldots \wedge V^{b_{11}} \varepsilon_{b_1 \ldots b_{11}} \\
& + 2\bar{\rho} \wedge \Gamma_{c_1 \ldots c_8}\psi \wedge V^{c_1} \wedge \ldots \wedge V^{c_8} \\
& - 84R^{\square} \wedge (i\bar{\psi} \wedge \Gamma_{a_1 \ldots a_5}\psi \wedge V^{a_1} \wedge \ldots \wedge V^{a_5} - 10A \wedge \bar{\psi} \wedge \Gamma_{ab}\psi \wedge V^{a} \wedge V^{b}) \\
& + \frac{1}{4}\bar{\psi} \wedge \Gamma^{a_1 a_2}\psi \wedge \bar{\psi} \wedge \Gamma^{a_3 a_4}\psi \wedge V^{a_5} \wedge \ldots \wedge V^{a_{11}} \varepsilon_{a_1 \ldots a_{11}} \\
& - 210\bar{\psi} \wedge \Gamma^{a_1 a_2}\psi \wedge \bar{\psi} \wedge \Gamma^{a_3 a_4}\psi \wedge V_{a_1} \wedge \ldots \wedge V_{a_4} \wedge A \\
& - 840R^{\square} \wedge R^{\square} \wedge A - \frac{1}{330}F_{a_1 \ldots a_4}F^{a_1 \ldots a_4}V^{c_1} \wedge \ldots \wedge V^{c_{11}} \varepsilon_{c_1 \ldots c_{11}} \\
& + 2F_{a_1 \ldots a_4}R^{\square} \wedge V_{a_5} \wedge \ldots \wedge V_{a_{11}} \varepsilon^{a_1 \ldots a_{11}} \Big\}
\end{aligned}$$

(continued)

Table 1 (Contd.)

ON-SHELL SOLUTION FOR THE CURVATURES :

$$R^a = 0$$

$$R^\square = F_{a_1\ldots a_4} V^{a_1} \wedge \ldots \wedge V^{a_4}$$

$$\rho = \rho_{ab} V^a \wedge V^b - \frac{1}{3}(i\Gamma^{a_1\ldots a_3}\psi \wedge V^{a_4} + \frac{1}{8}\Gamma^{a_1\ldots a_4 m}\psi \wedge V_m)F_{a_1\ldots a_4}$$

$$R^{ab} = R^{ab}_{\ .mn} V^m \wedge V^n + i\bar{\rho}_{mn}(\frac{1}{2}\Gamma^{abcmn} - \frac{2}{9}\Gamma^{mn[a}\delta^{b]c}$$

$$+ 2\Gamma^{ab[m}\delta^{n]c})\psi \wedge V_c$$

$$- \frac{7}{9}\bar{\psi} \wedge \Gamma_{mn}\psi F^{mnab}$$

$$+ \frac{55}{216}\bar{\psi} \wedge \Gamma^{abc_1\ldots c_4}\psi F_{c_1\ldots c_4}$$

PROPAGATION EQUATIONS

i) $\Gamma^{abc}\rho_{bc} = 0$

ii) $\mathfrak{D}_m F^{mc_1\ldots c_3} - \frac{1}{2\cdot 4!\cdot 7!}\,\varepsilon^{c_1\ldots c_3 a_1\ldots a_8} F_{a_1\ldots a_4} F_{a_5\ldots a_8} = 0$

iii) $R^{am}_{\ .bm} - \frac{1}{2}\delta^a_b R^{mn}_{\ .mn} - 3F^{ac_1\ldots c_3}F_{bc_1\ldots c_3}$

$$+ \frac{3}{8}\delta^a_b F^{c_1\ldots c_4}F_{c_1\ldots c_4} = 0$$

6. SUPERGROUP INTERPRETATION OF THE D = 11 CARTAN INTEGRABLE SYSTEM

In Section II we have discussed the possible equivalence of a Cartan integrable system with an ordinary supergroup.

Everything boils down to solving the system of algebraic equations relating the supergroup structure constants $C^\alpha{}_{\beta\gamma}$ with the components $K^{A(p)}_{\alpha_1\ldots\alpha_p}$ of the CIS-forms $\Theta^{A(p)}$. In the present section we solve this problem for the specific CIS of D = 11 supergravity, defined by Eqs. (3.11) and recalled in Table 3.

Since V , ω^{ab}, ψ are already one-forms and Eqs. (3.11) do already define a supergroup, all we have to do is to find a suitable decomposition of the three-form A in a basis of one-forms. Using a little bit of ingenuity we start with the following ansatz:

$$
\begin{aligned}
A = {} & B^{ab}\wedge V_a \wedge V_b + \alpha_1 B_{a_1 a_2}\wedge B^{a_2}{}_{\cdot a_3}\wedge B^{a_3 a_1} + \\
& + \alpha_2 B_{b_1 a_1 \ldots a_4}\wedge B^{b_1}{}_{\cdot b_2}\wedge B^{b_2 a_1 \ldots a_4} + \alpha_3 \epsilon_{a_1 \ldots a_5 b_1 \ldots b_5 m} B^{a_1 \ldots a_5}\wedge B^{b_1 \ldots b_5}\wedge V^m \\
& + \alpha_4 \epsilon_{a_1 \ldots a_6 b_1 \ldots b_5} B^{a_1 a_2 a_3 mn}\wedge B^{a_4 a_5 a_6}{}_{\cdot mn}\wedge B^{n_1 \ldots n_5} + i\beta_1 \bar{\psi}\wedge\Gamma^a \eta \wedge V_a \\
& + \beta_2 \bar{\psi}\wedge \Gamma^{ab}\eta \wedge B_{ab} + i\beta_3 \bar{\psi}\wedge\Gamma^{a_1 \ldots a_5}\eta\wedge B_{a_1 \ldots a_5}
\end{aligned}
\qquad (6.1)
$$

where B_{ab}, $B_{a_1 \ldots a_5}$ are two new skew-symmetric one-forms, η is a new spinorial one-form and α_1, α_2, α_3, α_4, β_1, β_2, β_3 are parameters. The structure of the supergroup is described by curvatures of the following type:

$$R^{ab} = d\omega^{ab} - \omega^{ac}\wedge\omega_c{}^b \qquad (6.2a)$$

$$R^a = \mathcal{D}V^a - \frac{i}{2}\bar{\psi}\wedge\Gamma^a\psi \qquad (6.2b)$$

$$\overset{\square}{R}{}^{a_1 a_2} = \mathcal{D}B^{a_1 a_2} - \frac{1}{2}\bar{\psi}\wedge\Gamma^{a_1 a_2}\psi \qquad (6.2c)$$

$$\overset{\square}{R}{}^{a_1 \ldots a_5} = \mathcal{D}B^{a_1 \ldots a_5} - \frac{i}{2}\bar{\psi}\wedge\Gamma^{a_1 \ldots a_5}\psi \qquad (6.2d)$$

$$\rho = \mathcal{D}\psi \qquad (6.2e)$$

$$
\begin{aligned}
\sigma = {} & \mathcal{D}\eta + i\delta\Gamma^a\psi\wedge V_a + \gamma_1 \Gamma_{ab}\psi\wedge B^{ab} \\
& + i\gamma_2 \Gamma_{a_1 \ldots a_5}\psi\wedge B^{a_1 \ldots a_5}
\end{aligned}
\qquad (6.2f)
$$

When we set $R^{ab} = R^a = R^{a_1 a_2} = R^{a_1 \ldots a_5} = \rho = \sigma = 0$ we obtain the Maurer-Cartan equations which are viable only if they satisfy the integrability condition dd = 0 (Jacobi identities). In our case the integrability of Eqs. (6.2a)-(6.2e) is self-evident: all we have to do is to check the integrability of Eq. (6.2f). At zero curvatures we obtain:

$$\mathcal{D}\mathcal{D}\eta = 0 = \frac{\delta}{2}\Gamma^a\psi\wedge\bar{\psi}\wedge\Gamma_a\psi - \frac{\gamma_1}{2}\Gamma^{ab}\psi\wedge\bar{\psi}\wedge\Gamma_{ab}\psi + \frac{\gamma_2}{2}\Gamma^{a_1\ldots a_5}\psi\wedge\bar{\psi}\wedge\Gamma_{a_1\ldots a_5}\psi \tag{6.3}$$

Using the Fierz decomposition of Table 5 of Chapter 2 we see that Eq. (6.3) is true only if

$$\delta + 10\gamma_1 - 720\gamma_2 = 0 \tag{6.4}$$

Equation (6.4) is the specific form taken in our case by condition (2.26). The explicit form of Eq. (2.27) is now worked out in the following way. We take the ansatz (6.1) and we compute dA at zero curvatures: $R^{ab} = R^a = R^{a_1 a_2} = R^{a_1 \ldots a_5} = \sigma = 0$. Imposing that the result be equal to $\frac{1}{2}\bar{\psi}\wedge\Gamma\psi\wedge V_a{}_a V_b$ by repeated use of Fierz identities we find the following two-fold solution for the parameters α_i, β_i, γ_i and δ:

$$\alpha_1 = \begin{pmatrix} \frac{4}{15} \\ -\frac{4}{15} \end{pmatrix};\ \alpha_2 = \begin{pmatrix} -\frac{5}{144} \\ \frac{5}{144} \end{pmatrix};\ \alpha_3 = \begin{pmatrix} \frac{1}{4!6!} \\ -\frac{1}{4!6!} \end{pmatrix};\ \alpha_4 = \begin{pmatrix} \frac{1}{2\cdot(72)^2} \\ -\frac{1}{2\cdot(72)^2} \end{pmatrix}$$

$$\beta_1 = \begin{pmatrix} 0 \\ 1 \end{pmatrix};\ \beta_2 = \begin{pmatrix} \frac{1}{2} \\ \frac{1}{5} \end{pmatrix};\ \beta_3 = \begin{pmatrix} \frac{1}{144} \\ \frac{1}{240} \end{pmatrix} \tag{6.5}$$

$$\gamma_1 = \begin{pmatrix} \frac{1}{5} \\ -\frac{1}{2} \end{pmatrix};\ \gamma_2 = \begin{pmatrix} \frac{1}{240} \\ -\frac{1}{144} \end{pmatrix};\ \delta = \begin{pmatrix} 1 \\ 0 \end{pmatrix}$$

Therefore we conclude that also D = 11 supergravity is a standard group manifold theory. The supergroup curvatures are the following:

$$R^{ab} = d\omega^{ab} - \omega^{ac}\wedge\omega_c{}^b \tag{6.6a}$$

$$R^a = \mathcal{D}V^a - \frac{i}{2}\bar{\psi}\wedge\Gamma^a\psi \tag{6.6b}$$

$$\rho = \mathcal{D}\psi \tag{6.6c}$$

$$\overset{\square}{R}{}^{a_1a_2} = \mathcal{D}B^{a_1a_2} - \frac{1}{2}\bar{\psi}\wedge\Gamma^{a_1a_2}\psi \tag{6.6d}$$

$$\overset{\square}{R}{}^{a_1\ldots a_5} = \mathcal{D}B^{a_1\ldots a_5} - \frac{i}{2}\bar{\psi}\wedge\Gamma^{a_1\ldots a_5}\psi \tag{6.6e}$$

$$\begin{aligned}\sigma = \mathcal{D}\eta + i\begin{pmatrix}1\\0\end{pmatrix}\Gamma_a\psi\wedge V^a + \begin{pmatrix}\frac{1}{5}\\-\frac{1}{2}\end{pmatrix}\Gamma_{ab}\psi\wedge B^{ab}\\ + i\begin{pmatrix}\frac{1}{240}\\-\frac{1}{144}\end{pmatrix}\Gamma_{a_1\ldots a_5}\psi\wedge B^{a_1\ldots a_5}\end{aligned} \tag{6.6f}$$

The action is the one given in Table 1: we replace A everywhere with its expression (6.1) in which the values (6.5) have been substituted.

Obviously the Lagrangian could have been determined by a direct application of the standard group-manifold method to the supergroups (6.6), without any reference to the Cartan integrable system C_1. It must be noted however that:

a) the Lagrangian written in terms of the supergroup potentials is gigantic and the cosmococycle equation would have been solvable only through the use of a computer;

b) the supergroups (6.6) introduce the novelty of a second Abelian spinorial generator Q'_α which is associated to the one-form η.

This very intringuing feature could not be guessed a priori.

7. CONCLUSIONS

D = 11 supergravity is the local theory of one of the two supergroups (6.6). The super Lie algebra is immediately read off from Eqs. (6.6) and it is given in Table 2. The $A_{\mu\nu\rho}$ field is not elementary, rather it is a non-linear combination of the one-form potentials.

$$B^{a_1a_2}{}_\mu\,,\ B^{a_1\ldots a_5}{}_\mu\,,\ V^a{}_\mu\,,\ \psi_\mu\,,\ \eta_\mu\,. \tag{7.1}$$

All the symmetries of the theory are generated by J_{ab}, P_a, Q, Q', $Z_{a_1a_2}$, $Z_{a_1...a_5}$ associated to ω^{ab}, V^a, ψ, η, $B_{a_1a_2}$, $B_{a_1...a_5}$ respectively. To determine the explicit transformations of all the fields under all the generators what we have to do is the following. Starting from Eq. (6.1) and taking the derivative we obtain:

$$
\begin{aligned}
R^{\square} = {} & \overset{\square}{R}_{ab}\, V^a \wedge V^b - 2\, B_{ab} \wedge R^a \wedge V^b + 3 \begin{pmatrix} \frac{4}{15} \\ -\frac{4}{15} \end{pmatrix} \overset{\square}{R}{}^{a_1a_2} \wedge B_{a_2}{}^{\cdot a_3} \wedge B_{a_3a_1} \\
& + \begin{pmatrix} -\frac{5}{144} \\ \frac{5}{144} \end{pmatrix} \overset{\square}{R}{}^{a_1 \ldots a_4 b_1} \wedge B_{b_1}{}^{\cdot b_2} \wedge B_{b_2 a_1 \ldots a_4} + \ldots + \\
& + i \begin{pmatrix} 0 \\ 1 \end{pmatrix} \bar{\rho} \wedge \Gamma^a \eta \wedge V_a - i \begin{pmatrix} 0 \\ 1 \end{pmatrix} \bar{\psi} \wedge \Gamma^a \sigma \wedge V_a + \ldots
\end{aligned}
\qquad (7.2)
$$

Comparing Eq. (7.2) with the on-shell curvatures given by Table 3 we can determine the structure of all the new curvatures

$$
\overset{\square}{R}_{ab} \quad , \quad \overset{\square}{R}_{a_1 \ldots a_5} \; , \; \sigma . \qquad (7.3)
$$

Once this is done we have the full set of rheonomic conditions and therefore we have the complete on-shell representation of the algebra. This programme is very straightforward but long and we postpone it to future work.

Table 2

SUPER LIE ALGEBRAS OF D = 11 SUPERGRAVITY

NORMALIZATION OF GENERATORS :

$$\omega^{ab}(iJ_{mn}) = \delta^{ab}_{mn} \; ; \; \psi_\alpha(Q_\beta) = \delta_{\alpha\beta} \; ; \; \eta_\alpha(Q'_\beta) = \delta_{\alpha\beta}$$

$$V^a(P_b) = \delta^a_b \; ; \; B^{a_1a_2}(Z_{b_1b_2}) = \delta^{a_1a_2}_{b_1b_2} \; ; \; B^{a_1\ldots a_5}(Z_{b_1\ldots b_5})\delta^{a_1\ldots a_5}_{b_1\ldots b_5}$$

COMMUTATION RELATIONS :

$$[J_{m_1m_2}, J^{n_1n_2}] = -4i\delta^{[n_1}_{[m_1} J^{\cdot n_2]}_{\cdot m_2]}$$

$$[J_{m_1m_2}, P^n] = -2i\delta^n_{[m_1}P_{m_2]}$$

$$[J_{m_1m_2}, Z^{n_1n_2}] = -4i\delta^{[n_1}_{[m_1} Z^{\cdot n_2]}_{m_2]}$$

$$[J_{m_1m_2}, Z^{n_1\ldots n_5}] = -10i\delta^{[n_1}_{[m_1} Z^{\cdot n_2\ldots n_5]}_{m_2]}$$

$$[J_{m_1m_2}, \binom{Q}{Q'}] = \frac{i}{4}\Gamma_{m_1m_2}\binom{Q}{Q'}$$

$$[P_n, P_m] = [Z_{m_1m_2}, Z_{n_1n_2}] = [Z_{n_1\ldots n_5}, Z_{m_1\ldots m_5}] =$$

$$= [P_n, Z_{m_1\ldots m_5}] = [Z_{m_1m_2}, Z_{n_1\ldots n_5}] = 0$$

$$[P_m, Q'] = [Z_{m_1\ldots m_5}, Q'] = [Z_{m_1m_2}, Q'] = 0$$

$$\{Q,Q\} = i\, C\Gamma^a P_a + i\, C\Gamma^{a_1a_2}Z_{a_1a_2} + i\, C\Gamma^{a_1\ldots a_5}Z_{a_1\ldots a_5}$$

$$\{Q',Q'\} = 0; \; [Q,P^a] = i\binom{1}{0}\Gamma_a Q' \; ; \; [Q, Z^{a_1a_2}] = \binom{1/5}{-1/2}\Gamma^{a_1a_2}Q'$$

$$[Q, Z^{a_1\ldots a_5}] = \binom{1/240}{-1/144}\Gamma^{a_1\ldots a_5}Q'$$

REFERENCES

1. E. Cremmer, B. Julia and J. Scherk, Phys. Lett. 76B:409 (1978).
2. E. Cremmer and B. Julia, Phys. Lett. 80B:48 (1978).
3. H. Nicolai, P.K. Townsend and P. van Nieuwenhuizen, Preprint "Comments on 11-dimensional supergravity", to be published in Nuovo Cimento Letters.
4. S. Coleman and J. Mandula, Phys. Rev. 159:1251 (1967).
5. P. van Nieuwenhuizen, private communication.
6. For an up to date review of the group manifold approach see R. D'Auria, P. Fré and T. Regge, Trieste preprint I.C./81/54 (1981), to be published in "Introduction to Supergravity", ed. J.G. Taylor, to be published by Cambridge University Press, and all references quoted therein.
7. R. D'Auria, P. Fré, E. Maina and T. Regge, Torino preprint IFTT 410 (1981), to be published in Annals of Physics.
8. A. D'Adda and T. Regge, unpublished notes.
9. R. D'Auria, P. Fré and T. Regge, Rivista del Nuovo Cimento 3:number 12 (1980).
10. R. D'Auria, P. Fré and A.J. Da Silva, Torino preprint IFTT 411 (1981), to be published in Nucl. Phys. B.
11. R. D'Auria and P. Fré, Nucl. Phys. B173:456 (1980).
12. R. D'Auria, P. Fré, E. Maina and T. Regge, Annals of Physics 135:237 (1981).
13. P. Fré, Nucl. Phys. B186:44 (1981).
14. S. Ferrara, "Unification of the Fundamental Particle Interactions", Eds. S. Ferrara, J. Ellis and P. van Nieuwenhuizen (Plenum Press, N.Y., 1980) p. 119;
    E. Cremmer, idem., p. 137.
15. R. D'Auria and T. Regge, Torino preprint IFTT 409 (1981) to be published in Nucl. Phys. B.

SUPERCURRENT AND ANOMALIES IN SUPER SYMMETRICAL YANG-MILLS THEORIES

Olivier Piguet *

Département de Physique Théorique
Université de Genève
1211 Genève 4, Switzerland

INTRODUCTION

The supercurrent[1] in N = 1 supersymmetric theories is a Lorentz axial vector superfield $V_\mu\ (x,\theta)$ whose components are the local currents $T_{\mu\nu}$, $Q_{\mu\alpha}$, $\bar{Q}_{\mu\dot{\alpha}}$, and R_μ associated with, respectively, translation invariance P, supersymmetry Q_α, $\bar{Q}_{\dot{\alpha}}$, and a chiral invariance R. The currents $T_{\mu\nu}$ and $Q_{\mu\alpha}$ are improved, which means that their traces T^μ_μ and $Q^\alpha_\mu\ \sigma^\mu_{\alpha\dot{\alpha}}$ vanish in the massless classical case. Moreover these traces lie in general in one and the same chiral multiplet S as the divergence of the axial current R_μ. As a consequence the chiral superfield S describes all the anomalies of the superconformal group [2,3]. In particular the dilatation anomalies as given by the Callan-Symanzik functions β,γ are directly related to the anomalies of the axial current R_μ. Now it can be shown [4] for any renormalizable supersymmetric model that the R-invariance is only softly broken by mass terms. This implies restrictions on the Callan-Symanzik functions β,γ. Such restrictions where already derived [5] in a simple case by using explicitly the renormalization properties special to supersymmetry. The connection with R-invariance was pointed out in ref. 6.

The purpose of this talk is to present a short description of the construction of the supercurrent and of its anomalies in N = 1 supersymmetrical Yang-Mills theories. A detailed discussion may be found in ref. 7. We begin with a general discussion of the supercurrent and of its relations with the Callan-Symanzik equation.

*Work supported in part by the Swiss National Science Foundation.

The results of ref. 7 are presented in the last section.

CONSTRUCTION OF THE SUPERCURRENT

The generators P_μ, Q_α, $\overline{Q}_{\dot\alpha}$ and R of translation, supersymmetry and chiral transformations obey the algebra

$$\{Q_\alpha, \overline{Q}_{\dot\alpha}\} = 2\sigma^\mu_{\alpha\dot\alpha} P_\mu$$
$$[Q_\alpha, R] = -Q_\alpha \quad , [\overline{Q}_{\dot\alpha}, R] = \overline{Q}_{\dot\alpha} \tag{1}$$

all other (anti-) commutators vanishing. Infinitesimal transformations of a superfield $\phi(x_\mu, \theta_\alpha \, \overline{\theta}_{\dot\alpha})$ are given by

$$[iP_\mu, \phi] := \delta^P_\mu \phi = \partial_\mu \phi$$
$$[iQ_\alpha, \phi] := \delta^Q_\alpha \phi = \left(\frac{\partial}{\partial\theta^\alpha} + i\sigma^\mu_{\alpha\dot\alpha}\overline{\theta}^{\dot\alpha}\partial_\mu\right)\phi$$
$$[iR, \phi] := \delta^R \phi = i\left(n + \theta^\alpha\frac{\partial}{\partial\theta^\alpha} - \overline{\theta}^{\dot\alpha}\frac{\partial}{\partial\overline{\theta}^{\dot\alpha}}\right)\phi \tag{2}$$
$$[iD, \phi] := \delta^D \phi = \left(d + x^\mu\partial_\mu + \frac{1}{2}\theta^\alpha\frac{\partial}{\partial\theta^\alpha} + \frac{1}{2}\overline{\theta}^{\dot\alpha}\frac{\partial}{\partial\overline{\theta}^{\dot\alpha}}\right)\phi$$

where the real numbers n and d are the "R-weight" and the dimension of ϕ. We have enclosed in eqs (2) the dilatation transformations D. P, Q, $\overline{Q}$, R and D are contained in the superconformal algebra [2] if the R-weight and the dimension of any chiral superfield are related by $n = -2/3\, d$, which will be assumed throughout this section.

Let us introduce for $A = P, Q, \bar{Q}, R$ or D the functional differential operators

$$W^A := -i \int dz\, \delta^A \phi(z) \frac{\delta}{\delta \phi(z)} \tag{3}$$

where z is a superspace point $(x, \theta, \bar{\theta})$ and dz is the integration measure dV, dS or $d\bar{S}$ for, respectively, general, chiral or antichiral superfields :

$$dV = d^4x\, DD\bar{D}\bar{D}, \quad dS = d^4x\, DD, \quad d\bar{S} = d^4x\, \bar{D}\bar{D}$$
$$D_\alpha = \frac{\partial}{\partial \theta^\alpha} - i\sigma^\mu_{\alpha\dot{\alpha}} \bar{\theta}^{\dot{\alpha}} \partial_\mu, \quad \{D_\alpha, \bar{D}_{\dot{\alpha}}\} = 2i\sigma^\mu_{\alpha\dot{\alpha}} \partial_\mu \tag{4}$$

The operators W^A obey the same algebra (1) as the generators. As a consequence the θ-dependent operator [3]

$$\hat{W}(\theta, \bar{\theta}) := W^R - i\theta^\alpha W^Q_\alpha + i W^{\bar{Q}}_{\dot{\alpha}} \bar{\theta}^{\dot{\alpha}} - 2\theta\sigma^\mu\bar{\theta}\, W^P_\mu \tag{5}$$

transforms as an x- independent superfield :

$$\left[i W^Q_\alpha , \hat{W}(\theta, \bar{\theta}) \right] = \frac{\partial}{\partial \theta^\alpha} \hat{W}(\theta, \bar{\theta}) \tag{6}$$

One can check [3,7] that there exists a local spinor superfield functional differential operator $w_\alpha(z)$ such that

$$\hat{W}(\theta, \bar{\theta}) = \int d^4x\, w(x, \theta, \bar{\theta}) \tag{7}$$

$$w(z) = D^{\alpha} w_{\alpha}(z) - \bar{D}_{\dot{\alpha}} \bar{w}^{\dot{\alpha}}(z)$$

$$\left\{ i W^{Q}_{\beta} , w_{\alpha}(z) \right\} = \delta^{Q}_{\beta} w_{\alpha}(z) \qquad \text{(7) (continued)}$$

$$\left\{ i W^{\bar{Q}}_{\dot{\beta}} , w_{\alpha}(z) \right\} = \delta^{\bar{Q}}_{\dot{\beta}} w_{\alpha}(z)$$

Supersymmetry and translation invariance of a theory is characterical by the Ward Identities (WI)

$$W^{Q}_{\alpha} \Gamma(\phi) = 0 \, , W^{\bar{Q}}_{\dot{\alpha}} \Gamma(\phi) = 0 \, , W^{P}_{\mu} \Gamma(\phi) = 0 \qquad (8)$$

where $\Gamma(\phi)$ is the generating functional of the one-particle-irreducible (1PI) Green's functions. These WI's can be generally proved[8]. It can also be proved for any supersymmetric model by using the methods of ref. 4 that $\bar{R}$-invariance is softly broken:

$$W^{R} \Gamma(\phi) \sim 0 \qquad (9)$$

where the symbol $\sim$ means that the equation holds in the deep Euclidean region of momentum space. Thus

$$\hat{W}(\theta,\bar{\theta}) \Gamma(\phi) \sim 0 \qquad (10)$$

It follows [3] from this last identity and eqs (7) that there exists an axial vector superfield $V_{\mu}(z)$ (dim. 3), a chiral superfield $S(z)$ (dim. 3) and a spinor superfield $B_{\alpha}(z)$ (dim. 7/2) such that (with $V_{\alpha\dot{\alpha}} = \sigma^{\mu}_{\alpha\dot{\alpha}} V_{\mu}$)

$$\bar{D}^{\dot{\alpha}} V_{\alpha\dot{\alpha}} \sim -2 w_{\alpha} \Gamma - 2 D_{\alpha} S + B_{\alpha} \qquad (11)$$

S and B_α are constrained by

$$\int dS\, S = \int d\bar{S}\, \bar{S} \tag{12}$$

$$D^\alpha B_\alpha = 0 \tag{13}$$

(For any field operator F, the generating functional of the 1PI Green's functions with F inserted is denoted by the symbol F itself).

Let us suppose that

$$B_\alpha = 0 \tag{14}$$

which must be proved for each model under consideration. Then eq. (11) defines V_μ as the supercurrent [1]. It follows indeed from eq. (11) with $B_\alpha = 0$ that the currents

$$\begin{aligned} Q_{\mu\alpha}(x) &= iD_\alpha V_\mu \Big|_{\theta=0} - i(\sigma_\mu \bar{\sigma}^\nu)_\alpha{}^\beta D_\beta V_\nu \Big|_{\theta=0} \\ T_{\mu\nu}(x) &= V_{\mu\nu} + V_{\nu\mu} - 2 g_{\mu\nu} V_\lambda{}^\lambda \end{aligned} \tag{15}$$

with $\quad V_{\mu\nu}(x) = [D_\alpha, \bar{D}_{\dot\alpha}]\, \sigma_\mu^{\alpha\dot\alpha} V_\nu \Big|_{\theta=0}$

are conserved and thus may be identified as the spinor current associated with supersymmetry and the energy - momentum tensor :

$$\partial^\mu Q_{\mu\alpha}(x) = i\, w^Q_\alpha(x)\,\Gamma$$
$$\partial^\mu T_{\mu\nu}(x) = i\, w^P_\nu(x)\,\Gamma \tag{16}$$

The right-hand-sides are local functional differential operators ("contact terms"), whose space-time integrals yield the variational operators (3) for $A = P, Q$. Eq. (11) with $B_\alpha = o$ implies also the broken conservation law of the axial current

$$R_\mu(x) = V_\mu(x)\Big|_{\theta=0} \tag{17}$$

and the trace identities for $T_{\mu\nu}$ and $Q_{\mu\alpha}$:

$$\partial_\mu R^\mu(x) \sim i\, w^R(x)\,\Gamma + i\left(DDS - \bar{D}\bar{D}\bar{S}\right)\Big|_{\theta=0}$$
$$Q_\mu{}^\alpha(x)\,\sigma^\mu_{\alpha\dot{\alpha}} \sim i\, w'_{\dot{\alpha}}(x)\,\Gamma + 2i\,\bar{D}_{\dot{\alpha}}\,\bar{S}\Big|_{\theta=0} \tag{18}$$
$$T_\mu{}^\mu(x) \sim i\, w'(x)\,\Gamma - \tfrac{3}{2}\left(DDS + \bar{D}\bar{D}\bar{S}\right)\Big|_{\theta=0}$$

which show that the respective anomalies are components of the chiral superfield S. Defining now the dilatation current

$$D_\mu(x) = x^\nu\, T_{\mu\nu}(x) \tag{19}$$

we get from eqs (16, 18) for $T_{\mu\nu}$ the anomalous conservation law

$$\partial^\mu D_\mu(x) \sim i w^D(x)\Gamma - \frac{3}{2}\left(DDS + \bar{D}\bar{D}\bar{S}\right)\Big|_{\theta=0} \qquad (20)$$

which yields after integration the anomalous dilation WI

$$W^D\Gamma \sim -\frac{3}{2} i \left[\int dS\, S + \int d\bar{S}\, \bar{S}\right] \qquad (21)$$

Let m_k, g_i denote the mass parameter and coupling constants of the theory, and N_ℓ the "counting operator" of the field ϕ_ℓ

$$N_\ell = \int dz\, \phi_\ell(z) \frac{\delta}{\delta\phi_\ell(z)} \qquad (22)$$

Expanding the dimension 4 insertion in the right-hand side of eq. (21) in the basis ($\partial_{g_i}\Gamma$, $N_\ell\Gamma$) and using the dimensional analysis identity

$$W^D\Gamma = i\sum_k m_k \partial_{m_k}\Gamma \qquad (23)$$

we obtain the Callan-Symanzik equation

$$\left[\sum_k m_k \partial_{m_k} + \sum_i \beta_i \partial_{g_i} - \sum_\ell \gamma_\ell N_\ell\right]\Gamma \sim 0 \qquad (24)$$

Relation between the coefficents β and γ will follow from the constraint (12) on the anomaly S, which is itself a consequence of the softly broken R-invariance (9, 10).

SUPERSYMMETRICAL YANG-MILLS THEORIES

Such a theory [9] involves a dimensionless gauge superfield $\phi = \phi^i\tau_i$, the τ's being the generators of the gauge group, and dimension 1 matter chiral superfields A_a in a unitary representation $\tau^i \to T^i_{ab}$. Faddeev - Popov ghost chiral superfields $c_+ = c_+^i \tau_i$, $c_- = c_-^i \tau_i$ of dimensions o and 1 must be intoduced and the theory is requested to be invariant under the BRS transformations

$$s\, e^{F(\phi)} = e^{F(\phi)} c_+ - \bar{c}_+ e^{F(\phi)}$$

$$s A_a = - c_+^i T^i_{ab} A_b \tag{25}$$

$$s c_+ = - c_+ c_+$$

$$s c_- = \bar{D}\bar{D}\phi$$

where

$$F(\phi) = \phi + \sum_{j \geq 2} a_j \phi^j \tag{26}$$

is an arbitrary function of ϕ, the a's playing the role of gauge parameters [10]. Introducing external superfields ρ, Y_a and Z coupled to the BRS transformations of ϕ, A_a and c , the BRS invariance is expressed through the Slavnov identity

$$\mathcal{S}(\Gamma) := \int dV \left[\frac{\delta\Gamma}{\delta\rho} \frac{\delta\Gamma}{\delta\phi} + \phi \left(\frac{\delta\Gamma}{\delta c_-} - \frac{\delta\Gamma}{\delta \bar{c}_-} \right) \right]$$

$$+ \int dS \left[\frac{\delta\Gamma}{\delta Z} \frac{\delta\Gamma}{\delta c_+} + \frac{\delta\Gamma}{\delta Y_a} \frac{\delta\Gamma}{\delta A_a} \right] + \text{c.c.} \sim 0 \tag{27}$$

(Mass terms are introduced for the gauge fields in order to avoid off-shell infrared singularities peculiar to their zero dimension. The Slavnov identity can however been shown to be only softly broken [10].) A consequence of the Slavnov identity is the "ghost equation"

$$\mathcal{G}\Gamma := \left[\frac{\delta}{\delta c_-} + \frac{2}{\alpha}\bar{D}\bar{D}DD\frac{\delta}{\delta\rho}\right]\Gamma \sim 0 \tag{28}$$

which defines the gauge parameter α.

A basis of dimension 4 insertions commuting with the Slavnov identity is provided by the application on Γ of the differential operators ∇_i

$$\{\nabla_i\} := \{\partial_g , \partial_{\lambda_i} , \partial_{a_j} , \mathcal{N}_\phi , \mathcal{N}_A , \mathcal{N}_+ , \mathcal{N}_-\} \tag{29}$$

where g is the gauge coupling constant, the λ's are the self-coupling constants of the matter fields and

$$\begin{aligned}
\mathcal{N}_\phi &= N_\phi - N_\rho + N_{c_-} + N_{\bar{c}_-} + 2\alpha\partial_\alpha \\
\mathcal{N}_A &= N_A + N_{\bar{A}} - N_y - N_{\bar{y}} \\
\mathcal{N}_+ &= N_{c_+} + N_{\bar{c}_+} - N_z - N_{\bar{z}} \\
\mathcal{N}_- &= N_\phi - N_\rho + N_{c_-} + N_{\bar{c}_-}
\end{aligned} \tag{30}$$

the N's being the counting operators (22). Let us remark that all of the ∇'s, $\mathcal{N}_-$ excepted, commute with the ghost equation (28). Expanding in this basis the insertion

$$\nabla_m \Gamma := \sum_k m_k \partial_{m_k} \Gamma \tag{31}$$

(where the m's are all the mass parameters of the theory) which commutes with both $\mathcal{S}$ and $\mathcal{G}$ (eqs (27, 28)) yields the Callan-Symanzik equation

$$\left[\nabla_m + \sum_i \beta_i \nabla_i\right]\Gamma \sim 0 \tag{32}$$

$$(\beta_{\mathcal{N}_-} = 0)$$

Let us start the construction [7] of a BRS invariant supercurrent First, a local functional differential operator w_α as defined by eqs (7) and commuting with the Slavnov identity is seen to be uniquely defined. Without writing it explicitly we note that it gives rise to R-transformations of weights

$$n = \left(0, -\frac{2}{3}, 0, -2\right) \text{ for } \left(\phi, A, c_+, c_-\right) \tag{33}$$

thus violating the rule $n = -2/3\ d$ for the chiral field c_-. Then it is shown that there exists a BRS invariant supercurrent V_μ and a BRS invariant chiral field S such that eq. (11) holds with $B_\alpha = 0$.

The connection with the Callan-Symanzik equation is made by choosing the following basis for the anomaly S :

$$\left\{ S_i : \int dS\, S_i + \int d\bar{S}\, \bar{S}_i = \nabla_i \Gamma \right\} \tag{34}$$

The expansion of S in this basis reads

$$S = \frac{2}{3}\sum_i \alpha_i S_i \tag{35}$$

$$\alpha_{S_-} = -\frac{4}{3} \quad ; \quad \alpha_i = O(\hbar)\ , \ i \neq S_-$$

(The coefficient of S_- is a consequence of the ghost equation (28).) The procedure leading to eqs (20-24) may now be applied. But now one obtains

$$"W^D"\Gamma \sim -i\sum_i \alpha_i \nabla_i \Gamma \tag{36}$$

where "W^D " is a dilatation operator with wrong dimensions for c_-, ϕ and ρ. However the correct dilatation operator is seen to be given by

$$W^D = "W^D" - 2i\mathcal{N}_- \tag{37}$$

Thus the Callan-Symanzik equation (32) is recovered with the identification $\alpha_i = \beta_i$ for $i \neq \mathcal{N}_-$.

It remains to use the constraint (12) on the anomaly. Recalling that the β's are of order $\hbar$ and noting that it is possible to define S_- in such a way that it fulfils the constraint (12) by itself, the restriction on the β's are easily worked out at the order $\hbar$. The result is (in the case of one matter field with selfcoupling λA^3) :

$$\beta_\lambda = -3\lambda\beta_{\mathcal{N}_A}\left(1 + O(\hbar)\right)$$
$$\beta_{\mathcal{N}_+} = O(\hbar^2) \tag{38}$$

The explicit relation at higher order will depend on the normalization conditions defining the parameters of the theory (c.f. refs 3, 6).

REFERENCES

1. S. Ferrara and B. Zumino, Nucl. Phys. B87:207 (1975).
2. S. Ferrara, Nucl. Phys. B77:73 (1974) ;
   P.H. Dondi and M. Sohnius, Nucl. Phys. B81:317 (1974).
3. T.E. Clark, O. Piguet and K. Sibold, Nucl. Phys. B143:445 (1978).
4. T.E. Clark, O. Piguet and K. Sibold, Nucl. Phys. B119:292 (1977) ;
   O. Piguet, M. Schweda and K. Sibold, Nucl. Phys. B168:337 (1980).
5. S. Ferrara, J. Iliopoulos and B. Zumino, Nucl. Phys. B77:413 (1974).
6 H.S. Tsao, Phys. Letters, 53B:381 (1974) ;
   O. Piguet and M. Schweda, Nucl. Phys. B92:334 (1975).
7. O. Piguet and K. Sibold, Nucl. Phys. B196:428,447 (1982).
8. T.E. Clark, O. Piguet and K. Sibold, Annals of Phys. 109:418 (1977) ;
   O. Piguet, M. Schweda and K. Sibold, Nucl. Phys. B174:183 (1980).
9. S. Ferrara and B. Zumino, Nucl. Phys. B79:413 (1974) ;
   A. Salam and J. Strathdee, Phys. Lett. 51B:353,475 (1974) ;
   S. Ferrara and O. Piguet, Nucl. Phys. B93:261 (1975).
10. O. Piguet and K. Sibold, Nucl. Phys. B197:257,272 (1982).

GRAND UNIFICATION IN EXTENDED SUPERGRAVITY

John Ellis

CERN, Geneva, Switzerland

INTRODUCTION

This paper is a review of attempts that have been made in the last couple of years to forge a connection between grand unified theories of elementary particle interactions and extended supergravities. My principal collaborators in this endeavour have been Mary K. Gaillard and Bruno Zumino[1-3], and as between us we have already written several reviews[4-6] of our approach, this paper is intended to emphasize recent developments at the expense of older results.

The motivations for seeking grand unification in extended supergravity are well known. Present grand unified theories[7] (GUTs) are clearly inadequate - one does not know which group to choose, GUTs do not explain the number of fermion generations, they have a hierarchy problem with elementary scalars and difficulties with strong CP violation, and even the minimal SU(5) GUT has at least 23 parameters, which we might like to reduce to 0, 1 or 3 depending on our religion. On the other hand, supersymmetry[8] (susy) is the only way of combining internal symmetries with Lorentz invariance in a non-trivial way, and is so beautiful that it must be true. It may even be useful[9], for example for "solving" the hierarchy problem by enabling one to "fix and forget" the Higgs potential. Recently proposed[10] schemes of this type only use a simple $N = 1$ rigid susy, which is not true "superunification" as the fermionic charges do not carry internal indices. To do this one should go to extended ($N > 1$) susy. Furthermore, if one is to include gravity (and what else is there left to do if one believes in GUTs?) one must invoke local susy. We are therefore led to try[1-3] to extract a GUT from an extended supergravity (ESG) theory[11]. These are the only theories of gravitational interactions with matter fields which are not known to have unrenormalizable infinities. Maybe ESGs also have uncontrollable infinities, but N = 4 Yang-Mills theories are already known[12] to avoid many infinite pitfalls, so that one has motivation to hope that some ESG may be "it".

THE SPECTRA OF EXTENDED SUPERGRAVITY THEORIES

There are[11] few of them: N = 2, 3, 4, 5, 6 and 7 = 8 with the particle spectra shown in Table 1. These all contain just one graviton, as many gravitinos as there are susy generators, and just the right number of spin one particles to gauge an internal SO(N) symmetry [or SO(6) × O(2) in the case of N = 6]. It has long been

Table 1. Multiplets in extended supergravities

Helicity \ N =	2	3	4	5	6	7 = 8
2	1	1	1	1	1	1 = 1
3/2	2	3	4	5	6	7 + 1 = 8
1	1	3	6	10	15 + 1	21 + 7 = 28
1/2	-	1	4	10 + 1	20 + 6	35 + 21 = 56
0	-	-	1 + 1	5 + 5	15 + 15	35 + 35 = 70
-1/2	-	1	4	1 + 10	6 + 20	21 + 35 = 56
-1	1	3	6	10	1 + 15	7 + 21 = 28
-3/2	2	3	4	5	6	1 + 7 = 8
-2	1	1	1	1	1	1 = 1

known[13] that the $N \leq 4$ theories could be gauged, but the gauging of $N = 5$ [14] to 8 [15] has only recently been demonstrated. Gauging introduces a dimensionless gauge parameter g, hitherto absent, and entails the introduction of a cosmological constant

$$\Lambda = O(g^2 m_P^4) \tag{1}$$

and the energy of the gauged theory is not even bounded below. Is gauging SO(N) necessary, desirable or useful? One may take the traditional point of view of the Swiss law that anything which is not forbidden is compulsory, or one may be so anticlerical as to refuse the offer of an arbitrary input parameter. Perhaps the SO(N) gauge interactions could be useful for "preconfining" states in the original supermultiplets of Table 1 so that the observed particles are composites, a necessity to be emphasized subsequently. However, confinement is conventionally associated with a non-Abelian gauge coupling rising to O(1) on some scale Λ_{NA} and giving bound states of radius $1/\Lambda_{NA}$. But the SO(N) gauge coupling is known[16] to have a vanishing β function at one-loop and it is suspected that this may continue to be true at the multiloop level[12]. In this case there would be no Λ_{NA} for these interactions, and the bound state scale could only be provided by the dimensional coupling $O(m_P)$ of the supergravitational interactions, which would therefore have to play a crucial rôle in the "preconfinement" mechanism. What of the associated cosmological constant (1)? The present failure of the Universe to expand exponentially means that the cosmological constant averaged over large distance scales must be very small:

$$\langle\Lambda\rangle_{\text{large}} < O(10^{-29})\text{gm/cc} \simeq 10^{-47}\ \text{GeV}^4 \tag{2}$$

If susy exists at all, it must be broken at some energy scale $\gtrsim$ 1 GeV, which guarantees that the cosmological constant must have been non-zero before symmetry breakdown, as only susy-invariant vacua have zero energy[17]. Indeed, it has been argued[18] that the cosmological constant has to exhibit fluctuations of order m_P^4 on very small distances if it is to be negligibly small (1) on very large distance scales. While the gauging[14,15] of higher N ESGs is very exciting, it is not obvious to us whether we should avail ourselves of the opportunity.

It is well known[19] that the supermultiplets of Table 1 cannot be used directly for grand unification. They do not contain enough spin one particles, since

$$SO(8) \not\supset SU(3) \times SU(2) \times U(1) \tag{3}$$

let alone even a minimal GUT group. In fact the maximal promising subgroup is

$$SO(8) \supset SU(3) \times U(1) \times U(1) \tag{4}$$

which does not contain candidates for $W^\pm$ bosons, so that they could only be interpreted as composite states. Similar difficulties occur with the fermion spectra in Table 1: they do not contain candidates for known fermions such as the μ, τ, b, etc. Another trouble with SO(N) ($N \neq 6,10,\ldots$) groups in general is that their representations available for fermions are necessarily real. GUTters[7] worry about this because they know that the observed fermions are complex: each generation is in a $(3,2) + 2(\bar{3},1) + (1,2) + (1,1)$ of $SU(3) \times SU(2)$. Furthermore, such a chiral representation is necessary if one is to prevent spin 1/2 fermions from acquiring masses $\geq O(m_X) >> O(m_W)$ in a natural way, i.e., by matching different helicity states with $|\text{helicity}| \leq J$ and identical internal symmetry properties to get a massive particle of spin J. Indeed, if one constructs the N = 8 ESG by dimensional reduction from the N = 1 theory in 11 dimensions in the most general way[20], one finds masses for particles of all spins $\leq 3/2$. This illustrates the dangers of real fermion representations, and may also suggest that one should seek an alternative formulation of the N = 8 theory if one wants a sufficiently non-trivial spectrum of low-mass fermions.

So far we have seen that the $W^\pm$, μ, τ and b do not appear in the spectra of Table 1, and in the ESG framework can therefore only be interpreted as composite states[21]. One may also encounter problems if one seeks to identify Higgs fields with the elementary spin-zero states in Table 1. It has been argued[18] in the context of quantum gravity that elementary scalars necessarily acquire masses $O(m_P)$ from their propagation through a foamy structure of space-time at short distances $O(1/m_P)$. Hence phenomenological scalar fields should probably also be composite. Furthermore, it has been suggested[18] that the space-time foam in a supergravity theory may be even more singular, to such an extent that space-time becomes fractal at very short distances with Hausdorff dimension $\gamma = 6$. If there is no cut-off on these foamy fluctuations, they will give logarithmically divergent masses to elementary spin 1/2 particles. In this case all observed spin 1/2 particles would also have to be composite. It is amusing to speculate on an extension of this argument. If the effective dimension of space-time at short distances were $\gamma = 11$, the resulting super-duper-foam might give masses $O(m_P)$ to all elementary particles of spin $\leq 3/2$, in striking coincidence with the results[20] of dimensional reduction from the N = 1 theory in 11 dimensions. Thus one might argue that all light particles of spin $\leq 3/2$ must necessarily be composite. This possibility will be pursued in the fourth section, but first we need to recall some more structure of ESGs.

HIDDEN SYMMETRIES AND DYNAMICS OF EXTENDED SUPERGRAVITY THEORIES

When they obtained the unbroken N = 8 ESG in four dimensions by dimensional reduction from the N = 1 theory in 11 dimensions,

Cremmer and Julia[22] noticed that the Lagrangian could be written in a form invariant under both a global non-compact $E_{7(+7)}$ symmetry and also a local SU(8) symmetry. Written in a manifestly gauge-invariant form, the theory contains 133 fundamental scalar fields in an adjoint representation of E_7. One can remove 63 scalars by making an appropriate choice of SU(8) gauge, leaving oneself with 70 physical scalar fields which parametrize the coset space $E_7/SU(8)$ [the maximal compact subgroup of non-compact $E_{7(+7)}$ is $SU(8)/Z_2$]. These 70 scalar fields are those shown in the last column of Table 1. The general structure of the theory is reminiscent of non-linear σ (or CP^{N-1}) models[23]: is has long been known[24] that one can write consistent Lagrangians with a non-compact global symmetry if the non-compact parts are realized non-linearly in terms of the scalar fields. The SU(8) gauge invariance of the Lagrangian is realized[22] using an SU(8)-covariant derivative

$$D^A_{\mu B} \equiv \delta^A_{\mu B}\partial_\mu + iQ^A_{\mu B} \tag{5}$$

with the $Q^A_{\mu B}$ vectors in an adjoint representation of SU(8) which do not have a kinetic term in the Lagrangian. Hence they can be eliminated using the equations of motion, and seem to contain a quadratic dependence on the fundamental scalar fields. This is strikingly similar to what happens in the CP^{N-1} model[23]:

$$\mathcal{L}_{CP^{N-1}} = -\Sigma^N_{i=1}(\partial_\mu - iv_\mu)z^*_i(\partial_\mu + iv_\mu)z_i \quad : \quad \Sigma^N_{i=1}|z_i|^2 = 1 \tag{6}$$

which manifestly possesses a local U(1) gauge invariance as well as the global SU(N) invariance, and whose equations of motion enable one to express

$$v_\mu(x) = \frac{1}{2}\Sigma^N_{i=1} z^*_i \overleftrightarrow{\partial}_\mu z_i \tag{7}$$

as a quadratic function of the fundamental scalar fields.

The 1/N expansion has been used to show[23] that in two dimensions quantum effects generate a kinetic term for the v_μ field, so that the gauge field becomes dynamical and generates a long range confining potential. The physical spectrum therefore contains bound states sitting in unitary representations of the global SU(N) groups[25]. A supersymmetric version of the two-dimensional CP^{N-1} model also contains a fermionic superpartner of the gauge field which also becomes dynamical. Furthermore, it has been shown recently[26] that the 1/N expansion in three dimensions yields similar results of a dynamical gauge boson and an N = 1 fermionic superpartner. This model is perhaps more interesting than the two-dimensional example because in this case the gauge boson has a physical helicity state. The infra-red properties of two- and three-dimensional models are however different from those in four dimensions, and it may well be that a purely bosonic CP^{N-1} model in four dimensions would not perform the same trick of generating dynamical gauge bosons. However, our four-dimensional supergravity theory of course contains fermion fields as well, and it has recently been argued[27] that a non-linear fermion theory in four dimensions can yield dynamical, physical, composite gauge bosons.

It therefore seems permissible to follow Cremmer and Julia in their conjecture[22] that the physical spectra of four-dimensional ESGs contain dynamically generated composite gauge bosons corresponding to a local SU(8) symmetry in the case of the N = 8 theory. Unless all susy is lost at the Planck mass, which does not happen in the two- and three-dimensional models, it is also plausible to conjecture[1] that at least some of the states in a supermultiplet containing the gauge fields $Q^A_{\mu B}$ will also become dynamical, physical, composite particles. Thus we suppose that all the observed and expected "elementary" particles: quarks, leptons, gauge bosons and Higgses, are all composites of the states seen in Table 1, which are now to be interpreted as preons bound on a distance scale of order $1/m_P$. It may be that the graviton is truly elementary: it is clearly a singlet of all the manifest and hidden internal symmetries of the ESG, and as such may escape whatever "preconfinement" mechanism removes the other preons from the low energy spectrum. We are therefore led to the geological stratigraphy of "elementary particles" laid out in Fig. 1, which is reminiscent of a wall of the Grand Canyon. However, the "elementarity" of the graviton is by no means essential: there are well known theorems[28] which guarantee that any massless spin two particle, even if composite, will have couplings identical with those of an elementary graviton at energies $\ll m_P$. Nevertheless, for definiteness we will pursue the scenario of Fig. 1.

GRAND UNIFICATION IN THE BOUND STATE SUPERMULTIPLET

Analogies with the known N = 1 and 2 cases, and other arguments[1-6], suggest that for general N the appropriate supermultiplet of bound states containing the vector fields $Q^A_{\mu B}$ (5) is

$$\left(\tfrac{3}{2}\right)^A,\ (1)^A_B,\ \left(\tfrac{1}{2}\right)^A_{[BC]},\ (0)^A_{[BCD]},\ \ldots\ \left(\tfrac{3-N}{2}\right)^A \qquad (8)$$

where the numbers in parentheses are helicities, and the lower indices are antisymmetrized up to N times, thereby fixing the lowest helicity. The supermultiplet obtained applying (8) to the case of N = 8 is shown in Table 2. A recent analysis[29] of embryonic Regge behaviour in the 2 ↔ 2 scattering amplitudes of the N = 8

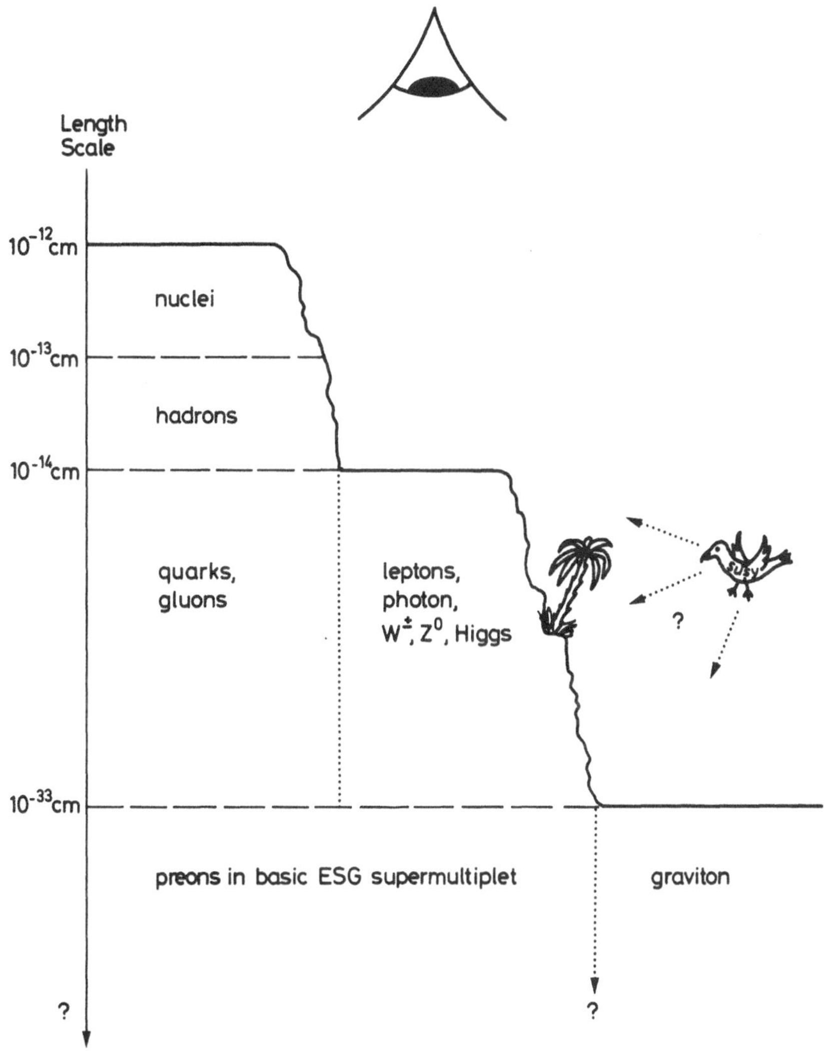

Fig. 1 The geological strata of "fundamental" particles, as seen by an observer from a large distance scale. At what level will the susy-bird land ?

ESG theory suggests the existence of Regge trajectories which could support physical massless states with the quantum numbers of Table 2, as well as of the original preon multiplet in Table 1. It is not possible to be sure that the residues of these Regge poles do not vanish on mass-shell, and the analysis of multiparticle scattering amplitudes would probably signal the existence of other trajectories, but this result may be a signal that we are on the right track.

The representations in parentheses in Table 2 are those obtained by tracing one upper and one lower index in (8). In contrast to theories with lower N, the spin one trace field $Q^A_{\mu A}$ does not have a a special symmetry rôle in the N = 8 theory[22]: the gauge symmetry is SU(8), not U(8). We therefore conjecture[1,2] that this field does not manifest itself as a low mass bound state, and it is

Table 2. The N = 8 supercurrent multiplet

Helicity	-3/2	-1	-1/2	0	+1/2	+1	+3/2	+2	+5/2
SU(8) content	$\overline{\underline{8}}$	$\underline{63}$ (+$\underline{1}$)	$\underline{216}$ (+$\underline{8}$)	$\underline{420}$ (+$\underline{28}$)	$\underline{504}$ (+$\underline{56}$)	$\overline{\underline{378}}$ (+$\underline{70}$)	$\overline{\underline{168}}$ (+$\overline{\underline{56}}$)	$\overline{\underline{36}}$ (+$\overline{\underline{28}}$)	$\overline{\underline{8}}$

natural[30] to extend this conjecture to at least some of the other parenthesized trace representations in Table 2. Neglecting trace representations in general involves breaking at least some supersymmetries. Is there a profound connection between the fact that susy is not manifest at very low energies O(1) GeV, and the absence of any reason for breaking susy if one uses bound states in supermultiplets with $N \leq 6$?

If one tries to identify the observed "elementary" particles with (at least a subset of) states in the supermultiplet (8), one immediately encounters a major problem of chirality. We recalled earlier on that GUTs want chiral fermion representations:

$$R(f_R) \neq R(f_L) \tag{9}$$

in order that $\bar{f}_R f_L$ not be a singlet of the GUT group, in which case every fermion could acquire a GUT-invariant mass $\geq m_X$. We are gratified that the multiplet of Table 2 is chiral, in contrast to the "preons" of Table 1. However, one can have too much of a good thing, and the representation (8) of Table 2 is not real or vector-like under the exact low energy gauge group SU(3) × U(1), which means that the normal rules of helicity matching would exclude any mass for many of these states[1].

There is also a problem with axial anomalies. Since SO(N) is a safe group, gauging it can never be upset by anomalies, but gauging the effective low-energy SU(N) is not necessarily safe, and the bound state fermions (8) in Table 2 do indeed seem to yield anomalies. Römer[31] has given a general formula for computing the axial anomaly due to a fermion in an (n,m) representation of the Lorentz group with internal group Casimir C(n,m):

$$C(n,m) \times \frac{(m+1)(n+1)}{6} \left[m(m+2) - n(n+2)\right] \frac{g^2}{32\pi^2} F^a_{\mu\nu} F^a_{\mu\nu} \varepsilon^{\mu\nu\rho\sigma} \qquad (10)$$

which gives the anomalies of Table 3 when it is applied to the traceless fermions in Table 2. Unfortunately, it is not clear what sense to make of this calculation. Although one can set up consistent gauge interactions for spin 3/2 particles, this is only possible if they are massive, and one has no consistent field theoretical formalism at all for the interactions of spin 5/2 particles. Furthermore, Coleman and Grossman[32] have given S-matrix arguments that only triangle diagrams containing massless spin 1/2 fermions can contribute to the anomaly. However, it is clear from Table 3 that restricting ourselves to these states does not help us.

Table 3. Anomalies from the bound state supermultiplet

Helicity	3/2	1/2	-1/2	-3/2	-5/2
Group factor	-1	+3	+75	+55	-1
Spin factor	+3?	+1	-1	-3?	-5??
Product	-3?	+3	-75	-165?	5??

We conclude on the basis of these arguments that the full supermultiplet (8) of Table 2 is too chiral. We must find some way either of discarding some helicity states, or else of adding in more supermultiplets. It is easy to convince oneself that the normal rules for giving SU(3) × U(1) invariant masses to fermions probably require an infinite number of complete supermultiplets[5,33] and this has been proven under certain restrictive assumptions[34]. We will turn in the fifth section to a possible scenario for an infinite set of supermultiplets. For now we focus in the alternative of discarding unwanted helicity states from Table 2.

It has been argued[1,2,35] that the only particles which can consistently have masses << m_P are a set forming a renormalizable field theory, which means that they can include no state with helicity > 1, that the only states with helicity one must be gauge vector bosons, and that the spin 1/2 fermions must be free of anomalies with respect to the low-energy effective gauge group. In addition to the conventional renormalizable interactions there may be non-renormalizable interactions whose couplings are inverse powers of the Planck mass. Such interactions would probably cause baryons to decay[36] and neutrinos to have masses[37] even independently of GUT interactions, and might help to explain the magnitudes and ratios of the u, d and electron masses[38]. We have looked[1,2] for subsets of the states in Table 2 which could form a phenomenologically acceptable low-energy gauge theory. Some popular GUT groups such as SO(10) and E_6 are not subgroups of SU(8) and so cannot be obtained. Also it turns out that any subsets of Table 2 whose fermions are vector-like with respect to SU(3) × U(1) and free of SU(6) or SU(7) anomalies in fact contain unacceptable completely real fermion representations. This philosophy therefore solves the problem of choosing the GUT group: it must be SU(5). Furthermore, the N = 8 bound state supermultiplet is the only representative of the set (8) which has a sufficiently rich fermion spectrum to accommodate three generations[1,2]. We are therefore led to

$$(N = 8 \text{ ESG}) \rightarrow SU(8) \rightarrow SU(5) \tag{11}$$

as the unique way of embedding grand unification in extended supergravity.

Our first example of a candidate superGUT was obtained[2] by taking the maximal set of chiral SU(5) anomaly-free left-handed fermions from among the traceless fermions in Table 2:

$$(\underline{45}+\overline{\underline{45}}) + 4(\underline{24}) + 9(\underline{10}+\overline{\underline{10}}) + 3(\overline{\underline{5}}+\underline{5}) + 9(\underline{1}) + 3(\overline{\underline{5}}+\underline{10}) \tag{12}$$

We were encouraged to see that this spectrum contained three generations of chiral SU(5) fermions, in accord with phenomenological prejudices based on the successful prediction[39] of the mass of the bottom quark and the constraint of three (or at the most four) light neutrinos allowed by cosmological nucleosythesis calculations[40]. It is also amusing that the spectrum (12) contains so many fermions which are real with respect to SU(5) that the reduced SU(5) β function is

$$\beta_{SU(5)} = +147\tfrac{1}{2} \text{ instead of } -55 + 4N_G \tag{13}$$

once one gets above all the particle thresholds. The β function (13) is so large that $g_{SU(5)}$ may become O(1) at energies $E \sim m_P$[1,41], as illustrated in Fig. 2. Getting such an increase in $g_{SU(5)}$ only requires $\langle\beta\rangle \approx +70$ in the energy range betweem m_X and m_P, so that the trick of Fig. 2 can be played even if some of the fermions in (12) have masses $O(m_P)$. It is heartening that there is no longer a need for non-perturbative supergravity effects to generate a small effective coupling constant $\alpha_{GUT} \equiv g^2_{SU(5)}/4\pi = O(10^{-2})$: this is taken care of by the renormalization group evolution between m_X and m_P. We already knew[42] that the consistency of GUTs required

$$\frac{1}{120} < \alpha_{em} < \frac{1}{170} \tag{14}$$

and now we have a qualitative explanation how it got there: largely thanks to the existence of many high-mass particles.

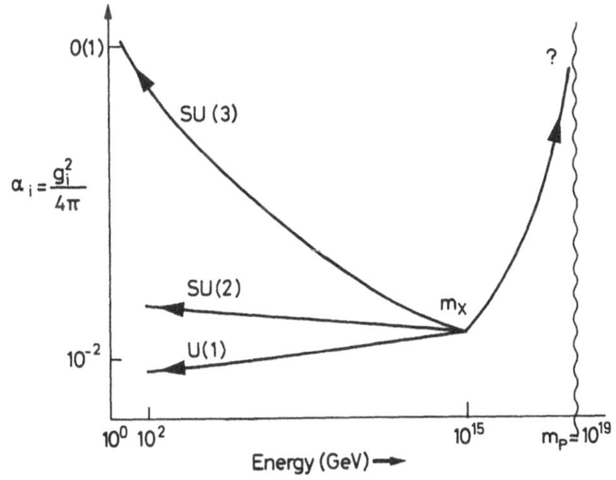

Fig. 2 Possible evolution of the GUT coupling constant at energy scales between m_X and m_P.

One can construct many alternative[30] superGUTs within the general philosophy outlined above. For example, one might postulate a set of bound state supermultiplets which enables one to use the normal helicity-matching rules to dispose of all unwanted states of low helicity, while tolerating the apparent existence of light states of high helicity. Alternatively, one might take non-maximal subsets of the traceless spin 1/2 fermions in Table 2, or perhaps include some particles from the trace representations. Here I would like to mention[3] some alternative superGUTs which accommodate some desirable variations on the GUT theme which have been proposed in the recent literature.

Global N = 1 supersymmetric GUTs[9,10] may "solve" the hierarchy problem thanks to the celebrated "no-renormalization" theorem[43]. In our original superGUT we broke all supersymmetries in the acts of discarding trace representations and taking the maximal renormalizable subset of the traceless states in Table 2. Alternatively one could keep an N = 1 susy in the effective theory, but it turns out[3] that the maximal susy GUT does not contain three generations. The maximal three-generation susy GUTs contain the following chiral supermultiplets

$$\text{with traces} \quad : \quad (\underline{45}+\overline{\underline{45}}) + 4(\underline{24}) + 3(\underline{10}+\overline{\underline{10}}) + 3(\overline{\underline{5}}+\underline{5}) + 6(\underline{1}) + 3(\overline{\underline{5}}+\underline{10}) \tag{15a}$$

$$\text{without traces:} \quad (\underline{45}+\overline{\underline{45}}) + 4(\underline{24}) + 2(\underline{10}+\overline{\underline{10}}) + 1(\overline{\underline{5}}+\underline{5}) + 3(\underline{1}) + 3(\overline{\underline{5}}+\underline{10}) \tag{15b}$$

which are sufficient to construct a viable susy GUT. Another recent GUT variation is to solve the strong CP problem by embedding a global, anomalous U(1) symmetry into the theory à la Peccei-Quinn[44] and breaking it spontaneously at an energy scale $O(m_X)$. Such a theory contains a very light ($m \sim 10^{-8}$ eV), very weakly interacting ($g_{\bar{f}f} \sim 10^{-15}$) axion which is essentially unobservable[45]. The minimal SU(5) example of such a GUT contains a complex $\underline{24}$ and two $\underline{5}$'s of Higgses in addition to the conventional $\overline{\underline{5}}$ + $\underline{10}$ generations of fermions. The SU(8) $\underline{420}$ of scalar fields contained in Table 2 are clearly sufficient to accommodate these Higgs fields, and hence a global Peccei-Quinn U(1) symmetry can be imposed on the superGUT if so desired. However, we do not yet see a natural way in which such a U(1) symmetry could emerge without being imposed by hand.

These examples of alternative superGUTs indicate that it is possible to extract a variety of phenomenologically interesting models using the general philosophy outlined previously. However, all these models beg the question of how to dispose of unwanted helicity states from the bound state supermultiplet (8) of Table 2, and we turn to this question in the next section.

POSSIBLE FATES OF UNWANTED HELICITY STATES[3]

It was emphasized in the previous section that the bound state supermultiplet (8) is highly chiral, and one cannot give masses to all the unwanted helicity states using the normal rules of matching states of $|\text{helicity}| \leq J$ with identical internal quantum numbers so as to get a massive state of spin J. So what happens to the unwanted helicity states of $|\text{helicities}| \geq 1$, $= 1/2$? And a related question: how do the dynamics select which (if any) renormalizable subset of particles from among the states available with $|\text{helicity}| \leq 1/2$? We can identify three distinct possibilities for the unwanted helicity states:

A - They were never bound in the first place.
B - Some of them do exist as massless states.
C - They all exist, and have found "partner" helicity states to "eat" and become massive.

After a few remarks about the first two possibilities we come to the final solution which we prefer[3].

A - If they had bound with low mass, the unwanted states would have made the theory unrenormalizable. Perhaps the unknown bound state equations which should bind the preons in these channels develop some related singularities that prevent the binding from occurring? There is no theorem that every composite channel must necessarily exhibit a bound state pole, and one would not discard the quark model if no obvious resonance were found in some obscure $\bar{q}q$ channel. There is indeed a theorem[46] that one cannot have massless bound states with $|\text{helicity}| > 1$ (or $> 1/2$) in theories with a Lorentz covariant energy momentum tensor (or conserved current). Unfortunately, gravity (and gauge) theories do not satisfy these conditions of Lorentz covariance, nor <u>a fortiori</u> do the ESGs of interest[47]. It therefore seems that the theorem may be too stringent to be useful, which is just as well as it would have forbidden our desired massless composite gauge bosons if it had been applicable !

B - Some unwanted helicity states may exist with zero (or low) mass if they have very small couplings. General theorems[48] on kinematical singularities in $2 \leftrightarrow 2$ scattering amplitudes imply[3] that the coupling constants of massless high-spin particles must contain inverse powers of masses. In our case these can only be m_P^{-n}, and hence such particles must have negligible interactions at low energies $E \ll m_P$. In general a $2 \leftrightarrow 2$ scattering amplitude

$$F(1+2 \leftrightarrow 3+4) = (\sqrt{s})^{\lambda_s} (\sqrt{-t})^{\lambda_t} (\sqrt{-u})^{\lambda_u} \hat{F}(s,t) \tag{16}$$

where F is dimensionless, s, t and u are the conventional Mandelstam variables, $\hat{F}$ is a kinematical singularity-free reduced amplitude, and

$$\lambda_s \equiv |\lambda_1+\lambda_2+\lambda_3+\lambda_4|, \quad \lambda_t \equiv |\lambda_1-\lambda_2-\lambda_3+\lambda_4|$$
$$\lambda_u \equiv |\lambda_1-\lambda_2+\lambda_3-\lambda_4| \tag{17}$$

The external kinematical factors in (16) clearly mean that $\hat{F}$ contains inverse powers of masses in general. Consider for example (1/2,-1/2) ↔ (3/2, 1/2) scattering: using (17) one finds that

$$F = (us)\hat{F} \tag{18}$$

and if one goes to a direct channel (photon or gluon?) pole 1/s in $\hat{F}$ one sees that its coefficient, and hence the (3/2,1/2,γ or g) coupling, must be $O(m_P^{-2})$. This line of argument works directly for all massless states of |helicity| > 1, and also for massless states of |helicity| = 1 unless they have the specific gauge form of couplings[3]. By this reasoning one can argue that all unwanted states of |helicity| $\geq$ 1 would have negligible couplings at low energies, but one cannot wish away unwanted |helicity| = 1/2 states in the same way, and so this is only a partial solution to our problem.

Even if the direct interactions of unwanted helicity states are undetectable at low energies, one might worry[3] that they could have an indirect cosmological effect. If they were once in thermal equilibrium and then decoupled because of their weakening interactions as the temperature fell, they would still be present in the Universe during cosmological nucleosynthesis. Their energy-momentum would necessarily accelerate the primordial expansion rate[2] and might thereby enhance the cosmological ^{4}He production rate above astrophysically acceptable limits. The amount of speed-up is[49]

$$\xi^2 - 1 = \frac{4}{43}(N_B+\frac{7}{8}N_F)\left(\frac{43}{4N_D}\right)^{4/3} \tag{19}$$

where N_B and N_F are the numbers of species of decoupled bosons and fermions, and N_D is the total number of residual interacting degrees of freedom at the epoch of decoupling. From primordial ^{4}He nucleosynthesis one deduces

$$\xi^2 - 1 < 0.15 \tag{20}$$

which is satisfied when (here we corect Ref. 3)

$$(N_B+\frac{7}{8}N_F) < 0.07\ N_D^{4/3} . \tag{21}$$

This is because a large value of N_D degrades (19) the energy of every unseen boson and fermion sufficiently for their total contribution to the Universe's energy density to be negligible. The condition (21) is satisfied for all the unwanted N_B and N_F if one puts into N_D all the desired low-mass states included for example in the candidate superGUT (12) with its associated Higgs scalars. Therefore the presence of all the unwanted helicity states in Table 2 as weakly-interacting states of low mass is compatible with the nucleosynthesis constraints, though it should again be emphasized that we do not know how to guarantee that all the unwanted |helicity| = 1/2 states have negligibly weak interactions at low energies.

C - Maybe all the unwanted helicity states have large masses ? To arrange this within the normal helicity-matching rules we need extra helicity states to "eat" the unwanted ones, and probably an infinite set of bound state supermultiplets. We propose[3] that this infinite set may be obtained from unitary representations of the global non-compact $E_{7(+7)}$ symmetry in the physical spectrum. As a word of caution it should be mentioned that the gauged version[15] of the N = 8 ESG has a local SO(8) x SU(8) symmetry, but not a compact $E_{7(+7)}$ symmetry. Therefore this proposal breaks down if it turns out that SO(8) is gauged. Why should one think that the physical spectrum might contain unitary representations of $E_{7(+7)}$? It has already been mentioned that in two-dimensional CP^{N-1} models the physical spectrum contains[25] unitary representations of the global SU(N) symmetry, and the same is also true in three-dimensional models[26]. There are also some indications[50] of an infinite spectrum in two-dimensional models based on a global non-compact SO(N,1) group, redolent of a unitary representation. So if $E_{7(+7)}$ is present in the Lagrangian it may well show up in the physical spectrum.

It seems[3] that the relevant $E_{7(+7)}$ representations may have the non-linear structure

$$\{\lambda_{max},R\} \times \Sigma_{n=0}^{\infty}\left[\phi_{[ABCD]}\right]^{\{n\}} \tag{22}$$

where $\{\lambda_{max},R\}$ denotes an SU(8) supermultiplet (e.g., $\{3/2,\bar{8}\}$ in the case of Table 2), the ϕ_{ABCD} are the 70-dimensional scalars of Table 1, and the n-fold product is completely symmetric. If SU(8) is broken down to SU(6) or a subgroup thereof, as we proposed in the fourth section, it turns out[3] that at every value of helicity present in the original supermultiplet the representation (22) contains an infinite number of examples of every representation of the unbroken subgroup. This enables us to find partner helicity states for every unwanted helicity state $\hat{R}(\lambda)$, for if we look at any λ': $|\lambda'| \leq \lambda$, we can always find in (22) another example of the same representation $\hat{R}$ with helicity λ'. The representations of $E_{7(+7)}$ contained in (22) with any given helicity are not irreducible: this need not surprise us since the SU(8) representations at any given helicity in the supermultiplet (8) of Table 2 are themselves not reducible. To proceed further we must understand better the appropriate infinite-dimensional unitary representations[51] of $E_{7(+7)}$. We expect that the reducibility is forced upon us by susy, in the same way as in Table 2. To see how this comes about, we need to know how to combine susy with the global non-compact symmetry into a more complete algebraic structure, and then look for its representations. We have started[3,52] by looking at the N = 4 theory which has SU(4) × SU(1,1) global and U(4) local symmetries. It contains complex scalar fields z in the coset space SU(1,1)/U(1), on which the SU(1,1) actions are

$$L_+ z = i, \quad L_- z = iz^2, \quad L_0 z = -iz \tag{23}$$

The SU(1,1) transformations do not commute with susy transformations:

$$[L_+, Q_L] = -i\bar{z}Q_L \quad [L_-, Q_L] = -izQ_L, \quad [L_0, Q_L] = iQ_L \tag{24}$$

which suggests that the full algebra contains generators of the form

$$\left(f_L(\bar{z}) + g_L(\bar{z})z\right) Q_L, \quad \left(f_R(z) + g_R(z)\bar{z}\right)\bar{Q}_R \tag{25}$$

This structure begins to look like a graded version of a Kac-Moody[53] algebra

$$[Q_a^u, Q_b^m] = i\, f_{abc} Q_c^{n+m} \;; \quad n,m \quad \mathbb{Z} \tag{26a}$$

obtained from a conventional Lie algebra of local charges

$$[Q_a(\theta), Q_b(\theta')] = f_{abc} Q_c(\theta)\ (\theta-\theta') \quad 0 \leq \theta \leq 2\pi \tag{26b}$$

by making the Fourier transform

$$Q_a^n \equiv \int_0^{2\pi} d\theta e^{in\theta} Q_a(\theta) \tag{27}$$

It will be interesting[52] to see whether "graded Kac-Moody algebras" exist in ESGs, and then see whether they play any important physical rôle.

Even if the helicity-matching works as we have proposed above, we have no indication why which preferred finite subset of helicity states should be left massless by the dynamics. However, at least the mechanism discussed here indicates that it is not impossible to imagine zapping all the unwanted helicity states up to large masses.

WHAT NEXT IN SUPERUNIFICATION ?

The central problem in these attempts to embed grand unification in extended supergravity using composite fields remains the disposal of the unwanted helicity states. There is clearly a need for more detailed dynamical studies like those being done to look at confinement in gauge theories. We need to know how (if at all) the dynamics selects which renormalizable effective theory at low energies $E \ll m_P$.

We must figure out whether it is necessary or useful to gauge the internal SO(N) symmetry, and also the mirror question whether the possible non-compact global symmetries play an important rôle. The connection with conventional GUTs is still very obscure, and in particular the question of which supersymmetries to break at what energy scales. All at 10^{19} GeV? Retain an $N = 1$ susy down to $O(10^2)$ GeV to "solve" the hierarchy problem in a susy GUT? At what scale[54] is susy no longer local ?

Clearly there is food for thought, but the stakes are high. To paraphrase Hawking[55], perhaps we have the ultimate theory, and all we have to do is learn how to solve it.

ACKNOWLEDGEMENTS

In addition to a big thank you for my principal collaborators on this subject, Mary K. Gaillard and Bruno Zumino, I also wish to acknowledge valuable insights gleaned from Murat Günaydin, A. Kabelschacht and D.V. Nanopoulos.

REFERENCES

1. J. Ellis, M.K. Gaillard, L. Maiani and B. Zumino, "Unification of the Fundamental Particle Interactions", ed. by S. Ferrara, J. Ellis and P. Van Nieuwenhuizen, (Plenum Press, N.Y., 1980), p. 69.
2. J. Ellis, M.K. Gaillard and B. Zumino, Phys. Lett. 94B:343 (1980).
3. J. Ellis, M.K. Gaillard and B. Zumino, LAPP preprint TH-44/ CERN TH.3152 (1981), to appear in Acta Physics Polonica.
4. J. Ellis, "First Workshop on Grand Unification", ed. by P. Frampton, S.L. Glashow and A. Yildiz, (Math. Sci. Press, Brookline, 1980), p. 287.
5. B. Zumino, Proc. 1980 Madison Int. Conf. on High Energy Physics, ed. by L. Durand and L.G. Pondrom, (A.I.P., N.Y., 1981), p. 964; and "Superspace and Supergravity", ed. by S.W. Hawking and M. Roček, (Cambridge University Press, Cambridge, 1981), p. 423.

6. M.K. Gaillard, talk presented at the Heisenberg Symposium, München 1981, LBL preprint LBL-13371 (1981).
7. J. Ellis, "Gauge Theories and Experiments at High Energies", ed. by K.C. Bowler and D.G. Sutherland, (Scottish Universities Summer School in Physics, Edinburgh, 1981), p. 201 and Lectures presented at the Les Houches Summer School, August 1981, LAPP preprint TH-48/CERN TH.3174 (1981), to appear in the proceedings;
P. Langacker, Phys. Rep. 72C:185 (1981); and
Talk presented at the Int. Symp. on Lepton and Photon Interactions at High Energies, Bonn 1981, Univ. of Pennsylvania preprint UPR-0180 T (1981).
8. Y.A. Gol'fand and E.O. Likhtman, J.E.T.P. Lett. 13:323 (1971);
D. Volkov and V.P. Akulov, Phys. Lett. 46B:109 (1973);
J. Wess and B. Zumino, Nucl. Phys. B70:39 (1974).
9. D.V. Nanopoulos, these proceedings.
10. S. Dimopoulos and H. Georgi, Nucl. Phys. B193:150 (1981);
N. Sakai, Zeit. für Phys. C11:153 (1982)
11. For a review, see
P. van Nieuwenhuizen, Phys. Rep. 68C:189 (1981).
12. L.V. Avdeev, O.V. Tarasov and A.A. Vladimirov, Phys. Lett. 96B:94 (1980);
M.T. Grisaru, M. Roček and W. Siegel, Phys. Rev. Lett. 45:1063 (1980);
L. Caswell and D. Zanon, Phys. Lett. 100B:152 (1980).
13. D.Z. Freedman and A. Das, Nucl. Phys. B120:221 (1977);
E. Cremmer and J. Scherk (unpublished).
14. B. de Wit and H. Nicolai, Nucl. Phys. B188:98 (1981); see also these proceedings.
15. B. de Wit and H. Nicolai, Phys. Lett. 108B:285 (1982).
16. T.L. Curtright, Phys. Lett. 102B:17 (1981)
17. B. Zumino, Nucl. Phys. B89:535 (1975).
18. S.W. Hawking, Talk presented at the Heisenberg Symposium, München 1981; and
S.W. Hawking, D.N. Page and C.N. Pope, Phys. Lett. 86B:175 (1979) and Nucl. Phys. B170 (FS1):283 (1980).
19. M. Gell-Mann, Talk presented at the 1977 Washington Meeting of the American Physical Society (unpublished).
20. S. Ferrara and B. Zumino, Phys. Lett. 86B:279 (1979) and references therein.
21. See also
T.L. Curtright and P.G.O. Freund, "Supergravity", ed. by P. Van Nieuwenhuizen and D.Z. Freedman, (North Holland, Amsterdam, 1979), p. 167.
22. E. Cremmer and B. Julia, Nucl. Phys. B159:141 (1979).
23. A. D'Adda, P. Di Vecchia and M. Lüscher, Nucl. Phys. B146:63 (1978) and B152:129 (1979);
E. Witten, Nucl. Phys. B149:285 (1979).
24. J. Ellis, Nucl. Phys. B21:217 (1970).

25. H. Haber, I. Hinchliffe and E. Rabinovici, Nucl. Phys. B172:458 (1980).
26. E.R. Nissimov and S.J. Pacheva, Comptes Rendus de l'Académie Bulgare des Sciences 32:1475 (1979); Lett. Math. Phys. 5:67, 333 (1980).
27. D. Amati, R. Barbieri, A.C. Davis and G. Veneziano, Phys. Lett. 102B:408 (1981); see
A.C. Davis, these proceedings.
28. S. Weinberg, "Lectures on Elementary Particles and Quantum Field Theory", ed. by S. Deser, M. Grisaru and H. Pendleton (M.I.T. Press, Cambridge, 1970).
29. M.T. Grisaru and H.J. Schnitzer, Brandeis University preprint (1981); see
M.T. Grisaru, these proceedings.
30. See however
J.-P. Derendinger, S. Ferrara and C.A. Savoy, Nucl. Phys. B188:77 (1981) and
C.A. Savoy, these proceedings;
J.E. Kim and H.S. Song, Seoul National University preprint (1981).
31. H. Römer, Phys. Lett. 83B:172 (1979).
32. S. Coleman and B. Grossman, private communication (1981).
33. M. Gell-Mann, unpublished (1980).
34. P.H. Frampton, Phys. Rev. Lett. 46:881 (1981)
35. M. Veltman, Acta Physica Polonica B12:437 (1981).
36. Ya.B. Zeldovich, Phys. Lett. 59A:254 (1976).
37. R. Barbieri, J. Ellis and M.K. Gaillard, Phys. Lett. 90B:249 (1980).
38. J. Ellis and M.K. Gaillard, Phys. Lett. 88B:315 (1979).
39. D.V. Nanopoulos and D.A. Ross, Phys. Lett. 108B:351 (1982).
40. G. Steigman, Bartol Research Foundation preprint BA-81-20 (1981) and references therein; also these proceedings.
41. M. Glück and E. Reya, Phys. Lett. 105B:30 (1981).
42. J. Ellis and D.V. Nanopoulos, Nature 292:436 (1981).
43. J. Wess and B. Zumino, Phys. Lett. 49B:52 (1974);
J. Iliopoulos and B. Zumino, Nucl. Phys. B76:310 (1974);
S. Ferrara, J. Iliopoulos and B. Zumino, Nucl. Phys. B77:413 (1974).
44. R.D. Peccei and H.R. Quinn, Phys. Rev. Lett. 38:1440 (1977) and Phys. Rev. D16:1791 (1977).
45. M. Dine, W. Fischler and M. Srednicki, Phys. Lett. 104B:199 (1981);
M.B. Wise, H. Georgi and S.L. Glashow, Phys. Rev. Lett. 47:402 (1981).
46. S. Weinberg and E. Witten, Phys. Lett. 96B:59 (1980).
47. M.K. Gaillard and B. Zumino, Nucl. Phys. 193:221 (1981).
48. J.P. Ader, M. Capdeville and H. Navelet, Nuov. Cim. 56A:315 (1968).

49. G. Steigman, K.A. Olive and D.N. Schramm, Phys. Rev. Lett. 43:239 (1979);
D.A. Olive, D.N. Schramm and G. Steigman, Nucl. Phys. B180: (FS2),497 (1981).
50. E. Rabinovici, unpublished (1980).
51. M. Günaydin and C. Saçlioğlu, Phys. Lett. 108B:180 (1982).
52. J. Ellis, M.K. Gaillard, M. Günaydin and B. Zumino, dormant interest (1981).
53. V. Kac, Math. U.S.S.R. Izvestiya Ser. Math. 32:1271 (1968);
R. Moody, Bull. Am. Math. Soc. 73:217 (1967) and J. of Algebra 10:211 (1968).
54. R. Barbieri, S. Ferrara and D.V. Nanopoulos, CERN preprint TH.3159 (1981).
55. S.W. Hawking, Lucasian Inaugural Lecture "Is the end in sight for theoretical physics?", (Cambridge University Press, Cambridge, 1980).

BOUND STATE REGGE TRAJECTORIES IN N=8 SUPERGRAVITY

M. T. Grisaru *
H. J. Schnitzer **

* California Institute of Technology, Pasadena, CA 91125
** Brandeis University, Waltham, MA 02254

We describe here some recent work[1] in which we undertook to investigate, based on the dynamics of N=8 supergravity, the existence of bound states in this system. We were motivated to some extent by the suggestion of Ellis, Gaillard, Maiani and Zumino[2] (EGMZ) that N=8 supergravity could make contact with current particle phenomenology if one postulated a certain set of massless (before symmetry breaking) bound states transforming according to the local SU(8) invariance group that the theory may possess[3]. In their original proposal EGMZ assumed that these bound states had the quantum numbers of a so-called "current multiplet":

$$(-3/2)_L, \ (-1)_L^A, \ \ldots . \ (5/2)_L^{AB} \ \ldots . + \text{(TCP conjugates)}.$$

The particles of SU(5) GUT were to be found among these, following a sequence of symmetry breakings. It has also been argued[6] that in order to make the EGMZ scenario viable, one might have to postulate an infinite tower of massless particles.

We have nothing to say about how realistic a theory one can build out of the bound states of N=8 supergravity. We believe we can say something about the existence of such bound states. We find fairly good evidence that "current multiplet" states do exist as two-preon bound states. Furthermore we believe that our methods, if used beyond the present stage of our work, would indicate the existence of infinite sets of multipreon bound states. However, our technology is not sufficiently developed to handle this issue quantitatively.

It is well known that finding bound states in field theory is difficult and it becomes even more so in a potentially divergent theory like supergravity. In fact, it is clear that standard methods will not work here: one such method is to extract out of the field theory diagrams a static potential to be used in conjunction with the Schrodinger equation. However, this is clearly senseless in a theory where the interacting particles are massless. Another method is to set up a Bethe-Salpeter equation for a 2-body amplitude which, for example, sums ladder diagrams, and to solve the homogeneous equation. However, the Bethe-Salpeter equation uses off-shell amplitudes and one runs immediately into the bad off-shell divergences of (super-) gravitational theories. We know, of course, that such theories are better behaved on-shell and this suggests using methods which deal exclusively with on-shell quantities. Such methods exist in the context of S-matrix theory where one uses unitarity and analyticity to set up integral equations for on-shell scattering amplitudes.

The object one studies is, for example, a partial-wave amplitude $f_j(s)$ which satisfies a nonlinear integral equation whose solution may exhibit a spin j, mass M bound state pole $f_j(s) \sim (s-M^2)^{-1}$. However, it is not always easy to set up and solve the equation sufficiently accurately to exhibit this bound state pole especially since the solution suffers from so-called CDD ambiguities. It turns out to be much more efficient and also much easier to find the desired information by studying the equation satisfied by analytic continuation to complex angular momentum f(s,J) of the partial wave amplitude[5,6]. In general, this function has s-dependent "Regge" poles

$$f(s,J) \sim \frac{\beta(s)}{J-\alpha(s)} \tag{1}$$

with "Regge residue" $\beta(s)$ and "Regge trajectory" $\alpha(s)$. If for some $s=M^2$ $\alpha(s)=j$, an integer or half integer, then

$$f(s,j) \sim \frac{\beta(M^2)}{(s-M^2)\alpha'(M^2)} \tag{2}$$

and this function, if it can be identified with the physical amplitude $f_j(s)$, exhibits a bound state of mass M and spin j provided $\beta(M^2)\neq 0$. Our procedure, in fact, is to look for Regge poles in N=8 supergravity. Some aspects of the complex angular momentum properties of supergravity are discussed in Ref. 6.

In the past ten years, considerable work has gone into studying Regge poles in Yang-Mills theories. One approach[7] consists of studying in leading logarithm approximation the high-momentum transfer behavior of sums of Feynman diagrams. This turns out to be

$$\sim \sum_i \gamma_i(s)\, t^{\alpha_i(s)}$$

where the α's are various Regge trajectories:

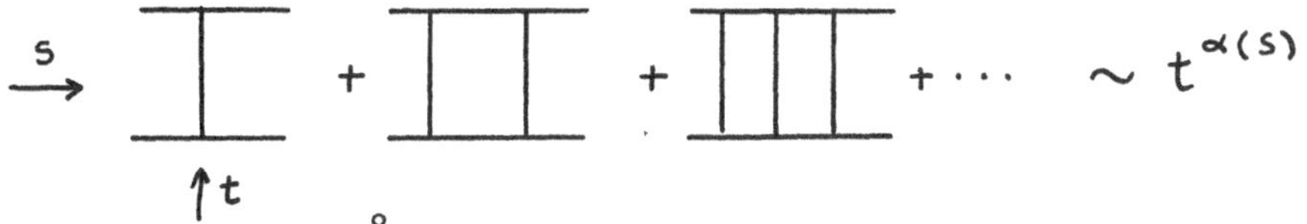

The second procedure[8] sets up and solves in a simple approximation the set of integral equations that the partial-wave helicity amplitudes $f(s,J)$ satisfy[5]. This requires only knowledge of the Born approximation and is quite easy. While in principle less reliable, in all Yang-Mills calculations where comparison is possible it has given results for Regge poles identical to those obtained by summing diagrams. In supergravity the first procedure, summing Feynman diagrams, is out of the question because computing diagrams beyond tree approximation is as yet much too difficult. The second procedure is easy to apply and while it is subject to some uncertainties, the experience with Yang-Mills theories makes us quite confident about the validity of our results.

We are considering the scattering of particles with spin and we look at the integral equation satisfied by the two-body helicity amplitudes

$$F_{\lambda_3\lambda_4,\lambda_1\lambda_2}(s,J) = V_{\lambda_3\lambda_4,\lambda_1\lambda_2}(s,J)$$

$$+ \sum_{\lambda_5\lambda_6} \int \frac{ds'}{s'-s}\, \rho(s')F_{\lambda_3\lambda_4\lambda_5\lambda_6}(s',J)F_{\lambda_5\lambda_6\lambda_1\lambda_2}(s',J)\ . \qquad (3)$$

The ingredients that go into these equations are certain assumptions of analyticity and boundedness which allow one to write partial wave dispersion relations valid for $\text{Re } J > N$, for some positive N

$$F(s,J) = \frac{1}{\pi}\int_{-\infty}^{s_-} \frac{ds'}{s'-s} A_L(s',J) + \frac{1}{\pi}\int_{s_+}^{\infty} \frac{ds'}{s'-s} A_R(s',J)\ . \qquad (4)$$

Here

$$A_R = |F|^2 + A_R^{\text{inelastic}}$$

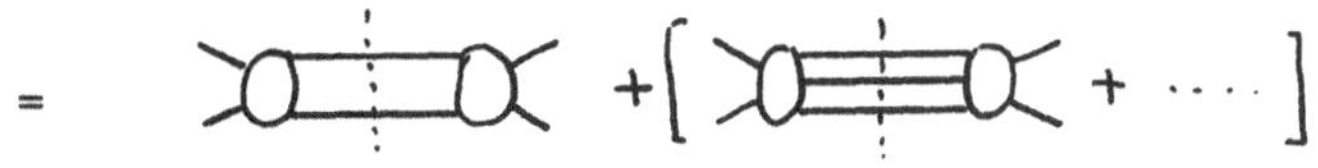

(5)

while A_L can, in principle, be obtained from cross-channel discontinuities[5]. We have denoted the A_L contribution by V and ignore for now $A_R^{\text{inelastic}}$.

V(s,J) can be obtained to any given order of perturbation theory from diagrams. In particular, if we keep only lowest order diagrams, we find

$$V_{\lambda_3\lambda_4\lambda_1\lambda_2}(s,J) \sim Q_{J-\lambda_m}(Z_o(s))v_{\lambda_3\lambda_4\lambda_1\lambda_2}(s) \qquad (6)$$

where $\lambda_m = \max(|\lambda_3-\lambda_4|,|\lambda_1-\lambda_2|)$ is an integer of half integer (we continue to complex J with fixed, real helicities) and Q a Legendre function of the second kind. We observe that the Q functions have poles at negative integral orders. If we continue the integral equation (3) to a neighborhood of $J = j_o = \lambda_m - n$ where n = 1,2,3, ... we find that elements of the matrix V have pole singularities. Let us denote $F_{\lambda'\lambda} \equiv F_{\lambda_3\lambda_4\lambda_1\lambda_2}$ and write the matrix equations (3), for J in the neighborhood of j_o, in the form

$$\begin{pmatrix} F_{ss} & F_{sn} \\ F_{ns} & F_{nn} \end{pmatrix} = \begin{pmatrix} v_{ss} & v_{sn}(J-j_o)^{-1/2} \\ v_{ns}(J-j_o)^{-1/2} & v_{nn}(J-j_o)^{-1} \end{pmatrix}$$

$$+ \int \frac{ds'}{s'-s} \, (\;F\;)\;(\;F\;)\;. \qquad (7)$$

We have introduced submatrices F_{nn}, F_{ns}, F_{sn}, F_{ss} depending on whether $\lambda' = |\lambda_3-\lambda_4|$, $\lambda = |\lambda_1-\lambda_2|$ are or are not greater than j_o (i.e., have "nonsense" or "sense" values[5]. In the neighborhood of j_o the potential matrix V breaks up as shown, the v's being polynomials in s, and one can solve the integral equation for $J \simeq j_o$ by ignoring the "small" potential V_{ss} compared to the other, "large" entries. The solution is, for example[8] (using the polynomiality of v and a particular subtraction philosophy)

$$F_{ss}(s,J) = v_{sn'}\left[\frac{K(s)}{J-j_o-v(s)K(s)}\right]_{n'n} v_{ns} \qquad (8)$$

where K(s) is a simple known integral and we find Regge poles with trajectories

$$\alpha(s) = j_o + (\text{eigenvalues of } v_{nn}) \times K \qquad (9)$$

equal in number to the rank of v_{nn}.

We see that information about the trajectories is contained entirely in v_{nn} while v_{sn}, v_{ns} give some information about the residues. In our work we have concentrated for simplicity on the trajectories only, which require that we calculate v_{nn}. In general, in the scattering of particles of spin S_1, S_2 nonsense values exist for

$j_o \leq (S_1+S_2) -1$. We look at two-body helicity amplitudes for the preons of N=8 supergravity with individual helicities 2, 3/2, 1 .. -2. The maximum value of $|\lambda_1-\lambda_2| = 4$ and therefore our methods may give Regge poles in the vicinity of $j_o = 3, 5/2, 2$ .. . For each value of j_o we compute from the Born approximation the matrix v_{nn}. We label our individual preons by SU(8) labels according to their helicity and global SO(8) content, i.e., (-2), $(-3/2)^A$, $(-1)^{AB}$ ... $(3/2)_A$, (2) and project v_{nn} onto irreducible representations of SU(8). The corresponding Regge poles will carry the quantum numbers of such representations.

For example, we find the Born approximation helicity amplitudes

$$F(2, -2; 2, -2) = \kappa^2 \frac{u^3}{st} \tag{10a}$$

$$F(1^{AB}, -\frac{1}{2}_{CDH}; \frac{3}{2}_E, -1^{FG}) = i\kappa^2 \sqrt{\frac{-st}{u}} \cdot$$

$$\cdot \left[\frac{\delta^{AB}_{DE}\delta^{FG}_{HC}+\ldots}{t} + \frac{\delta^{AB}_{HC}\delta^{FG}_{DE}+\ldots}{u} \right] \tag{10b}$$

where $\delta^{AB}_{DE} = \delta^A_D\delta^B_E - \delta^A_E\delta^D_B$, etc., and s,t,u are Mandelstam invariants. The graviton-graviton amplitude in (10b) has "nonsense" helicity values for j_o = 3,2,1 .... and after partial wave projection may contribute to SU(8) singlet Regge poles in the vicinity of such values. The amplitude in (10b) is "nonsense-nonsense" for $j_o = 1/2$ and will contribute to the v_{nn} matrix which gives Regge poles near this value, carrying (from the reduction of $(3/2)_E X(-1)^{FG}$ octet and 216 representations of SU(8).

For each value of j_o and each irreducible representation of SU(8) we compute v_{nn} and determine the number of Regge trajectories. We find Regge poles having the form

$$F(s,J) \sim \frac{1}{J-j_o-s\gamma} \tag{11}$$

passing through "right signature" points[5], j_o = 1/2, 1, 3/2, 2 .... for s=0 (i.e., apt to produce zero mass bound states with spins j_o). We find that for each irreducible bound state trajectory

$$(\lambda) \begin{matrix} [AB\ldots] \\ [LM\ldots] \end{matrix}$$

there exist trajectories corresponding to the same helicity forming "trace" representations

$$\delta^{[A \quad B\ldots]}_{[L \quad M\ldots]}(\lambda), \quad \delta^{[AB \quad \ldots]}_{[LM \quad \ldots]}(\lambda), \text{ etc.}$$

If we lump these together into <u>reducible</u> representations, we find a very simple picture: a set of trajectories corresponding to the original preons (which therefore "Reggeize"[5],

$$(-2),\ (-3/2)^{A},\ (-1)^{AB}\ \ldots$$

and another set

$$(-3/2)_{L},\ (-1)^{A}_{L},\ (-1/2)^{AB}_{L}\ \ldots$$

which corresponds precisely to the current multiplet of EGMZ (including traces!). We also find sets of "wrong signature" trajectories[5]:

$$(-3),\ (-5/2)^{A},\ (-2)^{AB}\ \ldots$$

$$(-5/2)_{L},\ (-2)^{A}_{L},\ (-3/2)^{AB}_{L}\ \ldots$$

$$(-2)_{LM},\ (-3/2)^{A}_{LM},\ (-1)^{AB}_{LM}$$

These do not produce bound states (the residues $\beta(s)$ necessarily vanish for values of s such that $\alpha(s)$ is physical) but may be relevant for the discussion of many-particle bound states.

In spite of some uncertainties in our approach (in particular, we cannot calculate the Regge residues reliably), we are fairly confident about the outcome of our calculations: Regge trajectories exist, and some of them correspond to the original preons, which in this sense are both elementary and composite. The others correspond to additional bound states. That binding occurs in spite of the small value of the coupling constant may seem strange, but we observe that very little binding energy is required when the constituents are massless. Just as in Yang-Mills theory, the value of the coupling parameter in the Lagrangian determines the slope of the Regge trajectories, but not their position.

We believe that multipreon bound states exist and could be discovered by studying equations similar to the ones we have used, but including now inelastic channels and many particle amplitudes. In general, one expects now nonsense values and therefore Regge poles near $j_0 \leq S_1 + S_2 + \ldots + S_n - (n-1)$. The technology for such a study is not very well developed but it may be possible to make some qualitative statements. For example, since the existence of two-body bound states has been settled, one may be able to approximate three-body scattering channels by the quasi two-body scattering of preons and current multiplet Reggeons. Since the highest spins available are 2 (for the graviton) and 5/2 (for the highest member of the current multiplet), these may produce Regge poles in the neighborhood of j_0's up to $|5/2-(-2)|-1 = 7/2$. Repeating the procedure, one could imagine finding trajectories in the neighborhood of arbitrary large values of j_0. What is lacking, of course, is the dynamical input, e.g., the Born approximation, which in the two-preon case gave us some quantitative information about binding forces and determined the fact that only two-body bound states with the quantum numbers of the current multiplet (among several other possibilities) may actually occur. However, one might still be able to put some restrictions on the spin and SU(8) content of possible multipreon bound states. In a scenario where infinite numbers of multiplets may be required (but so far in an unspecified manner) to give masses to unwanted particles[4] and make contact with GUT phenomenology, these restrictions would be most welcome. We hope to be able to say something on the subject in the near future.

The work reported here was supported in part by NSF Grant No. PHY 79-20801 and DoE Contract DE-AC03-76-ER03230-A005.

REFERENCES

1. M. T. Grisaru and H. J. Schnitzer, "Dynamical Calculation of Bound-state Supermultiplets in N=8 Supergravity," Phys. Lett. (to be published and in preparation).
2. J. Ellis, M. K. Gaillard, L. Maiani and B. Zumino, "Unification of the Fundamental Interactions," Eds. S. Ferrara, J. Ellis and P. van Nieuwenhuizen (Plenum Press, New York, 1980); J. Ellis, M. K. Gaillard and B. Zumino, Phys. Lett. 94B:343 (1980).
3. E. Cremmer and B. Julia, Phys. Lett. 80B:48 (1978); Nucl. Phys. B159:141 (1979).
4. P. Frampton, Phys. Rev. Lett. 46: 881 (1981); J. P. Derendinger, S. Ferrara and C. A. Savoy, Nucl. Phys. B188: 77 (1981).
5. P. D. Collins and E. J. Squires, Springer Tracts in Modern Physics, vol. 45, (Springer-Verlag, Berlin, 1968).
6. M. T. Grisaru, "(Super) Gravity in the Complex Angular Momentum Plane," in the Proceedings of the 1981 Nuffield Quantum Gravity Workshop, eds. M. Duff and C. Isham (Cambridge University Press, to be published).
7. B. M. McCoy and T. T. Wu, Phys. Rev. D12: 2357 (1376); D13: 1076 (1981); J. Bartels, Nucl. Phys. B151: 293, (1978); B175: 365 (1980).
8. M. T. Grisaru, H. J. Schnitzer and H. -S. Tsao, Phys. Rev. Lett. 30: 811 (1973); Phys. Rev. D8: 4498 (1973); D9: 2864 (1976); M. T. Grisaru, Phys. Rev. D13: 2916 (1976); D16: 1962 (1977); M. T. Grisaru and H. J. Schnitzer, Phys. Rev. D20: 784 (1979); D21: 1952 (1980).

NEW RESULTS IN CONFORMAL SUPERGRAVITY

B. de Wit

NIKHEF-H
Amsterdam

1. INTRODUCTION

The important role that local symmetries play in quantum field theory has generally been recognized. However, in specific cases the presence of such a symmetry may seem rather insignificant. For instance, consider a massive vector field described by the Proca lagrangian

$$L = -\tfrac{1}{4}(\partial_\mu V_\nu - \partial_\nu V_\mu)^2 - \tfrac{1}{2}m^2 V_\mu{}^2 \ , \tag{1}$$

which exhibits no local gauge invariance. Nevertheless, it is possible to introduce gauge transformations by simply redefining V_μ according to

$$V_\mu = A_\mu - m^{-1}\ \partial_\mu\phi \ . \tag{2}$$

This linear combination is left invariant under

$$\begin{aligned} \delta A_\mu(x) &= \partial_\mu \Lambda(x) \ , \\ \delta\phi(x) &= m\Lambda(x). \end{aligned} \tag{3}$$

The presence of these local gauge transformations is crucial to achieve the decomposition of the four degrees of freedom contained in V_μ into the three degrees of freedom of the gauge field A_μ and one of the new field ϕ. The field ϕ is called a <u>compensating field</u>. Compensating fields generally allow one to redefine fields that transform under gauge transformations as gauge invariant quantities; in those redefinitions ϕ occurs as the parameter of a field-dependent gauge transformation, whose variation compensates for the effect of the gauge transformation on the noninvariant fields. Indeed (2) is one example of such a redefinition, which in this form is due to Stueckelberg[1]. Obviously when one substitutes (2) into the lagrangian (1) the new lagrangian is invariant under (3). Of course it remains simply gauge equivalent to (1): the field ϕ can be removed by a choice of gauge (such as $\phi = 0$), or it can be reabsorbed into the definition of the remaining fields through a (uniform) field-dependent gauge transformation.

The introduction of gauge invariance through substitutions such as (2) may seem trivial at first sight. Nevertheless, there are sometimes good reasons to use the second formulation. For instance, one has the option of imposing a variety of gauge conditions, some of which may lead to softer quantum divergences in the corresponding Green functions. Another reason which is important in what follows, is that the second formulation is based on smaller multiplets: the presence of gauge invariance implies a higher degree of irreducibility of the multiplets in question. Note that it is not possible to find a local decomposition of the degrees of freedom contained in V_μ without introducing gauge invariance.

Within the context of gravity and supergravity it is also possible to use formulations with a high degree of symmetry which lead to invariants that are gauge equivalent to actions for theories with less symmetry. The maximal symmetry that one considers in that context is related to the gauge transformations of the (super)conformal algebra[2]. Formulations with this higher symmetry offer important advantages. They are helpful in mastering the complexities of extended supergravity: the additional symmetry severely restricts possible nonlinear terms in the transformation rules, and moreover the fundamental field representation contains fewer degrees of freedom. The latter is analogous to what happens in the example that we have discussed above, where the extra gauge symmetry allows one to decompose the fields into smaller multiplets with a higher degree of irreducibility. Many of the low-spin degrees of freedom that are required in order to construct specific Lagrangians correspond to compensating fields. In supergravity these compensating fields must be part of an entire supermultiplet, and in fact it is usually possible to have the appropriate compensators for different choices of supermultiplets. This then leads to inequivalent versions of Poincaré and de Sitter supergravity theories. The superconformal approach can thus be applied uniformly to all those theories, and clarifies their structure within one common framework; hence independent studies of all these theories are unnecessary.

We should emphasize that we employ the gauge equivalence only at the classical level, since the conformal symmetry is expected to be destroyed by quantum corrections. However, some of the extra symmetries may be helpful in understanding the dynamics of these complicated theories. This hope has been inspired by studies of two-dimensional models where extra (non-dynamical) symmetries play a useful role. This has led to the conjecture that a similar situation may arise in supergravity in four dimensions[3]. But we stress that even without immediate implications for the dynamics the extra symmetries are a useful tool for the construction of classical supergravity theories[4].

In this talk I will briefly review the essential features of conformal supergravity and the use of gauge-equivalent formulations in supergravity. Subsequently I will discuss some new results in N=1 supergravity in ten and four dimensions.

2. CONFORMAL SUPERGRAVITY

The symmetries of conformal supergravity are derived from the SU(2,2 | N) superalgebra, which is an extension to supersymmetry of the conventional conformal algebra[5]. The latter contains translations (P), Lorentz rotations (M), dilatations (D) and conformal boosts (K). When considering supersymmetry (Q) one finds that closure of the symmetry algebra requires the presence of special supersymmetry (S) and chiral U(N) transformations (in the case N=4 the chiral transformations may be restricted to those of SU(N)). These symmetries with their corresponding gauge fields have been listed in Table 1. Not all gauge fields are independent; those fields corresponding to M, K and S are expressed in terms of the remaining fields, and their structure is indicated generically in the Table.

The reasons for this dependence have been discussed in several references[2,6,7,8]. One obvious consequence is that the theory is based on a smaller number of degrees of freedom. If we count the gravitational degrees of freedom, related to the conformal P, M, D and K gauge fields, we find 45 (we always subtract gauge degrees of freedom); the restrictions on $\omega_\mu{}^{ab}$ and $f_\mu{}^a$ reduce this number to only 5, which is precisely the minimal number of field components needed to describe spin 2. A similar phenomenon takes place for the fermionic gauge fields $\psi_\mu{}^i$ and $\phi_\mu{}^i$, which represent 24N degrees of freedom. The fact that $\phi_\mu{}^i$ is not an independent field reduces this number to 8N, which is the minimal number to describe fermions with spin 3/2 (to describe fermions one generally needs twice as many degrees of freedom as for bosons). For the chiral gauge fields no constraints are necessary; gauge invariance already reduces their number of independent components to 3 for each field, which is the minimal number for describing spin 1.

Hence the constraints imposed on the conformal gauge fields lead to maximal irreducibility. However, these field representations are too small to qualify as true supermultiplets; it turns out that the conformal field representation contains additional low-spin degrees of freedom. The structure of these representations is completely

Table 1. Gauge symmetries and gauge fields of conformal supergravity; the last column indicates the characteristic terms for the dependent fields.

P	4	$e_\mu{}^a$	
M	6	$\omega_\mu{}^{ab}$	; dependent: $\omega \sim \partial e$
D	1	b_μ	
K	4	$f_\mu{}^a$	; dependent: $f \sim \partial\omega$
Q	4N	$\psi_\mu{}^i$	
S	4N	$\phi_\mu{}^i$	; dependent: $\phi \sim \partial\psi$
U(N)	N^2	$V_\mu{}^i{}_j$	

known for N≤4; beyond N=4 only the gauge fields corresponding to the superconformal algebra have been given. For a general discussion of the field representation of conformal supergravity we refer to refs. 2 and 8.

We will now demonstrate the use of conformally invariant formulations for spin 2. This is the direct analogue of the Stueckelberg mechanism discussed in the previous section. In D dimensions conformal gravity is based on $\frac{1}{2}(D+1)(D-2)$ degrees of freedom, to which we add a compensating scalar field ϕ. For such a field it is possible to construct an action invariant under all conformal symmetries. The corresponding lagrangian is

$$L = - e\, \phi\, \Box^c \phi \quad , \tag{4}$$

where $\Box^c$ denotes the conformally covariant d'Alembertian. The crucial point is that the conformal covariance of this operator requires the presence of the gauge field $f_\mu{}^a$ of the conformal boosts (K). Schematically we can write

$$\Box^c\phi = \Box^{grav}\phi - \tfrac{1}{2}(D-2)\, f_\mu{}^\mu\, \phi \ , \tag{5}$$

where $\Box^{grav}$ denotes the standard d'Alembertian covariantized with respect to general coordinate transformations, with some minor modification.

We now recall that $f_\mu{}^a$ depends on the other gravitational fields (see Table); it is proportional to the derivative of the spin connection $\omega_\mu{}^{ab}$, which in turn is proportional to the derivative of the

D-bein fields (more precisely, to the objects of anholonomity). Covariance under general coordinate transformations ensures that the trace of $f_\mu{}^a$ is simply proportional to the standard Riemann curvature scalar R(e), and we find

$$L = -\, e\, \phi\, \Box^{grav}\phi - \tfrac{1}{4}\, \frac{D-2}{D-1}\, e\, R(e)\phi^2 \, . \tag{6}$$

This lagrangian is gauge-equivalent to Einstein gravity. To see this we note that (6) is invariant under all conformal transformations, and in particular under local dilatations which act on ϕ according to

$$\phi(x) \to \phi'(x) = e^{\frac{1}{2}(D-2)\Lambda_D(x)}\, \phi(x) \, . \tag{7}$$

This local invariance allows one to impose a gauge condition

$$\phi = \kappa^{-1} \, , \tag{8}$$

where κ is a dimensionful constant. With this condition (6) takes the form of Einstein's lagrangian

$$L = -\, \tfrac{1}{4}\, \frac{D-2}{D-1}\, \kappa^{-2}\, e\, R(e) \, , \tag{9}$$

which shows that Newton's constant is proportional to κ^2. Henceforth we shall usually put κ equal to one.

Hence it is possible to write Poincaré gravity in a formulation with conformal invariance. Within that context the theory is based on two multiplets: the conformal gauge fields corresponding to $\frac{1}{2}(D+1)(D-2)$ degrees of freedom, and the compensating field representing one degree of freedom. The same procedure can be applied to supergravity, but in that case it is important to realize that the compensators are contained in supermultiplets. The choice of these supermultiplets is not unique; different multiplets may contain the appropriate compensating fields, and their addition may thus lead to inequivalent versions of the corresponding Poincaré or de Sitter supergravity theories. The difference is usually present in the off-shell sector of the theory: the equations of motion for the physical fields are then the same. The relevance of conformal supergravity should now be obvious; since all Poincaré and de Sitter supergravity theories are expected to correspond to a combination of conformal supergravity with an appropriate set of compensating multiplets, all these theories can be described within one common framework.

I should emphasize that the application of gauge-equivalent formulations has three important aspects. In the case of conformal supergravity these can be formulated as follows:

a. One must select a field representation of the superconformal gauge algebra to provide for the compensating fields.
b. For this representation one must be able to construct a superconformal invariant.

c. This invariant should then lead to a meaningful action from which the appropriate equations of motion will follow.

It is important to realize that these three aspects are logically independent; for instance, it is possible to find compensating multiplets for which the requirements b. or c. are not satisfied. Furthermore it is sometimes necessary to consider a whole hierarchy of compensators in order to describe a certain theory.

Let us now briefly review how the known field representations of supergravity are decomposed into superconformal fields and compensating multiplets. The N=1 superconformal fields[6] constitute 8+8 (bosonic + fermionic) degrees of freedom, to which one may add the well-known chiral multiplet as a compensator[9,10]. This then leads to the minimal set of auxiliary fields of N=1 supergravity, which is contained in a field representation of 12+12 degrees of freedom[11]. Another option is to add a complex linear multiplet[10] which contains 12+12 degrees of freedom; the resulting theory corresponds to the nonminimal form of N=1 supergravity based on 20+20 components[12]. For N=2 the decomposition is more complicated. The superconformal theory contains 24+24 degrees of freedom to which one adds a vector multiplet of 8+8 degrees of freedom as a compensator[7]. The result is the so-called minimal field representation of 32+32 components[7,13]. However, the addition of one compensating multiplet does not suffice in this case; although it is possible to define a superconformal invariant, this invariant does not qualify as a suitable supergravity action. Therefore one must introduce a second compensating multiplet. Two options are known: one may add the so-called nonlinear multiplet or the scalar multiplet. Both multiplets contain 8+8 degrees of freedom, so that one finds two inequivalent N=2 supergravity theories, both based on 40+40 field components[7,14]. Beyond N=2 no complete field representations are known for supergravity, except for the superconformal theory itself.

3. MAXWELL-EINSTEIN SUPERGRAVITY IN TEN DIMENSIONS

One way to understand extended supergravity theories is to study simple supergravity in higher space-time dimensions. The highest dimension in which it is possible to couple matter to N=1 supergravity is D=10; after dimensional reduction to D=4 one obtains N=4 supergravity coupled to matter. Simple supergravity in ten dimensions is based on the following physical fields: the zehnbein field $e_\mu{}^a$, the gravitino field ψ_μ, a scalar field ϕ, a spinor λ, and an antisymmetric tensor gauge field $A_{\mu\nu}$. The supersymmetric Maxwell theory contains a physical spinor and a vector, precisely as in four dimensions, whose fields are denoted by χ and A_μ, respectively. All spinor fields are Majorana-Weyl fields, i.e. they are both real and chiral.

It is possible to couple the supersymmetric Maxwell theory to supergravity by means of an iterative procedure known as the Noether method. On the other hand one may analyse the currents that describe the coupling of the supergravity fields to this matter theory. Both aspects have been studied in collaboration with E. Bergshoeff, M. de Roo and P. van Nieuwenhuizen[15]; the strategy of this work was to use the combined information obtained from these two separate approaches in order to acquire some further understanding of the off-shell structure of D=10 supergravity. For independent studies of ten-dimensional supergravity theories, see Ref. 16.

The currents of the supersymmetric Maxwell system that we have mentioned above consist of the energy-momentum tensor, i.e., the current that describes the coupling of the graviton to the supersymmetric Maxwell fields, and a large number of other quantities bilinear in those fields which can be found by considering successive supersymmetry transformations. One of these components is the Noether current of supersymmetry transformations which couples to the gravitino field ψ_μ. The current multiplet for the supersymmetric Maxwell system has been fully constructed in four and five dimensions[8,17]; recently also the ten-dimensional currents have been analyzed[18].

Knowledge of the multiplet of currents gives information about the off-shell structure of supergravity; since the currents form a true off-shell multiplet, the supergravity fields to which they couple form an off-shell multiplet as well. Some of the fields of this multiplet correspond to physical degrees of freedom of supergravity, whereas others are only auxiliary. However, it is important to realize that this multiplet does not necessarily encompass all fields of Poincaré supergravity; in fact the supersymmetric Maxwell system only couples to an irreducible subset. As we have discussed previously the presence of an irreducible submultiplet is an indication that a higher degree of gauge invariance is relevant. In four dimensions the currents of the supersymmetric Maxwell theory thus lead to the field representation of conformal supergravity. The fact that only the superconformal fields couple to this theory comes as no surprise, since the supersymmetric Maxwell lagrangian is locally scale invariant. The extra fields of Poincaré supergravity can be added in the form of compensating multiplets, but these fields will not couple to the supersymmetric Maxwell system. In higher dimensions one qualitative difference is that not all <u>physical</u> supergravity fields are among those that couple to the multiplet of currents[17]. Furthermore the supersymmetric Maxwell lagrangian is no longer locally scale invariant in dimensions different from D=4. But the essential point remains that the supersymmetric Maxwell theory couples to an irreducible subset of the fields of Poincaré supergravity.

Another aspect of having such an irreducible subset of fields is that it is not possible to construct an invariant action of the standard form (i.e. of second order in derivatives) on the basis of these fields. We understand why this is so from the examples discussed in the previous sections: it is for instance not possible to formulate Einstein's gravity action on the basis of the conformal degrees of freedom; as we have seen in section 2 one must introduce one extra compensating degree of freedom. Without adding degrees of freedom, it is only possible to have gravitational actions with higher-order derivatives.

If one compares some of the explicit supergravity couplings of the supersymmetric Maxwell theory in D=10 dimensions to the coupling as predicted by the analysis of the current multiplet, one discovers an obvious puzzle. One of the current components is proportional to $A_{[\mu}F_{\nu\rho]}$, where $F_{\nu\rho}$ is the Maxwell field strength; this current couples to a field $t_{\mu\nu\rho}$:

$$L_{\text{coupling}} \propto t^{\mu\nu\rho}(A_{[\mu}F_{\nu\rho]}) \ . \tag{10}$$

The field $t_{\mu\nu\rho}$ is an antisymmetric 3-rank tensor, which satisfies

$$\partial_\mu t^{\mu\nu\rho} = 0 \ . \tag{11}$$

On the other hand the field $t_{\mu\nu\rho}$ is not one of the physical fields of D=10 supergravity. Therefore the explicit Noether coupling procedure does not make reference to this field; instead it reveals the presence of a term

$$L_{\text{Noether}} \propto \partial^{[\mu}A^{\nu\rho]}\ (A_{[\mu}F_{\nu\rho]}) \tag{12}$$

in the lagrangian of D=10 Maxwell-Einstein supergravity. Although (10) and (12) have the same structure it is clear that $\partial_{[\mu}A_{\nu\rho]}$ cannot represent the auxiliary tensor field $t_{\mu\nu\rho}$; its divergence is not identically zero, but it is part of the field equation for the tensor gauge field $A_{\mu\nu}$ of D=10 supergravity.

A solution for this problem has been presented in ref. 15; namely, one may introduce $A_{\mu\nu}$ as a Lagrange multiplier in order to ensure that its divergence vanishes. In that case the linearized lagrangian of Maxwell-Einstein supergravity contains the terms

$$L = 3/4\ t_{\mu\nu\rho}{}^2 + 3/2\ A_{\mu\nu}(\partial_\rho t^{\mu\nu\rho}) + 3/4\ \sqrt{2}\ t^{\mu\nu\rho}(A_{[\mu}F_{\nu\rho]}) \ , \tag{13}$$

which lead to the field equations

$$\begin{aligned} &\partial_\mu t^{\mu\nu\rho} = 0 \ , \\ &t_{\mu\nu\rho} = \partial_{[\mu}A_{\nu\rho]} - \tfrac{1}{2}\sqrt{2}\ A_{[\mu}F_{\nu\rho]} \ . \end{aligned} \tag{14}$$

This solution is somewhat ad hoc, but it makes the results (10)-(12) compatible. However, a more important aspect of introducing the auxiliary field $t_{\mu\nu\rho}$ is that it allows us to write the D=10 supersymmetric Maxwell theory and its coupling to supergravity in a locally scale invariant way. Namely we have

$$L = L_{SG} + L_{Maxwell} \,, \tag{15}$$

with

$$L_{SG} = -\tfrac{1}{2}\, e\, R - \tfrac{1}{2}\, e\, \bar{\psi}_\mu \Gamma^{\mu\nu\rho} D_\nu \psi_\rho - \tfrac{1}{2}\, e\, \bar{\lambda} \not{D} \lambda - 9/16\, e\, (\partial_\mu \phi/\phi)^2$$
$$+ 3/4\, e\, t_{\mu\nu\rho}{}^2 - 3/2\, e\, \phi^{-3/4}\, t^{\mu\nu\rho}\, \partial_\mu A_{\nu\rho} + \ldots \,, \tag{16}$$

$$L_{Maxwell} = -\tfrac{1}{4}\, e\, \phi^{-3/4}\, F_{\mu\nu}{}^2 - \tfrac{1}{2}\, e\, \bar{\chi} \not{D} \chi$$
$$+ \tfrac{1}{4}\sqrt{2}\, e\, t^{\mu\nu\rho} \left(3\phi^{-3/4}\, A_\mu F_{\nu\rho} + \tfrac{1}{4}\, \bar{\chi} \Gamma_{\mu\nu\rho} \chi\right)$$
$$- \tfrac{1}{4}\, e\, \phi^{-3/8}\, \bar{\chi} \Gamma^\mu \Gamma^{\rho\sigma} F_{\rho\sigma} (\psi_\mu + 1/12\, \sqrt{2}\, \Gamma_\mu \lambda) + \ldots \,, \tag{17}$$

where we have suppressed some of the higher-order terms which can be found in ref. 15. The most conspicuous feature of (15)-(17) is the way in which the scalar ϕ occurs in the coupling to the Maxwell theory. Indeed ϕ plays a crucial role in achieving the scale invariant coupling; when we assign the canonical scaling weights

$$w(e_\mu{}^a) = -1, \quad w(\psi_\mu) = -\tfrac{1}{2}, \quad w(t_{\mu\nu\rho}) = -2,$$
$$w(A_\mu) = 0 \,, \quad w(\chi) = 9/2, \tag{18}$$

then the lagrangian (17) becomes <u>locally</u> scale invariant provided that we choose

$$w(\phi) = -8, \quad w(\lambda) = \tfrac{1}{2}. \tag{19}$$

The Weyl weights for $e_\mu{}^a$, ψ_μ and $t_{\mu\nu\rho}$ follow directly from the way in which these fields transform into one another under supersymmetry. The weight of χ is the canonical weight for a spinor in ten dimensions, for which the properly covariantized Dirac action is locally scale invariant. As usual, the photon field A_μ must be inert under local scale transformations, since electromagnetic gauge transformations and local scale transformations commute[2]. For the same reason the tensor gauge field $A_{\mu\nu}$ must also be inert under scale transformations. It is then clear why the tensor field $t_{\mu\nu\rho}$ is crucial to achieve a locally scale invariant coupling, since the field equation for $t_{\mu\nu\rho}$

$$t_{\mu\nu\rho} = \phi^{-3/4}\, \partial_{[\mu} A_{\nu\rho]} + \ldots \,, \tag{20}$$

is not compatible with the Weyl weight assignments. Of course, this is a direct consequence of the lack of scale invariance of the ten-dimensional supergravity lagrangian (16), which is also the reason why the field ϕ cannot be eliminated from the combined lagrangian (15) by means of a local scale transformation. Although the presence of ϕ is thus important to compensate for the lack of scale invariance of the generic terms in (17), it cannot be viewed as a compensating field in the strict sense.

In view of the scale invariance of the coupling (17) it should not come as a surprise that the full supersymmetry transformations of the Maxwell fields are compatible with the Weyl weight assignments (18) and (19). We have found

$$\begin{aligned} \delta A_\mu &= \tfrac{1}{2}\,\bar{\varepsilon}\Gamma_\mu\chi\,\phi^{3/8} \quad , \\ \delta\chi &= -\tfrac{1}{4}\,\Gamma.F\varepsilon\,\phi^{-3/8} \quad . \end{aligned} \tag{21}$$

The above results are somewhat surprising because the ordinary Maxwell theory is not locally scale invariant in ten dimensions. Supergravity evidently improves the situation by introducing a scalar field ϕ into the gravitational sector, whose coupling to the supersymmetric Maxwell theory reestablishes dilatational invariance.

It is known[18] that the ten-dimensional multiplet of currents leads to 5760+5760 degrees of freedom, which will constitute only a subset of a possible off-shell formulation of ten-dimensional Poincaré supergravity. The higher degree of irreducibility is indicative of a higher gauge symmetry, as we have been emphasizing in this talk. The result found above shows that conformal invariances must be part of this gauge symmetry. Therefore, there is hope that superconformal techniques can also be used succesfully to clarify the structure of higher-dimensional supergravity theories, and thus of higher-extended supergravity in four dimensions.

4. IMPROVED TENSOR GAUGE FIELDS

We now return to four dimensions where we consider an antisymmetric tensor gauge field $E_{\mu\nu}$, subject to gauge transformations

$$E_{\mu\nu} \to E'_{\mu\nu} = E_{\mu\nu} + \partial_{[\mu}\xi_{\nu]} \quad . \tag{22}$$

Its invariant field strength in curved space is

$$E^\mu = \tfrac{1}{2}i\; e^{-1}\;\varepsilon^{\mu\nu\rho\sigma}\partial_\nu E_{\rho\sigma} \quad . \tag{23}$$

It is well-known that tensor gauge fields are difficult to treat within the context of conformal symmetries, because the obvious candidate for a lagrangian

$$L = \tfrac{1}{2}\,e\,(E^\mu)^2 \quad , \tag{24}$$

is not locally scale invariant in four dimensions. The reason is that $E_{\mu\nu}$ must be inert under dilatations, as we have already pointed out in the previous section, to avoid a conflict between the gauge transformations (22) and local dilatations. Therefore, only the six-dimensional analogue of (24) is locally scale invariant

$$L_{D=6} = - 3/2 \; e \; (\partial_{[\mu} E_{\nu\rho]})^2 \; . \tag{25}$$

We should stress that this problem only arises for tensor gauge fields; for antisymmetric tensors without gauge invariance one can find locally scale invariant actions for arbitrary dimensions, provided the Weyl weight is $w = \frac{1}{2}(D-2)$. The corresponding lagrangian for such a tensor T_{ab} is

$$L = - \tfrac{1}{4} \; e \; T^{ab} \; D^c D_c \; T_{ab} + 2 \; (D-2)^{-1} \; e \; T^{ab} \; D_a D^c \; T_{cb} \; , \tag{26}$$

where D_a denotes the conformally covariant derivative[2].

There is a clear analogy with the Maxwell theory where we have only a scale invariant lagrangian in four dimensions. However, there is a simple way to improve such a situation by multiplying the gauge field lagrangian by a scalar field taken to some power, in such a way that the resulting lagrangian leads to a scale invariant action. In general this procedure is arbitrary, but in the context of supersymmetry and supergravity scalar fields are an intrinsic part of the theory, so that the compensation for the lack of scale invariance may occur in a natural way. Indeed, this is precisely what happens in the coupling of the supersymmetric Maxwell theory to supergravity in ten dimensions, as we have exhibited in section 3.

In supersymmetry the tensor gauge field occurs as part of a so-called tensor multiplet, which for N=1 supersymmetry contains a scalar L, a spinor φ and a tensor gauge field $E_{\mu\nu}$ [19]. From the fixed Weyl weight (w=0) of $E_{\mu\nu}$ it follows from the supersymmetry transformations that L has w=2 and φ has w=5/2. The field strength (23), L and φ are contained in a superfield $G(x,\Theta,\bar{\Theta})$, which is a so-called linear multiplet, i.e.

$$D^2 G = \bar{D}^2 G = 0 \; . \tag{27}$$

Its lowest Θ-component is the field L, whereas φ and E^μ occur in higher orders of Θ. The standard lagrangian for the tensor multiplet is of the form (modulo total derivatives)

$$L \propto \int d^4\Theta \; G^2 = - \tfrac{1}{2} \left\{ (\partial_\mu L)^2 + \bar{\varphi} \partial\!\!\!/ \varphi - (E^\mu) \right\} \; . \tag{28}$$

This is the supersymmetric extension of (24), and because of the Weyl weight assignments, the corresponding action is not locally scale invariant.

Although (28) is not scale invariant the tensor multiplet itself is a genuine superconformal multiplet, i.e. it can be coupled to the fields of conformal supergravity and constitutes a representation of the superconformal algebra. It is therefore a meaningful question whether an alternative action can be constructed for the tensor multiplet which is conformally invariant. This turns out to be the case, as was shown recently in collaboration with M. Roček[20]. In superspace this superconformal action takes the form

$$S \propto \int d^4x\, d^4\Theta \quad G\, \ell n[G/\phi\bar{\phi}] \ , \tag{29}$$

where ϕ is an arbitrary chiral superfield with w=1, and $\bar{\phi}$ its complex conjugate, which has been introduced to make the argument of the logarithm in (29) dimensionless. Note that this modification does not affect the action, since

$$\int d^4x\, d^4\Theta \quad G\, \ell n\, \phi = \int d^4x\, D^2\bar{D}^2(G\, \ell n\, \phi) = 0 \ , \tag{30}$$

because of (27) and the defining condition for a chiral superfield: $\bar{D}_\alpha\phi = 0$. The presence of the chiral field has further interesting aspects, which have been discussed in ref. 20. To verify the invariance of (29) under constant scale transformations is straightforward: the combination $d^4\Theta\, G$ is already scale invariant, whereas the scaling of G in the logarithm vanishes according to (30), since a constant is a special case of a chiral superfield.

The action (29) can be expressed in components, and we find

$$S \propto \int d^4x\, \{-\tfrac{1}{2}\, L^{-1}(\partial_\mu L)^2 - \tfrac{1}{2}\, L^{-1}\bar{\varphi}\not{\partial}\varphi + \tfrac{1}{2}\, L^{-1}(E^\mu)^2 + \tfrac{1}{4}i\, L^{-2}\, \bar{\varphi}\not{E}\gamma_5\varphi - 1/32\, L^{-3}(\bar{\varphi}\gamma_a\gamma_5\varphi)^2\} \ . \tag{31}$$

Indeed, appropriate powers of the scalar field L compensate for the lack of scale invariance of the generic terms in this action. Furthermore supersymmetry implies the presence of several interaction terms. Nevertheless, in spite of these terms, the action (31) is equivalent to the scale invariant action of a free massless chiral multiplet, as has been shown recently in ref. 21. Hence, (31) describes free massless spin-0 and spin-½ particles, precisely as its alternative form without local scale invariance (28). This result may be verified by explicit S-matrix calculations in perturbation theory (expanding L about some constant value).

The coupling of (31) to conformal supergravity starts from the observation that there exists another conformally invariant action

$$S \propto \int d^4x\, d^4\Theta\, GV \ , \tag{32}$$

where $V(x,\Theta,\bar{\Theta})$ is a real superfield subject to gauge transformations

$$V \to V' = V + i(\bar{\Lambda} - \Lambda) \ . \tag{33}$$

In Wess-Zumino gauge V is described in terms of a gauge field V_μ, a Majorana spinor Ψ, and an auxiliary field D. Comparison of (29) to (32) suggests that $\ell n[G/\phi\bar{\phi}]$ transforms as a vector multiplet. We have determined the explicit relation between the components in Wess-Zumino gauge, in the coupling to conformal supergravity. The correspondence is

$$
\begin{aligned}
V_\mu &= \hat{E}_\mu L^{-1} + \tfrac{1}{4}i\,\bar{\varphi}\gamma_\mu\gamma_5\varphi L^{-2} - \tfrac{1}{2}\,\bar{\psi}_\mu\varphi L^{-1} + A_\mu, \\
\Psi &= \not{D}\varphi L^{-1} - \tfrac{1}{2}\,(\not{D}L + i\hat{\not{E}}\gamma_5)\varphi L^{-2} + 1/8\,(\bar{\varphi}\gamma_a\gamma_5\varphi)\gamma_a\gamma_5\varphi L^{-3}, \\
D &= 2\,(\Box\sqrt{L})L^{-\frac{1}{2}} + \tfrac{1}{2}\,(\bar{\varphi}\not{D}\varphi - \hat{E}_\mu{}^2)L^{-2} - \tfrac{1}{2}i\,\bar{\varphi}\,\hat{\not{E}}\gamma_5\varphi L^{-3} \\
&\quad + 3/32\,(\bar{\varphi}\gamma_a\gamma_5\varphi)^2\,L^{-4},
\end{aligned}
\tag{34}
$$

where the vierbein $e_\mu{}^a$, the gravitino ψ_μ and the chiral U(1) gauge field A_μ are the fields of N=1 conformal supergravity; $\hat{E}^\mu$ is the conformally covariant generalization of the tensor field strength (23)

$$
\hat{E}^\mu = \tfrac{1}{2}i\;e^{-1}\,\varepsilon^{\mu\nu\rho\sigma}(\partial_\nu E_{\rho\sigma} - i\bar{\psi}_\nu\sigma_{\rho\sigma}\gamma_5\varphi - \tfrac{1}{2}\,L\bar{\psi}_\nu\gamma_\rho\psi_\sigma)\;. \tag{35}
$$

We now introduce the density formula corresponding to (32) for the superconformal coupling of a vector and a tensor multiplet[7]:

$$
L = e(D + \tfrac{1}{2}i\;\bar{\Psi}\gamma_5\gamma\cdot\psi)L - e\bar{\Psi}\varphi - eV_\mu E^\mu\;. \tag{36}
$$

Substitution of (34) into (36) then leads directly to the action (31) coupled to conformal supergravity.

One of the immediate consequences of this result is that it opens the way for using a tensor multiplet as a compensator for N=1 supergravity. The field representation thus consists of the fields of conformal supergravity combined with those of the tensor multiplet. To show that these 12+12 degrees of freedom are gauge equivalent to those of Poincaré supergravity, we observe that the gauge conditions

$$
\begin{aligned}
b_\mu &= 0\;, \\
L &= 1\;, \\
\varphi &= 0\;,
\end{aligned}
\tag{37}
$$

break the invariance under special conformal boosts (K), dilatations (D) and S supersymmetry. Hence we obtain a supergravity theory based on the vierbein field $e_\mu{}^a$, the gravitino field ψ_μ, the axial vector gauge field A_μ, and an antisymmetric tensor $E_{\mu\nu}$. The most conspicuous feature of this representation is that the chiral U(1) invariance of conformal supergravity is left intact. The reason is that the tensor multiplet does not contain a compensating field for this symmetry. The conformally invariant action for the tensor multiplet that we have constructed previously can now be used to find a Poincaré supergravity action. Imposing the gauge conditions (37) on the improved tensor multiplet action leads to

$$L = -\tfrac{1}{2}\, eR(e,\omega) - \tfrac{1}{2}\, \varepsilon^{\mu\nu\rho\sigma}\, \bar{\psi}_\mu \gamma_5 \gamma_\nu D_\rho(\omega)\psi_\sigma$$
$$- 3/2\, eA_\mu \hat{E}^\mu - \tfrac{3}{4}\, e(\hat{E}^\mu)^2 \;, \qquad (38)$$

Hence we have derived an alternative minimal version of N=1 Poincaré supergravity, which modulo a field redefinition coincides with the recently proposed formulation of Sohnius and West[22] (for related work, see also ref. 23) The chiral U(1) invariance of that theory is thus traced back to the fact that the tensor multiplet does not provide for a chiral compensator; the characteristic tensor-vector gauge invariant coupling $\varepsilon^{\mu\nu\rho\sigma} A_\mu \partial_\nu E_{\rho\sigma}$ follows directly from inserting the representation (34) for V_μ into the $V_\mu E^\mu$ term in the superconformal density formula (36). Also the transformation rules can be understood. The spinor φ of the tensor multiplet transforms under Q and S supersymmetry as

$$\delta\varphi = -(i\gamma_5 \not{D} L + \hat{\not{E}})\varepsilon - 2iL\gamma_5 \eta \;, \qquad (39)$$

where D_a is the superconformal derivative, and the spinors ε and η are the parameters of Q and S supersymmetry. To preserve the gauge condition $\varphi = 0$ we must consider only a specific linear combination of Q and S supersymmetry, chosen such that (39) vanishes subject to the conditions (37). This leads to a uniform decomposition rule

$$\delta_{\text{Poincaré}}(\varepsilon) = \delta_Q(\varepsilon) + \delta_S(\eta = \tfrac{1}{2} i\gamma_5 \hat{\not{E}}\varepsilon) \;. \qquad (40)$$

Hence to obtain the transformations of Poincaré supersymmetry one must uniformly add an S supersymmetry transformation with a field-dependent parameter specified by (40). Therefore the fields ψ_μ and A_μ acquire the following supersymmetry transformations:

$$\begin{aligned} \delta\psi_\mu &= 2(\partial_\mu - \omega_\mu^{ab}\sigma_{ab} + \tfrac{3}{4} i\gamma_5 A_\mu)\varepsilon - \tfrac{1}{2} i\gamma_\mu\gamma_5 \hat{\not{E}}\varepsilon, \\ \delta A_\mu &= i\bar{\varepsilon}\gamma_5\phi_\mu - i\bar{\psi}_\mu(\tfrac{1}{2} i\gamma_5 \hat{\not{E}}\varepsilon) \;, \end{aligned} \qquad (41)$$

where ϕ_μ is the gauge field of S supersymmetry which is subject to a constraint (cf. Table 1):

$$\phi_\mu = (\sigma^{\rho\sigma}\gamma_\mu - 1/3\, \gamma_\mu \sigma^{\rho\sigma})\,(\partial_\rho - \omega_\rho^{ab}\sigma_{ab} + \tfrac{3}{4} i\gamma_5 A_\rho)\psi_\sigma \;. \qquad (42)$$

The remaining fields $e_\mu{}^a$ and $E_{\mu\nu}$ are not affected by (42), since they are inert under S supersymmetry. The second term in the variation of A_μ is precisely a supercovariantization of ϕ_μ corresponding to the extra term present in the variation of ψ_μ. Both transformations can be compared to those of ref. 22.

We recall that there exist alternative versions of Poincaré supergravity, whose conformal decomposition has been briefly discussed in section 2. Classically these different formulations are entirely equivalent, but one may wonder whether this equivalence is preserved at the quantum level. This is clearly not the case because of the chiral anomaly[24], which is the same for both minimal formulations of N=1 Poincaré supergravity[20]. In the standard minimal formulation[11] this anomaly is harmless, since the chiral invariance has been eliminated by a compensating field. However, in the second minimal formulation the anomaly couples to the chiral U(1) gauge field, which leads to a violation of chiral symmetry, and thereby of supersymmetry[22], at the quantum level.

This shows the importance of exploring off-shell formulations for supergravity theories. Within the superconformal framework different variants of Poincaré (and de Sitter) supergravity are recognized as the result of different sets of compensating multiplets fused with conformal supergravity. Also their coupling to supersymmetric matter can be described in this context. The interesting aspect is not so much that many different variants of supergravity exist, but that quantum-mechanical considerations may ultimately dictate which formulation should be adopted. Clearly the subtle relation between off-shell formulations and quantum-mechanical consistency deserves further study.

REFERENCES

1. E.C.G. Stueckelberg, Helv. Phys. Acta 11:225 (1938).
2. For a review, see B. de Wit, in proc. of the 18th Winter School of Theoretical Physics, Karpacz, 1981, Gordon and Breach (to be published); in "Supergravity 1981", proc. of the Trieste Spring School in Supergravity, eds. S. Ferrara and J.G. Taylor, Cambridge Univ. Press (to be published).
3. E. Cremmer and B. Julia, Nucl. Phys. B159 (1979); J. Ellis, M.K. Gaillard, L. Maiani and B. Zumino, in "Unification of the Fundamental Particle Interactions", eds. S. Ferrara, J. Ellis and P. van Nieuwenhuizen, Plenum Press (1980).
4. For a recent example where extra symmetries play a crucial role in the construction of a lagrangian, see H. Nicolai and B. de Wit, this volume.
5. S. Ferrara, M. Kaku, P.K. Townsend and P. van Nieuwenhuizen, Nucl. Phys. B129:125 (1977).
6. M. Kaku, P.K. Townsend and P. van Nieuwenhuizen, Phys. Rev. D17:3179 (1978); P.K. Townsend and P. van Nieuwenhuizen, Phys. Rev. D19:3166 (1979).
7. B. de Wit, J.W. van Holten and A. Van Proeyen, Nucl. Phys. B167:186 (1980) (E: B172:543 (1980)); B184:77 (1981).
8. E. Bergshoeff, M. de Roo and B. de Wit, Nucl. Phys. B182:173 (1981).
9. M. Kaku and P.K. Townsend, Phys. Lett. 76B:54 (1978); A. Das, M. Kaku and P.K. Townsend, Phys. Rev. Lett. 40:1215 (1978).

10. W. Siegel and S.J. Gates, Nucl. Phys. B147:77 (1979).
11. S. Ferrara and P. van Nieuwenhuizen, Phys. Lett. 74B:333 (1978); K.S. Stelle and P.C. West, Phys. Lett. 74B:330 (1978).
12. P. Breitenlohner, Phys. Lett. 67B:49 (1977); Nucl. Phys. B124:500 (1977).
13. P. Breitenlohner and M.F. Sohnius, Nucl. Phys. B165:483 (1980); B178:151 (1981).
14. E.S. Fradkin and M.A. Vasiliev, Lett. Nuovo Cim. 25:79 (1979); Phys. Lett. 85B:47 (1979); B. de Wit and J.W. van Holten, Nucl. Phys. B155:530 (1979).
15. E. Bergshoeff, M. de Roo, B. de Wit and P. van Nieuwenhuizen, Nucl. Phys. B195:97 (1982).
16. A. Chamseddine, Nucl. Phys. B185:403 (1981); North Eastern Univ. preprint (1981).
17. P. Howe and U. Lindström, Phys. Lett. 103B:422 (1981).
18. E. Bergshoeff and M. de Roo, Leiden preprint (1982).
19. W. Siegel, Phys. Lett. 85B:333 (1979).
20. B. de Wit and M. Roček, preprint NIKHEF-H81-28 (1981).
21. S.J. Gates, M. Roček and W. Siegel, preprint CALT-68-868 (1981).
22. M.F. Sohnius and P.C. West, preprint ICTP 80-81/37 (1981).
23. S.P. Bedding and W. Lang, Max Planck preprint MPI-PAE/PTh 42/81 (198 M.F. Sohnius and P.C. West, talks at the London Nuffield Workshop (1981); P. Howe, K.S. Stelle and P.K. Townsend, CERN preprint TH.3179 (1981); K.S. Stelle, this volume.
24. S.M. Christensen and M.J. Duff, Phys. Lett. 76B:571 (1981); P. van Nieuwenhuizen, M.T. Grisaru, H. Römer and N.K. Nielsen, Nucl. Phys. B140:477 (1978).

UNITARY REALIZATIONS OF THE NON-COMPACT SYMMETRY GROUPS OF SUPERGRAVITY

M. Günaydin

CERN, Geneva, Switzerland

INTRODUCTION

In this talk I will be reporting on some work done in collaboration with Cihan Saclioḡlu on the oscillator-like unitary representations of the non-compact groups of extended supergravity theories[1,2] (ESGT). Our construction uses boson annihilation and creation operators whose transformation properties are the same as the vector fields in the corresponding supergravity theories.

In the first part of my talk I will briefly summarize how non-compact groups emerge in ESGTs[3,4] and motivate the study of their unitary representations following mainly the ideas of Ellis, Gaillard and Zumino[5,6]. After this motivation I will give the bosonic construction of the Lie algebras of these non-compact groups and the corresponding unitary representations. In the last part of my talk I will discuss the relevance of these representations to the attempts to extract a realistic grand unified theory (GUT) from N = 8 ESGT[5,6] and stress the point that with an additional assumption one may be able to obtain a realistic GUT based on SU(5) in N = 5 ESGT with a generation group U(1). I conclude with some comments on the infinite dimensional Lie superalgebras that contain the generators of supersymmetry and of the non-compact symmetry groups.

NON-COMPACT GROUPS OF EXTENDED SUPERGRAVITY THEORIES

In their important work on N = 8 ESGT, Cremmer and Julia have shown that in ESGTs for N = 5,6,8 the natural SO(N) invariance that can be extended to SU(N) via chiral dual transformations can be further enlarged to a non-compact invariance group on-shell[3]. The first non-compact invariance group of this type was found for the N = 4 theory by Cremmer, Ferrara and Scherk[7].

Under the action of the non-compact invariance group G the vector field strengths get transformed into their duals and together form a linear representation of G, whereas the scalar fields transform non-linearly as the coset space G/H where H is the maximal compact subgroup of G. The largest invariance group of these theories on-shell has the form $G_{global} \otimes H_{local}$ where the local invariance group H_{local} is isomorphic to (but not identical with) the maximal compact subgroup of G_{global}. The Fermi fields (s = 1/2 or s = 3/2) are all singlets under G_{global} and transform as some non-trivial linear representation of H_{local}. The graviton is a singlet of both G_{global} and H_{local}. The content of the fundamental fields that enter ESGTs (N = 4-8) in four dimensions and the corresponding groups G_{global} and H_{local} are listed below[4]:

Table 1

N	8	7	6	5	4
Spin 2	1	1	1	1	1
Spin 3/2	8	7+1	6	5	4
Spin 1	28	21+7	15+1	10	6
Spin 1/2	56	35+21	20+6	10+1	4
Spin 0	70	35+35	15+15	5+5	1+1
G_{global} Rank	$E_{7(+7)}$ 7	$E_{7(+7)}$ 7	SO*(12) 6	SU(5,1) 5	SU(4) × SU(1,1) 4
H_{local} Rank	SU(8) 7	SU(8) 7	U(6) 6	U(5) 5	U(4) 4

Cremmer and Julia constructed the N = 8 ESGT in four dimensions by dimensional reduction from the simple supergravity in eleven dimensions combined with duality transformations. The emergence of the exceptional group $E_{7(7)}$ as a symmetry of this theory is rather surprising. In fact N = 8 ESGTs in d space-time dimensions obtained from the eleven dimensional simple supergravity all have non-compact global invariance groups belonging to the E-series[8]. The Lie groups E_n are all finite dimensional for $n \leq 8$. The Lie algebra of E_9 is infinite dimensional and corresponds to a Kac-Moody extension of the Lie algebra of E_8[9]. The invariance groups of N = 8 ESGTs in various space-time dimensions d are listed below[8]:

Table 2

d	G_{global}	H_{local}
9	GL(2,R)	SO(2)
8	$E_{3(+3)} \equiv SL(3,R) \times SL(2,R)$	$SO(3) \times SO(2)$
7	$E_{4(+4)} \equiv SL(5,R)$	SO(5)
6	$E_{5(+5)} \equiv SO(5,5)$	$SO(5) \times SO(5)$
5	$E_{6(+6)}$	USp(8)
4	$E_{7(+7)}$	SU(8)
3	$E_{8(+8)}$	SO(16)

Note that the rank of H_{local} is in general not equal to the rank of G_{global}.

THE QUESTION OF BOUND STATES IN EXTENDED SUPERGRAVITY THEORIES

The non-compact global invariance groups in four dimensional ESGTs make their appearance only in those theories that contain fundamental scalar fields. These scalar fields sit in the coset space G/H and the "gauge fields" associated with H_{local} are all composites of the scalar fields as in the two dimensional CP^N models[10]. The Lagrangian for the scalar fields has the general form

$$\mathcal{L} = \mathrm{Tr}\,(V^{-1}D_\mu V)^2 \qquad (1)$$

where

$$D_\mu V \equiv \partial_\mu V - VA_\mu \quad \text{and} \quad A_\mu = V^{-1}\partial_\mu V$$

with V(x) denoting a matrix function of the scalar fields parametrizing the coset space G/H. The potential problem with ghosts due to the non-compactness of G is avoided by the gauging of its maximal compact subgroup[11].

Now as was pointed out by Gell-Mann the fundamental fields that enter the largest ESGT (N=8) in four dimensions do not have a rich enough structure to accommodate the basic fields of a realistic gauge theory of strong, weak and electromagnetic interactions[12]. Thus it was thought that some of the fields entering such a theory might have to be made composites of the fundamental fields of N = 8 ESGT in order to make contact with elementary particle physics[13,14]. The first important development in this direction came from the suggestion of Cremmer and Julia that the classically non-propagating composite gauge fields of $SU(8)_{local}$ in N = 8 ESGT may become dynamical on the quantum level[3]. Their suggestion was motivated by the analogy with the two-dimensional CP^N models whose study at large N limit shows that the composite gauge field develops a pole at $p^2 = 0$ in its propagator and hence becomes dynamical on the quantum level[10]. Nissimov and Pacheva have extended this analysis to the three-dimensional generalized non-linear sigma models and shown that in the large N limit these theories have a phase in which the composite gauge fields develop poles at $p^2 = 0$ in their propagators and become propagating[15,16].

The major attempt at making contact with the elementary particle physics came from Ellis, Gaillard and Zumino (originally together with Maiani) who postulated that in N = 8 ESGT in addition to massless gauge fields other massless bound states (fermionic and bosonic) form whose low energy effective theory is a grand unified theory (GUT) based on SU(5) with three generations of quarks and leptons[5,6]. All the "elementary" particles entering this GUT (quarks, leptons, gauge fields, Higgs fields) are all composites of the fundamental "preon" fields of N = 8 ESGT with a binding scale of order m_{Planck}. The only truly elementary particle at low energies is the graviton which is a singlet of the $E_{7(7)global} \otimes SU(8)_{local}$ symmetry of the theory.

Ellis, Gaillard and Zumino chose the supercurrent multiplet of bound states from which to construct a realistic GUT. In this multiplet the vector bound states transform like the adjoint representation of $SU(8)_{local}$. In addition to the particles needed for a realistic GUT this supercurrent multiplet contains many unwanted helicity states. These unwanted helicity states cannot be made supermassive in an SU(5) or SU(3) × SU(2) × U(1) invariant fashion without introducing an infinite set of supermultiplets of bound states[6,17,18]. In fact, an infinite set of supermultiplets seems to be necessary for a vector-like embedding of $SU(3)_{colour}$ in these theories[17-19]. Both of these are semi-phenomenological motivations for looking for an infinite set of bound states in the N = 8 ESGT.

On the other hand there are strong theoretical arguments indicating that in ESGTs with non-compact invariance groups there may be an infinite set of bound states formed. The strongest of these arguments is the analogy with two- and three-dimensional generalized σ models. We have already mentioned that in the two-dimensional generalized CP^N models with $G_{global} \times H_{local}$ invariance the composite gauge fields associated with H_{local} become dynamical on the quantum level. Haber et al. have shown that, in the large N limit, these gauge fields get confined and act as the binding force between the scalar fields, and the spectrum of theory consists of scalar bound states that transform like a linear representation of the parent global symmetry group G_{global}[20]. Similarly, Nissimov and Pacheva have shown that, in the three-dimensional supersymmetric generalized σ-models, the phase in which the composite gauge fields become dynamical massless "gluons" on the quantum level has a bound state spectrum consisting of equal mass s = 0 bosons and s = 1/2 fermions transforming like a linear representation of the global invariance group G_{global}[16].

Now, as was first mentioned by Zumino[18] and later elaborated by Ellis, Gaillard and Zumino[6], if the N = 8 supergravity theory has a phase in which the composite gauge fields of $SU(8)_{local}$ become dynamical and act as the binding force between the other fields, in such a way that the bound state spectrum corresponds to a linear representation of $E_{7(7)}$, then one must have an infinite set of bound states since all the non-trivial unitary representations of non-compact groups are infinite dimensional.

Moreover, as was reported in this workshop[21], Grisaru and Schnitzer have shown that under certain reasonable assumptions the scattering amplitudes in extended supergravity theories seem to reggeize, indicating again the possibility of an infinite set of bound states[22].

Further clues in this direction come again from the study of two-dimensional models. Makhankov and Pashaev[23], in their study of the non-linear Schrödinger equation with a non-compact U(1,1) invariance, find that the spectrum of soliton solutions is far richer than the compact case and suggest that this be understood in the language of unitary realizations of the non-compact invariance group. Similar results were obtained by Rabinovici in the case of σ models with a non-compact global invariance group[24].

All these theoretical arguments provide sufficient motivation for a study of the unitary representations of the non-compact groups of ESGTs.

THE BOSONIC CONSTRUCTION OF THE LIE ALGEBRAS
OF THE NON-COMPACT GROUPS OF EXTENDED SUPERGRAVITY THEORIES

The construction of the Lie algebras of non-compact groups using boson annihilation and creation operators is well known in the physics literature[25]. Consider, for example, a set of boson annihilations and creation operators a_i and $a_i^\dagger$ (i=1,...,n) satisfying the commutation relations

$$[a_i,a_j^\dagger] \equiv [a_i,a^j] = \delta_i^j \qquad i,j = 1,\ldots,n$$
$$[a_i,a_j] = 0 = [a^i,a^j] \qquad a_i^\dagger \equiv a^i \tag{2}$$

where the creation operators are denoted with upper indices. Then the bilinear operators $a^i a_j \equiv a_i^\dagger a_j$ generate the Lie algebra of U(n) under commutation. The diboson annihilation and creation operators $a_i a_j$ and $a^i a^j$ close into the set $a^i a_j$ under commutation and all together they generate the Lie algebra of the non-compact group Sp(2n,ℝ). If one uses two sets of boson operators $a_i(a^i)$ and $b_i(b^i)$ $i = 1,\ldots,n$ then the generators $I_i^j = a^j a_i + b_i b^j$ of U(n) can be extended to the Lie algebra of SP(2n,ℝ) by using symmetric diboson annihilation and creation operators:

$$S_{ij} = a_i b_j + a_j b_i \;, \quad S^{ij} = a^i b^j + a^j b^i \tag{3}$$

or to the Lie algebra of SO(2n)* by using antisymmetric diboson annihilation and creation operators[1,2]:

$$A_{ij} = a_i b_j - a_j b_i \;, \quad A^{ij} = a^i b^j - a^j b^i \tag{4}$$

Note that in either extension the rank of the Lie algebra does not change. The Lie algebras of SU(m,n) and SO(m,n) can be similarly constructed[2].

In Ref. 1 this standard construction has been extended to the case when the boson operators transform like the antisymmetric tensor representation of SU(n). Under the restriction that the Lie algebra of the non-compact group have the same rank as its maximal compact subgroup [U(n) or SU(n) as the case may be] one finds that such an extension yields a finite dimensional Lie algebra only for $n \leq 8$. For $n < 4$ this extended construction corresponds to the standard construction outlined above. For n = 4, 5,...,8 one obtains exactly the Lie algebras of the non-compact symmetry groups of ESGTs in four dimensions for N = 4,5,...,8. Let us now summarize this construction using an arbitrary number of pairs of boson annihilation and creation operators as was done in Ref. 2. The use of more than one pair of boson operators corresponds to taking direct sums of copies of the resulting Lie algebra. However, this trivial extension on the Lie algebra level makes it possible to construct much larger classes of unitary irreducible representations as will be explained in the next section.

Consider the antisymmetric boson operators

$$a_{ij}(K) = -a_{ji}(K) \quad \text{and} \quad b_{ij}(K) = -b_{ji}(K)$$
$$i,j = 1,\ldots,n \qquad K = 1,\ldots,N$$

that satisfy the commutation relations

$$\begin{aligned} \left[a_{ij}(K), a^{k\ell}(L)\right] &= \delta^{KL}(\delta_i^k\delta_j^\ell - \delta_i^\ell\delta_j^k) = \left[b_{ij}(K), b^{k\ell}(L)\right] \\ \left[a_{ij}(K), a_{k\ell}(L)\right] &= 0 = \left[b_{ij}(K), b_{k\ell}(L)\right] \end{aligned} \tag{5}$$

where the creation operators are denoted by upper SU(n) indices. Then the bilinear operators

$$I^i_j \equiv \vec{a}^{im}\cdot\vec{a}_{jm} + \vec{b}_{jm}\cdot\vec{b}^{im} \equiv \sum_{K=1}^{N}\sum_{m=1}^{n} a^{im}(K)a_{jm}(K) + b_{jm}(K)b^{im}(K) \tag{6}$$

generate the Lie algebra of U(n). Separating out the trace part Q generating the U(1) subgroup we can write it as:

$$I^i_j = T^i_j + \frac{1}{n}\delta^i_j Q, \quad Q = \vec{a}^{k\ell}\cdot\vec{a}_{k\ell} + \vec{b}_{k\ell}\cdot\vec{b}^{k\ell} \tag{7}$$

where T^i_j generate the SU(n) Lie algebra.

<u>n = 4</u>: In this case the Lie algebra of U(4) can be extended to the Lie algebra of SU(4) × SU(1,1) by the diboson annihilation and creation operators Q^- and Q^+ defined as

$$Q^+ = \frac{1}{4}\varepsilon_{ijk\ell}\,\vec{a}^{ij}\cdot\vec{b}^{k\ell} \qquad Q^- = \frac{1}{4}\varepsilon^{ijk\ell}\,\vec{a}_{ij}\cdot\vec{b}_{k\ell} \tag{8}$$

$$\begin{aligned} \left[Q, Q^{\mp}\right] &= \mp Q^{\mp} \\ \left[Q^-, Q^+\right] &= 2Q \end{aligned} \tag{9}$$

n = 5: Again using the SU(5) invariant totally antisymmetric ε tensor we can define diboson annihilation and creation operators

$$A_i = \frac{\sqrt{2}}{4} \varepsilon_{ijk\ell m} \vec{a}^{jk} \cdot \vec{b}^{\ell m}$$
$$A^i = \frac{\sqrt{2}}{4} \varepsilon^{ijk\ell m} \vec{a}_{jk} \cdot \vec{b}_{\ell m} \tag{10}$$

which close into the I^i_j and together form the Lie algebra of SU(5,1):

$$[A_i, A^j] = T^j_i - \frac{3}{10} \delta^j_i Q$$
$$[T^i_j, A_k] = -\delta^i_k A_j + \frac{1}{5} \delta^i_j A_k \tag{11}$$
$$[Q, A_i] = 4A_i$$

n = 6: In this case the diboson operators constructed using the ε tensor do not close into the U(6) subalgebra. To close the algebra one has to introduce pairs of boson annihilation and creation operators $\vec{v}(\vec{v}^\dagger)$ and $\vec{w}(\vec{w}^\dagger)$ that are singlet under SU(6). Then the diboson operators

$$A_{ij} = \frac{1}{4} \varepsilon_{ijk\ell mp} \vec{a}^{k\ell} \cdot \vec{b}^{mp} + \frac{1}{\sqrt{2}}(\vec{a}_{ij} \cdot \vec{v} + \vec{b}_{ij} \cdot \vec{w})$$
$$A^{ij} = \frac{1}{4} \varepsilon^{ijk\ell mp} \vec{a}_{k\ell} \cdot \vec{b}_{mp} + \frac{1}{\sqrt{2}}(\vec{a}^{ij} \cdot \vec{v}^\dagger + \vec{b}^{ij} \cdot \vec{w}^\dagger) \tag{12}$$

together with SU(6) and U(1) generators

$$T^i_j = \vec{a}^{im} \cdot \vec{a}_{jm} + \vec{b}^{im} \cdot \vec{b}_{jm} - \frac{1}{6} \delta^i_j (\vec{a}^{k\ell} \cdot \vec{a}_{k\ell} + \vec{b}^{k\ell} \cdot \vec{b}_{k\ell})$$
$$Q = \vec{a}^{k\ell} \cdot \vec{a}_{k\ell} + \vec{b}_{k\ell} \cdot \vec{b}^{k\ell} - 6\, \vec{v}^\dagger \cdot \vec{v} \quad v - 6\, \vec{w} \cdot \vec{w}^\dagger \tag{13}$$

form the Lie algebra of SO(12)*:

$$[A_{ij}, A^{k\ell}] = \frac{1}{2}(\delta^k_i T^\ell_j + \delta^\ell_j T^k_i - \delta^\ell_i T^k_j - \delta^k_j T^\ell_i) - \frac{1}{12}(\delta^k_i \delta^\ell_j - \delta^k_j \delta^\ell_i) Q$$
$$[T^i_j, A_{k\ell}] = -\delta^i_k A_{j\ell} - \delta^i_\ell A_{kj} + \frac{1}{3} \delta^i_j A_{k\ell} \tag{14}$$
$$[Q, A^{k\ell}] = 4A^{k\ell}$$

n = 7: In this case as well the diboson operators constructed from the antisymmetric tensor boson operators do not close into the U(7) generators. To close the algebra one needs to introduce pairs of boson operators $\vec{v}_i(\vec{v}^i)$, $\vec{w}_i(\vec{w}^i)$ transforming like the fundamental representation of SU(7). Doing this however makes the construction in this case equivalent to the case for n = 8. Referring the reader for details to Refs. 1 and 2 we now give the n = 8 construction.

n = 8: In this case the diboson annihilation and creation operators constructed using the ε tensor transform like the totally antisymmetric tensor representation of rank four under SU(8). Since this corresponds to a real representation $\underline{70}$ of SU(8) one has to take "self-conjugate" combinations of the diboson operators. Thus one finds that the diboson operators

$$V_{ijk\ell} = \vec{a}_{[ij}\cdot\vec{b}_{k\ell]} + \frac{1}{4}\varepsilon_{ijk\ell rstu}\vec{a}^{rs}\cdot\vec{b}^{tu} \tag{15}$$

where the bracket $[ijk\ell]$ denotes antisymmetrization of all four indices, close into the Lie algebra of SU(8):

$$\begin{aligned} [V_{ijk\ell}, V_{mnpq}] &= -\frac{1}{144}\varepsilon_{ijk\ell rstu}\varepsilon^{vstuabcd}\varepsilon_{abcdmnpq}T^r_v \\ [T^i_j, V_{mnpq}] &= \frac{1}{6}\varepsilon_{mnpqjrst}V^{irst} - \frac{1}{2}\delta^i_j\, V_{mnpq} \end{aligned} \tag{16}$$

and together form the Lie algebra of $E_{7(7)}$. The self-conjugacy of the 70 operators $V_{ijk\ell}$ is reflected in the identity

$$V^{\dagger}_{ijk\ell} \equiv V^{ijk\ell} = \frac{1}{4!}\,\varepsilon^{ijk\ell rstu}V_{rstu}$$

To summarize, we have the following representation content for the boson operators for various n and the resulting Lie algebras L:

				L
n = 4:	$a_{ij}(b_{ij})$	$\underline{6}$	of SU(4) →	SU(4) × SU(1,1)
n = 5:	$a_{ij}(b_{ij})$	$\underline{10}$	of SU(5) →	SU(5,1)
n = 6:	$a_{ij}(b_{ij}) \oplus v(w)$	$\underline{15}+\underline{1}$	of SU(6) →	SO(12)*
n = 7:	$a_{ij}(b_{ij}) \oplus v_i(w_i)$	$\underline{21}+\underline{7}$	of SU(7) →	$E_{7(7)}$
n = 8:	$a_{ij}(b_{ij})$	$\underline{28}$	of SU(8) →	$E_{7(7)}$

The boson operators have exactly the same representation content as the vector fields in extended supergravity theories and the Lie algebras (L) are those of the non-compact groups (G) of the respective theories as well. Moreover, under the action of the non-compact group generated by L, the boson operators $a_{ij}(b_{ij})$ get mixed with $b^{ij}(a^{ij})$ and together form a linear representation of G. The vector field strengths in ESGTs and their duals transform into each other under the action of the non-compact invariance group and form a linear representation in exactly the same fashion.

OSCILLATOR-LIKE UNITARY REPRESENTATIONS OF NON-COMPACT GROUPS OF SUPERGRAVITY

With the exception of $E_{7(7)}$ all the Lie algebras constructed above have a Jordan structure with respect to their subalgebras generating a maximal compact subgroup, i.e., they decompose as a vector space direct sum

$$L = L^- \oplus L^0 \oplus L^+ \qquad (17)$$

where

$$(L^+)^\dagger \cong L^- \qquad (L^0)^\dagger \approx L^0$$

and L^0 contains the generator Q of a U(1) factor group which gives the grading, i.e.,

$$\begin{aligned} [Q,L^+] &= L^+ \\ [Q,L^-] &= -L^- \\ [Q,L^0] &= 0 \end{aligned} \qquad (18)$$

The simple non-compact Lie groups which have a Jordan structure with respect to their maximal compact subgroups have been classified[26]. They are the groups SU(m,n), SO(n,2), Sp(2n,ℝ), SO(2n)* $E_{6(-14)}$ and $E_{7(-25)}$, where $E_{6(-14)}$ and $E_{7(-25)}$ have maximal compact subgroups SO(10) × SO(2) and $E_6 \otimes SO(2)$, respectively. The Lie algebras of all these groups can be constructed in terms of boson annihilation and creation operators and one can give a general construction of a class of unitary representations of these groups in the Fock space of the boson operators[1,2].

Consider the Fock space constructed from the tensor product of Fock spaces of individual boson operators. The vacuum state $|0\rangle$ will be a tensor product of the individual vacua $|0\rangle$

$$|0\rangle \equiv |0\rangle|0\rangle\ldots|0\rangle$$

which is annihilated by all the annihilation operators.

$$a_i(K)\ |0\rangle = 0 = b_i(K)\ |0\rangle \qquad K = 1,\ldots,N$$

Now choose a set of states $|\psi_A\rangle$ in our Fock space which is annihilated by all the diboson operators in the L^- space and transform as a certain representation of the maximal compact subgroup generated by L^0:

$$L^-|\psi_A\rangle = 0 \tag{19}$$

Then the infinite set of states generated by applying the operators L^+ on $|\psi_A\rangle$ form the basis of a unitary representation of the non-compact group G:

$$|\psi_A\rangle,\ \ L^+|\psi_A\rangle,\ \ L^+L^+|\psi_A\rangle,\ \ldots \tag{20}$$

If the states $|\psi_A\rangle$ are chosen such that they transform like an irreducible representation of the maximal compact subgroup then the corresponding representation of the non-compact group is also irreducible. The proof of this theorem is very simple and uses only the condition (19) and the Jordan structure of the Lie algebra L^1.

We should note that the Jordan structure does not guarantee the existence of states $|\psi_A\rangle$ annihilated by L^- space that transform irreducibly under the compact subgroup. For example, in the construction of the Lie algebra of SO(12)* in terms of antisymmetric tensor and singlet boson operators of SU(6) [see Eq. (12)], such states do not exist in the Fock space, whereas in the construction of SO(12)* in terms of boson operators transforming like the fundamental representation of SU(6) there exists an infinite set of such states. In cases when there does not exist any such set of states $|\psi_A\rangle$ annihilated by the L^- space the application of our method leads to reducible unitary representations.

To illustrate our construction we now give a detailed study of the corresponding representations of SU(5,1) using its Lie algebra given by Eq. (11). Now any of the states of the form

$$[a^{ij}(1)]^{n_1}[a^{k\ell}(2)]^{n_2}\ldots[a^{pq}(N)]^{n_N}|0\rangle$$

or of the form $\qquad i,j\ldots = 1,\ldots,5$

$$[b^{ij}(1)]^{n_1}[b^{k\ell}(2)]^{n_2}\ldots[b^{pq}(N)]^{n_N}|0\rangle$$

are annihilated by the operators $A^i = \varepsilon^{ijk\ell m}a_{jk}b_{\ell m}$ belonging to the L^- space. Any irreducible subset $|\psi_A\rangle$ of these states can be used as the starting representation of U(5) for the construction of a unitary irreducible representation (UIR) of SU(5,1).

Restricting ourselves to the case of only one pair of boson operators a and b and assuming that the annihilation operators transform as the representation $\underline{10}$ of SU(5) we find the following transformation properties of various states:

$$
\begin{aligned}
a^{jk}|0\rangle &\longrightarrow |(1,1,1,0,0)\rangle \\
a^{jk}a^{\ell m}|0\rangle &\longrightarrow |(2,1,1,1,1,)\rangle + |(2,2,2,0,0)\rangle \\
b^{jk}|0\rangle &\longrightarrow |(1,1,1,0,0)\rangle \\
&\cdots
\end{aligned} \tag{21}
$$

where $(m_1,\ldots,m_5)$ denotes a representation of U(5) whose Young Tableaux contain m_i boxes in the ith row. It is easy to see that any irreducible representation of SU(5) with a definite U(1) charge can be constructed by repeated application of a^{jk} followed by a suitable projection operator. Thus we can take any irreducible representation of SU(5) as the starting states for the construction of an irreducible representation of SU(5,1).

Starting from states $|\psi_A\rangle$ transforming like a representation (m_1,m_2,m_3,m_4,m_5) of U(5) that are annihilated by the operators A^i of the L^- space we can construct the states

$$(A_i)^k|\psi_A\rangle \qquad\qquad k = 1,2,\ldots$$

by repeated application of the operators A_i of the L^+ space. Under U(5) they transform like

$$
\begin{aligned}
(A_i)^k|\psi_A\rangle \overset{\approx}{\longrightarrow} (2k,k,k,k,k) \otimes (m_1,m_2,m_3,m_4,m_5) \\
k = 0,1,2,\ldots
\end{aligned} \tag{22}
$$

and form the basis of an irreducible unitary representation of SU(5,1).

In the case of the Lie algebra of $E_{7(7)}$ we do not have a Jordan structure with respect to the Lie algebra of its maximal compact subgroup SU(8). It decomposes as

$$L = T^i_j \oplus V_{ijk\ell} = 63 \oplus 70 \tag{23}$$

where T^i_j are the generators of SU(8) and $V_{ijk\ell}$ the non-compact generators. On a set of states $|\psi_A\rangle$ that are constructed by acting on the vacuum state with the creation operators a^{ij} $(i,j = 1,\ldots,8)$ and which transform irreducibly under SU(8) we can apply the non-compact generators to generate new states

$$|\psi_A\rangle, \quad V_{ijk\ell}|\psi_A\rangle, \quad VV|\psi\rangle, \ \ldots \tag{24}$$

Clearly these states form the basis of a unitary representation of $E_{7(7)}$. However, they are infinitely reducible. This can be seen as follows: the product of two copies of the V's contains an SU(8) singlet, i.e.,

$$V_{ijk\ell}V_{mnpq} = \frac{1}{2}\{V_{ijk\ell},V_{mnpq}\} + \frac{1}{2}[V_{ijk\ell},V_{mnpq}] = \\ = (1 + 720 + 1764)_{sym} + 63_{antisym} \quad (25)$$

This means that every irreducible representation of SU(8) that occurs in the infinite set of states (24) will reappear again after two applications of the V's. Thus the multiplicity of an SU(8) representation occurring in the set (24) is infinite. This implies that the unitary representation we have is infinitely reducible since the multiplicity of an irreducible representation of the maximal compact subgroup inside a UIR of a non-compact group is always less than or equal to its dimension[27].

There is a method due to Gell-Mann for constructing a class of UIRs of some non-compact groups on certain coset spaces of their maximal compact subgroups[28,29]. His method is applicable to $E_{7(7)}$ and is quite simple for determining the multiplicities of representations of SU(8). For example, one possible coset space on which to realize the UIRs of $E_{7(7)}$ is SU(8)/Sp(8). In this case the multiplicities of representations of SU(8) inside a UIR of $E_{7(7)}$ are determined by the number of Sp(8) singlets they contain. Unfortunately, this method cannot be applied to our construction simply because the boson operators we use transform linearly under SU(8) rather than non-linearly as a certain coset space of SU(8) satisfying Gell-Mann's criteria[29]. However, though reducible, our representations of $E_{7(7)}$ may still be of relevance for physical applications since, as explained in the next section, the compatibility of supersymmetry with non-compact invariance groups do in general imply reducible unitary representations.

GRAND UNIFIED THEORIES AND NON-COMPACT SYMMETRY GROUPS OF SUPERGRAVITY

In their attempt to extract an effective "low energy" GUT from the N = 8 ESGT, Ellis, Gaillard and Zumino have considered only the supercurrent multiplet of bound states from which to choose their "elementary" particles (quarks, leptons, etc.). The choice of this multiplet was dictated by the requirement that it contain the composite gauge fields of $SU(8)_{local}$. In addition to the "low energy" "elementary" particles this composite multiplet contains many unwanted helicity states. Since there seems to be no experimental evidence for the existence of these additional particles at low energies it was suggested that they be made superheavy at the order of Planck mass[30]. The same problem also arises in the scheme of Derendinger, Ferrara and Savoy who do not restrict themselves to the current supermultiplet alone[31]. To make these unwanted states massive in an SU(5) or even SU(3) × SU(2) × U(1) invariant fashion so as to be left with three families of chiral massless fermions, one needs to introduce an infinite set of composite supermultiplets[6,17]. The fact that one needs to introduce an infinite set of supermultiplets to get rid of the unwanted helicity states leads one to consider an alternative possibility: instead of taking both the gauge fields and the "matter fields" (quarks and leptons) for an effective low energy GUT from the same supermultiplet one takes the gauge fields from one supermultiplet and the matter fields from other composite supermultiplets. If such a scenario is adopted one need not go to the largest ESGT to obtain a realistic GUT as an effective low energy theory. Already the N = 5 ESGT with $SU(5,1)_{global} \times U(5)_{local}$ invariance may in principle be large enough to accommodate such a theory. If the gauge fields were taken from the supercurrent multiplet then in analogy with the CP^N models we would expect them to be singlets under the global invariance group G_{global}[32]. Thus if one goes outside the supercurrent multiplet, to be consistent one must then choose both the matter multiplets and the gauge fields from among the infinite set of bound states forming representations of G_{global}. Such a scenario however requires a resolution of an apparent conflict between supersymmetry and unitary realizations of G_{global}. As mentioned earlier the basic Fermi fields of ESGTs are all inert under G_{global}. Thus one would naïvely expect only the bosonic bound states with integer helicity to form unitary representations of G_{global}. But if supersymmetry is valid at any level then it would imply that these infinite towers of bosonic bound states have fermionic partners and together form infinite towers of supermultiplets. This in turn means that the fermionic bound states must also form unitary representations of the non-compact G_{global}. One may ask whether there is a super invariance group of these theories that contains the non-compact symmetry group as well as the supersymmetry transformations and transforms the bosonic and fermionic bound states into each other. The fact that there is no simple finite dimensional supergroup whose even subgroup contains $E_{7(7)}$ as a factor group makes one suspect that such a group may be infinite dimensional. In fact, when one takes the commutator of non-compact symmetry generators and supersymmetry generators one obtains new generalized supersymmetry generators[6]. Thus starting from the Lie algebra of the non-compact group and supersymmetry generators one can generate an infinite dimensional superalgebra with generalized momenta and generalized supercharges[6,33]. The existence of this infinite dimensional superalgebra does not automatically ensure the compatibility of unitary realization of G_{global} on the bound states and supersymmetry. One has to show that one can realize this infinite dimensional algebra unitarily[33]. For the class of UIRs given in the previous section this compatibility can be shown easily[33] as follows: in the infinite set of generalized supersymmetry generators thus obtained, there are some that commute with the non-compact generators belonging to the L^- space. These operators F_α carry half integer helicity $|\lambda| = 1/2$. Thus by applying on the initial set of states $|\psi_A\rangle$ annihilated by the L^- space with these generalized supergenerators F_α we can create half-integer helicity states annihilated by the L^- space[34]:

$$L^- F_\alpha |\psi_A\rangle = F_\alpha L^- |\psi_A\rangle = 0$$

By repeated application of the $F\alpha$'s we can create half-integer and integer spin helicity states that form a supermultiplet. All of these states are annihilated by the operators in the L^- space and hence can be used as the starting states for constructing oscillator-like unitary representations. Diagramatically we have

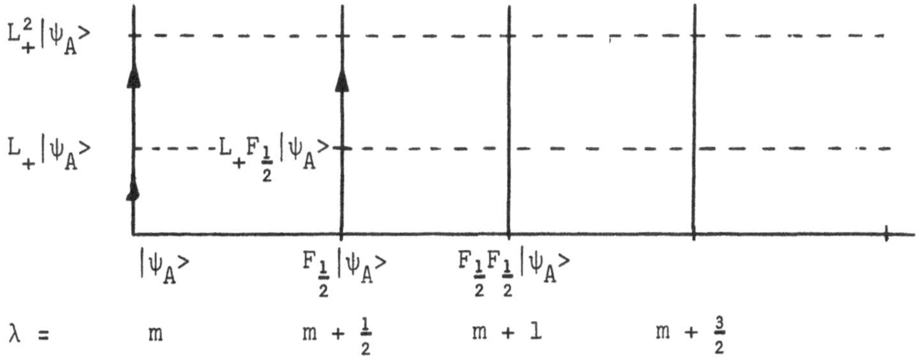

At each helicity λ we have an infinite tower of states forming a unitary representation of G_{global}. This shows that the unitary realizations of the non-compact symmetries of ESGTs on the bound states can be made compatible with supersymmetry. However, the resulting unitary representations will in general not be irreducible. This is because even if the initial set of states $|\psi_A\rangle$ transform irreducibly under the maximal compact subgroup H of G_{global}, the states $F_\alpha|\psi_A\rangle$, $F_\alpha F_\alpha|\psi_A\rangle$,... etc., are in general reducible and consequently the unitary representations of G_{global} obtained by repeated application of L^+ operators on these states will also be reducible. The infinite dimensional group that contains G_{global} and the supersymmetry transformations will transform all these states into each other. It would be interesting to know if one such unitary supermultiplet will be rich enough to accommodate all the elementary particles of a realistic GUT.

Before concluding let us remind ourselves once more that the crucial problem in the attempts to extract a realistic GUT from ESGTs is to show that there are indeed bound states formed that correspond to the "elementary" particles observed so far. If they are formed and fall into representations of $G_{global} \times H_{local}$ invariance of these theories one must then show that one can break these larger symmetries as well as the supersymmetries so as to be left with the observed low energy symmetries[6] and particles. One would also like to know what this breaking implies for the infinite dimensional superalgebra incorporating supersymmetry and the global invariance group. On the other hand as was reported in this meeting in the largest ESGT (N=8) one has the option of gauging the SO(8) invariance at the price of losing $E_{7(7)}$global invariance while still preserving supersymmetry[35,36]. If the bound states do indeed form representations of $E_{7(7)}$ what would happen to them when one turns on the SO(8) gauge couplings? Do they become unbound or do the non-singlets of SO(8) get confined? Will there be any trace of $E_{7(7)}$ left? Hopefully by the time of the next meeting we will have answers to some of these problems.

ACKNOWLEDGEMENTS

I would like to thank I. Bars, E. Cremmer, J. Ellis, S. Ferrara, M. Gell-Mann, F. Gürsey, S. MacDowell, H. Nicolai, C. Saçlioḡlu, J. Schwarz and B. Zumino for many enlightening discussions. Part of this work was done at Yale University, Seattle Summer Institute and Aspen Centre for Physics. I thank these institutions for their hospitality and the Deutsche Forschungsgemeinschaft for a travel grant.

REFERENCES

1. M. Günaydin and C. Saçlioḡlu, Bonn University preprint HE-81-11 (1981), to be published in Phys. Lett. B.
2. M. Günaydin and C. Saçlioḡlu, CERN preprint TH.3209 (1981).
3. E. Cremmer and B. Julia, Phys. Lett. 80B:48 (1978); Nucl. Phys. B159:141 (1979).
4. For a review of the symmetries of ESGTs see
   E. Cremmer, Lectures at "Spring School on Supergravity" at Trieste (1981) and Summer Institute in Theoretical Physics in Seattle (1981), Ecole Normale Supérieure preprint LPTENS 81/18 , Paris, (1981);
   B. Julia, Talk presented at the Johns Hopkins Workshop on Particle Theory (1981), Ecole Normale Supérieure preprint LPTENS 81/14 (1981).
5. J. Ellis, M.K. Gaillard, L. Maiani and B. Zumino, "Unification of the Fundamental Particle Interactions", ed. by S. Ferrara, J. Ellis and P. van Nieuwenhuizen, Plenum Press, N.Y. (1980);
   J. Ellis, M.K. Gaillard and B. Zumino, Phys. Lett. 94B:343 (1980).

6. J. Ellis, M.K. Gaillard and B. Zumino, LAPP preprint TH-44/ /CERN TH.3152 (1981) to appear in Acta Physica Polonica;
J. Ellis, "First Workshop on Grand Unification", ed. by P. Frampton, S.L. Glashow and A. Yildiz, Math. Sci. Press, Brookline, Mass., 1980, p. 287 and this proceedings;
M.K. Gaillard, Talk presented at the Heisenberg Symposium, München 1981, LBL preprint LBL-13371 (1981);
B. Zumino, Proc. 1980 Madison Int. Conf. on High Energy Physics, ed. by L. Durand and L.G. Pondrom, A.I.P., N.Y. (1981), p. 964.
7. E. Cremmer, S. Ferrara and J. Scherk, Phys. Lett. 74B:61 (1978).
8. E. Cremmer, in "Unification of the Fundamental Particle Interactions", ed. by S. Ferrara, J. Ellis and P. van Nieuwenhuizen Plenum Press, N.Y. (1980), pp 137-155;
J.H. Schwarz, Phys. Lett. 95B:219 (1980)
9. V. Kac, Math. U.S.S.R. Izvestiya Ser. 32:1271 (1968);
R. Moody, Bull. Am. Math. Soc. 73:217 (1967) and J. of Algebra 10:211 (1968).
10. A. D'Adda, P. Di Vecchia and M. Lüscher, Nucl. Phys. B146:63 (1978)
11. K. Cahill, Phys. Rev. D18:2930 (1978);
E. Cremmer and J. Scherk, Phys. Lett. 74B:341 (1978);
B. Julia and J.F. Luciani, Phys. Lett.90B:270 (1980).
12. M. Gell-Mann, Talk at the 1977 Washington Meeting of the American Physical Society, unpublished.
13. M. Gell-Mann, Talk at the Aspen Workshop on Octonionic Quantum Mechanics (1978), unpublished.
14. T. Curtright and P.G.O. Freund, "Supergravity", ed. by P. van Nieuwenhuizen and D.Z. Freedman, North Holland, Amsterdam (1979), p. 197.
15. E.R. Nissimov and S.J. Pacheva, Comp. Rendus de l'Acad. Bulgare des Sciences 32:1475 (1979).
16. E.R. Nissimov and S.J. Pacheva, Lett. Math. Phys. 5:67,333 (1981).
17. M. Gell-Mann, unpublished (1980).
18. B. Zumino, "Superspace and Supergravity", ed. by S.W. Hawking and M. Ročec, Cambridge Univ. Press, Cambridge (1981), p. 423.
19. P. Frampton, Phys. Rev. Lett. 46:881 (1981).
20. H.E. Haber, I. Hinchliffe and E. Rabinovici, Nucl. Phys. B172:458 (1980).
21. M.T. Grisaru, these proceedings
22. M.T. Grisaru and H.J. Schnitzer, Brandeis Univ. preprint (1981).
23. V.G. Makhankov and O.K. Pashaev, Dubna preprints JINR E2-81-264 (1981) and E2-81-540 (1981).
24. E. Rabinovici, unpublished. Private communication through J. Ellis and B. Zumino.

25. For an extensive list of references on the subject see, B.G. Wybourne, "Classical Groups for Physicists", J. Wiley & Sons, New York (1974).
26. J.A. Wolf, Journ. Math. and Mech. 13:489 (1964).
27. H. Hecht and W. Schmid, Inv. Math. 31:129 (1975).
28. M. Gell-Mann, private communication.
29. For an outline of Gell-Mann's method see R. Hermann, "Lie Groups for Physicists", Benjamin, New York (1966), p. 182.
30. J. Ellis, M.K. Gaillard and B. Zumino consider other scenarios as well. See Ref. 6.
31. J.-P. Derendinger, S. Ferrara and C.A. Savoy, CERN preprint TH.3025 (1981), unpublished and Nucl. Phys. B188:77 (1981).
32. E. Cremmer, private communication.
33. J. Ellis, M.K. Gaillard, M. Günaydin and B. Zumino, in preparation.
34. Note that in this case the Fock space is enlarged and the vacuum is the direct product of the vacua of bosonic and fermionic operators.
35. H. Nicolai, these proceedings
36. B. De Wit and H. Nicolai, CERN preprint TH.3208 (1981).

TOWARDS UNIFICATION WITH JUST FUNDAMENTAL FERMIONS

A.C. Davis

CERN, Geneva, Swizterland

Fermions are an essential constituent in Nature. Could they be the only fundamental constituents? For instance, could all other particles required by low energy phenomenology, e.g., gauge bosons and Higgs scalars, be bound states of just fermions? Here I show how this appealing suggestion might be realized.

First I discuss a class of four-dimensional fermionic Lagrangians with a local gauge invariance, but no explicit gauge field. In this class of Lagrangians a classical gauge symmetry induces dynamical gauge fields at the quantum level. The resulting effective action is equivalent to a standard gauge theory at energies less than an ultra-violet cut-off Λ. To give an unambiguous meaning to the theory a cut-off is required, and it cannot be removed with our present understanding of the theory. However, to establish the existence of dynamical gauge bosons renormalizability is not crucial due to a decoupling between the massless sector and a spectrum at Λ. A QED-like model, non-Abelian-like model and an Abelian-Higgs-like model are successively discussed. For non-Abelian groups the construction requires: i) flavour (or families) since, ultimately, the theory has to be infra-red free; ii) for non-simple groups a unification of all gauge couplings at a common Landau pole mass of the generated effective theory. This work is the result of a collaboration with D. Amati, R. Barbieri and G. Veneziano and is described in Ref. 1.

To discuss unified field theories it is necessary to consider the mechanism of symmetry breaking. In theories with just fermions an appealing possibility is that of 'tumbling'. That is, if a fermion bilinear, which is not a scalar of the gauge group, acquires a non-zero vacuum expectation value (VEV) the gauge symmetry will be dynamically broken. I present an explicit example of a 'tumbling' gauge theory in two dimensions. The model is a non-Abelian generalization of the CP^{n-1} model with fermions, having a global U(n) and local U(ℓ) symmetry. This model allows a study of dynamical Higgs mechanisms and shows the difference between dynamical and elementary Higgs bosons. This was done in collaboration with A. D'Adda and P. Di Vecchia[2]. I discuss some interesting features of the model, in particular the problem of vacuum alignment. This is in progress, in collaboration with M. Peskin.

Finally I briefly discuss the possibility and problems of constructing a unified theory in four dimensions with just fermions.

There have been previous attempts to generate gauge bosons from fermions[3], trying to convert a global symmetry of the classical theory into a local symmetry of the quantum theory. This approach has met with limited success[4] and has to circumvent a no-go theorem[5]. Alternatively, in the two-dimensional (bosonic) CP^{n-1} models[6] a local gauge invariance (but no explicit gauge field) classically is converted into a dynamical gauge field quantum mechanically.

Following this second line of approach, consider the four-dimensional, fermionic Lagrangian

$$\mathcal{L} = i\bar{\psi}\,\partial\!\!\!/\,\psi - i\bar{\psi}\gamma_\mu\psi\ \frac{\bar{\psi}\overleftrightarrow{\partial}_\mu\psi}{\bar{\psi}\psi} \tag{1}$$

where ψ is a Dirac field*). Now $\mathcal{L}$ is invariant under a local U(1) transformation, $\psi(x) \to e^{i\alpha(\alpha)}\psi(x)$. It is scale invariant and contains no dimensionless parameter. However, it is not obviously renormalizable. (That it contains no dimensionless parameter might, ultimately, make it well-behaved in the ultra-violet, as has recently been suggested for a similar, scalar theory[8].)

For (1) to be well-defined $\bar{\psi}\psi$ should have a non-zero VEV. A correct treatment may well result in $<\bar{\psi}\psi> \neq 0$. However, with present techniques we are unable to treat Lagrangians with such a high degree of non-linearity. Thus we introduce auxiliary fields to make (1) quadratic, and to give $\bar{\psi}\psi$ a non-zero VEV. Hence we consider the theory defined by the (Euclidean) generating functional

$$Z = \int \mathcal{D}\bar{\psi}\,\mathcal{D}\psi\,\mathcal{D}A_\mu\,\mathcal{D}\lambda_\mu\,\mathcal{D}\lambda\ \exp - \int d^4x \left[\bar{\psi}(\partial\!\!\!/ + i\,A\!\!\!/)\,\psi + \right. \tag{2}$$

*) This Lagrangian was first considered by Sugawara[7] in a similar context. We have only recently become aware of this paper. I wish to thank a member of a seminar audience in Copenhagen for informing me of it.

$$+ i\lambda(\bar{\psi}\psi - v) + i\lambda_\mu(v A_\mu - \bar{\psi}\, i \overleftrightarrow{\partial}_\mu \psi)] \tag{2 cont.}$$

After elimination of the auxiliary gauge field, A_μ, and Lagrange multipliers, λ_μ and λ, this Lagrangian reduces to (1), supplemented by the condition $\bar{\psi}\psi = v$. (The VEV for $\bar{\psi}\psi$ could be introduced through a potential without modifying our conclusions.) We introduce an ultra-violet cut-off, Λ, and, to preserve gauge invariance, use the Pauli-Villars regularization.

Integrating over fermion fields gives

$$Z = \int \mathcal{D}A_\mu\, \mathcal{D}\lambda_\mu\, \mathcal{D}\lambda \exp - S_{eff} \tag{3}$$

where

$$S_{eff} = \int d^4x\, (i\, \lambda_\mu A_\mu v - i\, \lambda v) - \mathrm{tr}\log(\not{\partial} + \lambda_\mu \partial_\mu + i\lambda) \tag{4}$$

Solving by saddle-point approximation in the remaining fields yields the Lorentz invariant solution

$$\langle \lambda_\mu \rangle = \langle A_\mu \rangle = 0; \quad \langle i\lambda \rangle = m \tag{5}$$

where m is the generated fermion mass given by

$$v = -\mathrm{tr} \int \frac{d^4p}{(2\pi)^4} \frac{1}{(i\not{p}+m)} \tag{6}$$

Expanding around the saddle-point solution we find the Green functions. These are:

$$\frac{\delta^2 S_{eff}}{\delta A_\mu \delta A_\nu} = \langle A_\mu A_\nu \rangle = (q_\mu q_\nu - q^2 g_{\mu\nu}) \frac{1}{12\pi^2} \log\left(\frac{\Lambda^2}{q^2+4m^2}\right) \tag{7}$$

This is just the standard QED vacuum polarization with $e = 1$. To determine whether or not A_μ has acquired dynamics we have also to consider $\langle \lambda_\mu \lambda_\nu \rangle$ and $\langle \lambda_\mu A_\nu \rangle$. These are:

$$\frac{\delta^2 S_{eff}}{\delta \lambda_\mu \delta \lambda_\nu} = \langle \lambda_\mu \lambda_\nu \rangle = g_{\mu\nu}\, O(\Lambda^4) + (q_\mu q_\nu - q^2 g_{\mu\nu})\, O(\Lambda^2) \tag{8}$$

showing, upon rescaling $\ell_\mu = \lambda_\mu \Lambda$, that it is a non-propagating field for all $q^2 \ll \Lambda^2$. Finally

$$\frac{\delta^2 S_{eff}}{\delta \ell_\mu \delta A_\nu} = \langle \ell_\mu A_\nu \rangle = O\left(\frac{m}{\Lambda}\right) (q_\mu q_\nu - q^2 g_{\mu\nu}) \tag{9}$$

because of a cancellation between the tree and one-loop contributions. Whence, because of the invariant form of (9), A_μ rigorously describes a massless photon in interaction with fermions of mass m. Finally,

$$\frac{\delta^2 S_{eff}}{\delta\lambda\delta\lambda} = \langle\lambda\lambda\rangle \sim O(\Lambda^2) \tag{10}$$

Thus, for all $q^2 \ll \Lambda^2$, $\lambda(x)$ is a non-propagating field. From gauge invariance there is no λ-A_μ mixing.

We should now do a systematic study of higher point functions. An inspection of diagrams with insertions of 'heavies' (λ_μ and λ) suggests that the effective action is equivalent to that of standard QED for $q^2 \ll \Lambda^2$, deviations being of order q/Λ. The only quantities for which 'heavies' give renormalizations of O(1) are the fermion mass and the vacuum polarization, (7). Thus there is a decoupling between the 'heavies' and the light spectrum. Upon rescaling we obtain the standard $-(1/4)F_{\mu\nu}F^{\mu\nu}$ kinetic energy term for A_μ, with renormalized charge

$$e_R^2|_{q^2=0} = \left[\frac{1}{12\pi^2}\log\left(\frac{\Lambda^2}{m_R{}^2}\right) + O(1)\right]^{-1} \tag{11}$$

at the one-loop level. O(1) includes the 'heavy' insertions. Similarly, $m_R = O(m)$. Equation (11) shows that the cut-off, Λ, should be identified with the position of the Landau pole for the effective QED Lagrangian.

We could also consider (1) with many fermion flavours and do a $1/N_f$ expansion. The saddle-point approximation is more justified in this case. 'Heavy' insertions are further suppressed in this case by powers of $1/N_f$.

Finally we mention that Lagrangians such as (1) have no Noether current - the standard construction leads to 0 = 0, as in the CP^{n-1} model. This is due to having an auxiliary gauge field initially. The quantum mechanical generation of kinetic terms for A_μ gives rise to a non-trivial conserved current.

Generalizing our construction to non-Abelian groups we consider a Lagrangian with a local U(n) invariance,

$$\mathcal{L} = i\bar{\psi}^{i\alpha}\partial\!\!\!/\,\psi_{i,\alpha} + \bar{\psi}^{i,\alpha}\, W\!\!\!\!/_{\alpha}{}^{\beta}\psi_{i,\beta} \tag{12}$$

where

$$W_\mu = \tfrac{1}{2}M_\mu M^{-1} + \text{h.c.} \tag{13}$$

$$(M_\mu)^\beta{}_\alpha = -\psi^{Ti\beta}C^{-1}\, i\partial_\mu\, \psi_{i\alpha},\quad M^\beta{}_\alpha = \psi^{Ti\beta}\, C^{-1}\, \psi_{i\alpha} \tag{14}$$

$\psi_{i\alpha}$ is a (left-handed) Weyl spinor; $i = 1,\ldots,f$ is the flavour index and the gauge index is $\alpha = 1,\ldots,n$. Equation (12) is invariant under

$$\psi_{i\alpha}(x) \rightarrow U_{\alpha}{}^{\beta}(x)\, \psi_{i\beta}(x); \quad U(x)\, U^{+}(x) = U^{+}(x)\, U(x) = \mathbb{1} \tag{15}$$

To gauge a subgroup G of U(n) then

$$W_{\mu} = \sum_{a\epsilon U(n)} \lambda^{a} Tr(W_{\mu}\lambda^{a}) \rightarrow \sum_{b\epsilon G} \lambda^{b} Tr(W_{\mu}\lambda^{b}) \tag{16}$$

where λ^{a} are the generators of U(n) in the fundamental representation and λ^{b} span the subgroup, G.

For a vector-like theory we can proceed by introducing

$$\lambda^{\alpha}{}_{\beta}(M^{\beta}{}_{\alpha} - v\delta^{\beta}{}_{\alpha}) \tag{17}$$

and use Pauli-Villars regularization. For a theory with chiral fermions (17) would break the gauge group down to a vector subgroup. We can get around this problem by being more subtle. Also, there are indications from the Abelian case that we did not really need a VEV as long as we were off-shell. However, there are problems regularizing chiral fermions[9].

For vector-like theories we can establish the dynamical generation of massless gauge bosons. At low energies their interactions with fermions will be the same as in standard gauge theories. Thus,

$$\alpha(q^{2})^{-1} = \left(\frac{f}{6\pi} - \frac{11n}{12\pi}\right) \left[\log \frac{\Lambda^{2}}{q^{2}} + O(1)\right], \tag{18}$$

whence the effective coupling constant will be O(1) for q^{2} $O(\Lambda^{2})$. Hence, for the theory to be meaningful at low energies ($q^{2} << \Lambda^{2}$), it must be infra-red free such that α decreases with decreasing q^{2}. This requires the number of flavours to be greater than the number of colours so that the fermionic contribution to the β function dominates over that of gluons. Furthermore, for non-simple groups the couplings have to be unified at Λ. This plays the role of a universal Landau pole for the effective theory. At low energies couplings develop according to their respective β functions. Our 'Landau pole unification' scheme is similar to that suggested in Ref. 10 on the basis of different motivations.

A realistic theory necessitates the treatment of broken gauge theories. This can be done within our construction. For instance, consider the Abelian Lagrangian

$$\mathcal{L} = i\bar{\psi}\partial\!\!\!/\psi + i\bar{\chi}\partial\!\!\!/\chi + \lambda(\bar{\psi}\chi\bar{\chi}\psi - v^{2}) - \tag{19}$$

$$- \left(\frac{e_{1}\bar{\psi}\gamma_{\mu}\psi + e_{2}\bar{\chi}\gamma_{\mu}\chi}{e_{1}+e_{2}}\right) \left(\frac{\bar{\psi}i\overleftrightarrow{\partial}_{\mu}\chi}{\bar{\psi}\chi} + \frac{\bar{\chi}i\overleftrightarrow{\partial}_{\mu}\psi}{\bar{\chi}\psi}\right)$$

where ψ,χ are Weyl spinors, one left- and one right-handed. Equation (19) has a local U(1) invariance

$$\psi \to e^{ie_1\alpha(x)}\psi; \quad \chi \to e^{ie_2\alpha(x)}\chi$$

and a global U(1) symmetry

$$\psi \to e^{ic}\psi, \ \chi \to e^{-ic}\chi$$

When $\bar{\psi}\chi$, $\bar{\chi}\psi$ acquire non-zero VEV, two possible breaking patterns are possible:

$$\left.\begin{array}{l} e_1 = e_2 \\ \\ e_1 \neq e_2 \end{array}\right\} \ U(1)_{local} \otimes U(1)_{global} \left\{\begin{array}{l} \supset U(1)_{local} \\ \\ \supset U(1)'_{global} \end{array}\right.$$

where $U(1)_{global}$ is a combination of the original $U(1)_{local}$ and $U(1)_{global}$ symmetries.

To proceed we make (19) quadratic in the fields as previously. When $e_1 = e_2$ A_μ again describes a massless photon and the 'heavies' decouple. In addition we obtain a Goldstone boson from chiral symmetry breaking.

For $e_1 \neq e_2$ we obtain

$$\frac{\delta^2 S_{eff}}{\delta A_\mu \delta A_\nu} = \langle A_\mu A_\nu \rangle = \left(\frac{e_1^2+e_2^2}{12\pi^2}\right)(q_\mu q_\nu - q^2 g_{\mu\nu}) \log\left(\frac{\Lambda^2}{q^2+4m^2}\right) + \frac{1}{4\pi^2} g_{\mu\nu} m^2 (e_1-e_2)^2 \log\left(\frac{\Lambda^2}{q^2+4m^2}\right) \tag{20}$$

Thus the 'photon' acquires a mass of

$$m_\gamma^2 = 3m^2 \frac{(e_1-e_2)^2}{e_1^2+e_2^2}\left[1 + 0\left[\left(\log\left(\frac{\Lambda^2}{q^2}\right)\right)^{-1}\right]\right] \tag{21}$$

i.e., $m_\gamma^2 \propto$ generated fermion mass.

Here, the Goldstone boson has been eaten to give rise to the longitudinal component of the photon. This is just the Higgs mechanism. All left-over scalars are of $O(\Lambda)$. We do not find a light Higgs scalar, though this is not required for renormalizability in this Abelian example.

So what next? There are several natural extensions to the scheme discussed here. Gravity has already been included in such a construction by Amati and Veneziano[11] and will be discussed here by Veneziano[12]. When gravity is included the cut-off, Λ, naturally becomes the Planck mass. Alternatively, we could investigate a non-linear realization of supersymmetry in such a scheme, in a manner similar to that of Volkov and Akulov[13]. Arguably the most natural, and maybe the most ambitious, extension is to try to construct a phenomenologically realistic model based on our 'Landau pole unification' scheme. For fermionic Lagrangians such as ours, we have to discuss the breaking of the (non-Abelian) gauge symmetry. Now, if a fermion bilinear, which is not a singlet of the gauge group, acquires a non-zero VEV the gauge group will be broken[14,15]. If several energy scales result then this is known as tumbling[15]. It would be natural to try to merge our scheme with that of tumbling. However, this would be plagued with the usual technical problem of trying to find a gauge invariant regularization for chiral fermions[9].

I am going to be more modest and discuss a two-dimensional toy model of a 'tumbling' gauge theory*). There are several motivations for going over to two dimensions. (This is demonstrated by the fact that several speakers at this meeting have ventured into the playground of two dimensions - or lower.) In two dimensions one can do explicit calculations. This can help gain insight and confidence for the four-dimensional world [e.g., a study of the CP^{n-1} model with fermions[6] helped elucidate the U(1) problem[16]]. Recently it has been shown how to regularize chiral fermions in two dimensions[17]. In our case we can write down an explicit Lagrangian which undergoes 'tumbling' and display the difference between dynamical and elementary Higgs mechanisms. (Tumbling in two dimensions has also been studied by Banks et al.[17], though the tumbling picture was never realized in Ref. 17. We differ from Ref. 17 in that our model has a quartic interaction.)

The model I am going to discuss is a non-Abelian generalization of the CP^{n-1} model with fermions. It has a global U(n) symmetry, represented by the index $i = 1,\ldots,n$, and a local $U(\ell)$ symmetry, represented by the index $\alpha = 1,\ldots,\ell$. In this model the scalar Z fields take their values on a Grassmannian manifold, i.e.,

$$\bar{Z}^{i\alpha} Z_{i\,\beta} = \frac{n}{2f}\,\delta^{\alpha}{}_{\beta} \tag{22}$$

where f is a coupling constant. The Lagrangian is

$$\mathcal{L} = \bar{D}_{\mu}{}^{\alpha}{}_{\beta}\bar{Z}^{i\beta} D_{\mu\alpha}{}^{\gamma} Z_{i\gamma} + i\lambda_{\alpha}{}^{\beta}(\bar{Z}^{i\alpha}Z_{i\,\beta} - \frac{n}{2f}\,\delta^{\alpha}{}_{\beta}) + \tag{23}$$

*) I use the word 'tumbling' rather loosely in what follows for want of a better word.

$$+ \bar{\psi}^{i}_{L\alpha} D^{\alpha}_{-\beta} \psi_{Li}{}^{\beta} + \bar{\psi}^{i\alpha}_{R} D_{+\alpha}{}^{\beta} \psi_{Ri\beta}$$
$$- \frac{1}{\sqrt{n}} [\bar{\psi}^{i}_{L\alpha} \psi_{Ri\beta} \bar{\phi}^{\alpha\beta} + \bar{\psi}^{i\alpha}_{R} \psi_{Li}{}^{\beta} \phi_{\alpha\beta}] + \frac{1}{g} \phi_{\alpha\beta} \bar{\phi}^{\beta\alpha} \quad (23)\ \text{cont.}$$

where

$$D_{\mu\alpha}{}^{\beta} = \partial_{\mu} \delta_{\alpha}{}^{\beta} + \frac{i}{\sqrt{n}} A_{\mu\alpha}{}^{\beta} \quad (24)$$

$\lambda_{\alpha}{}^{\beta}$ imposes the constraint (22). ϕ, $\bar{\phi}$ and A_{μ} are auxiliary fields. [Equation (23) has already been linearized and written in a form convenient for a 1/n expansion],

$$\phi_{\alpha\beta} = \frac{g}{\sqrt{n}} \bar{\psi}^{i}_{L\alpha} \psi_{Ri\beta} \quad (25)$$

and

$$D_{\pm} = D_{0} \pm i\, D_{1} \quad (26)$$

In (23) the left- and right moving fermions are in different representations of the gauge group - ψ_R in (ℓ), ψ_L in $(\bar{\ell})$. (We could have taken them in the same representations. This is considered in Ref. 2.) The left- (right-) mover couples only to $D_-(D_+)$.

The field $\phi_{\alpha\beta}$ is a Lorentz scalar and a scalar of U(n) allowing a 1/n expansion. However, it is not a scalar under the $U(\ell)$ gauge group. Thus, if it acquires a non-zero VEV the gauge group will be broken.

Now (23) has a hidden symmetry. Since ψ_L couples only to D_- and ψ_R to D_+, A_- and A_+ can be two independent fields as far as the fermionic sector is concerned. Thus, (23) has a $U(\ell) \otimes U(\ell)$ symmetry, which is broken by the gauge field in the bosonic sector. It is also broken by a non-Abelian anomaly. This anomaly plays a crucial role in the properties of the model, as will be shown later.

Integrating over ψ and Z we obtain the effective action

$$S_{eff} = n\,\mathrm{trlog}(-D^2 + \frac{i}{\sqrt{n}} \lambda_{\alpha}{}^{\beta}) - n\,\mathrm{trlog} \begin{pmatrix} D_{+\alpha}{}^{\beta} & -\frac{1}{\sqrt{n}} \phi_{\alpha\beta} \\ -\frac{1}{\sqrt{n}} \bar{\phi}^{\alpha\beta} & D^{\alpha}_{-\beta} \end{pmatrix} \quad (27)$$
$$+ \frac{1}{g} \int d^2x\, \phi_{\alpha\beta} \bar{\phi}^{\beta\alpha} + i \int d^2x\, \lambda_{\alpha}{}^{\beta} \delta_{a}{}^{\beta} \sqrt{n}/2f$$

Solving in a 1/n expansion gives

$$<\phi_{\alpha\beta}> = \sqrt{n}\, M_{\alpha\beta} \tag{28}$$

and

$$M_{\alpha\beta}\, M^{\beta\gamma} = \delta_{\alpha}{}^{\gamma}\, M^2 \tag{29}$$

where M is the generated fermion mass, obtained by dimensional transmutation of g. From (28) and (29) we see that ϕ has acquired a non-zero VEV. Thus, the gauge group is broken. However, (29) does not tell us the direction of symmetry breaking. There is a manifold of vacuum states, related by a chiral transformation, and to determine the alignment of the vacuum we have to go beyond leading order. This is in progress[18].

Expanding in 1/n we obtain

$$\begin{aligned} S^{(2)}_{eff} = \int d^2x \Big\{ & (\partial_\mu A_{\nu\alpha}{}^{\beta} - \partial_\nu A_{\mu\alpha}{}^{\beta})^2 + (\partial_\mu B_{\nu\alpha}{}^{\beta} - \partial_\nu B_{\mu\alpha}{}^{\beta})^2 \\ & + 4M^2 B_{\mu\alpha}{}^{\beta} B_{\mu\beta}{}^{\alpha} + |\partial_\mu \phi_{-\alpha}{}^{\beta}|^2 + |\partial_\mu \phi_{+\alpha}{}^{\beta}|^2 + 4M^2 (\phi_{+\alpha}{}^{\beta})^2 \\ & - 4M\partial_\mu\, \phi_{-\alpha}{}^{\beta} B_{\mu\beta}{}^{\alpha} + 4iM\epsilon_{\mu\nu}\, \partial_\mu\, A_{\nu\alpha}{}^{\beta} \phi_{+\beta}{}^{\alpha} \Big\} \end{aligned} \tag{30}$$

where the fields diagonalizing the spectrum are:

$$B_{\mu\alpha}{}^{\beta} = \tfrac{1}{2}(A_{\mu\alpha}{}^{\beta} - M_{\alpha\rho} \frac{A^{\rho}_{\mu\gamma}}{M^2} M^{\gamma\beta}) \tag{31}$$

and we redefine

$$A_{\mu\alpha}{}^{\beta} = \frac{1}{2}\, (A_{\mu\alpha}{}^{\beta} + M_{\alpha\rho} \frac{A^{\rho}_{\mu\gamma}}{M^2} M^{\gamma\beta}) \tag{32}$$

$$\phi_{-\alpha}{}^{\beta} = \frac{i}{2M}(\phi_{\alpha\gamma} M^{\gamma\beta} - M_{\alpha\gamma}\, \bar{\phi}^{\gamma\beta}) \tag{33}$$

these are would-be Goldstone bosons from chiral symmetry breaking.

$$\phi_{+\alpha}{}^{\beta} = \frac{1}{2M}(\phi_{\alpha\gamma} M^{\gamma\beta} + M_{\alpha\gamma}\, \phi^{\gamma\beta}) \tag{34}$$

The gauge bosons B_μ have acquired a mass, whilst A_μ remain massless. Since the unitary matrix $M_{\alpha\beta}$ does not appear in $S^{(2)}$ we do not know the unbroken subgroup.

To analyze the spectrum of states we may take

$$M_{\alpha\beta} = M\, \delta_{\alpha\beta} \tag{35}$$

The unbroken subgroup is then $O(\ell)$ [similar considerations can be made for the other subgroups of $SU(\bar{\ell})$, e.g., $S_p(\ell)$]. Here the components of ϕ_- not in $O(\ell)$ are eaten by B_μ in a dynamical Higgs mechanism. Defining

$$B_{\mu\ \alpha}{}^{\beta} = B^{i}_{\mu}(t^{i})_{\alpha}{}^{\beta},\ A_{\mu\alpha}{}^{\beta} = A_{\mu}{}^{a}(\tau^{a})_{\alpha}{}^{\beta} \tag{36}$$

$$\phi_{-\alpha}{}^{\beta} = B^{i}(t^{i})_{\alpha}{}^{\beta} + A^{a}(\tau^{a})_{\alpha}{}^{\beta},\ \phi_{+\alpha}{}^{\beta} = C^{s}(T^{s})_{\alpha}{}^{\beta} \tag{37}$$

where T^{s} are the generators of $SU(\ell)$ and

$$\tau^{a} = (T^{a}-T^{a*})\ \tfrac{1}{2},\ t^{i} = (T^{i}+T^{i*})\ \tfrac{1}{2}, \tag{38}$$

τ^{a} spanning the $O(\ell)$ algebra. The index a (i) runs over the unbroken $O(\ell)$ subgroup [broken part of $U(\ell)$]. Finally we obtain

$$S^{(2)} = \int d^{2}x \left\{ F^{a\ 2}_{\mu\nu} + G^{i\ 2}_{\mu\nu} + 4M^{2}W^{i2}_{\mu} + (\partial_{\mu}\ A^{a})^{2} + (\partial_{\mu}C^{s})^{2} + 4M^{2}\ C^{s2} + 4iMA^{a}\varepsilon_{\mu\nu}\partial_{\mu}A^{a}_{\nu} \right\} \tag{39}$$

where

$$W^{i}_{\mu} = B^{i}_{\mu} - 2M\partial_{\mu}B^{i} \tag{40}$$

and $F^{a}_{\mu\nu}$, $G^{i}_{\mu\nu}$ are the kinetic terms for A^{a}_{μ} and W^{i}_{μ} respectively. From (39) we see that the B^{i} fields have been eaten by W^{i}_{μ} and disappear from the physical spectrum. Thus, we have a massive W^{i}_{μ} and massive C^{s} field. The pseudo-scalar A^{a} couples to the unbroken gauge field A^{a}_{μ} via a non-Abelian anomaly. This is a generalization of the U(1) anomaly and here gives mass to the entire A^{a} multiplet. In the absence of this anomaly term the A^{a} would be Goldstone bosons from chiral symmetry breaking. This term would be absent if we had elementary Higgs bosons. The presence of such an anomaly term signals the difference between dynamical and elementary Higgs mechanisms. The unbroken gauge fields also acquire a mass. This screens the long-range force, as in the massless Schwinger and CP^{n-1} models. It is a consequence of there being massless fermions in the original Lagrangian. The implications of our results are being investigated and could be generalized to more realistic models.

Perhaps the most interesting, and bizarre, feature of the model is that, to O(1) in the 1/n expansion, we are unable to determine the unbroken subgroup. There is a manifold of vacuum states related by a chiral transformation. To determine the alignment of the vacuum we have to calculate the vacuum energy to more orders in 1/n. It is interesting to speculate that there may be some hidden symmetry protecting the vacuum state to this order. This is in progress[18].

We can consider other representations for the left- and right-moving fields. The representations have to be chosen such that anomalies in the gauged current are cancelled between the left- and right-handed representations. There will be an anomaly in the ungauged current analogous to the one found here, which will give mass to would-be Goldstone bosons. Such models could also provide insight into the role of the anomaly in tumbling. This is currently under investigation.

Finally, I wish to consider constructing a realistic model with just fermions. For instance, considering the 5 and $\overline{10}$ representations of SU(5) we can generalize (12) - (14),

$$M_\mu = -\chi^{T\alpha\beta} C^{-1}\, i\, \partial_\mu \psi_\beta \qquad\qquad M = \chi^{T\alpha\beta} C^{-1} \psi_\beta$$

and

$$W_\mu = M_\mu M^{-1} + \text{h.c.}$$

where ψ and χ are left-handed Weyl spinors in the 5 and $\overline{10}$ representation of SU(5). If M acquires non-zero VEV then the SU(5) group will be broken. Problems here arise with the regularization of chiral fermions. Clearly it would be interesting if we could construct unification models with just fermions. This is a very ambitious programme. However, the models I have presented indicate that it is not completely impossible. Much work is needed in this direction, but the results could be rewarding.

ACKNOWLEDGMENTS

I wish to take this opportunity to thank my collaborators Daniele Amati, Riccardo Barbieri, Gabriele Veneziano, Alessandro D'Adda, Paolo Di Vecchia and Michael Peskin for making the many enlightening, fruitful and often frustrating hours we have spent collaborating so enjoyable.

REFERENCES

1. D. Amati, R. Barbieri, A.C. Davis and G. Veneziano, Phys. Lett. 102B:408 (1981).
2. A. D'Adda, A.C. Davis and P. Di Vecchia, CERN preprint TH-3189 (1981).
3. J.D. Bjorken, Ann. Phys. 24:174 (1963);
   I. Bialynicki-Birula, Phys. Rev. 130:465 (1963);
   G.S. Guralnik, Phys. Rev. 136:1404 (1964);
   T. Eguchi, Phys. Rev. D14:2755 (1976);
   H. Terazawa, Y. Chikashige and K. Akama, Phys. Rev. D15:480 (1977);
   C. Bender, F. Cooper and G.S. Guralnik, Ann. Phys. 109:165 (1977);
   K. Shizuya, Phys. Rev. D21:2237 (1980).

4. T. Banks and A. Zaks, Nucl. Phys. B184:303 (1981).
5. S. Weinberg and E. Witten, Phys. Lett. 96B:59 (1980).
6. A. D'Adda, P. Di Vecchia and M. Lüscher, Nucl. Phys. B146:63 (1978); B152:125 (1979);
E. Witten, Nucl. Phys. B149:285 (1979).
7. H. Sugawara, Proc. Tokyo Conference (1978).
8. V. de Alfaro, S. Fubini and G. Furlan, Phys. Lett. 97B:67 (1980).
9. H.B. Nielsen and M. Ninomiya, Nucl. Phys. B185:20 (1981); preprints NBI-81-1 and RL-81-052 (1981);
A. Casher, private communication;
F. Englert, private communication.
10. G. Parisi, Phys. Rev. D11:909 (1975);
L. Maiani, G. Parisi and R. Petronzio, Nucl. Phys. B136:115 (1978);
N. Cabibbo, L. Maiani, G. Parisi and R. Petronzio, Nucl. Phys. B158:295 (1979).
11. D. Amati and G. Veneziano, Phys. Lett. 105B:358 (1981).
12. G. Veneziano, these proceedings.
13. D.V. Volkov and V.P. Akulov, Zh. ETF. Pis. Red. 16:621 (1972); Phys. Lett. 46B:109 (1973).
14. M.E. Peskin, unpublished.
15. S. Raby, S. Dimopoulos and L. Susskind, Nucl. Phys. B169:373 (1980).
16. For a review see P. Di Vecchia, Proc. Schladming Inst. (1980); Acta Physica Austriaca Suppl. XXII:341 (1980).
17. T. Banks, Y. Frishman and S. Yankielowicz, Nucl. Phys. B191:493 (1981).
18. A. D'Adda, A.C. Davis, P. Di Vecchia and M.E. Peskin, in preparation.

BARYON STABILITY AND NEUTRINO OSCILLATIONS

Donald H. Perkins

Department of Nuclear Physics
University of Oxford
Keble Road, Oxford, England

1. PROTON LIFETIME EXPERIMENTS

Baryon instability is discussed here because it constitutes a crucial test of grand unified gauge theories. There is, at the present time, neither evidence that baryons are unstable nor any good reasons to suppose that they are not. Conservation of baryon number B or lepton number L must be associated with an invariance principle. Thus, charge conservation is associated with gauge invariance in electromagnetism and the existence of a long range field (i.e. massless photons). If B (or L) were absolutely conserved, we would expect a long-range field coupled to baryon number, and this should show up in Eötvös experiments as a difference in gravitational attraction for objects of the same inertial mass but different baryon number (i.e. of different elements). No such effects are observed. Several experiments - designed originally to do other things - happen to provide limits on proton stability, and we discuss these briefly before turning to the experiments dedicated to the search for nucleon decay.

1.1 EXISTING LIMITS

1.1.1 Biological Limit

As pointed out many years ago by M. Goldhaber, existence of life sets a strong limit on nucleon stability. If all the decay products produce ionisation and the mean lifetime of the proton is τ years the body dose is:

$$\text{Body dose} = \frac{6.10^{18}}{\tau} \text{ rad yr}^{-1}$$

$$1 \text{ rad} = 100\text{ergs g}^{-1} = 6.10^{7} \text{ MeV g}^{-1}$$

Natural body dose = 0.12 rad yr^{-1} (cosmic rays + radioactivity)

Permissible body dose = 5 rad yr^{-1}

Probably 500 rads yr^{-1}, that is 100 x permissible body dose, would have serious effects, long term, so $\tau > 10^{16}$ years, already long compared with the known age of the galaxy ($\sim 4.10^{10}$ yr).

1.1.2 Geochemical Limit

The basic idea is to try to find rare, long-lived isotopes from small underground ore samples. Such isotopes might be formed as a result of nucleon decay, an example being I^{129} from Te^{130}:-

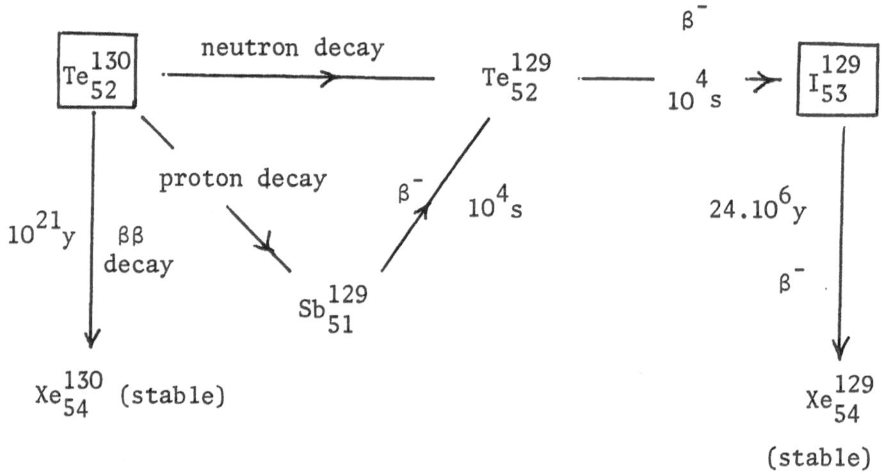

If Tellurium 130 loses either a proton or neutron, and the daughter nucleus is left unexcited (the probability as computed by shell model calculations is $\sim$40%), Te^{129} decaying to I^{129} would be formed. The lifetime of I^{129} is long enough to integrate the decay of Te^{130} over a long period, yet short compared with the age of the solar system - so there is no primordial I^{129}. Existing limits using mass-spectroscopic methods of measuring the I^{129} abundance give $\tau > 10^{25}$ years[1]. Possibly in the future one can extract, accelerate and detect single iodine ions, using stripping techniques to remove organic molecules of mass 129, and magnet spectrograph and time of flight methods to get the precision required. For $\tau = 10^{30}$ yrs, 1Kg Te^{130} will provide 10^3 atoms of

I^{129}. A weakness of this technique is that it depends on geological evolution over $\sim 10^6$ years, for example in comparing samples from different depths, subjected to different cosmic ray backgrounds.

1.1.3 Radiochemical Limit

A different approach is to detect the decay of a short-lived radioisotope produced as a result of baryon decay. Since the isotope is chosen to have a short lifetime, a large sample of ore is required to give any effect. An example is nucleon decay in K^{39} followed by p or n emission, resulting in Ar^{37} which decays by K-capture to Cl^{37}. Again, shell plus evaporation model calculations give $\sim 20\%$ for the probability

$$\boxed{{}^{39}_{19}K} \xrightarrow[\text{+ p, n emission}]{\text{n, p decay}} \boxed{{}^{37}_{18}Ar} \xrightarrow[\text{K capture}]{\text{50d}} \boxed{{}^{37}_{17}Cl}$$

that nucleon decay will be followed by the clean removal of just one nucleon. Using 1.7 ton KAc, Evans, Fireman and Steinberg[2] working at 4400mwe ((Homestake Mine) found a production rate of <1 Ar^{37} atom/day and hence $\tau > 2.10^{26}$ yr. These authors have proposed an experiment with 450 tons KAc at 9000mwe depth, where the cosmic background will be <0.1 Ar^{37} atoms/day, so that they expect to reach a limit $\tau > 5.10^{29}$ years.

In summary, radiochemical and geochemical methods have placed limits on the nucleon lifetime of $\tau > 10^{26}$ years and can probably attain a sensitivity up to $\tau \sim 10^{30}$ years. However, there is no way they can prove that nucleons decay, since what is observed is an "excess" of a rare isotope, but this excess could have arisen in many other ways. This drawback more than offsets the advantage that the "signal" is independent of the nucleon decay mode.

1.1.4 Cosmic Ray Limit

The most sensitive limit on baryon stability derives from an experiment deep underground in a South African gold mine (9000mwe), designed to detect the interactions of atmospheric neutrinos. The apparatus[3] consisted of flash tubes and liquid scintillator and could detect a stopping μ^+ by its decay with $2\mu s$ lifetime. 6 $\mu^+ \to e^+$ decays were observed in the detector; all could be attributed to neutrino reactions in the surrounding rock, $\nu_\mu + N \to \mu^+ + X$. If, however, they are attributed to proton decay, they yield the limit

$$\tau_p > 10^{30} \text{ yrs.} \qquad (1)$$

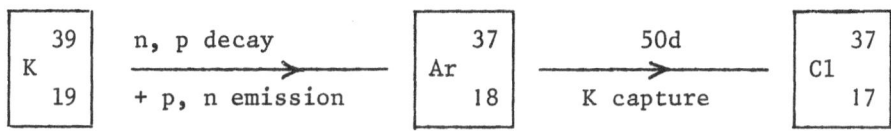

This limit depends on assumptions about p (or n) decay modes, namely that μ^+ arise as products in ∿15% of decays. For the SU(5) models to be discussed below, μ^+ will be produced at this rate because π^+ are important products (50% of proton decays) and half the π^+ will decay to μ^+. The limit (1) holds also for the Pati-Salam model, where decay modes involving neutrinos (e.g. $p \rightarrow \pi^+ + 3\nu$) and pions dominate. However, the above result is not generally model independent.

1.2 THEORETICAL PREJUDICE

There is no question that the present spate of experimental activity, searching for nucleon decay, neutron oscillations etc. stems from the theoretical work on grand unification schemes for the fundamental interactions, over the last few years.

Fig.1 shows the dependence on q, the 4-momentum transfer, of the running coupling constants g and g' of the SU(2) and U(1) gauge groups of the electroweak interaction and that, g_s, of the colour SU(3) strong interaction. So g and g' represent the coupling of leptons to the gauge bosons W^+, W^-, Z^o and γ of the SU(2) x U(1) model, g_s the coupling of quarks to the gauge bosons (gluons) of the strong interactions. The Abelian coupling g' increases with q, while the non-Abelian g and g_s couplings decrease. To within the experimental errors, the three couplings coincide at a unification mass $M_x \sim 10^{14}$GeV. If the three interactions are unified at this energy, then most of the distinction between quarks and leptons disappears, and indeed the "lepto-quark" bosons X and Y of charge 4/3 and 1/3, can mediate proton decay (Fig.2), transforming a quark to a lepton or quark to antiquark and giving, for example $p(=duu) \rightarrow e^+\pi^o(=e^+u\bar{u})$. From this diagram of X (or Y) exchange between quarks, one can see that the matrix element contains a propagator term $(q^2 + M_x^2)^{-1}$ which at the low q^2 of proton decay gives a rate factor $\propto M_x^{-4}$. Thus, on dimensional grounds the proton lifetime can be written

$$\tau_p = c \cdot \frac{M_x^4}{M_p^5} \qquad (2)$$

where c is a dimensionless constant containing numerous factors (integrals over quark wavefunctions etc.), with a value of order unity (i.e. 10^{-2} to 10^{+2}). If we take c = 1, $M_x = 3.10^{14}$GeV one finds $\tau \sim 10^{32}$ yrs. The "best" lifetime estimate available gives[4]

$$\tau_p = 10^{31\pm2} \text{ yrs} \qquad (3)$$

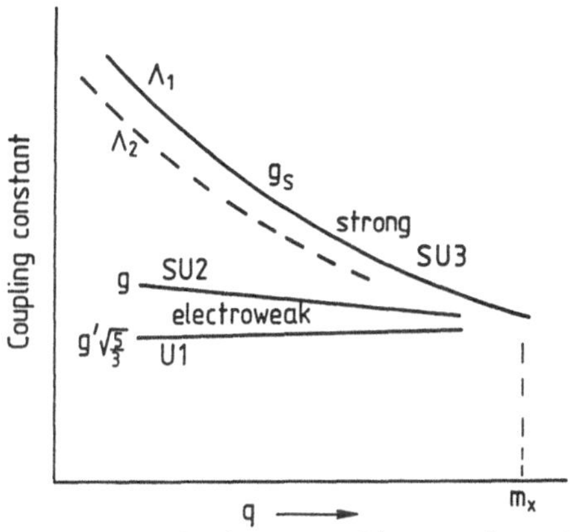

Fig.1 The q dependence of the couplings g' and g of the electroweak theory, and g_s of QCD. They appear to coincide at $q = M_X \sim 3.10^{14}$GeV.

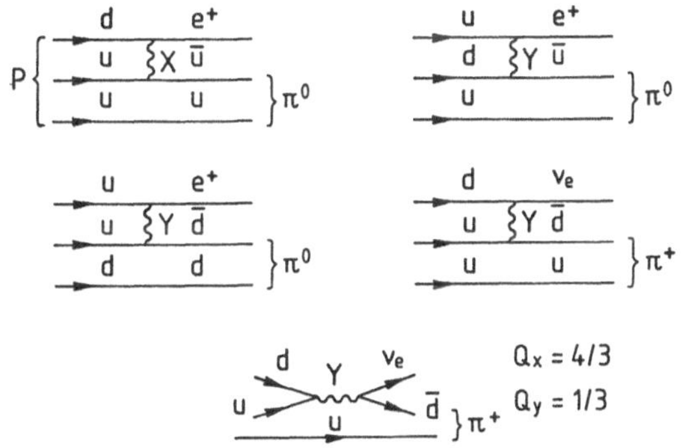

Fig.2 Possible diagrams representing nucleon decay via the leptoquark bosons X, Y of the grand unified theories.

However, we should recognise that it depends on numbers, the values of which are not well known and which can change very dramatically. As an example, the unification mass M_x is determined mostly by the rate of decrease of $\alpha_s(=g_s^2)$, the strong coupling, with q^2. In leading order QCD $\alpha_s = 12/(33-2f)\ell n q^2/\Lambda^2$ where f is the number of quark flavours, and the parameter Λ sets the scale of strong interactions. The result (3) assumed Λ = 400MeV, before the most recent results from the CERN EMC muon scattering experiment gave $\Lambda \sim$ 100MeV. Obviously $M_x \propto \Lambda$, hence if Λ changes by a factor 4, τ_p changes by 4^4 = 256. So theoretical estimates - even if the theory is right - need experimental input which leaves very large uncertainties.

The actual decay modes expected are of course of absolutely crucial importance to any detection system.

Table 1.1 gives a list of typical branching ratios based on SU(5) as the symmetry of the unified SU(3) x SU(2) x U(1) interaction. According to this model, u and d quarks appear in the same SU(5) generation as ν_e and e, while s and c quarks are partnered with ν_μ and μ. Thus a nucleon consisting of u and d quarks is expected to decay to e^+ + hadrons, not μ^+ + hadrons. The expected dominance of the decays $p \rightarrow \pi^o e^+$, $n \rightarrow \pi^- e^+$ had dictated the design of water Cerenkov detectors tailored to detection of electromagnetic showers.

TABLE 1.1

Typical Nucleon Decay Branching Ratios (SU(5))

$p \rightarrow \pi^o e^+$	34%	$n \rightarrow \pi^- e^+$	51%
$p \rightarrow \rho^o e^+$	16%	$n \rightarrow \rho^- e^+$	26%
$p \rightarrow \eta\ e^+$	12%		
$p \rightarrow \omega\ e^+$	21%		
$p \rightarrow \bar{\nu}_e X_{NS}$	13%	$n \rightarrow \bar{\nu}_e X_{NS}$	20%
$p \rightarrow \bar{\nu}_e X_S$		$n \rightarrow \bar{\nu}_e X_S$	
$\rightarrow \bar{\nu}_\mu X$	4%	$\rightarrow \bar{\nu}_\mu X$	3%
$\rightarrow e^+ X_S$		$\rightarrow e^+ X_S$	
$\rightarrow \mu^+ X$		$\rightarrow \mu^+ X$	

1.3 BACKGROUND CONSIDERATIONS

A 1000 ton detector contains 6.10^{32} nucleons, and if $\tau \sim 10^{32}$ years, only 6 nucleon decays will occur in a whole year, some by virtually undetectable modes. In contrast, at the earth's surface, nearly 10^{12} cosmic ray muons cross the detector, and these will produce background processes which can completely swamp any signal.

To reduce background as far as possible, experiments must be carried out underground. Fig.3 shows the total muon flux per year through a 10 x 10 x 10m cubic detector, as a function of depth, obtained by numerically integrating the measured differential muon spectrum[5]. Depth is quoted in metres of water equivalent (mwe): since "standard rock" has density $\sim$3, the depth in mwe is about equal to the actual depth in feet. The muon flux varies from $\sim$1 per second in the shallow mine experiments (1600mwe) to $\sim 10^{-3}$ per second in the Mont Blanc road tunnel (5000mwe).

1.3.1 Neutron Background

The muons themselves produce tracks traversing the entire detector and can hardly be mistaken for nucleon decays. The important background is contributed by energetic neutrons generated in nuclear cascades produced in inelastic muon-nucleon scattering.

First we observe from Fig.3 that the muon absorption length is about 700 times the nuclear mean free path (100gm cm^{-2}) so that hadrons are in equilibrium with, and proportional to, the muon flux at any particular depth. An estimate of neutron production has been made by A. Grant[6] using a Monte Carlo program to generate hadron cascades originating from muon interactions. These calculations give numbers for the energy spectrum of neutrons, either accompanied by a muon or other hadrons, or isolated in the sense that no other particle crosses the detector (again, a 10 x 10 x 10m cube). They also provide a spectrum of pions produced in hadron cascades. Whereas there are no useful measurements of fast neutron rates deep underground, there are measurements of the rate of stopping (and decaying) μ^+. For h > 1000mwe, the π^+ stop rate comes mainly from decay of π^+, the production rate of which can therefore be deduced and used to check the Monte Carlo predictions.

The results show that: (i) the total number of neutrons of E > 100MeV and the total number of $\pi^\pm$ are about equal, at $1.5.10^{-2}$ particles per crossing muon, per 100g cm^{-2} of material traversed (ii) the number of <u>isolated</u> neutrons of E > 1GeV is 2.10^{-3} per muon per 100g cm^{-2}. Thus one can estimate that the number of isolated neutrons capable of giving relativistic secondaries, with a visible energy of 0.9 ± 0.2GeV is <<0.25 of the neutrino background rate

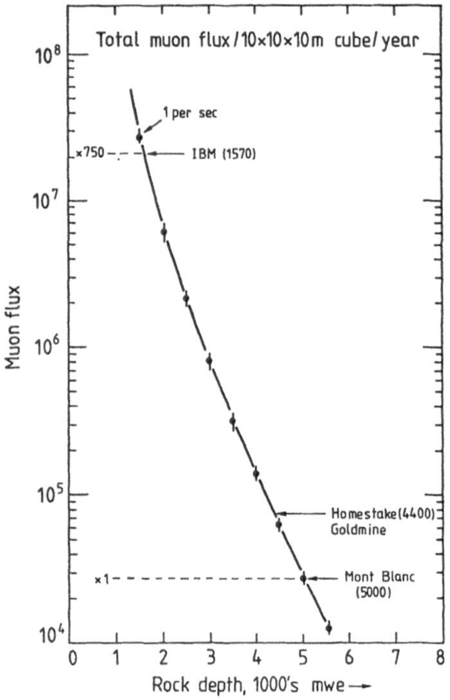

Fig.3 Cosmic ray muon flux through a 10m cube as a function of rock depth in meters of water equivalent (mwe).

(see below) for depths h > 1600mwe. Hence, provided one goes deep, neutron interactions are much less important than neutrino background.

This result depends on making sure that the neutron is isolated. If one misses an accompanying muon, the rate goes up by a factor 15. If one misses both the muon and any accompanying hadron, the rate will go up by a factor 120. The result also depends somewhat on detector size. Although the results are quoted for a 10 x 10 x 10m cube, they apply for almost any detector with a linear dimension large compared with the nuclear mean free path.

1.3.2 Neutrino Background

This background is irreducible (except by going to the Moon and thus eliminating cosmic ray production of pions in the atmosphere) and the principal limitation on baryon decay experiments.

The magnitude of the neutrino flux is quite well known; the pions (and kaons) produced in the stratosphere decay to muons and neutrinos, so that the measured muon flux is directly related to that of neutrinos. The neutrino spectrum can be approximated by a power law:-

$$\phi(\nu_\mu+\bar{\nu}_\mu) = 0.05\ E^{-2.74}\mathrm{cm}^{-2}\mathrm{st}^{-1}\mathrm{s}^{-1}\ \mathrm{GeV}^{-1} \qquad (E > 1\ \mathrm{GeV}) \qquad (4)$$

$$\phi(\nu_e+\bar{\nu}_e) \simeq 0.5\ \phi(\nu_\mu+\bar{\nu}_\mu)$$

and can be considered known to ±30%. The ν_e, $\bar{\nu}_e$ component originates from Ke3 and μ decay in the atmosphere.

From these fluxes and the known charged (CC) and neutral current (NC) cross-sections one obtains the following rate estimates:-

Table 1.2 Neutrino Background Rates

			Rate per 10^3 ton per year
CC	$\nu_\mu + \bar{\nu}_\mu$	$E = 0.9 \pm 0.2$GeV	27
	$\nu_e + \nu_e$		13
NC	$\nu_\mu + \bar{\nu}_\mu$	$E_{hadron} = 0.9 \pm 0.2$GeV	4
	$\nu_e + \bar{\nu}_e$		2
		TOTAL	46

A total of about 50 events have visible energy 0.9 ± 0.2GeV for a 1000 ton detector in one year's run. (The total number of neutrino events of $E_{vis} > 1$GeV is about 3 times this figure.) If all these events simulated proton decay, it follows that the maximum detectable lifetime would be $\tau(\max) \sim 10^{31}$ years and indeed a 30 ton detector, providing 2 neutrino and 2 proton decay events per year, which could not be distinguished, would be big enough. Clearly, the only justification for building a larger detector is that one hopes to cut down the background by a substantial factor, either by:

(a) improving the energy resolution - very difficult in practice, since very massive detectors with very high resolution are extremely expensive, or

(b) identifying the decay configuration, e.g. $n \to e^+\pi^-$ by having good track pattern recognition in the detector. This can in principle reduce background by a large factor, even 100, so that lifetimes as large as 10^{33} years might be measurable.

As an example of what might be achieved, Fig.4 shows a plot from the Irvine-Michigan-BNL proposal[7]. From events in the Gargamelle PS neutrino experiment, those compatible with $\nu_\mu + N \to \mu^- + \pi^+ + N$ were selected and the energy ratio $E_\pi/(E_\pi + E_\mu)$ plotted against the angle $\theta_{\pi\mu}$ between μ^- and π^+. We would expect a similar plot for the neutron decay ($n \to e^+\pi^-$) background reaction $\bar{\nu}_e + N \to e^+ + \pi^- + N$, since the PS neutrino spectrum is rather similar to the cosmic ray neutrino spectrum. Fig.4 shows the number of events expected in a 2 year run of their proposed detector. Nucleon decays (e.g.$n \to e^+\pi^-$) should exhibit a back-to-back configuration of the two secondaries, modified by Fermi motion of the neutron in the nucleus, and should be contained inside the dotted semi-circle. Out of 100 background events none actually falls into this allowed region, justifying - at least on paper - the factor 100 mentioned above.

1.4 FUTURE EXPERIMENTS

The requirements of a detector for nucleon decay can be summarised as follows:-

(1) It must be large enough to contain the decay products. For water Cerenkov detectors, oriented towards the decay mode $p \to e^+\pi^o$, the electromagnetic showers must be contained inside the fiducial volume, and their spatial extent will be several radiation lengths, i.e. several metres of water. Hence, a detector of linear dimension 5-10 metres is required.

(2) It must be massive. 3000 tons for 1 year would produce 1 event for lifetime 10^{33} years.

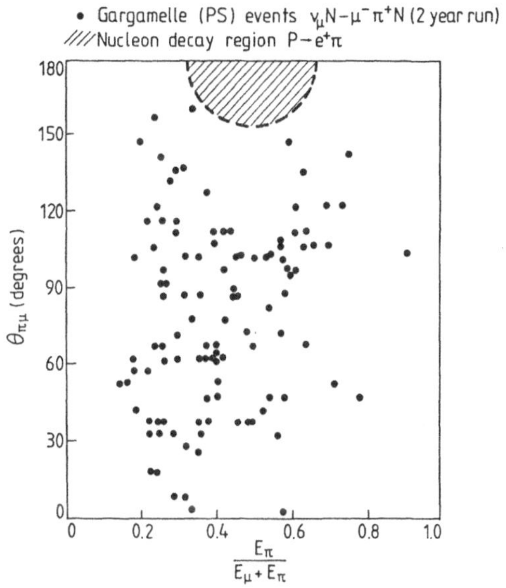

Fig. 4 Division of energy and lepton-pion angle in Gargamelle (PS) neutrino events of the type $\nu_\mu N \rightarrow \mu^- \pi^+ N$. The number of events is equal to that expected as neutrino background in a 2-year run of the IMB proton decay experiment (from IMB proposal[7]).

(3) It should be deep underground, h > 1500mwe to eliminate neutron background.

(4) It should have good energy resolution, good particle identification and track recognition.

Fig.5 summarises the timescales and rates involved. New experiments are aiming at the so far unexplored interval $10^{30} < \tau < 10^{33}$ years. For $\tau > 10^{33}$ years, background will be a major problem, and one would be reduced to comparing observed numbers of events with the predicted background with various cuts on the data. Eventually one will reach the point where one is taking differences of two large numbers (observed - predicted) with the signal/noise increasing only as $M^{0.5}$, so that a factor 10 on the lifetime implies a 100-fold increase in detector mass. This "neutrino brick wall" presumably occurs in the region of $\tau > 10^{33}$ years.

1.4.1 Water Cerenkov Devices

The Irvine-Michigan-BNL group[7] and the Harvard-Wisconsin-Purdue group[8] have proposed large water Cerenkov detectors, with useful (fiducial) masses on the order of 1000 - 4000 tons, or more.

In water, relativistic secondaries - and particularly electrons in electromagnetic showers from the decay modes $n \rightarrow e^+\pi^-$ and $p \rightarrow e^+\pi^o$ - will produce Cerenkov light at the Cerenkov angle in water (42°), which is transmitted in a cone about the particle trajectory and reaches the surface of the water as a ring of light. Fig.6 shows examples from the IMB proposal. The light is detected by a photomultiplier array placed at the surface (IMB) or in the water volume (HWP). Since the typical time jitter over the photocathode area of a 5" tube is ∿2ns, timing can also be used to estimate the direction of the light cone and thus, the track orientation.

Some of the factors involved in the water Cerenkov method are illustrated in Fig.7. The Cerenkov spectrum has the form $d\lambda/\lambda^2$, but the photoelectron yield depends on this, the quantum efficiency of the photocathode and attenuation of light in water. Even for pure (reverse osmosis) water, the number of photoelectrons is at least a factor 10 smaller than the number of Cerenkov photons in the visible region. The photoelectron yield has to be multiplied by the effective fractional surface area covered by photocathode, which is of order 4%. The resultant number of "hits" for a 20 x 20 x 20m cube with a surface array of 2400 5" photomultipliers is given in Table 1.3. The resolution could be improved by a factor 2, with 4 times the number of PM's, but this would treble the cost of the experiment.

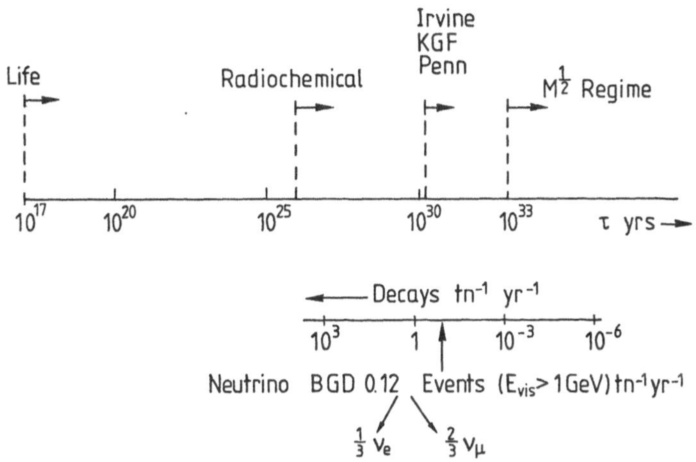

Fig.5 Timescales and backgrounds in proton decay.

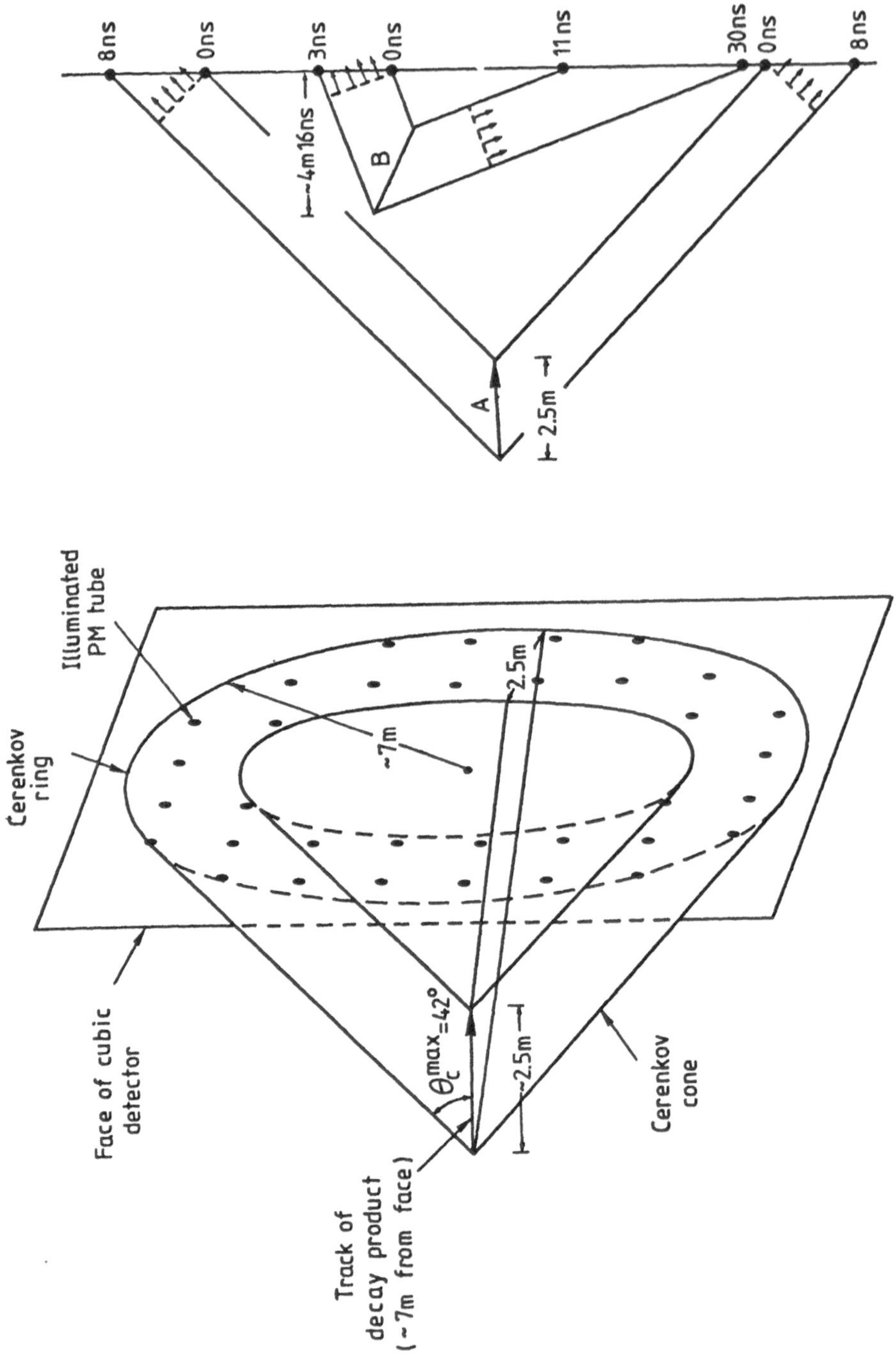

Fig.6 Typical Cerenkov light cones from relativistic particles in water detector (from IMB proposal).

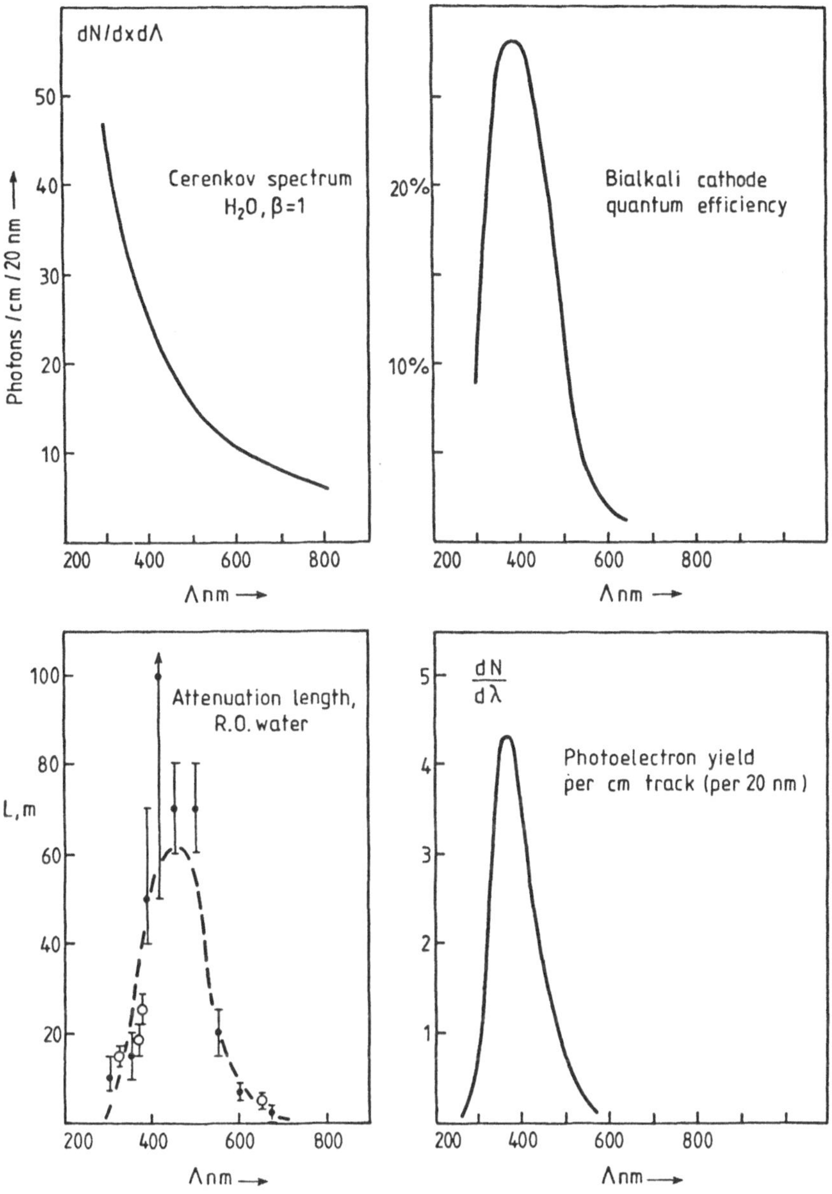

Fig.7 Cerenkov spectrum in water, photomultiplier quantum efficiency, and light attenuation as a function of wavelength. The photoelectron yield after 10m traversal of purified water is the product of factors from the preceding plots.

Table 1.3 Photomultiplier Hits for Various Decays

20 x 20 x 20m Cube with 2400 5" PM's on Surface

Decay Mode	# Hits	Energy Resolution (rms)
$p \to e^+\pi^o$	210	±16%
$n \to e^+\pi^-$	150	±19%
$p \to \mu^+\pi^o$	160	±18%
$n \to \mu^+\pi^-$	90	±24%

The analysis of the data from the water Cerenkov devices involves reconstruction of light cones, tracks and vertices and has to be done on-line by computer.

1.4.2 Calorimeter Devices

The other types of device proposed for detecting nucleon decay are solid calorimeters, in which it is planned to record the tracks of the charged secondary products by means of proportional wire or drift chambers, streamer chambers or flash tubes, arranged in planar arrays separated by iron plates. The main advantages of a solid calorimeter device is that the radiation length is short, a few cms rather than a metre, so that a decay event is spread over less than one metre linear dimension, and tracks can be easily reconstructed.

I mention as examples, two calorimeter devices. The Minnesota group[9] has built a 30 ton prototype calorimeter consisting of proportional tubes embedded in taconite (iron ore + cement) - see Fig.8. The density is $\sim 2.7 gcc^{-1}$, radiation length 6.7cm and nuclear mean free path 50 cm. The energy resolution for this device is rather poor (±30%) and the granularity is rather coarse, but it will serve to give experience for construction of a full-scale device. It is operating in the Soudan iron mine in northern Minnesota.

The Milano-Torino-Frascati group[10] are building a 150-ton calorimeter consisting of extruded plastic tubes operated in the streamer mode, with cathode strip read out in two orthogonal directions (Fig.9), layers of plastic tubes being separated by 1cm iron plates. A computer simulation of a $p \to e^+\pi^o$ decay is shown in Fig.10. It will be operated in the Mont-Blanc road tunnel (h = 5000mwe).

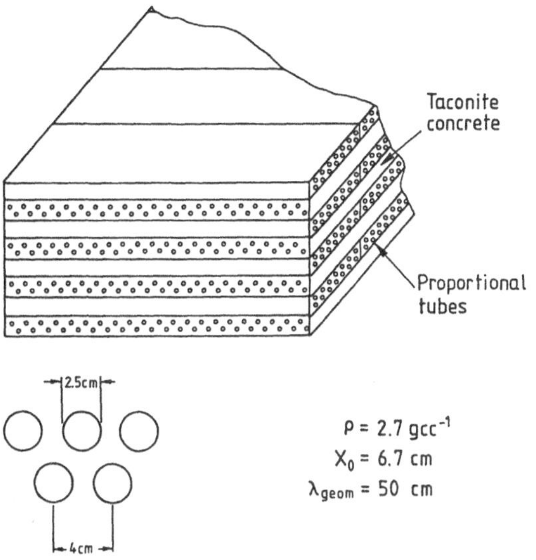

Fig.8 Taconite (iron + concrete) calorimeter of U. of Minnesota experiment employing proportional tube detectors.

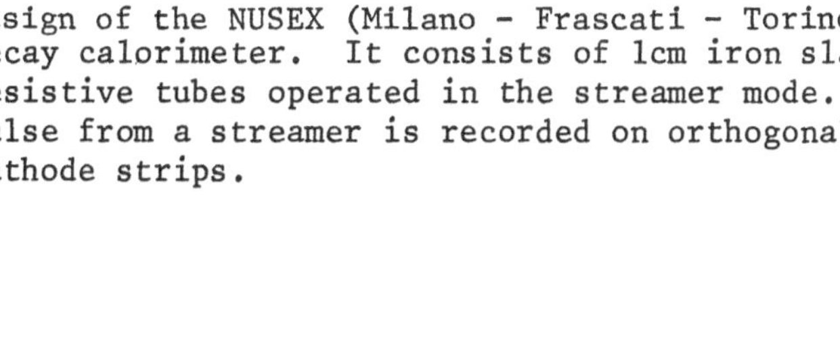

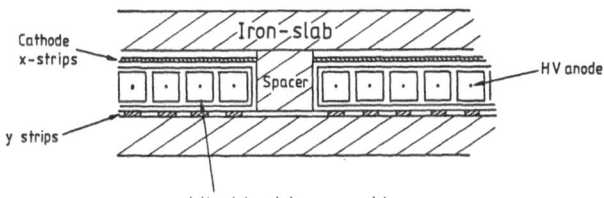

Fig.9 Design of the NUSEX (Milano - Frascati - Torino[10]) nucleon decay calorimeter. It consists of 1cm iron slabs with resistive tubes operated in the streamer mode. The ion pulse from a streamer is recorded on orthogonal layers of cathode strips.

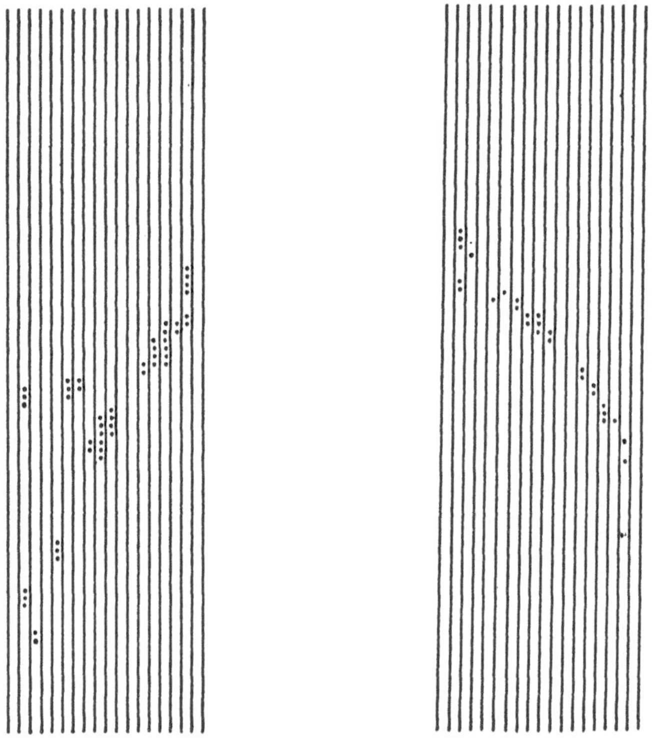

Fig. 10 Simulation of $p \rightarrow e^+\pi^0$ decay in NUSEX calorimeter. (90° Stereo Views).

One of the significant advantages of calorimeters is that test modules can be built small (∿1m dimension) and tested in accelerator neutrino beams, so that background problems can be studied in great detail. Compared with the water detectors, the mass one can achieve for a given sum of money is about a factor 10 smaller. Obviously, it is most important to build and operate both types of detector and to learn by experience of the different techniques which is more suitable. Unfortunately, this learning process will take years.

Even if the detectors now being constructed do not find evidence for nucleon decay, it is important to emphasise that totally unexpected new phenomena might turn up. This is the first time that physicists have operated detectors of hundreds of tons capable of observing very rare phenomena in conditions of very low background.

1.5 NEUTRON-ANTINEUTRON OSCILLATIONS

In the SU(5) grand unification scheme, baryon decay obeys the selection rule $\Delta B = \Delta L (=1)$. It is also possible to consider the process $\Delta B = 2$, that is, neutrons transforming to antineutrons via a 6 quark coupling (see, for example, Marshak & Mohapatra[11]). The rate of this spontaneous transition $\Gamma(n\bar{n}) < 10^{-22}$eV, by the arguments leading to τ(proton) $\sim 10^{32}$ years. If neutrons were free and in remote space, the $N\bar{N}$ transitions would be like those in the $K^o\bar{K}^o$ system. In real life, N, $\bar{N}$ states are not degenerate, because of the effects of magnetic fields, etc. which split the degeneracy by $\Delta E \gg \Gamma$ ($\Delta E \sim 10^{-11}$eV for free N, $\bar{N}$ in the earth's field). Thus, in practice the mixing angles are small. For a short observation time $t \ll \Gamma^{-1} = 10^5$ sec, the antineutron intensity in a free neutron beam has the simple form

$$I(\bar{N}) = I(N).(\Gamma/\Delta E)^2 \quad (5)$$

By using magnetic shielding to reduce ΔE, and assuming cooled thermal neutrons from a reactor observed over a 10m length, rates of a few events per year are calculated for a free oscillation time $\Gamma^{-1} = 10^5$ secs. An experiment is now being carried out at ILL, Grenoble[12]. The signal due to $\bar{N}$ annihilation after dumping the beam in a target is large (∿2GeV energy release). The main background comes from cosmic rays. Of course, $\Delta B = 2$ transitions are not predicted in the sense that proton decay is predicted, and the question of their existence is an experimental one.

2. NEUTRINO MASS AND NEUTRINO OSCILLATIONS

2.1 NEUTRINO MASS MEASUREMENTS

The most sensitive way to determine the ν_e mass is to observe the end-point of tritium β-decay, an allowed transition with a maximum electron energy E_o of only 18.6keV. The spectrum has the form

$$N(p) = \text{const.} f(z,E)\ p^2(E_o-E)\ \sqrt{(E_o-E)^2-m_\nu^2} \qquad (6)$$

where p,E are the momentum and kinetic energy of the electron, f(z,E) is the nuclear Coulomb factor which can be calculated exactly. If we plot $[N(p)/p^2f(z,E)]^{\frac{1}{2}}$ against E, a straight line should result (Kurie plot) as in Fig.11(a). For $m_\nu = 0$, the plot cuts the axis at $E=E_o$, while for finite m_ν, the plot turns over and cuts the axis vertically at $E=E_o-m_\nu$. So, in principle, the shape of the spectrum at the end-point provides a measure of m_ν. In practice, one has to fold in experimental resolution of the magnetic spectrometer, etc. which introduces a tail and makes the shape for $m_\nu=0$ and finite m_ν not so very different (Fig.11(b)).

Everything hinges on (i) counting rate (statistics) and (ii) resolution. Suppose one aims at a resolution ΔE, so one is concerned with the shape in the region $E_o-\Delta E$. Furthermore, to achieve a resolution of ΔE, the source size and thickness, magnet apertures etc. have to be matched and this introduces a further factor $(\Delta E)^2$ or $(\Delta E)^3$, so that the count rate varies as $(\Delta E)^5$ or $(\Delta E)^6$.

To try to combat the low rate near the end point, Bergkvist[13] developed a long source with a potential difference down it, so that β particles emitted with the same energy will be accelerated by different potentials from different points on the source, and traverse different path lengths in the magnetic field, such that they will reach the same momentum slit (see Fig.12). The required potential distribution along the source is determined experimentally by use of an electron gun. The calibration and resolution of the spectrometer was determined by use of Tm^{170} γ-source, giving K-conversion electrons of unique energy (22.9keV). From the Tm^{170} profile (determined by measuring the counting rate as a function of the Helmholtz coil current) the energy resolution was found to be ±0.12% (40eV). The source used by Bergkvist consisted of tritium gas absorbed in an aluminium substrate, and energy loss fluctuations of the electrons leaving the surface were measured to be 14 ± 5eV, yielding a total resolution of 55eV.

There were many other complications in the interpretation of the experiment. For example, in the β-decay

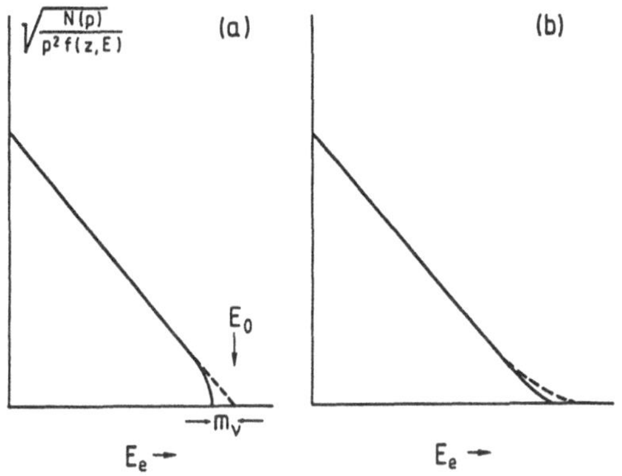

Fig.11 (a) Kurie plot of Tritium β-decay, showing the effect of finite neutrino mass.
(b) Actual shape of plot when experimental resolution is folded in.

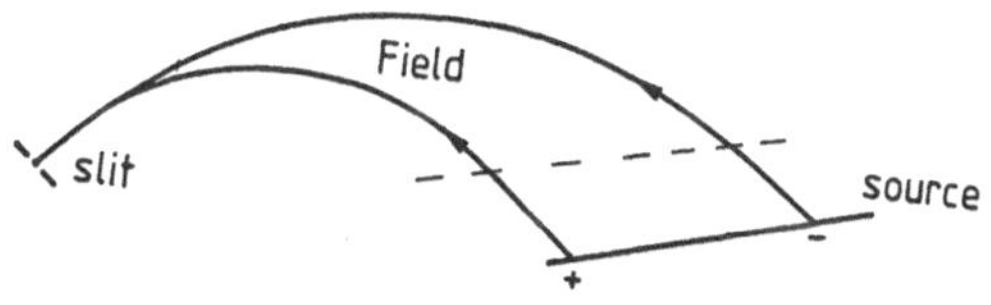

Fig.12 The acceptance of the β-source can be improved by extending it over a line and applying a suitable potential gradient, so that electrons emitted with the same energy from different points arrive at the momentum slit.

$$H^3 \rightarrow (He^3)^+ + e^- + \nu_e \qquad (7)$$

the He^+ ion can be left in the ground (1s) state or an excited (2s) state, with an energy difference of 40.5eV. Calculations showed that ∿70% of the ions are found in the ground state. When this fact is taken into account, the effective line width is increased to 70eV.

The ITEP experiment of Lyubimov et al[14], the first results of which were reported in 1976 by Tretyakov et al,[15] used a similar technique to that of Bergkvist, but the source consisted of an organic material (valine) with H^3 replacing H atoms. One of the problems with this source is that, when one triton in the valine molecule decays, the molecule breaks up and tritium atoms are released and form a background of decays in the spectrometer, proportional to the time. The resolution of the Moscow experiments was comparable with that of Bergkvist, but the source strength employed and observation time were much greater.

The Kurie plots for these two experiments are shown in Fig.13. Figs.13(a) and 13(b) are the results of Bergkvist et al[13] and Tretyakov et al,[15] both consistent with $m_\nu = 0$. The more recent data of Lyubimov et al[14] is given in (c) and (d). Fig.13(c) shows the combined result of 8 out of a total of 16 runs, indicating a finite mass. For each of the 16 runs, a most probable neutrino mass was computed, and the distribution is given in Fig.13(d), together with that expected for $m_\nu = 0$ and $m_\nu = 34$eV. These latest results appear to prove that the neutrino mass is finite and of order 40eV.

Improvements in these experiments seems possible. For example, it is feasible to design spectrometers with intrinsic resolution of 10^{-4}(=2 eV for tritium). With such precision, gas scattering in the spectrometer becomes a problem and thus, high vacuum is needed, and this suggests the use of solid tritium sources at low temperature (2.5°K) and with correspondingly low vapour pressure. Electrostatic methods are also being investigated. But such experiments are bound to be very difficult and to take a long time to give results.

The importance of the neutrino mass problem is not restricted to particle physics. It has profound cosmological significance. For years, it has been known that the motional kinetic energy of visible matter in the universe is an order of magnitude larger than the gravitational binding energy. However, if galaxies possess large neutrino haloes, these can correct the above mismatch if the neutrino mass is finite. Presumably in the "big bang", neutrinos were produced with similar number density to photons (the 3°K background radiation today). The observed ratio of baryons to photons is $B/\gamma \sim 10^{-9}$. So, a mass of 10^{-9} x 1GeV = 1eV for a

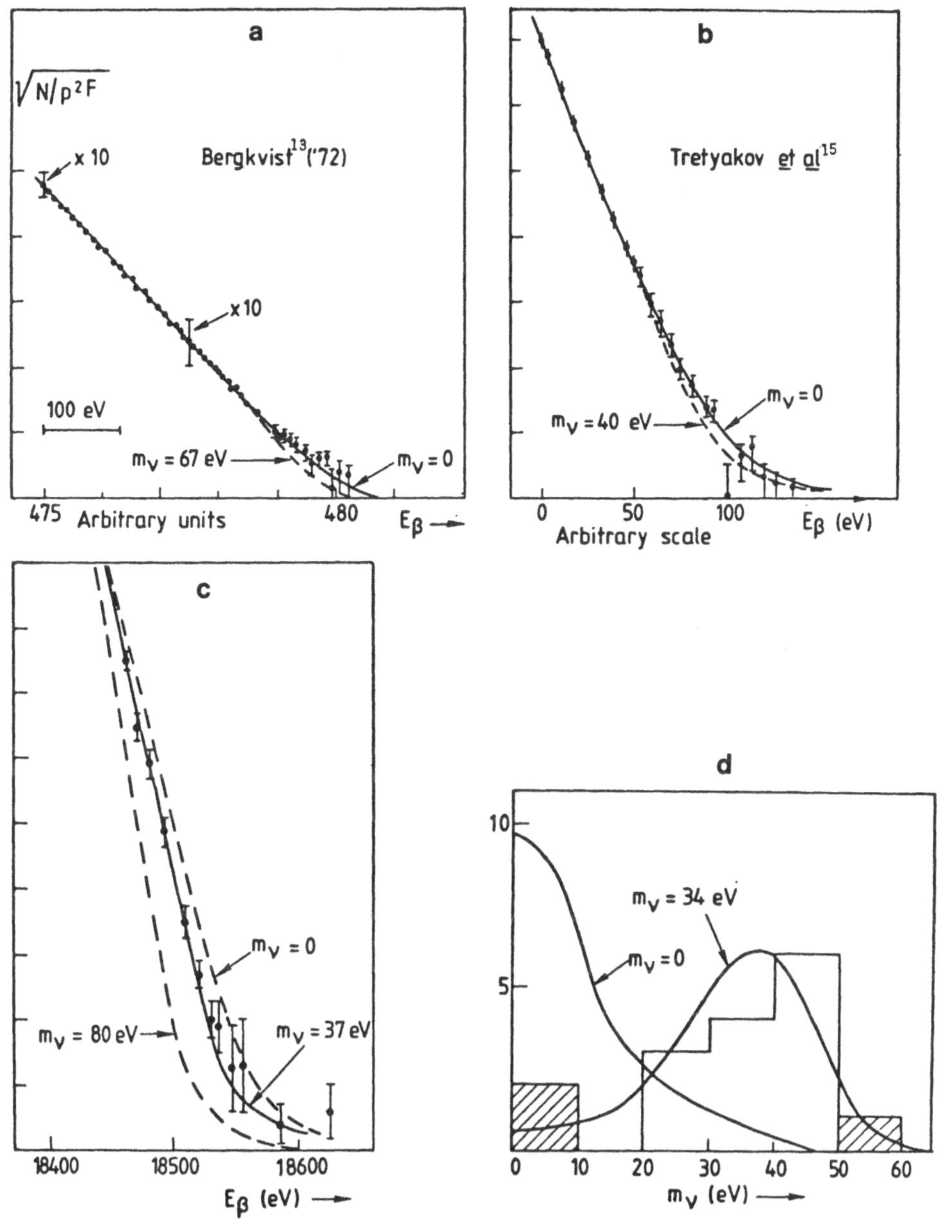

Fig.13 Kurie plots from recent observations[13,14,15] of the Tritium β-spectrum.

neutrino would appreciably increase the gravitational energy from the halo. Indeed, $m_\nu \sim 40eV$ is what astrophysicists estimate is required to close the universe, converting it from one which expands indefinitely to one that may even oscillate. By the same argument values of $m_\nu > 60eV$ are ruled out because that would conflict with the fact that the universe has expanded to the present scale.

The theoretically expected values of neutrino mass or mass differences are unknown and one can only make guesses. In the electroweak SU(2) × U(1) theory, neutrinos are exclusively left-handed and thus $m_\nu = 0$. This is also true in the SU(5) model of unification with minimal Higgs structure. With more complex Higgs structure, a small neutrino mass is expected (usually estimated at $\gtrsim 10^{-5}eV$). For the O(10) unifying group, a finite neutrino mass is expected, with a value proportional to that of the charge 2/3 quark in the same generation. For example one could have

$$m_{\nu_e} \sim 10^{-2} eV$$

$$m_{\nu_\mu} \sim 1eV$$

$$m_{\nu_\tau} \sim 30eV$$

compared with the observed limits of < 60eV, 1MeV and 250MeV respectively. In neutrino mixing, $\nu_\mu \leftrightarrow \nu_\tau$ is favoured over $\nu_\mu \leftrightarrow \nu_e$, in analogy with quark (c,s) mixing.

2.2 NEUTRINO OSCILLATIONS

The possibility that lepton number is not conserved, and that one type of neutrino (say ν_e) could oscillate into another type (say ν_μ) was proposed long ago by Pontecorvo[16] and Maki _et al_[17]. Considered as something of a curiosity at that time, it later received considerable impetus from the so-called solar neutrino problem. More recently, with the development of interest in gauge theories, it was realised that there were no good theoretical reasons for believing that baryon and lepton number should be conserved.

The theoretical treatment of neutrino oscillations is an almost exact analogue of the temporal development of a neutral kaon beam. Neutrino _mass eigenstates_ with well defined proper frequency are denoted ν_i (i=1,2,3.....). The _weak intereaction eigenstates_ are denoted ν_α (where α=e, μ, τ.......) and they are superpositions of the mass eigenstates. If the total lepton number $L = L_e+L_\mu+L_\tau+......$ is conserved, the two are related by a unitary transformation

$$|\nu_\alpha> = U_{\alpha i} |\nu_i> \tag{8}$$

where $U_{\alpha i}$ is a unitary matrix, and the probability of the transformation $\nu_\alpha \leftrightarrow \nu_\beta$ is given by

$$P|\nu_\alpha \to \nu_\beta| = |\sum_{i=1}^{i=n} U_{\alpha i} U^\dagger_{\beta i} \exp(-iE_i t)|^2 \tag{9}$$

where $E_i = p + m_i^2/2p$.

If the states ν_i are to be spatially coherent, they must be eigenstates of the momentum operator, so that they have energies which are slightly different, depending on the mass m_i, and where we assume $m_i << p$.

In order to illustrate the main features of the oscillation phenomenon, let us consider the simplest case of two types of neutrino, say ν_e and ν_μ. Then U is a 2 x 2 matrix which can be specified by one parameter, the mixing angle

$$\begin{pmatrix} \nu_\mu \\ \nu_e \end{pmatrix} = \begin{pmatrix} \cos\theta & \sin\theta \\ -\sin\theta & \cos\theta \end{pmatrix} \begin{pmatrix} \nu_1 \\ \nu_2 \end{pmatrix} \tag{10}$$

where

$$\begin{aligned} \nu_1(t) &= \nu_1(0)\, e^{-iE_1 t} \\ \nu_2(t) &= \nu_2(0)\, e^{-iE_2 t} \end{aligned} \tag{11}$$

$$E_{1,2} = p + \frac{m^2_{1,2}}{2p}$$

Let us start with $\nu_\mu(0)=1$ and $\nu_e(0)=0$ at t=0. So inverting (10) we get

$$\begin{aligned} \nu_2(0) &= \nu_\mu(0) \sin\theta \\ \nu_1(0) &= \nu_\mu(0) \cos\theta \end{aligned} \tag{12}$$

and from (10) and (11)

$$\frac{\nu_\mu(t)}{\nu_\mu(0)} = \cos^2\theta\, e^{-iE_1 t} + \sin^2\theta\, e^{-iE_2 t}$$

$$\frac{I_\mu(t)}{I_\mu(0)} = \left|\frac{\nu_\mu(t)}{\nu_\mu(0)}\right|^2 = \cos^4\theta + \sin^4\theta + \sin^2\theta\cos^2\theta[e^{i(E_2-E_1)t} + e^{-i(E_2-E_1)t}]$$

$$= 1 - \frac{\sin^2 2\theta}{2}\left[1 - \cos(E_2-E_1)t\right]$$

$$= 1 - \sin^2 2\theta . \sin^2\left[\frac{(E_2-E_1)t}{2}\right] \qquad (13)$$

If we write $(E_2-E_1) = \Delta m^2/2p$ where $\Delta m^2 = m_2^2 - m_1^2$ is the difference in mass squared, $p=E/c$ where E is the average energy, L the distance over which the beam has evolved, then

$$P(\nu_\mu \to \nu_e) = 1 - P(\nu_\mu \to \nu_\mu)$$

$$= \sin^2 2\theta . \sin^2\left(\frac{\Delta m^2 L}{4E}\right) \qquad (14)$$

$$= \sin^2 2\theta .\ \sin^2\left(\frac{1.27\ \Delta m^2 L}{E}\right) \qquad (15)$$

In (15), the numerical constant is that which applies if Δm^2 is in $(eV/c^2)^2$, L is in metres and E in MeV. Eqn.(15) shows that the relative intensity of ν_e and ν_μ oscillates with an amplitude depending on the mixing angle θ and a frequency or wavelength depending on the mass difference. This variation is shown in Fig.14. The crucial quantity is $\Delta m^2 L/E$, which must be greater than 0.25 if oscillating effects above the 10% level are to occur. In all practical experiments, one is dealing with a beam of neutrinos with an energy spread. Thus, if $\Delta m^2 L/E \gg 1$, the spectrum - averaged result in the case of 2 neutrinos would lead to an effective value $\langle P/\sin^2 2\theta\rangle = 0.5$.

The case of an arbitrary number n of neutrino states is more complicated. Usually it is assumed that one mass difference, and hence one oscillation frequency, dominates, when a formula like (15) will again hold, except that the coefficient outside the oscillating term is a product of mixing angles. In the limit $\Delta m^2 L/E \gg 1$, the spectrum - integrated intensity of any one component ν_α will have an average value 1/n, if there are n types of neutrino in total.

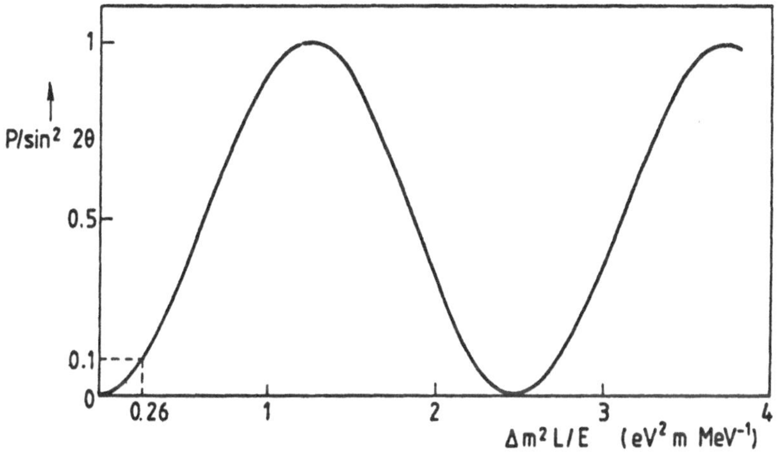

Fig.14 Oscillation amplitude as a function of mass difference, observation length and neutrino energy.

Table 2.1 Sensitivity for Various Neutrino Sources

Source	Mean energy	L(m)	Lower limit on Δm(eV)
Solar ν_e	100 keV	10^{11}	10^{-6}
Atmospheric ($\phi_{\nu_\mu} \sim 2\phi_{\nu e}$)	0.5 GeV	10^7	10^{-3}
Reactors $\bar{\nu}_e$	3 MeV	10	10^{-1}
Meson Factories (ν_μ,ν_e)	30 MeV	10-100	10^{-1}
Accelerators (ν_μ)	1-30 GeV	10^2-10^4	10^{-1}

Oscillation experiments can in principle be done with various sources. In the above table, we list these, together with the typical observation lengths L which can be achieved, and the typical mass differences to which they are sensitive (assuming the best possible experimental conditions, small statistical errors or large mixing angles).

The atmospheric neutrinos are generated by cosmic rays, and as indicated previously, the reaction rate is about 0.2 ton^{-1} yr^{-1}. Using the earth's diameter as the oscillation distance, one could for example observe any asymmetry of events containing muons, upwards as compared with downwards. Even for a 1000 ton detector operated for several years, it would be difficult to establish any oscillations of amplitude <10%. It is a useful fall-back possibility for nucleon decay detectors, since no artificial sources could match the mass resolution attainable.

2.3 REACTOR EXPERIMENTS

Nuclear reactors are a prolific source of antineutrinos $\bar{\nu}_e$, and the effect of any oscillations ($\nu_e \rightarrow \nu_\mu$, ν_τ.....) would be to reduce the number of charged-current reactions (e.g. $\bar{\nu}_e + p \rightarrow n + e^+$) but to leave the number of neutral-current reactions ($\bar{\nu}_e$, $\bar{\nu}_\mu$, $\bar{\nu}_\tau + d \rightarrow n + p + \bar{\nu}_e$, $\bar{\nu}_\mu$, $\bar{\nu}_\tau$) unchanged. Since the reactor spectrum extends only to a few MeV energy, any $\bar{\nu}_\mu$ or $\bar{\nu}_\tau$ particles will be below threshold for the charged-current process. Therefore, experimental evidence for oscillations depends on comparison of the absolute number of charged-current reactions with that expected in absence of oscillations, or on measurement of the ratio of charged to neutral current cross-sections and comparison with theory, for which all the required parameters are well known. Because of the low neutrino energy (2-6MeV), and the source-detector distances employed (5-10m), these experiments are sensitive to mass differences $\Delta m^2 \sim 0.1(eV/c^2)^2$

The Irvine experiment at the Savannah River Reactor[18] employed a deuterium target and He^3-filled neutron counters to detect the charged and neutral current reactions,

		$E_{(threshold)}$	
CC:	$\bar{\nu}_e + d \rightarrow n + n + e^+$	4MeV	(16)
NC:	$\bar{\nu}_e + d \quad n + p + \bar{\nu}_e$	2.2MeV	(17)

Note the difference in threshold energies, and recall that the reactor spectrum peaks at ∿0.5MeV and falls off very rapidly with increasing energy. The expected ratio of the two reaction rates must depend on the shape of the spectrum. Reines et al[18] observed a spectrum-averaged ratio at L=11m:-

$$r_{expt} = \frac{\sigma\ (CC)}{\sigma\ (NC)} = 0.167 \pm 0.093 \tag{18}$$

compared with the expected numbers (in the absence of oscillations)

$$r_{th} = 0.42 \qquad \text{(Avignone)}$$
$$= 0.44 \qquad \text{(Davis-Vogel)} \tag{19}$$

where the two estimates correspond to slightly different forms calculated for the spectrum., Thus the ratio of ratios was

$$R = \frac{r_{expt}}{r_{theory}} = 0.40 \pm 0.22 \quad \text{(Avignone)}$$
$$= 0.38 \pm 0.21 \quad \text{(Davis-Vogel)} \tag{20}$$

significantly less than unity, and claimed as evidence in favour of oscillations, with $\Delta m^2 \sim (1eV/c^2)^2$.

Another reactor experiment has been carried out at the ILL (Grenoble) reactor[19] in which the observed rate for the reaction

$$\bar{\nu}_e + p \rightarrow n + e^+ \tag{21}$$

was compared with that expected theoretically. In this case, the products are detected by a two-fold delayed coincidence of the γ-rays from e^+ annihilation and those from neutron capture in gadolinium. With L=8.7m they find, averaged over the spectrum

$$\frac{\sigma_{exp}}{\sigma_{th}} = 0.87 \pm 0.14 \quad \text{(Davis - Vogel)} \tag{22}$$

consistent with the unity value expected in absence of oscillations. The ratio (22) as a function of energy is shown in Fig.15. Boehm *et al*[19] deduce from these results $\Delta m^2 < 0.15(\mathrm{eV}/c^2)^2$, assuming maximal mixing.

The results (20) and (22) are not compatible - see Fig.16 - and it is clear that more experiments are required, and additionally better information on the spectrum. So far, spectra have all been calculated assuming a plausible model for the fission fragment distribution; recently, direct observations of the β spectrum from fission fragments have been carried out and preliminary results support the Davis - Vogel fluxes.

2.4 COSMIC AND SOLAR NEUTRINO EXPERIMENTS

Other evidence quoted in favour of neutrino oscillations comes from cosmic ray and solar neutrino observations.

In deep underground experiments, the Case-Witts-Irvine collaboration[20] and the India-Japan-Durham collaboration [21] compared the absolute rate of charged-current events produced by muon neutrinos generated by π and K decay in the atmosphere with that expected from cross-sections measured at accelerators. The events occurred in rock surrounding the detectors, which only observed the secondary muons (differentiated from atmospheric muons by a zenith angle cut, $\theta > 45^o$). The observed rates were somewhat smaller than expected. For example, Crouch *et al*[21] found

$$\frac{\bar{\sigma}(\text{predicted})}{\bar{\sigma}(\text{observed})} = 1.60 \pm 0.4$$

However, since what is observed are muons produced in rock by a neutrino flux which is not known to better than 30%, from which atmospheric muon effects have to be subtracted, any discrepancy between the above result and the unit value can hardly be significant.

Oscillations have also been suggested to account for the discrepancy between the observed and expected solar neutrino flux[22].

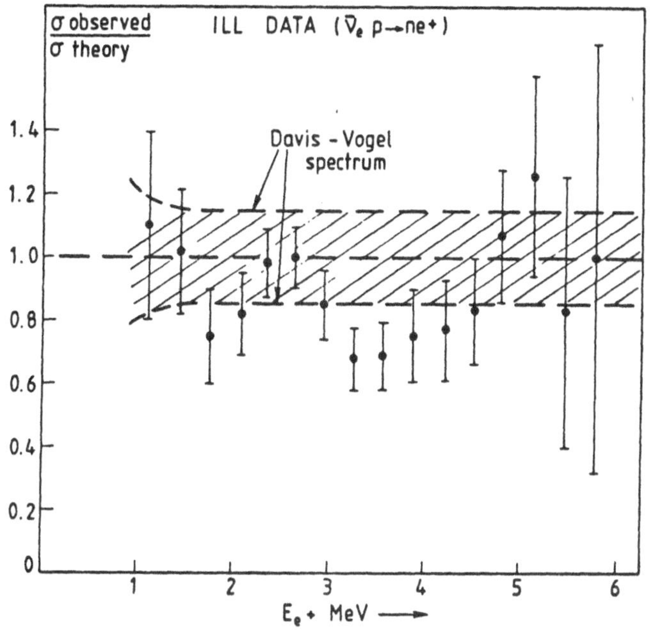

Fig.15 ILL data[19] on the ratio of observed to expected rates for $\bar{\nu}_e p \to ne^+$.

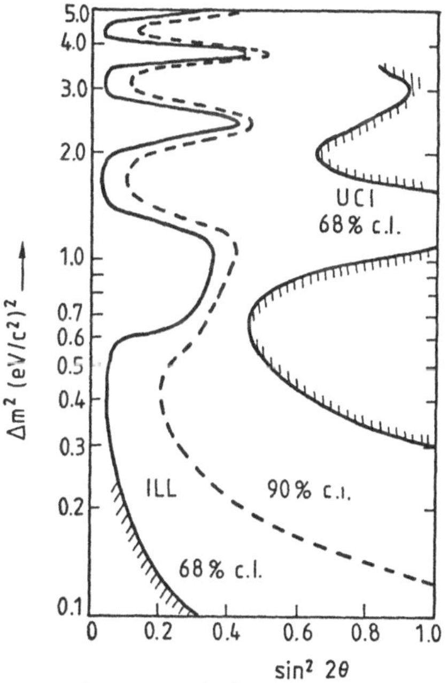

Fig.16 65% and 90% confidence limits from the Savannah River[18] and Grenoble[19] reactor experiments.

The Davis experiment[23] relies for detection on the reaction $\nu_e + Cl^{37} \to Ar^{37} + e^-$ which is above threshold for neutrinos from the main reaction ($p+p \to d+e^+ + \nu_e$) and relies on those from β decay of Be^7 and B^8 made in side reactions. The standard solar model predicts a rate of 7.8 ± 1.5 SNU (solar neutrino units) compared with an observed rate of 2.2 ± 0.4 SNU. The solution to this discrepancy has to await the development of Gallium and Indium detectors sensitive to the pp reaction, and it is quite wrong to conclude that the "solar neutrino problem" is yet evidence for neutrino oscillations.

2.5 MESON FACTORY EXPERIMENTS

Meson factories generate intense low energy pion beams (no kaons) and produce reactions via the decays at rest

$$\pi^+ \to \mu^+ + \nu_\mu \tag{23}$$

$$\mu^+ \to e^+ + \nu_e + \bar{\nu}_\mu \tag{24}$$

The experimental approach at LAMPF[24] was to use a 6 ton water Cerenkov counter to detect relativistic electrons ($e^\pm$) from the possible charged-current reactions

$$\left.\begin{array}{l} \nu_e d \to ppe^- \quad (25) \\ \bar{\nu}_e d \to nne^+ \quad (26) \end{array}\right\} D_2O$$

$$\bar{\nu}_e p \to ne^+ \qquad H_2O \tag{27}$$

If no oscillations occur, then from (24) only the reaction (25) is allowed. Hence, one expects a signal from D_2O but none from H_2O. They measured

$$R_{H_2O} = \bar{\nu}_e / \mu^+ \text{ decay} = -0.001 \pm .061 \tag{28}$$

and put a limit $\Delta m^2 < 0.9(\text{eV})^2$ for the transition $\bar{\nu}_\mu \leftrightarrow \bar{\nu}_e$. From the heavy water they found

$$R_{D_2O} = \nu_e/\mu^+ \text{ decay} = 1.1 \pm 0.4 \tag{29}$$

Assuming CP invariance (28) shows that $\nu_\mu \leftrightarrow \nu_e$ is excluded, so that the result (29) then sets limits on mixing $\nu_e \leftrightarrow \nu_\tau$, with $\Delta m^2 < 2.6(\text{eV})^2$. However, the errors are such that this result is only valid if the mixing is large, with $\sin^2 2\theta > 0.3$.

2.6 HIGH ENERGY ACCELERATORS

The results from high energy accelerators, as well as the meson factory experiment, are summarised in Table 2.2. The beams from high energy accelerators are formed from sign-selected pion, and to a lesser extent kaon, decay in flight. They are dominantly ν_μ or $\bar{\nu}_\mu$ with a small admixture, typically 0.5%, of ν_e, $\bar{\nu}_e$ from Ke3 decay. Thus, they are very sensitive to the transition $\nu_\mu \leftrightarrow \nu_e$, since even a 1% effect would double the number of charged current ν_e events with electron secondaries.

In experiments to date - not optimised for detecting oscillations but for maximum flux - the distances L from the source to the detector have been determined by the amount of earth or steel required to absorb background muons, which is proportional to the muon energies in pion decays in flight, and hence to the neutrino beam energy. Hence L/E is practically the same in all PS, SPS or FNAL experiments. In future experiments, L/E values can be 1-2 orders of magnitude larger.

The second entry in Table 2.2 relating to Gargamelle experiments at the CERN PS several years ago, still provides the most stringent limits on the amplitude squared for the transition $\nu_\mu \leftrightarrow \nu_\tau$ or $\nu_e \leftrightarrow \nu_\tau$. Since the tau neutrinos were below threshold for the charged current reaction

$$\nu_\tau + N \rightarrow \tau^- + X \tag{30}$$

the oscillation $\nu_\mu \leftrightarrow \nu_\tau$ might be detected as a decrease in the absolute number of charged current ν_μ events. But this absolute number is determined by fluxes and structure function integrals and cannot be computed from first principles.

The 15' bubble chamber experiments at FNAL relied on a comparison of the observed ratio of charged current events containing electron and muon secondaries, with that computed from the ν_e and ν_μ fluxes in the absence of oscillations. The reaction (30) results in (almost prompt) electron secondaries through the decay

$$\tau^- \rightarrow e^- + \bar{\nu}_e + \nu_\tau \tag{31}$$

with a 15% branching ratio. The observed e^-/μ^- ratio in neutrino running then sets the limit quoted. For antineutrino reactions, a better limit can be obtained, since in the charged current reaction

$$\bar{\nu}_e + N \rightarrow e^+ + X \tag{32}$$

Table 2.2 Accelerator Oscillation Experiments

$P(\nu_\alpha \to \nu_\beta)$

Experiment	$\bar{E}$	Test	90% CL or rate	(L/E) m MeV^{-1}	Δm^2(90%CL, $\theta=\pi/4$)
LAMPF (ref.24)	50MeV	$\bar{\nu}_\mu \to \bar{\nu}_e$	<0.10	0.3	$<0.9(eV)^2$
GGM PS (ref.25)	2GeV	$\nu_\mu \to \nu_e$	<.0013	0.02	$<1(eV)^2$
		$\bar{\nu}_\mu \to \bar{\nu}_e$	<.0014		$<1(eV)^2$
FNAL 15' (ref.26)	30GeV	$\nu_\mu \to \nu_\tau$	<.025	0.01	$<6(eV)^2$
(ref.27)	30GeV	$\bar{\nu}_\mu \to \bar{\nu}_\tau$	<.0075	0.01	$<6(eV)^2$
ν Emulsion (ref.28)	30GeV	$\nu_\mu \to \nu_\tau$	<.013	0.01	$<3.5(eV)^2$
BEBC SPS NBB (ref.29)	80GeV	$\nu_\mu \to \nu_e$	<.005	0.01	$<55(eV)^2$
GGM SPS WBB (ref.30)	30GeV	$\nu_\mu \to \nu_e$	<.004	0.01	$<2(eV)^2$
		$\nu_\mu \to \nu_\tau$	<.03	0.01	$<10(eV)^2$
LAMPF (ref.24)	50MeV	$\nu_e \to \nu_e$	1.1 ± 0.4		$<2.6(eV)^2$
GGM PS (ref.25)	2GeV	$\nu_e \to \nu_e$	0.92 ± 0.21		$<1(eV)^2$
BEBC SPS (ref.29)	80GeV	$\nu_e \to \nu_e$	1.04 ± 0.15		$<55(eV)^2$

the positron retains the bulk of the energy (the distribution varies as $dN/dy \propto (1-y)^2$ where $(1-y) = E_{e^+}/E_{\bar{\nu}}$). On the other hand production of τ^+ and subsequent decay as in (31) results in lower energy positrons. Hence, the low y events are very sensitive to a small $\bar{\nu}_\tau$ admixture. These, and the CERN SPS experiments, find no evidence for the transitions $\nu_\mu \leftrightarrow \nu_\tau$, but are only sensitive to mass differences such that $\Delta m^2 > 5(eV)^2$.

Lastly, absolute calculations of ν_e charged-current rates (based on e-μ universality and the known ν_μ cross-sections) are sensitive to the transition $\nu_e \leftrightarrow \nu_\tau$. The most reliable limit is set by a BEBC SPS experiment with a narrowband neutrino beam, where the ν_e flux from Ke3 decay was known to within 7%. Based on 70 e^- events, the experiment found $P(\nu_e \rightarrow \nu_e) = 1.04 \pm 0.15$. Because of the limited statistics and the beam energy, it would be sensitive only to $\Delta m > 7$ eV and mixing amplitudes, squared, exceeding 20%.

If we, optimistically, accepted a mass of order 30eV for ν_τ as proposed by cosmologists, and smaller masses for ν_μ and ν_τ, then the Δm^2 limits of present experiments are not important for the transition $\nu_e \leftrightarrow \nu_\tau$, or $\nu_\mu \leftrightarrow \nu_\tau$ and we could conclude that the mixing angles are small ($\theta < 0.2$ radians). If we make no assumptions about masses, then it seems that future progress will depend on

i) Accelerator experiments with much higher statistics, better measured fluxes and with larger values of L/E. However, regarding the last, only one order of magnitude could be obtained on the limiting value of $\Delta m^2 (<0.1(eV)^2)$. Such experiments are our best hope if mixing angles are small (say of order the Cabibbo angle). A number of such experiments are planned for the future.

ii) Experiments without accelerators, using the naturally occurring solar and atmospheric neutrinos and the large base lines provided by the size of the Earth or Earth-Sun distance. These would be sensitive to very small Δm^2 values but large mixing amplitudes would be required because event rates are inevitably low.

REFERENCES

1. E.W. Hennecke et al., Phys. Rev. D11:1378 (1975).
2. R. Steinberg and J. Evans, Proc. Neutrino Conf. '77 Vol.2:321 (1977).
3. J. Learned, F. Reines and A. Soni, Phys. Rev. Lett. 43:907 (1979).
4. J. Ellis et al., Nucl. Phys. B176:61 (1980).
5. M.G.K. Menon, Proc. of Neutrino Conf. '76, Aachen (1976).
6. A. Grant, CERN internal note EF/ALG (1979).
7. IMB proposal, D. Sinclair et al., presented at Workshop on Grand Unification, Ann Arbor, Michigan (1981).
8. Harvard-Purdue-Wisconsin proposal, C. Wilson et al., presented at Workshop on Grand Unification, Ann Arbor, Michigan (1981).
9. Minnesota proposal; J. Bartelt et al., Minnesota-ANL-Oxford preprint COO-1764-410, ANL-HEP-PR-81-12, OUNP-42-81 (1981).
10. Milano-Frascati-Torino proposal, G. Battistoni et al., Milano preprint (1979).
11. R.E. Marshak and R.N. Mohapatra, Phys. Lett. 91Bà222 (1980); Phys. Rev. Lett. 44:1316 (1980).
12. M. Baldo-Ceolin, 7th Trieste Conf. on Particle Physics, (1980).
13. K.E. Bergkvist, Nucl. Phys. B39:317 (1972).
14. V.A. Lyubimov et al., Phys. Lett. 94B:266 (1980).
15. E.F. Tretyakov et al., Proc. Neutrino Conf. '76, Aachen (1976); DESY preprint 81/012 (1981).
16. B. Pontecorvo, Sov. Phys. JETP 26:984 (1968).
17. Z. Maki et al., Prog. Th. Phys. 28:870 (1962).
18. F. Reines, H.W. Sobel and E. Pasierb, Phys. Rev. Lett. 45:1307 (1980).
19. F. Boehm et al., Phys. Lett. 97B:310 (1980).
20. M.F. Crouch et al., Phys. Rev. D18:2239 (1978).
21. M.R. Krishnaswamy et al., Proc. Roy. Soc. A323:489 (1971).
22. J. Bahcall and S.C. Frautschi, Phys. Lett. 29B:623 (1969).
23. R. Davis, S.C. Evans and B.T. Cleveland, Proc. Neutrino Conf., Purdue (1978).
24. S.E. Willis et al., Phys. Rev. Lett. 44:522 (1980).
25. J. Blietschau et al., Nucl. Phys. B133:205 (1978).
26. A.M. Cnops et al., Phys. Rev. Lett. 40:144 (1978).
27. B.P. Roe (FIMS collaboration) private communication (1980).
28. T. Kondo et al., FERMILAB-preprint, Conf.-80/92-EXP (1980).
29. H. Deden et al., Phys. Lett. 98B:310 (1981).
30. N. Armenise et al., CERN preprint EP/80-226 (1980).

THE UNIVERSE STRIKES BACK

Gary Steigman

Bartol Research Foundation
University of Delaware
Newark, Delaware 19711 U.S.A.

There is a theory which states that if ever anyone discovers exactly what the Universe is for and why it is here, it will instantly disappear and be replaced by something even more bizarre and inexplicable. *

INTRODUCTION

The mysteries of particle physics are conventionally probed at large accelerators. At present, much of the fascinating new phenomenology predicted by the currently fashionable crop of Unified, Grand Unified and Super Unified theories is too rare or at too high energy to be studied at accelerator facilities. To remedy this doleful situation, imaginative and ambitious new experiments in exotic terrestrial settings are being planned and built. It must be granted, however, that the most radical approach to date has been that which attempts to utilize the

* Douglas Adams in "The Restaurant at the End of the Universe" (Pan Books Ltd., London).

entire Universe as a laboratory.

The Universe is very big and very old; there is a lot of space and time available for rare events. The early Universe was very hot and very dense; high energy collisions occurred frequently in such an environment. Cosmological considerations have been most profitably employed in providing constraints on new models of particle physics (for reviews of this rapidly burgeoning field of interdisciplinary research see Steigman 1979, Dolgov and Zeldovich 1981, Kolb and Turner 1981). Occasionally there has been feedback and new developments in particle physics have suggested new solutions to old problems in cosmology (e.g. baryon nonconservation and the universal baryon asymmetry; massive neutrinos, or other "inos"; and the dark matter or "missing mass" problem).

At a superficial level, unified theories appear beautifully simple and esthetically elegant; under closer scrutiny, however, problems and complications often emerge. Do similar skeletons lurk in the cosmological closet? The growing symbiotic relationship between particle physics and cosmology requires that we search out and expose any problems or complications which may have been swept under the rug.

The simplest hot, big bang model - the "standard" model - has been remarkably successful in providing a theoretical framework for the interpretation of such cosmological observations as the black body radiation and the abundances of the light elements. The concordance of the theoretical predictions with the observational data suggests that the standard model may provide a correct description of the Universe when it was much younger, smaller, hotter and denser than it is today. Are there, nonetheless, inadequacies in this model which might cause us to pause in our headlong rush to use the Early Universe as a Cosmic Accelerator? There are, indeed, some nagging, persistent problems which, in our current euphoria, we tend to overlook or sweep aside. Their elucidation is the goal of this talk.

There are a series of Cosmological Puzzles which are well known to a limited coterie of cosmologists. In presenting them here to a broader audience, I lay no claim to originality. It is hoped that there may be some value in considering these problems from a fresh perspective. The juxtaposition of what may seem, superficially, to be unrelated issues will reveal, through a careful but simple analysis, a common ingredient. No solutions will be proposed; I am aware of none. It is my modest goal that more of my colleagues become aware of, and understand the nature and significance of, these problems. The three puzzles outlined here are described and discussed in greater detail in subsequent sections.

Flatness-Longevity Puzzle

Why has the Universe evolved for ~10 billion years without becoming "curvature dominated"? Why was the Early Universe so nearly "flat"? We will see that these two questions are really one; they may be rephrased as, Why is our Universe so "long-lived" and why, at present, it is "young"?

Horizon-Homogeneity Puzzle

Why is the Universe homogeneous on scales which were causally disjoint? Since the Universe was never as old as it is today, there are regions we can observe which have never been able to exchange information. How, then, did they "know" to be at the same temperature and density?

Cosmological Constant Puzzle

Why is the cosmological constant, Λ, (or, equivalently, the energy density of the vacuum) so very, very small? Observations limit Λ, in dimensionless units, to $\Lambda \lesssim 10^{-120}$. Is $\Lambda = 0$? If so, why and how?

THE EVOLUTION OF UNIVERSES

For homogeneous and isotropic Universes, the description of their evolution is particularly simple. Such Robertson-Walker-Friedman models fall into two categories: Open models which expand forever; Closed models which expand for a finite time and then collapse. These two possibilities are illustrated in Figure 1 which displays the scale factor R as a function of time. R will be carefully defined in a moment; for the present it is sufficient to think of R as the distance between test particles (e.g. galaxies or clusters of galaxies). For the closed model ($k>0$; the 3-space curvature k will be defined below) there is a special time t_* when the expansion stops. Although not as dramatically apparent, t_* also has significance for open models; t_* is the epoch at which an open model becomes "curvature dominated"; for $t \gtrsim t_*$, $R \propto t$. In both cases ($k \gtrless 0$) then, the Universe is Curvature Dominated (CD) for $t \gtrsim t_*$.

Within each category, open or closed, there are an infinite number of possible Universes. For the closed models this is illustrated in Figure 2. Model A quickly expands to its maximum extent and then collapses; models B and C (etc.) live longer before suffering the same fate. This leads us to ask an important, often overlooked, question: What determines the longevity (t_*) of a Universe? What is it that fixes the timescale t_* for a Universe to become Curvature Dominated?

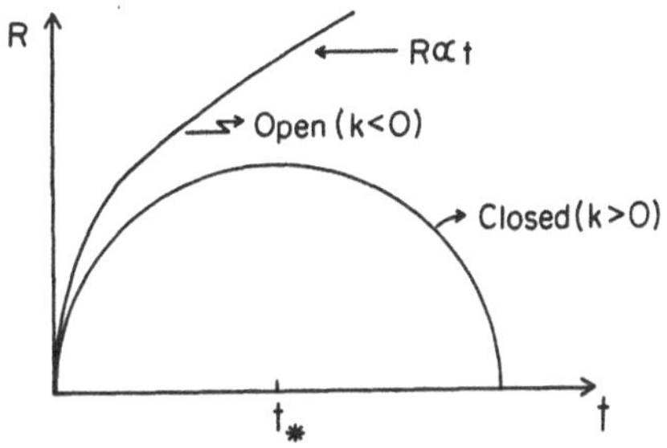

Fig. 1. The Evolution of Universes. The Scale Factor R is shown as a function of the Cosmic Time t for two representative models: Open (k<0), Closed (k>0). The epoch t_*, when the Universe becomes Curvature Dominated, is indicated.

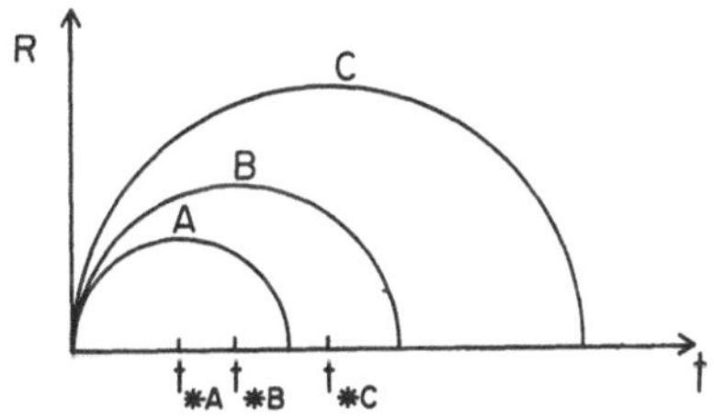

Fig. 2. A sample of Closed Universes. The evolution of models with different lifetimes ($\sim t_*$) is shown.

The Planck Scale

In cosmology, the Planck scale (involving G, $\hbar$ and c) is the only scale in town; of course, when the particle content (in particular, particle masses) is specified, other scales are available. The Planck mass is: $M_P=(\hbar c/G)^{1/2}=2\times10^{-5}$g; the corresponding temperature or energy scale is $kT_P=M_Pc^2=1.2\times10^{-19}$GeV. The Planck length is $l_P=\hbar c/kT_P=1.6\times10^{-33}$cm; the Planck time $t_P=l_P/c=0.5\times10^{-43}$s. In terms of these "natural" units for cosmology, our Universe is very old and very cold.

$$\frac{t_o}{t_P} \approx 10^{60} ; \frac{T_o}{T_P} \approx 10^{-32}. \tag{1}$$

If our Universe is curvature dominated, it could have only occurred recently so that $t_\star \gtrsim t_o \approx 10^{60} t_P$. Our Universe is exceedingly long-lived.

In our search for what distinguishes those R-W-F Universes which evolve quickly ($t_\star \approx O(t_P)$) from those which are long-lived, we seek a quantity which differs from model to model. That quantity must be time independent; otherwise we could be misled by comparing different models at different times. Our telltale clue must also be scale independent; it must not depend on the size of the volume employed to define it. As a preliminary step in our search, let us first look at the "Geometry" of the homogeneous, isotropic (RWF) models.

Geometry

A unique metric, the Robertson-Walker metric, describes the homogeneous, isotropic models.

$$ds^2 = c^2dt^2-a^2(t)[\frac{dr^2}{(1-kr^2)} + r^2(d\theta^2+\sin^2\theta d\phi^2)] \tag{2}$$

In equation (2), the scale factor a(t) describes the overall expansion (or contraction) of the Universe; r, θ, ϕ are the comoving coordinates which label the positions of those particles which expand with the average expansion of the Universe; the constant k is the 3-space curvature. The positive curvature (k>0) models are called closed or spherical; such models are finite: $r \leq r_{max}=k^{-1/2}$. The negative curvature models (k<0) are called open or hyperbolic; these models are infinite in extent ($r \to \infty$). The special, zero curvature (k=0) models are called flat; the flat models are unbounded ($r \to \infty$). We shall see later that the flat models are, in some sense, a pathological limit of both the open and closed models.

It is convenient to redefine the scale factor and the comoving radial coordinate in the R-W metric.

$$ds^2=c^2dt^2-R^2(t)[d\Theta^2+F^2(\Theta)(d\theta^2+\sin^2\theta d\phi^2)]. \tag{3}$$

The new radial comoving coordinate Θ is defined by

$$d\Theta^2 = \frac{dr^2}{(1-kr^2)} . \tag{4}$$

And,

$$k > 0: \quad R(t) = \frac{a(t)}{(k)^{\frac{1}{2}}} ; \quad F(\Theta) = \sin\Theta, \tag{5a}$$

$$k < 0: \quad R(t) = \frac{a(t)}{(-k)^{\frac{1}{2}}} ; \quad F(\Theta) = \sinh\Theta, \tag{5b}$$

$$k = 0: \quad R(t) = a(t) \quad ; \quad F(\Theta) = \Theta. \tag{5c}$$

The equation of motion for photons (or any massless particles) is: ds=0. Light emitted at t_1 and absorbed at t_2 will have travelled a comoving distance equal to

$$\Theta_{12} = \int_1^2 \frac{cdt}{R(t)}. \tag{6}$$

The "Particle Horizon", $\Theta_H(t)$, is obtained from equation (6) by taking the limits: $t_1 \to 0$(or, $t_1 \to t_P$) and $t_2 \to t$. The Particle Horizon defines the size, in comoving coordinates, of that part of the Universe over which information could, in principle, have been transmitted. It should not be confused with the less frequently used "Event Horizon", $\Theta_E(t)$, which is obtained from equation (6) by taking the limits: $t_1 \to t$ and $t_2 \to \infty (k<0)$ or $2t_*(k>0)$. The Event Horizon provides the size, in comoving coordinates, of that part of the Universe with which communication in the future will be, in principle, possible.

From the R-W metric it is possible to determine the proper volume at time t encompassed by the comoving radial coordinate .

$$V(\Theta,t)=2\pi R^3(t)\,\Phi(\Theta). \tag{7}$$

$\Phi(\Theta)$ is the "coordinate volume" out to comoving coordinate Θ .

$$k>0:\ \Phi = \Theta - \sin\Theta\cos\Theta \tag{8a}$$

$$k<0:\ \Phi = \sinh\Theta\cosh\Theta - \Theta, \tag{8b}$$

$$k=0:\ \Phi = 2\Theta^3/3. \tag{8c}$$

Notice that closed (k>0) models have a finite coordinate volume: $\Phi_{max} = \Phi(\Theta = \pi) = \pi$, so that $V_{max}(t) = 2\pi^2R^3(t)$.

The proper volume out to the particle horizon is $V_H(t)= V(\Phi_H(t),t) = 2\ R^3(t)\Phi_H(t)$ where $\Phi_H(t) = \Phi(\Theta_H(t))$.

Entropy And Extremely Relativistic Particles

In the absence of entropy producing processes (e.g. First

Order Phase Transitions), the Universe expands adiabatically and the entropy in a comoving volume, $S=S(\Theta)$ is conserved (i.e.: S is time independent). In a comoving volume V, the dimensionless entropy is

$$\frac{S}{k} = \left(\frac{P+\varepsilon}{kT}\right)V, \tag{9}$$

where P is the pressure and ε the energy density. The entropy is dominated by extremely relativistic (ER) particles for which $P=\varepsilon/3$ so that

$$\frac{S}{k} \approx \frac{4}{3}\left(\frac{\varepsilon_{ER}V}{kT}\right) = \frac{4}{3}\left(\frac{\langle E\rangle_{ER}}{kT}\right) N_{ER} \approx 4N_{ER} \tag{10}$$

In equation (10), $\langle E\rangle_{ER}$ is the average energy per ER particle, $N_{ER}=n_{ER}V$ is the number of ER particles in the volume V (n_{ER} is the number density of ER particles) and, we have used the approximate result that $\langle E\rangle_{ER} \approx 3kT$.

For an adiabatically expanding Universe, then, the number of ER particles in a comoving volume is conserved; N_{ER} is a candidate for the quantity we are seeking to distinguish among possible Universes. N_{ER} won't do, however, since it is an <u>extensive</u> quantity (i.e. it is proportional to the size of the comoving volume chosen). The quantity we seek is, nontheless, related to N_{ER}. To better understand what that quantity is, let us first specialize to the case of a closed cosmology. We have already learned that N_{ER} is time independent so that $N_{ER}(\Theta_{max}) = N_{ER}^{TOT}$ is the total number of ER particles in the entire closed, finite Universe.

$$N_{ER}^{TOT} = n_{ER}(t)V(\Theta_{max},t) = 2\pi^2R^3(t)n_{ER}(t). \tag{11}$$

For closed Universes, then, we may use N_{ER}^{TOT} to distinguish among different Universes. But, what we have just learned is easily generalized to open models.

The proper volume $V(\Theta,t)$ factorizes into a time-dependent part ($\propto R^3(t)$) and a coordinate-dependent part ($\propto \Phi(\Theta)$). The quantity $\mathcal{N}=N_{ER}/\Phi$ is coordinate and time-independent; $\mathcal{N}$ is the landmark we have been seeking. N is the number of ER particles per unit coordinate volume; $\mathcal{N}$ is defined for open models (where $N_{ER}\to\infty$) as well as closed models (where $N_{ER}^{TOT}=\pi\mathcal{N}$). Since, up to factors of order unity, the number density of ER particles is

$$n_{ER} \approx \left(\frac{kT}{\hbar c}\right)^3 = \frac{1}{\ell_p^3}\left(\frac{T}{T_p}\right)^3 , \tag{12}$$

we may write for $\mathcal{N}$

$$\mathcal{N} = \frac{N_{ER}}{\Phi} \approx \left[\left(\frac{R}{\ell_p}\right)\left(\frac{T}{T_p}\right)\right]^3 \tag{13}$$

Here and subsequently we are ignoring numerical factors of order unity (such as 2π).

It should be noted that $\mathcal{N}$ is not an observable. An observable such as the number of ER particles within the (particle) horizon is related to $\mathcal{N}$ by

$$N_H^{ER}(t) = \mathcal{N}\Phi_H(t); \; \Phi_H(t)=\Phi(\Theta_H(t)). \tag{14}$$

To see how it is that $\mathcal{N}$, the coordinate density of ER particles, is what we require to discriminate among possible cosmologies we must discuss the dynamics of the RWF models.

Dynamics

The evolution of a RWF model is described by the time dependence of the Hubble parameter.

$$H^2(t) = \left[\frac{1}{a}\left(\frac{da}{dt}\right)\right]^2 = \frac{8\pi G\rho}{3} - \frac{kc^2}{a^2} , \tag{15a}$$

$$H^2(t) = \left[\frac{1}{R}\left(\frac{dR}{dt}\right)\right]^2 = 8\pi G\rho \pm \frac{c^2}{R^2} . \tag{15b}$$

Equations (15) follow upon substitution of the RW metric in the Einstein equations; $\rho = \varepsilon/c^2$ is the mass-energy density. Instead of considering the evolution of the (unobservable) scale factor, R(t), we may use equation (13) to transform equation (15b) into an equation describing the temperature evolution.

$$[H(t)t_P]^2 = \frac{8\pi}{3} G\rho t_P^2 \mp \left(\frac{1}{N^{2/3}}\right) \left(\frac{T}{T_P}\right)^2 = \left[\frac{t_P}{T} \left(\frac{dT}{dt}\right)\right]^2 \quad (16)$$

In equation (16) each term has been made dimensionless by multiplying by t_P^2.

Radiation Dominated Evolution

When the Universal mass-energy density is dominated by ER particles,

$$\rho = \rho_{ER} \approx \frac{\hbar}{c}\left(\frac{kT}{\hbar c}\right)^4, \quad (17)$$

so that equation (16) may be rewritten as

$$[H(t)t_P]^2 \approx \left(\frac{T}{T_P}\right)^4 \mp \frac{1}{N^{2/3}} \left(\frac{T}{T_P}\right)^2 . \quad (18)$$

When the first term on the right hand side of equation (18) is larger than the second term, the Universe is Radiation Dominated (RD); when the second term is larger, the Universe is Curvature Dominated (CD).

It is now easy to define more precisely the epoch t_*, the epoch when the Universe changes from RD to CD. The epoch t_* is when the two terms on the right hand side of equation (18) are equal.

$$\frac{T_*}{T_P} \approx \frac{1}{N^{1/3}} \; ; \; \frac{t_*}{t_P} \approx N^{2/3} . \quad (19)$$

Cosmological Discrimination

It is now clear that all Universes are not created equal! It might be "natural" to expect that N, the coordinate density of ER particles, would be of order unity. Were that the case, the Universe would evolve from RD to CD in a time of order the Planck time. This is clearly not the case for our own Universe where $T_{*} \lesssim T_0 \approx 3K$ implies $N \gtrsim 10^{95}$.

It is N which determines whether or not a Universe will be Long-Lived; Universes require large N for longevity. Notice, here, that the spatially flat (k=0) cosmology is pathological in the sense that it corresponds to $N=\infty$. If we had a theory which told us how the magnitude of N is determined we would know whether or not a universe will be long-lived. Such a theory must be capable of explaining why $N \gtrsim 10^{95}$ rather than $N \approx O(1)$. [Could such a theory yield $N=\infty$?]

Spatial Flatness

The bonus provided by the preceding analysis is that it exposes a connection between the longevity puzzle (why is $t_0 \approx 10^{60} t_P$ rather than $t_0 \approx t_P$) and the spatial flatness puzzle. Notice from equation (15) that for k=0 (the spatially flat model), $3H^2 = 8\pi G_\rho$. It is convenient to define, for arbitrary k, the density parameter Ω,

$$\Omega = \Omega(t) \equiv \frac{8\pi G_\rho(t)}{3H^2(t)} \quad . \tag{20}$$

If Ω is close to unity, the Universe is very nearly flat. From equations (13), (15), (18) and (20) it follows that

$$|\Omega - 1| = [N^{2/3} (\frac{T}{T_P})^2 \mp 1]^{-1} \quad . \tag{21}$$

At the Planck epoch ($T \simeq T_P$),

$$|\Omega_P - 1| \simeq (N^{2/3} \mp 1)^{-1} \simeq (\frac{t_*}{t_P} \mp 1)^{-1} . \tag{22}$$

For a Universe to be long-lived ($t_* >> t_P$), N must be very large ($N \simeq (t_*/t_P)^{3/2}$); for N very large, the Universe is very flat during the Planck epoch.

$$|\Omega_P - 1| \simeq N^{-2/3} << 1 . \tag{22'}$$

For example, our Universe is very long-lived ($t_* \gtrsim t_0 \simeq 10^{60} t_P$) so that N must be very large ($N \gtrsim 10^{90}$). It follows, then, that at the Planck epoch, our Universe was very, very flat: $|\Omega_P - 1| \lesssim 10^{-60}$. This required no fine-tuning, no adjustment of quantities to sixty decimal places; if Ω_P had not been so close to unity, our Universe would not have been so long-lived. Of course, we are faced with the chicken and the egg problem: What made N so large?

We may use the issue of spatial flatness to establish an observational criterion for distinguishing "young" universes from "old" universes. At any time t when the temperature is T, a universe is young if $t \lesssim t_*$ ($T \gtrsim T_*$) and is old if $t \gtrsim t_*$ ($T \lesssim T_*$). For a young universe, then, $\Omega \simeq 1$; Ω differs noticeably from unity in an old universe. At present our Universe seems to be young but Ω_0 is poorly determined. Notice that if $N >> 10^{90}$, our Universe would be very young (<u>and</u> very long-lived); in this case it will be, from an observational point of view, indistinguishable from a flat (k=0) Universe.

It is particularly simple to solve for the evolution of the Universe during RD epochs. The time-dependence of the comoving coordinate of the horizon, $\Theta_H(t)$, may be found and the coordinate volume within the horizon, $\Phi_H(t) = \Phi(\Theta_H)$, calculated. The (in principle observable) number of ER particles within the horizon, $N_H(t) = N\Phi_H(t)$, is, for RD epochs

$$N_H \approx N\left(\frac{t}{t_*}\right)^{3/2} \approx N\left(\frac{T_*}{T}\right)^3. \tag{23}$$

Recall that $N^{1/3} \approx (t_*/t_P)(T_*/T_P)$ so that

$$N_H \approx 10^{87}(t_{10}T_3)^3. \tag{24}$$

In equation (24), t_{10} is the age of the Universe in units of 10 billion years and T is the temperature in units of 3K. During RD epochs the flatness relation becomes

$$|\Omega-1| \approx \left(\frac{T_*}{T}\right)^2 \approx \frac{t}{t_*} \approx \left(\frac{N_H}{N}\right)^{2/3}. \tag{25}$$

Once again, we see from equations (23) and (25) that a young Universe is flat; we also note that if a Universe is young then $N_H < N$. From yet another, albeit equivalent, point of view we are led to conclude that N is very very large: $N \gtrsim N_H \approx 10^{87}$.

Matter Dominated Evolution

Our preceding discussion has concentrated on the evolution of universes whose energy density is dominated by ER particles. How are the analysis and the results changed if, as appears to be the case for the present and recent evolution of our Universe, the density is dominated by nonrelativistic (NR) particles? Such "matter-dominated" (MD) evolution introduces another parameter into our cosmology. Suppose the density is dominated by NR particles of mass M whose abundance relative to the ER particles (e.g. photons) is η,

$$\rho_{NR} = Mn_{NR} = M\eta n_{ER}. \tag{26}$$

Neglecting factors of order unity (as we have been doing throughout), the ratio of the density in NR particles to that in ER particles is

$$\frac{\rho_{NR}}{\rho_{ER}} \approx \left(\frac{Mc^2}{kT}\right)\eta \equiv \frac{T_{eq}}{T}. \tag{27}$$

The temperature T_{eq} ($kT_{eq}=Mc^2\eta$) defines the epoch at which $\rho_{NR}\approx\rho_{ER}$; for $T>T_{eq}$ the Universe is RD, for $T<T_{eq}$ the Universe is MD. During MD epochs the evolution is described by

$$(Ht_P)^2 \approx \left(\frac{T_{eq}}{T_P}\right)\left(\frac{T}{T_P}\right)^3 \mp \frac{1}{N^{2/3}}\left(\frac{T}{T_P}\right)^2 . \qquad (28)$$

The epoch at which the Universe becomes CD now depends on T_{eq} as well as N.

$$\frac{T_*}{T_P} \approx \frac{1}{N^{2/3}}\left(\frac{T_P}{T_{eq}}\right) \; ; \; \frac{t_*}{t_P} \approx \left(\frac{T_{eq}}{T_P}\right)N \qquad (29)$$

Solving for the evolution during MD epochs we find for the number of ER particles within the horizon,

$$\frac{N_H(t)}{N} \approx \frac{t}{t_*} \approx \left(\frac{T_*}{T}\right)^{3/2} \qquad (30)$$

For the density parameter $\Omega(t)$, we find,

$$|\Omega-1| \approx \left(\frac{t}{t_*}\right)^{2/3} \approx \frac{T_*}{T} \approx \left(\frac{N_H}{N}\right)^{2/3} . \qquad (31)$$

From equations (29)-(31) it follows that

$$\frac{t}{t_P} \approx \left(\frac{T_{eq}}{T_P}\right) N_H \; ; \; \frac{T}{T_P} \approx \left(\frac{T_P}{T_{eq}}\right)\frac{1}{N_H^{2/3}} , \qquad (32a)$$

$$N_H^{1/3} \approx 10^{29}\, t_{10} T_3 \; ; \; \frac{T_{eq}}{T_P} \approx 10^{-26}\, t_{10}^{-2}\, T_3^{-3} . \qquad (32b)$$

As in the RD case, N_H is very large at present; for our Universe the epoch of equal densities is far from the Planck epoch. It is instructive to illustrate these results with a few examples.

Nucleons. If we live in a nucleon dominated Universe with $M_N \approx 10^{-19} M_P$ and $\eta \approx 10^{-9}$ then $T_{eq} \approx 10^{-28} T_P$ in rough agreement with (32b).

Massive Neutrinos. If, instead, our Universe is dominated by massive, relic neutrinos with $M\nu \approx 10\text{eV} \approx 10^{-27} M_P$ and $\eta \approx 1$ (Schramm and Steigman 1981) then $T_{eq} \approx 10^{-27} T_P$, also in good agreement with (32b).

Grand Unified Monopoles. Grand Unified Theories predict the existence of superheavy ($M_m \approx 10^{-3} M_P$), magnetic monopoles (Preskill 1979). Such Grand Unified Monopoles are expected to be cosmologically abundant (Preskill 1979), $n \approx 10^{-11}$. In this case, $T_{eq} \approx 10^{14} T_P$ in gross violation of (32b). This is the "monopole problem"; a Universe dominated by Grand Unified Monopoles evolves much too rapidly. A comment is in order here. It is often stated that such monopoles "close" the Universe. This is nonsense. Their presence or absence has no influence whatever on the geometry of the Universe; such monopoles simply (and devastatingly) speed up the evolution of the Universe.

Our analysis of the geometry and dynamics of the Universe has led us to an understanding of (but not a solution to) the Flatness-Longevity Puzzle. Possible Universes are distinguished by the value of a dimensionless quantity N which is independent of cosmic time. N is the coordinate density of ER particles; for a closed cosmology N is roughly the total number of ER particles in the Universe. The longevity of a Universe is fixed by the magnitude of N. A natural choice might have been $N \approx O(1)$; such a Universe evolves very quickly: $t_* \approx t_P$. For our Universe, which is clearly long-lived ($t_* \approx t_0 \approx 10^{60} t_P$), N must be exceedingly large: $N \gtrsim 10^{90}$. We have also seen that a long-lived Universe was very flat at the Planck epoch: $\Omega \approx 1 \pm N^{-2/3}$. Flatness and longevity are really two sides of the same coin; N is the ingredient common to both. During its early evolution, when a Universe is young, the number of ER particles within the horizon is small: $N_H < N$; in its youth a Universe is flat: $\Omega \approx 1 \pm (N_H/N)^{2/3}$.

All we have done, however, is to replace what seemed to be several puzzles by one unanswered question: What is it that determines N and, for our Universe, why is N so very large? We will return to discuss a brilliantly imaginative but apparently inadequate answer to this question shortly. First, though, let us turn to the Horizon-Homogeneity Puzzle.

THE HORIZON-HOMOGENEITY PUZZLE

On sufficiently large scales the Universe is observed to be homogeneous; this is true for the large scale distribution of galaxies, of radio sources, for the x-ray background and, most impressively, for the microwave background (black body radiation). This, in itself, is not a puzzle. What is a puzzle is that these observations reveal that regions of the Universe which could never exchange information were, at the same time, at the same temperature and density. For example, we observe the black body radiation coming from different directions in the sky, from regions which, when those photons were emitted, had never been in causal contact. How, then, did those causally disjoint regions "know" to be at the same temperature? Was it a cosmic conspiracy? Let us examine this question more quantitatively.

In a finite interval of time a photon, or any other carrier of information, can travel only a finite distance in comoving coordinates. For example, suppose a photon is emitted at t_e and observed later at t_o; it will have travelled a distance in comoving coordinates given by

$$\Theta_{eo} = \int_{t_e}^{t_o} \frac{cdt}{R(t)} = \frac{1}{N^{1/3}} \int_{T_o}^{T_e} \left(\frac{1}{Ht_P}\right) \frac{dT}{T_P} \; . \tag{33}$$

The size of the region which could have been in causal contact when the photon was emitted at t_e is given by

$$\Theta_{ie} = \int_{t_i}^{t_e} \frac{cdt}{R(t)} = \frac{1}{N^{1/3}} \int_{T_e}^{T_i} \left(\frac{1}{Ht_P}\right) \frac{dT}{T_P} \; , \tag{34}$$

where $t_i \to 0$ (or, $t_i \to t_P$). For convenience and with no loss of generality we may consider the case where $T_0 \ll T_e \ll T_{eq}$ so that $H \approx H_{MD}$. Let us compare Θ_{eo} with Θ_{ie}.

$$\left(\frac{\Theta_{eo}}{\Theta_{ie}}\right)_{MD} \approx \left(\frac{T_e}{T_o}\right)^{1/2} - 1 \approx \left(\frac{T_e}{T_o}\right)^{1/2} . \tag{34'}$$

Since $\Theta_{eo} > \Theta_{ie}$, we receive signals from regions which, at the time of emission had never been in causal contact.

There is an interesting, if trivial, solution to the horizon-homogeneity problem which has nothing to do with causality. Recall that the evolution of Universes is determined by a local quantity N (for MD evolution the local parameter $T_{eq} \alpha M\eta$ also enters). Once N is fixed, the time evolution of the temperature, density, etc. is uniquely determined. If the unknown physics we seek for the determination of N is microscopic (i.e. local) then each piece of the Universe, even those which are causally disjoint, evolves identically and the Universe is guaranteed to be homogeneous. A solution to the Flatness-Longevity Puzzle may, as a bonus, provide us with a solution to the Horizon-Homogeneity Puzzle.

THE INFLATIONARY SCENARIO

It is generally believed, today, that at very high energies all interactions are unified into a highly symmetric Grand Unified Theory. During the early evolution of the Universe, as the Universe cools below the "GUT" temperature $T_G \approx 10^{-5} T_P$, the symmetry will be broken spontaneously. Guth (1981) pointed out that if this transition is a First Order Phase Transition, several of the puzzles we have been discussing may be solved. Guth's point is that as T drops below T_G, most of the Universe may be trapped in the "False Vacuum" which has an enormous energy density $\varepsilon_V \approx kT_G(kT_G/\hbar c)^3$. Such a Universe may supercool to $T_{min} \ll T_G$ before the transition is completed. When the transition terminates an enormous "latent heat" will be released, possibly accounting for the large value of N inferred for our Universe.

There are difficulties with Guth's scenario and, below, we will call attention to two new problems, one possibly minor, the other more serious, perhaps fatal. But before criticizing the Inflationary Scenario, let us see how it has the potential to solve our puzzles.

For the early evolution of the Universe, Guth (1981) and those who have followed his lead have neglected (justifiably) the energy density in NR particles. Guth and others have also neglected the curvature term; this, we will see below, is inconsistent with the Inflationary Scenario. Let us, too, for the moment, neglect the curvature term. Then the evolution of the Universe is described by

$$(Ht_P)^2 = t_P[H_{RD}^2 + H_{VD}^2] \approx \left(\frac{T}{T_P}\right)^4 + \left(\frac{T_G}{T_P}\right)^4 . \tag{35}$$

In Figure 3 we sketch H_{RD} and H_{VD} versus the temperature. It is assumed that the coordinate density of ER particles, N_i, is small initially. At first the evolution is RD with $T \propto t^{-1/2}$. This continues until $t = t_G \approx (T_P/T_G)^2 t_P$ when $T \approx T_G$. Thereafter, the vacuum energy dominates and the Universe expands exponentially

$$\frac{T}{T_G} \approx \exp\left(\frac{-t}{t_G}\right) , \tag{36a}$$

$$\frac{R}{\ell_P} \approx N_i^{1/3} \left(\frac{T_P}{T_G}\right) \exp\left(\frac{t}{t_G}\right) . \tag{36b}$$

The Universe "Supercools" to some very low temperature $T_{MIN} << T_G$ before the transition is completed and the latent heat, the energy of the vacuum, is released heating the Universe and increasing N_i to N_f. Energy conservation provides an approach to calculating this reheating: $dE + PdV = 0$. For $E = (\varepsilon_R + \varepsilon_V)V$, $P_R = 1/3\varepsilon_R$ and $P_V = -\varepsilon_V$, we find

$$-d\varepsilon_V = d\varepsilon_R + \frac{4}{3}\varepsilon_R \frac{dV}{V}. \tag{37}$$

It is generally assumed that when the transition occurs, dV=0 so that $\varepsilon_{Rf} \approx \varepsilon_V$ and $T_f \approx T_G$. If so, then $R_f = R_i$ and

$$\left(\frac{N_f}{N_i}\right)^{1/3} = \left(\frac{R_f}{R_i}\right)\left(\frac{T_f}{T_i}\right) \approx \frac{T_G}{T_{min}}. \tag{38}$$

With sufficient supercooling ($T_{MIN} \ll T_G$) an enormous entropy

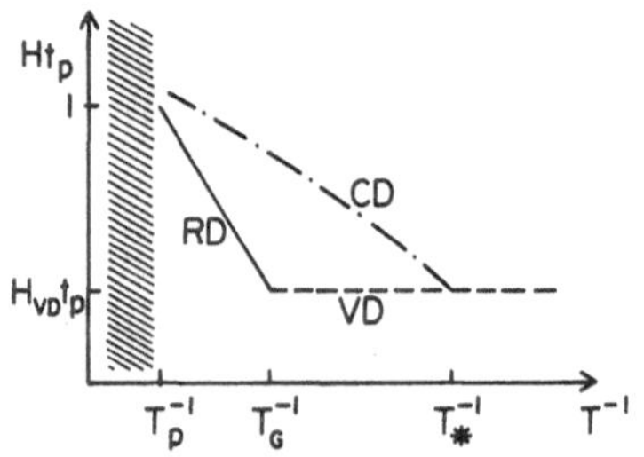

Fig. 3. The Hubble parameter in dimensionless form (Ht_p) is shown as a function of temperature for three limiting cases: Radiation Dominated (RD), Curvature Dominated (CD), Vacuum Dominated (VD).

may be created ($N_f \gg N_i$). To see what is actually required, though, let us assume that $N_i \approx O(1)$. Then, for $N_f \gtrsim 10^{90}$, the transition must not complete itself before the temperature drops to $T_{MIN} \lesssim 10^{-30} T_G \approx 10^{-3}$K!

Personally, I am skeptical about the assumption, made above, that the Universe does not expand when the transition occurs. Just before the transition, the Universe is expanding exponentially rapidly ($R \propto e^{t/t_G}$) and the temperature is dropping exponentially ($T \propto e^{-t/t_G}$). It must take a finite time for the energy released from the vacuum to be thermalized. In the time required to establish the new equilibrium state, surely the Universe must have expanded. To demonstrate how the previous results may be effected, let us parameterize our ignorance by

$$\frac{dV}{V} = \frac{3\beta}{4} \frac{d\varepsilon_R}{\varepsilon_R} \tag{39}$$

In this case we find

$$\frac{V_f}{V_i} = \left(\frac{R_f}{R_i}\right)^3 = \left(\frac{\varepsilon_{Rf}}{\varepsilon_{Ri}}\right)^{3\beta/4} = \left(\frac{T_f}{T_i}\right)^{3\beta}, \tag{40a}$$

$$\left(\frac{T_f}{T_G}\right)^4 \approx \frac{1}{1+\beta}, \tag{40b}$$

$$\left(\frac{N_f}{N_i}\right)^{1/3} = \left(\frac{T_f}{T_i}\right)^{1+\beta} = \left[\left(\frac{1}{1+\beta}\right)^{1/4} \frac{T_G}{T_{min}}\right]^{1+\beta} \tag{40c}$$

Although the Universe is reheated to a lower temperature, more entropy is generated when $\beta \neq 0$.

In any case, the latent heat released in a cosmological first order phase transition could provide an explanation for why N is so large and, in this manner, "solve" the Flatness-Longevity Puzzle. This scenario is also capable of coming to grips with the Horizon-Homogeneity Puzzle. For $T \ll T_G$, the comoving coordinate of the horizon is

$$\Theta_H = \frac{1}{N_i^{1/3}} \int_T^{T_p} \left(\frac{1}{Ht_p}\right)\frac{dT}{T_p} \approx \frac{2}{N_i^{1/3}}\left(\frac{T_p}{T_G}\right). \tag{41}$$

For N_i "small", Θ_H is large. We may now compare Θ_H with Θ_{eo} (see eq. (33)). For $T_e < T_{eq}$, so that the MD approximation for Θ_{eo} is valid, we find

$$\frac{\Theta_{eo}}{\Theta_H} \approx \left(\frac{N_i}{N_f}\right)^{1/3} \left(\frac{T_G}{T_{eq}}\right)\left(\frac{T_{eq}}{T_o}\right)^{1/2} \approx 10^{25}\left(\frac{N_i}{N_f}\right)^{1/3} . \qquad (42)$$

Provided that $N_f >> 10^{75} N_i$, there is no horizon problem ($\Theta_{eo} << \Theta_H$). For example, if $N_i \approx O(1)$ and $N_f \gtrsim 10^{90}$ then, $\Theta_{eo} \lesssim 10^{-5} \Theta_H$.

A CURVATURE DOMINATED INFLATIONARY UNIVERSE

The analysis outlined above is not internally consistent. In Guth's scenario, N_i is small and is inflated to its final large value $N_f \gtrsim 10^{90}$ (or, $N_f >> 10^{90}$). If, indeed, N_i is small (say, $N_i \approx O(1)$), then neglect of the curvature term is unjustified. Equation (35) should be replaced by

$$(Ht_p)^2 = t_p^2 [H_{RD}^2 + H_{CD}^2 + H_{VD}^2] \approx \left(\frac{T}{T_p}\right)^4 \pm \frac{1}{N^{2/3}}\left(\frac{T}{T_p}\right)^2 + \left(\frac{T_G}{T_p}\right)^4 . \qquad (43)$$

The Universe starts Radiation Dominated ($T \gtrsim T_p$) and expands until $T_\star \approx N^{-1/3} T_p$ when it becomes Curvature Dominated. If $T_\star > T_G$ ($N \lesssim 10^{15}$), the Universe is Curvature Dominated before it becomes Vacuum Dominated. This may be fatal. For a closed Universe (the minus sign in eq. (43)), the expansion and recollapse is completed before the Universe can ever inflate. Although the effect of curvature domination may be less dramatic for an open Universe, the inflationary scenario will, at the very least, have to be modified.

A CD Universe expands more rapidly than a RD or VD Universe (see Figure 3); this will effect the time scale for completion of the first order phase transition. Different fine tuning will now be required if the Universe is to supercool. For $N \approx O(1)$, the Universe becomes CD when T drops below T_p; the Universe remains

CD until it becomes VD at a temperature $T_V \gtrsim N^{1/3} T_G^2 / T_P \gtrsim 10^9 \mathrm{GeV} \ll T_G$. Below T_V the "standard" inflationary scenario is recovered.

There is yet another problem for the inflationary scenario. Ellis and Steigman (1980) have noted that Grand Unified Interactions cannot establish equilibrium in a RD Universe until the temperature drops below $\sim 3\times 10^{15}$GeV. If N is small, the Universe would have been CD at this temperature. Let us see how the Ellis-Steigman result is modified in a CD Universe.

As a criterion for whether or not equilibrium can be established we will compare the reaction rate, Γ, with the expansion rate H_{CD}. Neglecting factors of order unity we have,

$$\Gamma = n\sigma v \approx \left(\frac{kT}{\hbar c}\right)^3 \alpha^2 c \left(\frac{\hbar c}{kT}\right)^2 \approx \frac{\alpha^2}{t_P}\left(\frac{T}{T_P}\right), \tag{44a}$$

$$H_{CD} \approx \frac{1}{t_P}\left[\frac{1}{N^{1/3}}\left(\frac{T}{T_P}\right)\right]. \tag{44b}$$

So that

$$\frac{\Gamma}{H_{CD}} \approx \alpha^2 N^{1/3} \approx \left(\frac{N}{10^9}\right)^{1/3}. \tag{45}$$

For $N \lesssim 10^9$, Grand Unified Interactions are never in equilibrium in a CD Universe. Furthermore, when the Universe becomes VD for $T<T_V$,

$$\frac{\Gamma}{H_{VD}} \approx \alpha^2 N^{1/3}\left(\frac{T}{T_V}\right) < \frac{\Gamma}{H_{CD}} . \tag{46}$$

A Universe which emerges from the Planck epoch with low entropy (small N) will quickly become curvature dominated. Those Universes which are spatially closed (k>0) will recollapse before they ever inflate. Either our Universe is spatially open (k<0) or, it emerged from the Planck epoch generously endowed with enormous entropy (beware of Quantum Gravity!).

VANISHING COSMOLOGICAL CONSTANT

I have even less to offer on the issue of the cosmological constant than on the solutions of the other puzzles.

Einstein's equations may be modified by the inclusion of a constant term Λ, the cosmological constant, which has the same dimensions as H^2. Since, at present the cosmological term does not dominate the evolution of the Universe

$$\Lambda \lesssim H_0^2 \approx \frac{1}{t_0^2} \approx \frac{10^{-120}}{t_P^{\;2}} \tag{47}$$

In dimensionless units, Λt_P^2 is zero to better than 120 decimal places. Is Λ identically zero? If so, how did it get that way and, how did it stay that way? This is not a frivolous issue. Λ may be related to the energy density of the vacuum.

$$\Lambda t_P^2 \approx G\rho_\Lambda\, t_P^2 \sim \left(\frac{T_\Lambda}{T_P}\right)^4 . \tag{48}$$

We might expect that $T_\Lambda \approx T_P$ or T_G or M_W. Instead, we see from equation (47) that $T_\Lambda \lesssim 10^{-30} T_P \approx 10^{-2}$eV!

SUMMARY

Grand Unified Theories and Big Bang Comologies can boast of some impressive esthetic and experimental successes. Much observational and experimental work remains before we can be fully confident we're on the right track. The present euphoria is hardly justified by a large body of hard data. Under such circumstances it is incumbent on us to look for the blemishes as well as the beauty in our theories. The current excitement surrounding the use of cosmology as a probe of grand unified theories makes this critical re-examination imperative. Here, we've discussed some nagging cosmological puzzles whose ultimate solutions may prove to be trivial or profound.

The "natural" scale in cosmology is the Planck scale. Our Universe has lived for more than 10^{60} Planck times; why? At present the universal mass density is close to the critical density ($\Omega \approx 1$) but, during the Planck epoch $\Omega_P \approx 1 \pm 10^{-60}$; how was such fine tuning achieved? The Universe is observed to be homogeneous on scales which were never in causal contact; is this a cosmic conspiracy? Although we have proposed no answers to these questions, we've tried to show that they may all be related. There is a quantity, N, the coordinate density of ER particles which determines the evolution of a Universe. Universes with small N evolve quickly, those with large N are long-lived. A Universe is young if the number of ER particles within the horizon N_H is small compared to N. Young Universes are spatially flat: $\Omega \approx 1 \pm (N_H/N)^{2/3}$. During the Planck epoch, $N_H \approx O(1)$ so that a Universe destined to be long-lived (large N) will have been very flat ($\Omega_P \approx 1 \pm N^{-2/3}$). The problem still remains, though: What determines N? It should be noted that N is a local quantity and may follow from some microphysics in analogy, say, to the determination of the universal baryon asymmetry. If N is determined by local physics then all parts of the Universe, even those which are causally disjoint, may be endowed with the same value of N. If so, there is no longer a homogeneity problem; all parts of the Universe evolve identically. The issue of horizons would be moot.

The Inflationary Scenario was proposed (Guth 1981) as a solution to these puzzles. If it is to have a chance of working, we must live in a spatially open Universe; a spatially closed Universe recollapses before it has a chance to inflate. There are other difficulties with the scenario which suggest the Universe may have emerged from the Planck era (quantum gravity?) with its present entropy. Even so, what physics determines whether the

Universe will be open or closed?

It is far from clear if the puzzles discussed here pose any obstacles to the approach to particle physics via cosmology. However, until they are better understood and our cosmology more firmly established, we should be cautiously skeptical as well as cautiously optimistic.

ACKNOWLEDGMENTS

In the course of the work described here I have benefitted from discussions with many colleagues. I am particularly indebted to M. Einhorn, A. Guth, V. Petrosian and M. Sher. Part of this work was done while I was a visitor at the Institute for Theoretical Physics and I thank the director, W. Kohn, for hospitality. This work is supported at Bartol by DOE contract DE-A502-78ER05007.

REFERENCES

Dolgov, A. D. and Zeldovich, Ya. B., 1981, Rev. Mod. Phys., 53:1

Ellis, J. and Steigman, G., 1980, Phys. Lett., 89B:186.

Guth, A., 1981, Phys. Rev., D23:347.

Kolb, E. W., and Turner, M.S., 1981, Nature, in press.

Preskill, J. P., 1979, Phys. Rev. Lett., 43:1365.

Schramm, D. N. and Steigman, G., 1981, Ap. J., 243:1.

Steigman, G.,1979, Ann Rev. Nucl. Part. Sci., 29:313.

SUPERSYMMETRY IN NUCLEI

F. Iachello

Kernfysisch Versneller Instituut, Rijksuniversiteit Groningen, Nederland and Physics Department, Yale University, New Haven, Connecticut 06520

INTRODUCTION

Most applications of supersymmetry ideas deal with problems in elementary particle and gravitational physics[1]. Several of these applications have been discussed at this Conference. In my talk, I will discuss some applications of supersymmetry ideas to nuclear physics. In particular, I will show that experimental examples of supersymmetry appear to have been found in the spectra of complex nuclei. The supersymmetries encountered in nuclei link together collective and single particle excitations. They are based on an algebraic approach to nuclear spectra in which the collective excitations are treated as a system of interacting bosons. When treated alone the collective excitations have three possible dynamical symmetries, all of which have been found experimentally. These symmetries are of the ordinary type, since the corresponding symmetry operations transform bosons into bosons. If the single particle excitations are added, one is led to consider a system of interacting bosons and fermions. The symmetries of this system are of the super (or graded) type, since the corresponding symmetry operations contain terms which transform bosons into fermions and viceversa. There is a large number of symmetries of this type possible, two of which have been observed experimentally. In the first part of my talk, I shall review the ordinary symmetries of the interacting boson model, while in the second part I shall discuss the possible supersymmetries of the interacting boson-fermion model.

THE INTERACTING BOSON MODEL

In the interacting boson model[2], low-lying collective states of nuclei are described in terms of six dynamical bosons[3], one s-boson with angular momentum J=0 and a d-boson with J=2, assigned to a six-dimensional representation of the group U(6). These bosons have a dual interpretation: (i) a classical (geometric) interpretation, and (ii) a quantum (particle) interpretation. The classical interpretation is obtained by associating to the bosons classical variables. The most natural way to do this is by introducing[4,5] variables in the coset space $U(6)/U(5) \otimes U(1)$. This space is a five-dimensional (complex) space and it leads to a characterization of nuclei in terms of shapes (liquid drop model[6]). In the quantum (or particle) interpretation, the bosons are assumed to be correlated pairs of nucleons with J=0 and J=2 respectively[7], similar to the Cooper pairs of the electron gas. In first approximation, no distinction is made between proton and neutron pairs. In comparing with experiment, better results are obtained if proton (π) and neutron (ν) pairs are treated separately, thus giving to the bosons a two-dimensional label, called F-spin[7]. In this talk, I shall neglect the difference between proton and neutron pairs and speak only of nucleon pairs (F-spin invariance). Furthermore, in first approximation, only pairs in the valence shells are taken into account and a given nucleus with an even number of protons and neutrons is thus treated as a system of N bosons, where N is the sum of the number of proton (N_π) and neutron (N_ν) pairs outside the major closed shells, $N=N_\pi+N_\nu$. In order to take into account the particle-hole conjugation, this number is counted from the nearest closed shell, that is, if the shell is more than half full, N is the number of hole pairs.

In order to analyze the algebraic structure of this model, it is convenient to introduce boson creation $d^\dagger_\mu, s^\dagger$, ($\mu=0, \pm1, \pm2$) and annihilation (d_μ, s) operators, altogether denoted by $b^\dagger_\alpha$ (b_α), a = 1,....., 6. The 36 generators of U(6) can then be written as $G_{\alpha\alpha'}=b^\dagger_\alpha b_{\alpha'}$. All operators in the model are built out of generators G. For example, the Hamiltonian H, containing at most two-body interactions, is

$$H = H_0 + \sum_{\alpha\alpha'} \varepsilon_{\alpha\alpha'} b^\dagger_\alpha b_{\alpha'} + \tfrac{1}{2} \sum_{\alpha\alpha'\beta\beta'} u_{\alpha\alpha'\beta\beta'} b^\dagger_\alpha b^\dagger_{\alpha'} b_\beta b_{\beta'} . \tag{2.1}$$

This can be rewritten as

$$H = H_0 + \sum_{\alpha\alpha'} \varepsilon'_{\alpha\alpha'} G_{\alpha\alpha'} + \tfrac{1}{2} \sum_{\alpha\alpha'\beta\beta'} u_{\alpha\alpha'\beta\beta'} G_{\alpha\beta} G_{\alpha'\beta'} . \tag{2.2}$$

In both Eqs. (2.1) and (2.2), H_0 is assumed to be invariant under U(6). If the coefficients $\varepsilon'_{\alpha\alpha'}$, $u_{\alpha\alpha'\beta\beta'}$ would also be consistent with U(6), all low-lying levels of a given even-even nucleus would be degenerate. The splitting of the levels is achieved by choosing appropriate values of the coefficients $\varepsilon'_{\alpha\alpha'}$, $u_{\alpha\alpha'\beta\beta'}$.

DYNAMICAL SYMMETRIES

In general, for arbitrary values of the coefficients $\varepsilon'_{\alpha\alpha'}$, $u_{\alpha\alpha'\beta\beta'}$, the eigenvalue problem for H must be solved numerically. However, in some special cases, this can be done analytically. For a problem with group structure G, analytic solutions can be found whenever the Hamiltonian, H, can be written in terms only of Casimir invariants of a chain of subgroups of G

$$G \supset G' \supset G'' \supset \dots \quad (3.1)$$

These special cases correspond to dynamical symmetries of H. In the present case $G \equiv U(6)$ and there are three and only three possible chains[8,9,10] of subgroups that respect the J=0 and J=2 assignments of the pairs and contain the rotation group SO(3),

$$U(6) \begin{cases} U(5) \supset SO(5) \supset SO(3) \supset SO(2) & \text{(I)} \\ SU(3) \supset SO(3) \supset SO(2) & \text{(II)} \\ SO(6) \supset SO(5) \supset SO(3) \supset SO(2)\ , & \text{(III)} \end{cases} \quad (3.2)$$

where SU(3) is realized in the six-dimensional representation. For any of the three chains, the Hamiltonian is directly diagonalizable. States are then characterized by the labels appropriate to each group chain and the eigenvalues can also be written in terms of those labels. For example, for the group chain III, the labels are[10]

$$\left| \begin{matrix} U(6) \supset & SO(6) \supset & SO(5) \supset & SO(3) \supset & SO(2) \\ N\ , & \sigma\ , & \tau(\nu_\Delta)\ , & L\ , & M_L \end{matrix} \right\rangle . \quad (3.3)$$

Here only one number N, σ and τ characterizes the irreducible representations of U(6), SO(6) and SO(5) respectively, since these are totally symmetric. A delicate group theoretical problem arises in the reduction from SO(5) to SO(3), since SO(5) is not fully reducible with respect to SO(3). An extra label, called ν_Δ in Ref. 10, is needed. Similar labelling problems arise for the other group chains, as discussed in Refs. 8 and 9. However, since these extra labels do not appear in the energy eigenvalues, they will not be discussed further in this talk.

The energy eigenvalues corresponding to the three group chains of Eq. (3.2) can be written as[8,9,10]

$$E(N,n_d,v,n_\Delta,L,M_L)=E_0+E_1N+E_2N^2+\varepsilon n_d+\alpha\ \tfrac{1}{2}\ n_d(n_d-1)\ + \\ +\beta(n_d-v)(n_d+v+3)+\gamma[L(L+1)-6n_d]\ , \quad \text{(I)}$$

$$E(N,\lambda,\mu,K,L,M_L)=E_0+E_1N+E_2N^2+\ (\tfrac{3}{4}\kappa+\kappa')\ L(L+1)\ - \quad (3.4) \\ -\ \kappa[\lambda^2+\mu^2+\lambda\mu+3(\lambda+\mu)]\ , \quad \text{(II)}$$

$$E(N,\sigma,\tau,\nu_\Delta,L,M_L)=E_0+E_1N+E_2N^2+A\ \tfrac{1}{4}(N-\sigma)(N+\sigma+4)+$$
$$+\frac{B}{6}\ \tau(\tau+3)+C\ L(L+1)\ . \qquad \text{(III)}$$

Dynamical symmetries described by energy formulas of the type (2.6) also appear in other fields of physics. A notable example in elementary particle physics is Gell-Mann-Ne'eman SU(3), where the chain of groups is

$$SU(3) \supset SU(2) \otimes U(1)\ , \qquad (3.5)$$

leading to the Gell-Mann-Okubo mass formula[11]

$$E(I,I_3,Y) = a+bY+c[\tfrac{1}{4}Y^2 - I(I+1)]. \qquad (3.6)$$

A consequence of Eq. (3.4) is that the energies of all states in a given representation N of U(6) (a given even-even nucleus) are given in terms of few parameters. In particular, there are four parameters ε, α, β, γ which describe the excitation energies in the case I, two parameters κ, κ' in the case II, and three parameters A, B and C in the case III. Several examples of nuclei whose spectra can be described by the dynamical symmetries I, II or III have been found. Three examples, one for each of the three symmetries, are shown in Figs. 1, 2 and 3.

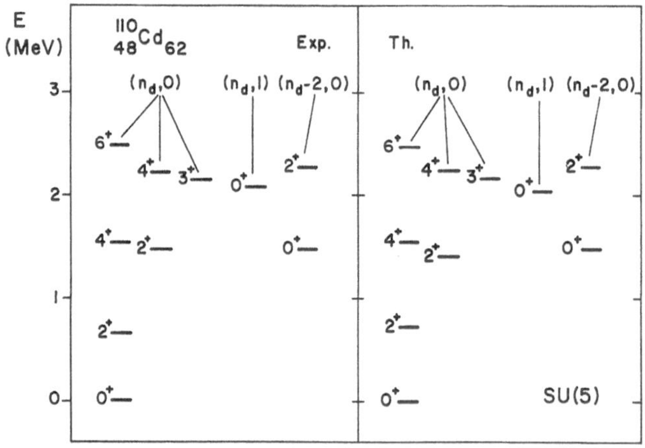

Fig. 1. An example of the dynamical symmetry I of the interacting boson model, the nucleus $^{110}_{48}Cd_{62}$.

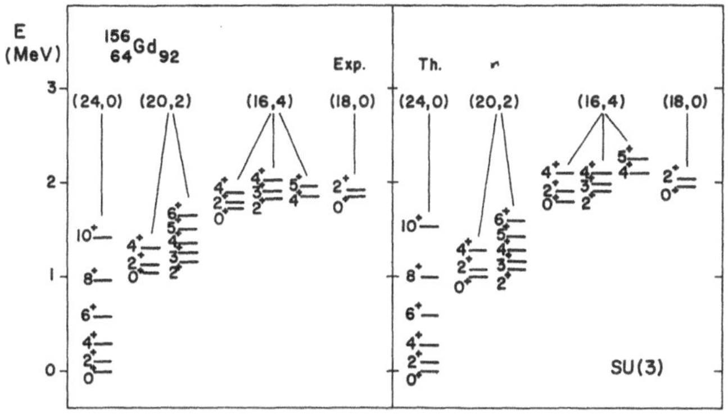

Fig. 2. An example of the dynamical symmetry II of the interacting boson model, the nucleus ${}^{156}_{64}Gd_{92}$.

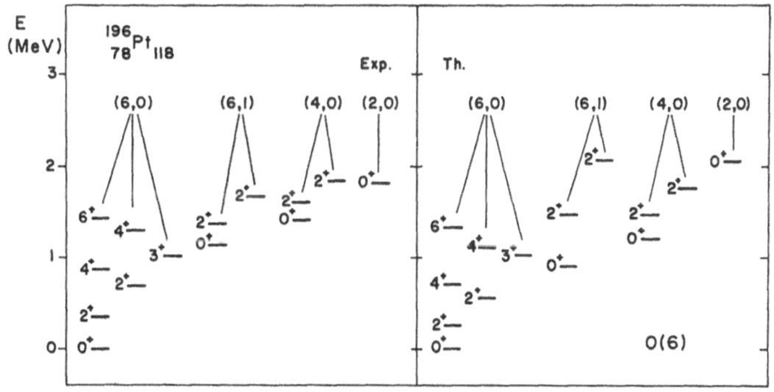

Fig. 3. An example of the dynamical symmetry III of the interacting boson model, the nucleus ${}^{196}_{78}Pt_{118}$.

THE INTERACTING BOSON-FERMION MODEL

The interacting boson model describes collective states in nuclei where all particles are paired together to J=0 and J=2. The study of its algebraic structure suggested the occurrence of three possible dynamical symmetries, all of which have been observed experimentally. There exist, however, also states in nuclei where some particles are explicitly unpaired. In nuclei with an odd number of protons or neutrons (odd-even nuclei) at least one particle must be necessarily unpaired. In nuclei with an even number of protons and neutrons there are states (usually above an excitation energy of ~1.8 MeV) with two unpaired particles (two-quasiparticle states[12]). In order to describe these states, one must introduce, in addition to the collective degrees of freedom (s-d bosons), the degrees of freedom of the unpaired particles (fermions)[13], Fig. 4. The resulting model has been called interacting boson-fermion model. In order to study the possible dynamical symmetries of this model, it is convenient to introduce, in addition to the boson creation ($b_\alpha^\dagger$, α=1,...,6) and annihilation (b_α) operators, a set of creation ($a_i^\dagger$, i=1,...,m) and annihilation (a_i) operators for fermions, where m labels the dimension of the fermionic space. In a given nucleus, the unpaired particles occupy the single particle shell-model levels appropriate to that nucleus. For example, for nuclei in the major shell 50-82, the single particle levels are $1g_{7/2}$, $2d_{5/2}$, $1h_{11/2}$, $2d_{3/2}$

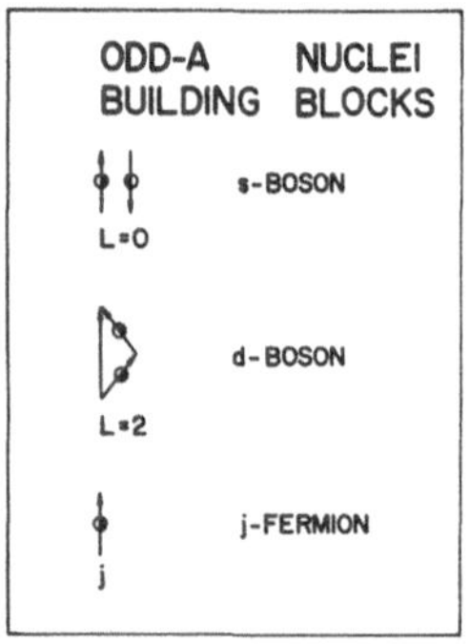

Fig. 4. Pictorial representation of the building blocks of odd-even nuclei in the interacting boson-fermion model.

and $3s_{1/2}$. Each single particle level with angular momentum j has degeneracy (2j+1). Thus, $m = \sum_j (2j+1)$, the sum going over all values of j in a major shell. (For the shell 50-82, m=32). In the same way in which the boson operators $b^\dagger_\alpha$ form a representation of U(6) with generators $G_{\alpha\alpha'} = b^\dagger_\alpha b_{\alpha'}$, the fermion operators $a^\dagger_i$ form a representation of U(m) with generators $G_{ii'}=a^\dagger_i a_{i'}$. In order to distinguish these two groups, it is convenient to place a superscript B to $U^{(B)}(6)$, and a superscript F to $U^{(F)}(m)$. The Hamiltonian for the mixed system of N bosons and M fermions, and containing at most two-body interactions can be written in the form

$$H = H^{(B)} + H^{(F)} + V^{(BF)} , \tag{4.1}$$

where $H^{(B)}$ is the same as before,

$$H^{(B)}=H_0+ \sum_{\alpha\alpha'} \varepsilon_{\alpha\alpha'} b^\dagger_\alpha b_{\alpha'} + \tfrac{1}{2} \sum_{\alpha\alpha'\beta\beta'} u_{\alpha\alpha'\beta\beta'} b^\dagger_\alpha b^\dagger_{\alpha'} b_\beta b_{\beta'} , \tag{4.2}$$

and $H^{(F)}$ and $V^{(BF)}$ are given by

$$H^{(F)}=H_0'+ \sum_{ii'} \eta_{ii'} a^\dagger_i a_{i'} + \tfrac{1}{2} \sum_{ii'kk'} v_{ii'kk'} a^\dagger_i a^\dagger_{i'} a_k a_{k'} , \tag{4.3}$$

$$V^{(BF)}= \sum_{\alpha\alpha' ii'} w_{\alpha\alpha' ii'}\, b^\dagger_\alpha b_{\alpha'} a^\dagger_i a_{i'} . \tag{4.4}$$

In Eqs. (4.2) and (4.3), H_0 and H_0' are invariant under $U^{(B)}(6)$ and $U^{(F)}(m)$ respectively and the coefficients $\eta_{ii'}, v_{ii'kk'}, w_{\alpha\alpha'ii'}$ are chosen as to conserve the angular momenta both of the fermions and of the bosons. Since, as stated above, the bilinear products $G^{(B)}_{\alpha\alpha'}=b^\dagger_\alpha b_{\alpha'}$ generate the group $U^{(B)}(6)$, while the bilinear products $G^{(F)}_{ii'}=a^\dagger_i a_{i'}$ generate the group $U^{(F)}(m)$, the algebraic structure of the model Hamiltonian, Eqs. (4.1)-(4.4), is that of the group $U^{(B)}(6) \otimes U^{(F)}(m)$. In general, no further symmetry is present, and the eigenvalue problem for the mixed system must be solved numerically. However, it is possible to extend the idea of dynamical symmetries to mixed systems of bosons and fermions, thus finding analytic solutions to the corresponding eigenvalue problem[14].

There are two ways in which the idea of dynamical symmetries can be extended to mixed systems of bosons and fermions. The first relies on the isomorphism of some Lie algebras and leads to symmetries which have been called "spinor" symmetries. This generalization, although new, makes still use of ordinary Lie algebras. The second, more interesting generalization, introduces explicitly graded (or super) algebras, thus leading to true supersymmetries. I will now briefly review both spinor- and supersymmetries of the interacting boson-fermion model.

SPINOR AND RELATED SYMMETRIES

Suppose that the boson part, H_B, of H, Eq. (4.1), has a dynamical symmetry described by the group chain

$$G^{(B)} \supset G'^{(B)} \supset G''^{(B)} \supset \ldots. , \qquad (5.1)$$

and the fermion part, H_F, a dynamical symmetry described by the group chain

$$G^{(F)} \supset G'^{(F)} \supset G''^{(F)} \supset \ldots , \qquad (5.2)$$

If the boson and fermion chains are accidentally isomorphic, then it is possible to form diagonal subgroups of $G^{(B)} \otimes G^{(F)}$ that transform simultaneously the bosons and the fermions. States can be labelled by representations of these groups and energy eigenvalues can be written in terms of the eigenvalues of the Casimir invariants of the appropriate groups. This procedure is illustrated best by an example. Consider the case in which the unpaired nucleons occupy only one single particle level with j=3/2. This situation is encountered, for example, in odd proton nuclei with proton number, Z, $76 \leq Z \leq 80$, where the odd protons occupy the level $2d_{3/2}$. For this situation, one must consider a system of bosons with J=0,2 and fermions with J=3/2, with group structure $U^{(B)}(6) \otimes U^{(F)}(4)$. Suppose now that the boson part, H_B, of H is well described by the dynamical symmetry III of Eq. (3.2),

$$U^{(B)}(6) \supset SO^{(B)}(6) \supset SO^{(B)}(5) \supset SO^{(B)}(3) \supset SO^{(B)}(2) . \qquad (5.3)$$

Then, since fermions with J=3/2 can be described by the group chain[15]

$$U^{(F)}(4) \supset SU^{(F)}(4) \supset Sp^{(F)}(4) \supset SU^{(F)}(2) \supset SO^{(F)}(2) , \qquad (5.4)$$

one can exploit the isomorphisms

$$SU(4) \approx SO(6) ,$$

$$Sp(4) \approx SO(5) , \qquad (5.5)$$

$$SU(2) \approx SO(3) ,$$

to form diagonal subgroups of $U^{(B)}(6) \otimes U^{(F)}(4)$ that transform simultaneously bosons and fermions. These are the spinor groups[16]

$$Spin(6) \approx SU(4) ,$$

$$Spin(5) \approx Sp(4) , \qquad (5.6)$$

$$\text{Spin}(3) \approx SU(2) \ .$$

The generators of these groups are written down explicitly in Ref. 14. A possible chain of subgroups for the mixed system of bosons and fermions is now

$$U^{(B)}(6)\otimes U^{(F)}(4)\supset SO^{(B)}(6)\otimes SU^{(F)}(4)\supset \text{Spin}(6)\supset \text{Spin}(5)\supset \text{Spin}(3)\supset \text{Spin}(2), \tag{5.7}$$

and a complete labelling of the representations is

$$\begin{array}{ll} U^{(B)}(6) & [N] \\ U^{(F)}(4) & \{M\} \\ SO^{(B)}(6) & \Sigma \\ \text{Spin}(6) & (\sigma_1,\sigma_2,\sigma_3) \\ \text{Spin}(5) & (\tau_1,\tau_2),\ \nu_\Delta \\ \text{Spin}(3) & (J) \\ \text{Spin}(2) & (M_J)\ . \end{array} \tag{5.8}$$

As in Sect. 3, an extra label is needed in the reduction from Spin(5) to Spin(3). This label, called ν_Δ, does not enter in the energy eigenvalues and it will not be discussed further. It should be noted that, while the $U^{(B)}(6)$ representations in Eq. (5.8) are totally symmetric

$$[N] \equiv \overbrace{\square\square\square \ldots \square}^{\text{N-boxes}} \ , \tag{5.9}$$

the $U^{(F)}(4)$ representations are totally antisymmetric

$$M \equiv \left.\begin{array}{c} \square \\ \square \\ \square \\ \vdots \\ \square \end{array}\right\} \text{M-boxes}, \tag{5.10}$$

according to the bosonic and fermionic nature of $U^{(B)}(6)$ and $U^{(F)}(4)$ respectively.

By writing H, Eq. (4.1), in terms of the Casimir invariants (at most quadratic) of all groups appearing in the chain Eq.(5.7), one can then find an analytic solution to the eigenvalue problem for the mixed system of bosons and fermions[17,14]

$$E(N,M,\Sigma,(\sigma_1,\sigma_2,\sigma_3),(\tau_1,\tau_2),\nu_\Delta,J,M_J) = E_0+E_1N+E_2N^2+E_3M+E_4M^2+E_5MN -$$

$$- \frac{A_1}{4}\Sigma(\Sigma+4) - \frac{A_2}{4}[\sigma_1(\sigma_1+4)+\sigma_2(\sigma_2+2)+\sigma_3^2] + \frac{B}{6}[\tau_1(\tau_1+3)+\tau_2(\tau_2+1)]+CJ(J+1). \quad (5.11)$$

Similar techniques can be used to construct other analytic solutions to the eigenvalue problem for mixed systems of bosons and fermions. In fact, since the accidental isomorphisms of all Lie algebras are known, it is possible to list all cases for which solutions are possible. I will give a part of this list in Sect.7.

SUPERSYMMETRIES

The spinor symmetries of the previous section describe states in a given nucleus (fixed values of N and M). One may inquire whether or not it is possible to imbed the group $G^{(B)} \otimes G^{(F)}$ into a larger group such that its representations would comprise several nuclei, thus leading to a further generalization of the idea of dynamical symmetries. Since the representations of this larger group must contain both bosonic (even M) and fermionic (odd M) states, it cannot be an ordinary Lie group. In fact, it must be a supergroup. Let this supergroup be G^*. If the Hamiltonian describing the mixed systems of bosons and fermions can be written in terms of Casimir invariants of a chain of subgroups of G^*

$$G^* \supset G' \supset G'' \supset \ldots\ldots , \quad (6.1)$$

this chain can be used to label states in several nuclei and a dynamical supersymmetry would arise. In Eq. (6.1), G', G", ... could be either ordinary Lie groups or supergroups.

For the applications described here, the appropriate supergroups are the unitary supergroups $U(n|m)$. The representation theory of supergroups has been discussed by several authors[18,19]. Here it is convenient to use the notation of Ref.18. Representations are labelled by Young supertableaux, similar in appearance to the tableaux of ordinary U(n) groups,

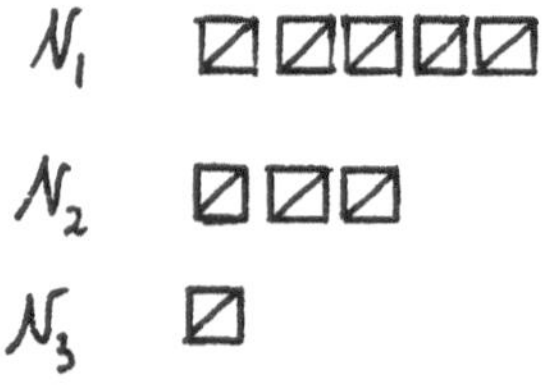

(6.2)

where $\mathcal{N}_1$, $\mathcal{N}_2$,...., is the number of boxes in the first, second, ..., row. However, the tableaux have a very different meaning from the ordinary tableaux. In order to see the difference, consider the Young supertableaux ⧄⧄ , obtained as the supersymmetrized product of two fundamental representations ⧄ . This tableau indicates symmetrization of indices for the purely bosonic subspace, but antisymmetrization of indices for the purely fermionic subspace. In the applications discussed here, where the bosonic degrees of freedom are completely symmetrized, while the fermionic are completely antisymmetrized, the representations which appear are the totally supersymmetric denoted by

$$\{\mathcal{N}\} \equiv \overbrace{⧄⧄\ldots⧄}^{\mathcal{N}\text{-boxes}} \ . \qquad (6.3)$$

For the same applications, it is convenient to realize the superalgebra $U(n|m)$ in terms of creation and annihilation operators. Introducing boson $b^{\dagger}_{\alpha}(b_{\alpha})$ $(\alpha=1,\ldots,n)$ and fermion $a^{\dagger}_{i}(a_{i})$ $(i=1,\ldots,m)$ operators as above, the generators of the superalgebra $U(n|m)$ can be written as[20]

$$\begin{array}{lll}
G^{(B)}_{\alpha\alpha'} = b^{\dagger}_{\alpha}b_{\alpha'} & & n^2 \\
G^{(F)}_{ii'} = a^{\dagger}_{i}a_{i'} & & m^2 \\
F^{\dagger}_{\alpha i} = b^{\dagger}_{\alpha}a_{i} & & mn \\
F_{i\alpha} = a^{\dagger}_{i}b_{\alpha} & & mn \\
\hline
 & & (m+n)^2 \ .
\end{array} \qquad (6.4)$$

The Bose sector of $U(n|m)$ is then $U^{(B)}(n) \otimes U^{(F)}(m)$. Branching rules for the decomposition of representations of $U(n|m)$ into representations of $U^{(B)}(n) \otimes U^{(F)}(m)$ are given in Ref. 14.

In order to show a concrete example of a supersymmetry, I will now return to the case discussed in detail in the previous section and embed the chain (5.7) into the supergroup $U(6|4)$,

$$\begin{array}{l}
U(6|4) \supset U^{(B)}(6) \otimes U^{(F)}(4) \supset SO^{(B)}(6) \supset \\
\supset Spin(6) \supset Spin(5) \supset Spin(3) \supset Spin(2) \ .
\end{array} \qquad (6.5)$$

The complete labelling of the states now includes a quantum number $\mathcal{N}$ which characterizes the totally supersymmetric representations $[\mathcal{N}]$ of $U(6|4)$. Here $\mathcal{N}$ is the total number of bosons plus fermions, $\mathcal{N}=N+M$. The number M is restricted to be $0 \leq M \leq 4$[14]. Assuming now that the Hamiltonian, Eq. (4.1), is written in terms of Casimir operators (at most quadratic) of the chain (6.5) leads to the energy formula[14]

$$E(\mathcal{N},N,M,\Sigma,(\sigma_1,\sigma_2,\sigma_3),(\tau_1,\tau_2),\nu_\Delta,J,M_J) =$$
$$= E_0+E_1N+E_2N^2+E_3M+E_4M^2+E_5MN+E_6\mathcal{N}+E_7\mathcal{N}^2 -$$
$$- \frac{A_1}{4}\Sigma(\Sigma+4) - \frac{A_2}{4}[\sigma_1(\sigma_1+4)+\sigma_2(\sigma_2+2)+\sigma_3^2]+\frac{B}{6}[\tau_1(\tau_1+3)+\tau_2(\tau_2+1)]+CJ(J+1). \tag{6.6}$$

Since $M=\mathcal{N}-N$, two of the parameters appearing in Eq. (6.6) can be eliminated to give

$$E(\mathcal{N},N,\Sigma,(\sigma_1,\sigma_2,\sigma_3),(\tau_1,\tau_2),\nu_\Delta,J,M_J) = E_0+E_1'N+E_2'N^2+E_3'\mathcal{N}+E_4'\mathcal{N}^2+E_5'\mathcal{N}N -$$
$$- \frac{A_1}{4}\Sigma(\Sigma+4) - \frac{A_2}{4}[\sigma_1(\sigma_1+4)+\sigma_2(\sigma_2+2)+\sigma_3^2]+\frac{B}{6}[\tau_1(\tau_1+3)+[\tau_2(\tau_2+1)]+CJ(J+1). \tag{6.7}$$

According to Eq. (6.7) the excitation energies of a given nucleus (fixed N and $\mathcal{N}$) are given by the last four terms. This is the same as in the case of a spinor symmetry, Eq.(5.11). However, a consequence of the supersymmetry is that the coefficients A_1, A_2, B and C must be the same for all nuclei belonging to the same supermultiplet $[\mathcal{N}\}$. In other words, even-even and odd-even nuclei are linked together by the same energy formula. In addition, only certain representations of $U^{(B)}(6) \otimes U^{(F)}(4)$ are allowed, corresponding to the decomposition of representations of $U(6|4)$ into representations of $U^{(B)}(6) \otimes U^{(F)}(4)$[14].

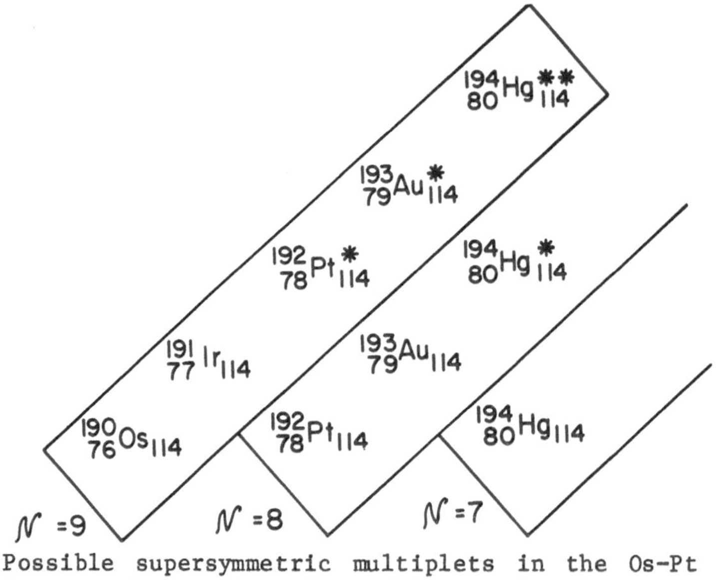

Fig. 5. Possible supersymmetric multiplets in the Os-Pt region. $\mathcal{N}$ denotes the total number of boson plus fermions. Excited configurations are labelled by one or two stars.

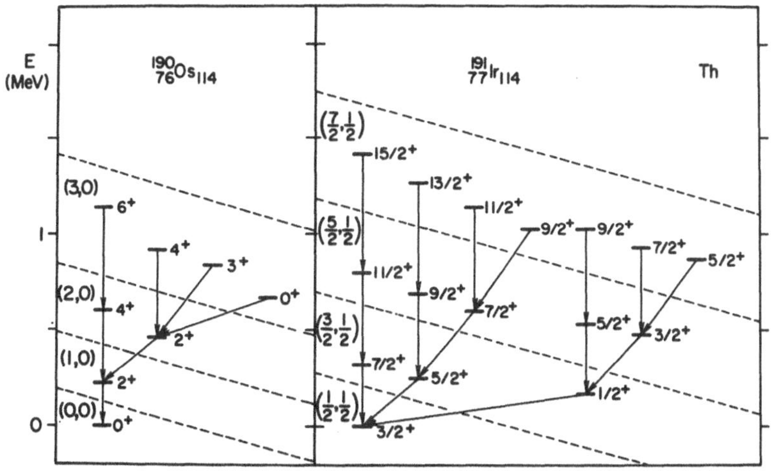

Fig. 6. Theoretical spectra of the pair of nuclei $^{190}_{76}Os_{114}$ and $^{191}_{77}Ir_{114}$ obtained using Eq. (6.7) with (B/6) = 40 keV and C = 10 keV for both nuclei. The numbers in parenthesis denote the Spin(5) labels (τ_1,τ_2). All states shown belong to the maximum allowed representation of Spin(6), σ_1=N, σ_2=σ_3=0 in the even-even nucleus and σ_1=N+$\frac{1}{2}$, σ_2=σ_3=$\frac{1}{2}$ in the odd-even nucleus. For describing the excitation energies of these states the parameters A_1 and A_2 are not needed.

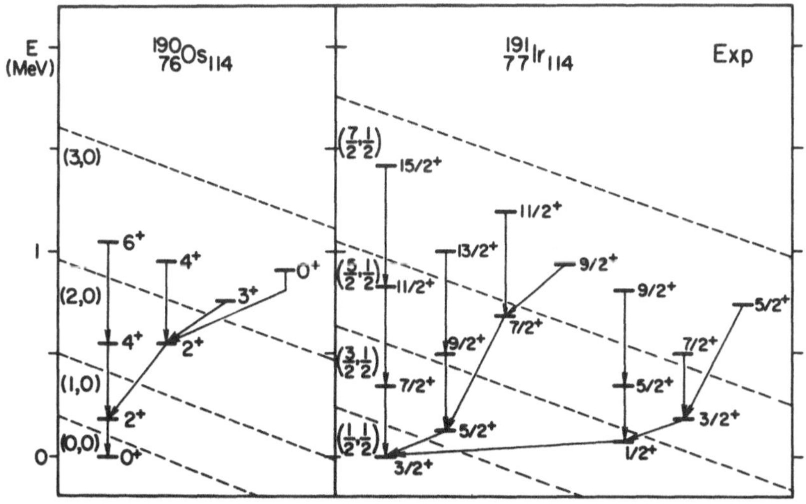

Fig. 7. Experimental spectra of the pair of nuclei $^{190}_{76}Os_{114}$ – $^{191}_{77}Ir_{114}$. The number in parenthesis denote the Spin(5) labels (τ_1,τ_2).

Examples of the supersymmetry U(6|4) have been found in the spectra of nuclei in the Pt-Os region. If the supersymmetry applies, one expects multiplets of nuclei belonging to representation of U(6|4) and with excitation energies described by Eq. (6.7). One of these multiplets would include, for example, the even-even nucleus ${}^{190}_{76}Os_{114}$, the odd-even nucleus ${}^{191}_{77}Ir_{114}$, the two-quasiparticle states in the nucleus ${}^{192}_{78}Pt_{114}$, denoted by a star, the three quasiparticle states in the nucleus ${}^{193}_{79}Au_{114}$, also denoted by a star, and the four quasiparticle states in the nucleus ${}^{194}_{80}Hg_{114}$, denoted by a double star, Fig. 5. The excitation energies for the first two nuclei in the multiplet, ${}^{190}_{76}Os_{114}$ and ${}^{191}_{77}Ir_{114}$ are shown in Fig. 6.

In Fig. 7 the observed excitation energies of the same two nuclei, ${}^{190}_{76}Os_{114}$ and ${}^{191}_{77}Ir_{114}$ are shown. A comparison between Figs. 6 and 7 shows that the spectra of these two nuclei can be described, to a good approximation, by the U(6|4) supersymmetry. The breaking of the supersymmetry (i.e. the average deviation from the energy formula (6.7)) appears to be ≈ 18%. Several other examples of U(6|4) supersymmetry have been reported in the literature[14,21,22].

It should be noted that the occurrence of dynamical supersymmetries allows one to construct analytic solutions not only for excitation energies but also for all other observables, such as electromagnetic transition rates, static moments and intensities of transfer reactions[14]. Several tests of the U(6|4) supersymmetry including these observables have been performed[14,21,22,23]. With the exception of some reactions involving the transfer of one nucleon, all tests appear to indicate that the supersymmetry U(6|4) is present in the spectra of nuclei in the Os-Pt region with a breaking of ≈ 30%. It is not clear at present whether the deviations observed in one nucleon transfer reactions are related to the form of the transfer operator used to calculate the reaction process or to the supersymmetry itself.

CLASSIFICATION OF SPINOR SYMMETRIES AND SUPERSYMMETRIES OF THE INTERACTING BOSON-FERMION MODEL

The U(6|4) chain of Eq. (6.5) provides one example of spinor- and supersymmetries encountered in nuclei. This is based on bosons with SO(6) symmetry and fermions with J=3/2. However, in the purely bosonic case, all three symmetries, U(5), SU(3) and SO(6), have been experimentally observed. One may therefore inquire whether or not other spinor- and supersymmetries are possible in the interacting boson-fermion model. This problem admits an infinite denumerable set of solutions. A partial list of the solutions which may be of interest in the study of nuclear spectra is given in Table I.

Table I

Boson Symmetry	Fermion Angular Momenta			
	j=1/2	j=3/2	j=5/2	j=7/2
U(5)	1			
		1		
	1		1	1
	1	1	1	1
SU(3)	1			
	1	1		
	1	1	1	
	1	1	1	1
SO(6)	1			
		1		
	1		1	1
	1	1	1	1

Table I. Partial list of spinor and supersymmetries in the interacting boson-fermion model. For each boson symmetry the possible values of the fermion angular momenta are indicated to the right.

The spinor and supersymmetries associated with the boson symmetries SU(5) and SO(6) are all of the type discussed in the previous section. For example the case of U(5) and j=1/2 gives rise to the group chain

$$U(6|2) \supset U^{(B)}(6) \otimes U^{(F)}(2) \supset U^{(B)}(5) \otimes SU^{(F)}(2) \supset SO^{(B)}(5) \otimes$$
$$\otimes \; SU^{(F)}(2) \supset SO^{(B)}(3) \otimes SU^{(F)}(2) \supset Spin(3) \supset Spin(2) \; . \quad (7.1)$$

Examples of this supersymmetry have been found recently by Vervier and Janssens[24]. Similarly, the case of SO(6) and j=1/2 gives

$$U(6|2) \supset U^{(B)}(6) \otimes U^{(F)}(2) \supset SO^{(B)}(6) \otimes SU^{(F)}(2) \supset SO^{(B)}(5) \otimes$$
$$\otimes \; SU^{(F)}(2) \supset SO^{(B)}(3) \otimes SU^{(F)}(2) \supset Spin(3) \supset Spin(2) \; . \quad (7.2)$$

It also appears that experimental examples of this chain have been found[25].

A new and interesting algebraic structure is brought in by the spinor and supersymmetries associated with the boson group $SU^{(B)}(3)$. A possible way to combine bosons and fermions here is to split the total angular momentum of the fermions $\vec{J}$ into an orbital, $\vec{L}$, and a spin part, $\vec{S}$, i.e. $\vec{J} = \vec{L}+\vec{S}$. If the angular momentum content of the orbital part is that of an irreducible representation of SU(3)[26], a supersymmetry will arise. For example, fermions with j=1/2 and j=3/2 can be thought of as composed of L=1 and S=1/2. Since L=1 gives rise to the (1,0) representation of SU(3)[26], one can construct the supersymmetry

$$U(6|6) \supset U^{(B)}(6) \otimes U^{(F)}(6) \supset SU^{(B)}(3) \otimes SU^{(F)}_L(3) \otimes SU^{(F)}_S(2) \supset$$
$$SU(3) \otimes SU^{(F)}_S(2) \supset SO(3) \otimes SU^{(F)}_S(2) \supset Spin(3) \supset Spin(2). \tag{7.3}$$

The corresponding energy formula, similar to Eq. (6.7), is now

$$E(\mathcal{N},N,(\lambda_B,\mu_B),(\lambda,\mu),K,L,J,M_J) =$$
$$= a'_0+a'_1\mathcal{N}+a'_2\mathcal{N}^2 +a'_3\mathcal{N}N+a'_4N + a'_5N^2 +$$
$$+ \alpha\ C(\lambda_B,\mu_B) + \beta\ C(\lambda,\mu) + \gamma\ L(L+1) + \delta\ J(J+1)\ . \tag{7.4}$$

The excitation spectrum is given in terms of the four parameters α,β,γ and δ. No experimental examples of this type of symmetry have been found yet. However, since many (purely bosonic) spectra appear to be well described by the $SU^{(B)}(3)$ symmetry, it is quite likely that examples of this type of symmetry will be found soon.

CONCLUSIONS

Several experimental examples of supersymmetric situations appear to have been found in the spectra of complex nuclei. These supersymmetries place in the same multiplet collective (bosonic) and single particle (fermionic) degrees of freedom. Thus, it appears that supersymmetries are relevant to physics and that graded Lie groups and algebras are useful for practical applications.

In concluding my talk, I would like to stress once more the profound difference between supersymmetries and normal symmetries. In a normal symmetry, for example isospin symmetry, ground or excited states in several nuclei are linked together, Fig. 8. These states have either integer of half-integer angular momenta. In a supersymmetry, ground or excited states in several nuclei are linked together, Fig. 9. However, the states which belong to the same representation of the supergroup have different angular momenta and they comprise both integer and half-integer values.

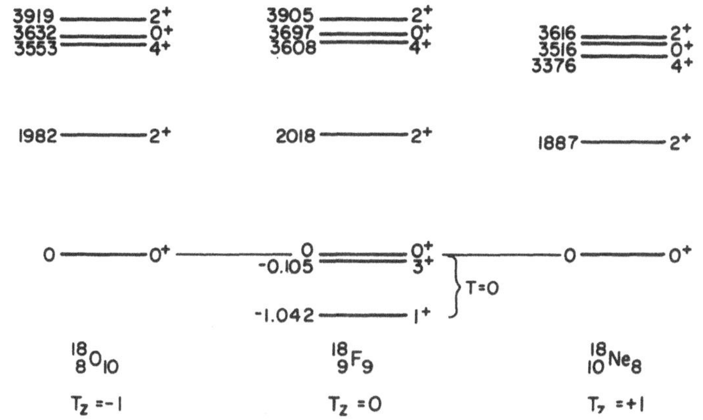

Fig. 8. Isospin symmetry in light nuclei: the triplet $^{18}_{8}O_{10}$ - $^{18}_{9}F_{9}$ - $^{18}_{10}Ne_{8}$, T = 1, T_Z = -1,0,+1. All energies are in keV.

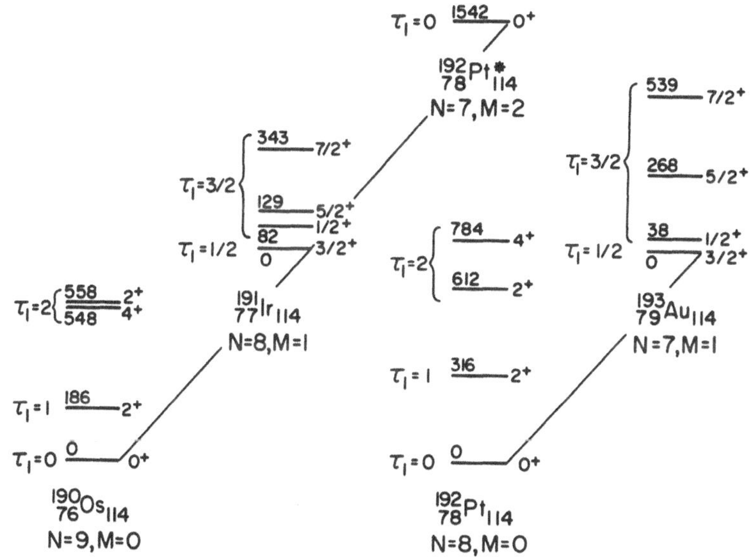

Fig. 9. Supersymmetry U(6|4) in heavy nuclei: a part of the $\mathcal{N}$ = 9 multiplet $^{190}_{76}Os_{114}$ (M=0), $^{191}_{77}Ir_{114}$ (M=1), $^{192}_{78}Pt^*_{114}$ (M=2) and a part of the $\mathcal{N}$ = 8 multiplet, $^{192}_{76}Os_{116}$ (M=0), $^{193}_{79}Au_{114}$ (M=1). All energies are in keV.

Another aspect worth emphasizing is that the supersymmetry observed in nuclei is not a "fundamental" symmetry, in the sense that it applies to some effective degrees of freedom. In fact, the collective bosons are correlated pairs of fermions. Although the correlations may be so large that the bosons effectively lose their memory of being fermion pairs, the ultimate description must certainly involve purely fermionic models.

ACKNOWLEDGMENTS

I wish to thank A.B. Balantekin and I. Bars, without whom most of the developments reported here would not have been possible, and all the experimentalists who have helped us understanding various aspects of supersymmetry in nuclei, in particular, J. Cizewski, M. Harakeh, M. Vergnes, J. Vervier and J. Wood. This work was in part supported under the Department of Energy Contract No. EY-76-C-02-3074 and in part under financial support from the Stichting FOM.

REFERENCES

1. For a review see P. Fayet and S. Ferrara, Phys. Rep. 32C:249 (1977).
2. For a review see Interacting Bosons in Nuclear Physics, ed. F. Iachello (Plenum Press, N.Y., 1979).
3. A. Arima and F. Iachello, Phys. Rev. Lett. 35:1069 (1975).
4. A.E.L. Dieperink, O. Scholten and F. Iachello, Phys. Rev. Lett. 44:1974 (1980).
5. D.H. Feng, R. Gilmore and S.R. Deans, Phys. Rev. C23:1254 (1981).
6. For a description of this model see A. Bohr and B.R. Mottelson, in Nuclear Structure, vol. 2 (Benjamin, N.Y., 1975).
7. A. Arima, T. Otsuka, F. Iachello and I. Talmi, Phys. Lett. 66B:205 (1978).
8. A. Arima and F. Iachello, Ann. Phys. (N.Y.) 99:253 (1976).
9. A. Arima and F. Iachello Ann. Phys. (N.Y.) 111:201 (1978)
10. A. Arima and F. Iachello, Ann. Phys. (N.Y.) 123:408 (1979).
11. S. Okubo, Prog. Theor. Phys. 27:949 (1962).
12. L.S. Kisslinger and R.A. Sorenson, Mat. Fys. Medd. Dan. Vid. Selsk. 32:No. 9 (1960).
13. F. Iachello and O. Scholten, Phys. Rev. Lett. 43:679 (1979).
14. A.B. Balantekin, I. Bars and F. Iachello, Phys. Rev. Lett. 47:19 (1981);
A.B. Balantekin, I. Bars and F. Iachello, Nucl. Phys. A370:284 (1981).
15. B.H. Flowers, Proc. Roy. Soc. A212:248 (1952).
16. R. Gilmore, Lie Groups, Lie Algebras and some of their Applications (eds. J. Wiley and Sons, New York, 1974), p. 111.
17. F. Iachello, Phys. Rev. Lett. 44:772 (1980).

18. A.B. Balantekin and I. Bars, *J. Math. Phys.* 22:1149; 22:1810 (1981).
19. V.G. Kac, *in* Differential Geometric Methods in Mathematical Physics, eds. K. Bleurer, H.R. Petry and A. Reetz (Springer Verlag, Berlin 1978);
Y. Ne'eman and S. Sternberg, *Proc. Nat. Acad. Sci.* (USA) 77:3127 (1980);
P.H. Dondi and P.D. Jarvis, *Zeit. f. Phys.* C4:201 (1980);
M. Scheunert, W. Hanm and V. Rittenberg, *J. Math. Phys.* 18:155 (1977);
M. Marcu, *J. Math. Phys.* 21:1277 (1980).
20. P.G. Freund and I. Kaplansky, *J. Math. Phys.* 17:228 (1976).
21. J. Wood, *Phys. Rev.* C24:1788 (1981).
22. J. Vervier, *Phys. Lett.* 100B:383 (1981);
J. Vervier, r. Holzmann, R.F. Hanssens, M. Loiselet and M.A. van Hove, *Phys. Lett.* 105B:343 (1981).
23. Y. Iwasaki, E.H.L. Aarts, M.N. Harakeh, R.H. Siemssen and S.Y. van der Werf, *Phys. Rev.* C23:1477 (1981);
M. Vergnes, G. Rotbard, J. Kalifa, G. Berrier-Rousin, J. Vernotte, R. Seltz and D.G. Burke, *Phys. Rev. Lett.* 46:584 (1981);
M.N. Harakeh, P. Goldhoorn, Y. Iwasaki, J. Łukasiak, L.W. Put, S.Y. van der Werf and F. Zwarts, *Phys. Lett.* 97B:21 (1980);
J.A. Cizewski, D.G. Burke, E.R. Flynn, R.E. Brown and J.W. Sunier, *Phys. Rev. Lett.* 46:1264 (1981).
24. J. Vervier and R.V.F. Janssens, *Phys. Lett.* B to be published (1981).
25. J. Vervier, private communication (1981).
26. J.P. Elliott, *Proc. Roy. Soc. Ser.* A245:128 (1958); A245:562 (1958).

AXIONS

J.-M. Frère

Faculté des Sciences, Université Libre, Bruxelles
and
CERN, Geneva, Switzerland

STRONG CP VIOLATION

Considering a coloured fermion f^a (a = 1,2,3) coupled to QCD, we first write the usual Lagrangian, using standard notations:

$$\mathcal{L} = -\frac{1}{4} G_{\mu\nu}^{a} G^{\mu\nu a} + \bar{f}(i\not{D} - m)f \tag{1}$$

This simple expression has to be modified in view of the existence of topologically inequivalent (classical) solutions for the gauge field, leading to inequivalent vacua. 't Hooft[1] has shown that these effects could be taken into account by adding to (1) the term

$$\theta \mathcal{L}_{CP} = \theta \frac{G^2}{32\pi^2} \cdot G_{\mu\nu}^{a} \tilde{G}^{a\mu\nu}$$
$$\tilde{G}^{a\mu\nu} = \frac{1}{2} \varepsilon^{\mu\nu\rho\sigma} G_{\rho\sigma}^{a} \tag{2}$$

Let us mention that this term is usually omitted, because it can be rewritten as a total divergence. The non-trivial topological solutions, however, correspond to potentials which decay like 1/r, and thus give non-vanishing contributions to the surface integral. We also remark that $\mathcal{L}_{CP}$ is odd under both the P and CP reflexions [each of its terms contains four vectors (derivatives) with different indices].

The term (2) can be rotated away when m vanishes in (1). We know indeed that in such a case, $\mathcal{L}$, although formally invariant under γ_5 rotation, is in fact altered due to the Adler anomaly[2]

$$f \to (1 + i\gamma_5\alpha)f$$

$$\delta \mathcal{L} = \frac{\partial}{\partial x^\mu}\left(\frac{\partial \mathcal{L}}{\partial\left(\frac{\partial f}{\partial x^\mu}\right)} \cdot \delta f\right) = -\alpha\frac{\partial}{\partial x^\mu}(\bar{f}\gamma^\mu\gamma_5 f) = -2\alpha\,\mathcal{L}_{CP} \tag{3}$$

If m vanishes, α can thus be chosen to rotate away the unwanted apparent CP violation. In the presence of several fermion flavours, it is sufficient for one of them to be massless to authorize a rotation (3) and avoid the "strong CP" problem completely.

In the more realistic case where none of the fermionic masses vanishes the strong CP problem has to be faced. In order to investigate its physical consequences, it is still more convenient to perform a γ_5 rotation to bring the CP violating term completely into the quark mass matrix. For three quark flavours one thus finds (2) to be equivalent to:

$$-i\theta \; \frac{m_u m_d m_s(\bar{u}\gamma_5 u + \bar{d}\gamma_5 d + \bar{s}\gamma_5 s)}{m_u m_d + m_d m_s + m_s m_u} \tag{4}$$

This term has been shown[3] to imply an electric dipole moment for the neutron at a level incompatible with experiment, unless:

$$\theta < 10^{-9} \tag{5}$$

This limit on the parameter θ can be accepted as such; alternatively one can look for a rationale for such a small value of the angle. The fact that θ can be completely rotated away in the absence of "current" mass for one fermion leads one to consider the case where such a mass only arises via a mechanism of spontaneous symmetry breaking, as in the models of weak interactions.

THE PECCEI-QUINN SYMMETRY AND THE AXION

The possibility of exploiting a spontaneous symmetry breaking mechanism to eliminate the CP violating θ angle was stressed by Peccei and Quinn[4]. While not possible in the minimal Weinberg-Salam model this mechanism can be realized in more extended versions; it, however, leads to the presence of a nearly massless boson, the "axion".

Let us first consider the minimal Weinberg-Salam model, with:

$$f_L = \begin{pmatrix} u \\ d \end{pmatrix}_L \quad \phi = \begin{pmatrix} \phi^+ \\ \phi^0 \end{pmatrix} \quad \tilde{\phi} = \begin{pmatrix} \phi^{0*} \\ -\phi^- \end{pmatrix} \quad \langle\phi^0\rangle \equiv v \tag{6}$$

We focus on the part of the Lagrangian containing the Yukawa coupling between the ϕ field and the fermions:

$$\mathcal{L}_Y = \frac{m_d}{v} \bar{f}_L \phi d_R + \frac{m_u}{v} \bar{f}_L \tilde{\phi} u_R + h.c. \tag{7}$$

Masses are generated when ϕ^0 develops its vacuum expected value v. In order to allow the elimination of the θ angle, $\mathcal{L}$ must be formally invariant under a γ_5 rotation. For convenience, we perform at the same time a phase rotation so that only the right-handed fields will be affected; starting with

$$\begin{aligned} d_R &\to e^{i\alpha_d} d_R \\ \phi &\to e^{-i\alpha_d} \phi \\ u_R &\to e^{i\alpha_u} u_R \end{aligned} \tag{8}$$

we find that formal invariance of $\mathcal{L}$ requires $\alpha_u = -\alpha_d$. We thus see that, while a γ_5 rotation is possible, it does not allow the removal of the CP violating θ term: under (8) the Lagrangian indeed receives an extra contribution through the Adler anomaly:

$$\delta\mathcal{L} \sim (\alpha_u + \alpha_d) \; G^{\mu\nu a}\tilde{G}_{\mu\nu a} = 0$$

The remedy to this failure is quite obvious, and entails a dedoubling of the field ϕ: we now write

$$\begin{aligned} \mathcal{L} &= \frac{m_u}{v_1} \bar{f}_L \tilde{\phi}_1 u_R + \frac{m_d}{v_2} \bar{f}_L \phi_2 d_R + h.c. \\ &\langle\phi_1^0\rangle = v_1 \qquad \langle\phi_2^0\rangle = v_2 \end{aligned} \tag{9}$$

This allows for independent rotations of α_u and α_d in (8) provided that:

$$\phi_1 \to e^{i\alpha_u}\phi_1 \qquad \phi_2 \to e^{-i\alpha_d}\phi_2 \tag{10}$$

So far we have only inspected the Yukawa part of the Lagrangian. We note however that the validity of (10) for arbitrary α_u and α_d implies that the Higgs potential cannot depend upon the relative phase of ϕ_1 and ϕ_2, so that the Lagrangian (before symmetry breakdown) possesses an extra symmetry - the Peccei-Quinn symmetry.

We still have to check what the situation becomes after symmetry breaking. In other words, does the minimum of the action coincide with the CP conserving solution ? Peccei and Quinn[4] have shown that this was indeed the case. A simple argument showing this is due to Voloshin and Zakharov[5]. Assuming a naîve Lagrangian $\mathcal{L}$ which conserves CP, and arbitrary (complex) vacuum expectation values v_1 and v_2, we first perform a chiral transformation on the fermion fields to make all the masses real. The only CP violating term in the Lagrangian is then

$$(\theta - \Sigma\alpha_i)\cdot \mathcal{L}_{CP}$$

Taking the vacuum expectation value of the effective action in a vacuum state which is by definition CP even, we see that the linear contribution must vanish. The leading contribution (for small $\theta_{eff} = \theta - \Sigma\alpha_i$) being quadratic, the CP conserving solution thus corresponds to a local extremum, as requested.

The strong CP problem is thus solved in principle, but we now have to face the phenomenological implications of the extended Weinberg-Salam model, in particular the axion[6], a nearly massless boson associated with the breaking of the Peccei-Quinn symmetry. Temporarily forgetting the fermions, we concentrate on the neutral Higgs scalars. With the symmetry (10) with $\alpha_u = +\alpha_d$ is associated the Noether current:

$$J^\mu = \phi_1^{0+}\partial_\mu\phi_1^0 - \phi_2^{0+}\partial_\mu\phi_2^0 \qquad (11)$$

After introducing the vacuum expectation values v_1 and v_2 we identify a Goldstone boson:

$$g \sim v_1 \,\mathrm{Im}\phi_1 - v_2\mathrm{Im}\phi_2 \qquad (12)$$

We know however that another combination, namely $(v_1\mathrm{Im}\phi_1 + v_2\mathrm{Im}\phi_2)$ is associated with a would-be Goldstone boson, which will be absorbed to form the longitudinal part of the Z : we thus have to orthogonalize those two states and we identify the (until now massless) axion:

$$h' = \frac{v^2}{v}\cdot \mathrm{Im}\phi_1 - \frac{v^1}{v}\cdot \mathrm{Im}\phi_2$$
$$v^2 = v_1^2 + v_2^2 \qquad (13)$$

The coupling of h' to the fermions is

$$\frac{h'}{v}\left[x\, \bar{u}m_u\gamma_5 u + \frac{1}{x}\bar{d}m_d\gamma_5 d\right] \qquad (14)$$

where $x \equiv v_2/v_1$ and the W boson mass is given for the present Higgs structure by $M^2 = 1/2(g^2v^2)$.

The presence of strong interactions responsible for the appearance of quarks' constituent masses implies the existence of more (pseudo) Goldstone bosons ($\pi,\eta,\ldots$). In the standard way, the high mass of the η is attributed to the strong Adler-Bell-Jackiw anomaly. We know from (13) that h' also couples to that anomaly; a mixing between h' and η results, and the light particle usually referred to as the "axion", now denoted by h, couples to an anomaly-free combination of quarks. The simultaneous presence of "current" and "dynamical" quark masses further mixes h and π. A detailed calculation can be found in Ref. 7.

One obtains, for N generations:

$$\begin{aligned} \xi_{\pi} &= \xi \cdot B_{\pi} \\ B_{\pi} &= \xi \cdot \left[N \frac{m_d - m_u}{m_u + m_d} \left(x + \frac{1}{x}\right) - x = \frac{1}{x} \right] \\ \xi_{\tilde{\eta}} &= (N-1)\left(x + \frac{1}{x}\right) \qquad [\tilde{\eta} \text{ is the SU(2) singlet}] \\ \xi &= 1.9\ 10^{-4} \end{aligned} \tag{15}$$

while the mass of the axion is given by:

$$m_{axion} = 23 \text{ keV } \frac{N}{2}\left(x + \frac{1}{x}\right) \tag{16}$$

In the above we have assumed that all the "up" quarks were coupled to ϕ_1, and all the "down" quarks to ϕ_2. This requirement is not compulsory, but has the advantage of leading automatically to flavour conservation in the neutral Higgs exchanges.

Further possibilities arise however when we turn to the leptons. The charged leptons could in principle be coupled with the coefficient $-x$ or $1/x$, or even not be connected to the "axion" at all.

Assuming that the couplings to all fermions are known, the decay rate into two photons is estimated via the usual triangular diagram:

$$\tau(h \to 2\gamma) = \frac{\tau(\pi^0 \to 2\gamma) \left(\frac{m_\pi}{m_h}\right)^5 3z(1+z)^{-2}}{\left(\frac{3z}{z+1} - N_1\right)} \tag{17}$$

where $z \equiv m_u/m_d$; N_1 = number of charged leptons coupled with $(-x)$, the others being coupled with $1/x$.

We may now summarize the various ways in which the axion interacts with ordinary matter.

i) Light quarks: interacts essentially via π or η mixing [Eq. (15)]

ii) Leptons: not really constrained, possibly ($-x$ or $1/x$)

iii) Two photons: [Eq. (17)]. Exact value depends upon ii)

iv) Heavy quarks: Yukawa coupling: $\sim m_q/m_W$ (x or $1/x$) [Eq. (14)]

EXPERIMENTAL EVIDENCE

These four types of physical probes have been used to investigate the possible existence of the axion. Starting with the coupling to light quarks, we first mention reactor experiments, where axions are assumed to be produced in nuclear transitions. The complicated structure of the source however makes flux estimates (based on a comparison with γ rates attributed to magnetic transitions) very uncertain. A critical discussion is to be found in Ref. 8.

A much clearer experiment is based on a pure nuclear transition, occurring in the de-excitation of ^{137}Ba from the state $11/2^-$ to the state $3/2^+$. This is essentially a probe of the neutron-axion coupling. Except for huge values of x or 1/x, which would forbid the transition on a pure energetical basis (the mass of the axion being then too big), but would imply a strong Yukawa coupling to quarks, the following limit was recently obtained (Ref. 9).

$$0.4 \leq x \leq 1.4 \tag{18}$$

Remaining in the domain of light quarks, beam dump experiments present in principle a very sensitive test. The situation is however plagued with some uncertainties in the treatment of the hadronic interactions. In the simplest case, where the axion produced in a hadronic collision is detected via its eventual re-interaction with hadronic matter, a typical limit is given by [10]

$$\sigma_{\text{production}} \cdot \sigma_{\text{interaction}} < 10^{-67} \sim 10^{-68}\ \text{cm}^4 \tag{19}$$

If one assumes that all the interaction takes place through pion mixing, this requires from Eq. (15) $B_\pi^2 < 0.12 \sim 0.04$. This limit is however not reliable: mixing with η (and possibly with other pseudoscalars) has to be added coherently. Ellis and Gaillard[11] have, for instance, considered a mechanism based on the incoherent production of $u\bar{u}$ and $d\bar{d}$ pairs; this leads to the replacement of B_π^2 above by $(B_\pi^2 + 1/3(B_\eta^2))(B_\pi^2 + 2/3(B_\eta^2))$.

We now turn to electron beam dump experiments. The interpretation is here quite reliable[12], but as expected the sensitivity is much lower. Furthermore we have already noticed that the leptonic couplings are not really constrained at the Lagrangian level; for this reason, leptonic tests should not be directly compared to hadronic ones.

Assuming (1/x) coupling, a SLAC experiment (electron beam dump and detection by leptonic interaction) expects $5.5/x^4$ events, but sees none[13]. An experiment based on production via bremsstrahlung and detection in e^+e^- or $\gamma\gamma$ gives[14]

$$1.2 < x < 9.2 \qquad (\text{or } \; x < 0.074, \text{ or } \; x > 70) \tag{20}$$

for 1/x coupling, the first region being excluded for a coupling $\sim$ (-x).

In view of the above, the clearest way to rule out the standard axion would certainly be to probe the Yukawa coupling to heavy flavours (iv). We begin by an indirect test for this, and consider K decays. The experimental limit for the branching ratio of K^+ into pion and axion[15] is $1.4 \; 10^{-6}$ (and could be improved to $3 \; 10^{-7}$). Kaons are expected to decay into axions via pion or eta mixing. (In the case of charged kaons, the latter would not be suppressed by the $\Delta I = 1/2$ rule.) Conflicting estimates have been made for this process (see Refs. 6,7 and 15). In addition to this, a short-distance contribution[16] to the process is due to the graphs:

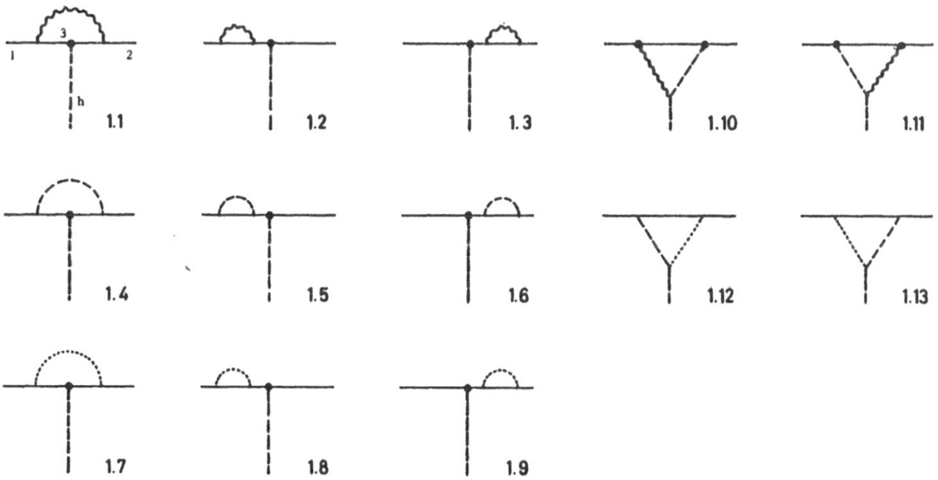

Fig. 1 Wavy line: $W^{\pm}$ propagator. Dotted line: unphysical charged scalar (ghost)(massless in Landau gauge). Dashed line: physical massive charged scalar (internal lines), axion (external lines). Solid line: fermions, 1,2,3 are quark labels, 3 being the (heavy) internal quark.

A detailed calculation of these contributions[17] shows that due to the pseudoscalar nature of the axion coupling, the graphs with intermediary physical charged scalar bosons are not negligible; the result depends strongly upon its mass m_H. If we consider only the short distance contribution we get:

$$B.R(K \to \pi h) \sim 0.8\ 10^{-6}\{xA_1(m_c) + x^3A_2(m_c) + \frac{m_t^2}{m_c^2}(s_2^2 + s_1s_3)[xA_1(m_t) + x^3A_2(m_t)]\}^2 \tag{21}$$

The functions A_1 and A_2 are represented in Fig. 2. From the number of parameters entering (21), it is quite clear that no definitive conclusion can be drawn from here. Let us just mention that for $m_H \sim 20$ GeV the contribution from the charmed quark is suppressed, but not the much bigger top contribution. Alternatively, for m_H between 50 and 150 GeV, cancellations are possible between those two terms (and the "soft" contribution from pion and eta mixing).

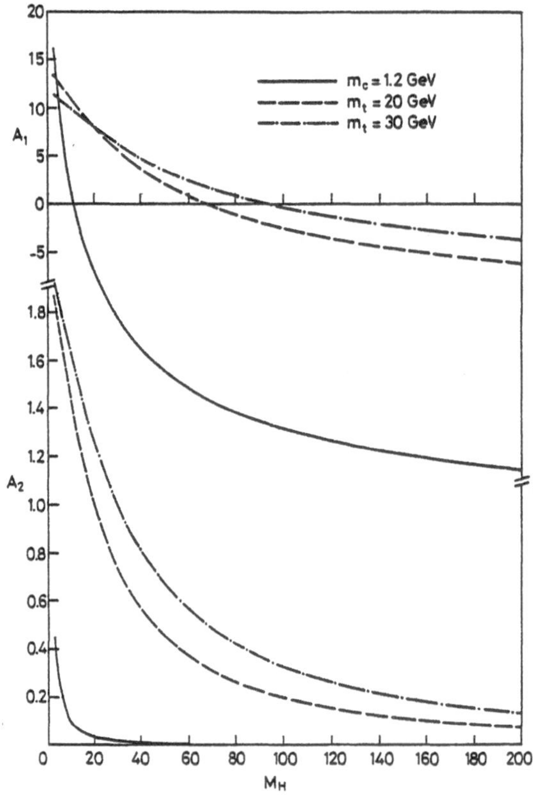

Figure 2

Strong cancellations are needed for the standard axion to survive this test; this cannot however be used reliably to exclude its existence. (We note that a better test would be provided by the similar decay of B into K and axion, where only the t quark contributes significantly.)

A clean and direct test of heavy quarks coupling is provided by ψ decays. We expect[6]

$$B. = \frac{\Gamma(\psi \to h\gamma)}{\Gamma(\psi \to e^+e^-)} = x^2 G_F \cdot m^2 c/(\sqrt{2} \cdot \pi \cdot \alpha) \qquad (22)$$

Recent results from the Crystall Ball collaboration[18] put the limit

$$B. < 3\ 10^{-5} \quad (90\%\ \text{C.L.}) \qquad (23)$$

which leads to $x < 1$ (a similar result with upsilon decays would be the best way to exclude the axion right away). Combining the two cleanest experimental limits (18) and (23) we see that, excepting extremely big or small values, the x parameter is already restricted to be

$$0.4 < x < 1 \qquad (24)$$

Furthermore no indication exists that values in that range would correspond to zeros of the production or interaction cross-sections in hadronic beam dump experiments and considerable cancellations are requested to reconcile it with K decays. We can thus conclude that the "minimal" axion scheme seems to be strongly disfavoured by present experiments. It should however be kept in mind that the above considerations rely on a basic assumption: namely, that "natural" flavour conservation is realized by coupling <u>all</u> up quarks to one Higgs boson, and all down quarks to another one. More "baroque" models could be envisaged (at the price of a fine tuning of some more parameters) and would possibly account for some discordant results[19].

ALTERNATIVE SOLUTIONS

The experimental situation prompts us to turn to some alternative solutions of the strong CP problem.

a) If we assume that $\theta = 0$ at the level of the strong interaction for some (yet unknown) reason, the question arises whether radiative corrections due to weak interactions would modify this by an unacceptable amount. Ellis and Gaillard[20] have shown that in the $SU(2) \times U(1)$ scheme the first (potentially) divergent contributions would arise only at the 14th order.

b) $SU(2)_L \times SU(2)_R \times U(1)$ models are frequently advocated in this context[21], since corrections are then finite and calculable. In order to avoid any confusion we should however mention that the axion type solution of the CP problem is also possible in left-right symmetrical models[22]. The reason why the scalar fields $\phi(1/2,1/2)$ have to be doubled is however different here (otherwise, the Peccei-Quinn symmetry forbids Cabibbo angles).

c) In theories based on dynamical symmetry breakdown the problem could be solved by an automatic alignment of the various breakings of symmetry[23]; no satisfactory model exists however.

d) The axion could be made heavy enough to avoid its production at present energies[24].

e) Finally another way out consists of transposing the solution of the CP problem to new sectors of the particle spectrum, weakly coupled to the known ones. Several attempts have been made in this direction[25]. We will develop somewhat the most economical[26].

THE SUPERLIGHT AXION

Going back to Eq. (14) we see that the Yukawa coupling of the axion to quarks is determined by the ratio of a quark mass to the vacuum expectation value of some scalar field. The latter is directly related in the minimal model considered so far to the mass of the weak bosons.

The possibility of generalizing the model has been considered in Ref. 7b, with the conclusion that this would usually only make the situation worse: if more fields contribute by their vacuum expectation values x to the W mass, the quantity v in Eq. (14) now has to satisfy

$$g^2\left(\frac{v^2}{2} + x^2\right) = M^2$$

and the axion-quark coupling actually becomes stronger. Only when the v.e.v. of the supplementary scalars does not take part in the W mass [which means that they must be singlets under $SU(2) \times U(1)$], while they still play a role in the Peceei-Quinn symmetry, is this conclusion modified.

This is precisely the case considered in Ref. 26. Adding to the model a complex scalar field S, we supplement the previously considered Higgs potential by

$$V_{DFS} = \lambda\, \phi_1 \phi_2^+ S^2 + V(S^2) \tag{25}$$

The Peccei-Quinn symmetry is preserved provided:

$$\begin{aligned} &\phi_1 \to e^{i\alpha}\phi_1 \qquad \phi_2 \to e^{i\beta}\phi_2 \qquad S \to e^{i\gamma}S \\ &2\gamma + \alpha - \beta = 0 \end{aligned} \tag{26}$$

The Noether current associated to (26) is now, instead of (11), given by:

$$J^\mu = \alpha\, \phi_1^+ \partial_\mu \phi_1 + \beta\, \phi_2^+ \partial_\mu \phi_2 + \gamma S^+ \partial_\mu S \tag{27}$$

Assuming <S> = X, and orthogonalizing as previously with respect to the longitudinal component of the Z field, we obtain the new form of the axion:

$$h' = \frac{v_2\ \mathrm{Im}\phi_1 - v_1\ \mathrm{Im}\phi_2 + X\ \mathrm{Im}S}{\sqrt{v_1^2 + v_2^2 + X^2}} \tag{28}$$

From Eq. (28), we note immediately that when $X \gg v_1, v_2$, the axion is essentially constituted by the imaginary part of the scalar field S, which is decoupled from quarks and gauge bosons. In this limit, the direct coupling of h' to matter is thus reduced by a factor v/X, namely by the ratio of the two scales of symmetry breaking. Modulo this suppression factor, Eqs. (15), (16) and (17) stay otherwise unchanged.

Three questions then arise:

i) How big must X be to agree with experimental data ?
ii) Is there any rationale to the second scale of symmetry breaking associated with X ?
iii) Is there any way to still detect the axion ?

While a suppression of a few orders of magnitude of the axion coupling would be sufficient to reconcile it with existing earth based experiments, astrophysical bounds provide us with stringent limitations on the ratio v/X. The evolution of red giants is in fact very sensitive to the presence of light, long-lived particles. Such particles, produced in the dense core of the star, would indeed escape and take away a quantity of energy.

This would lead to a cooling of the core and forbid further burning of helium, shortening the lifetime of the red giant stage beyond observational limits. In the case of interest, we focus on the axions produced by the Primakoff effect (production by leptons has also been considered, but we know that the axion coupling to leptons is not really constrained). This process can be suppressed in two ways:

i) due to the unavailability of photons of sufficient energy in the stellar core, whose temperature is estimated to be 10^8 ^{0}K ($\sim$ 10 KeV).

ii) due to the axion coupling to two photons [Eq. (17)]. We know that the coupling is proportional to the 5th power of the mass.

The first of these limitations was actually effective for the "classical" axion, with a mass of a fraction of an MeV. However, as soon as the mass drops below that value, the production of axions becomes much too important, (and raises exponentially with decreasing energy). Even in view of the uncertainty on the coefficient of Eq. (17) due to the coupling of leptons, little can be gained. Only the second mechanism can then be called into play. As was shown in Ref. 27, the axion mass then has to be either > 200 keV, or < 10^{-2} eV ! This corresponds to X = <S> > 10^9 GeV. Clearly this is suggestive of embedding the superlight axion mechanism in a grand unification scheme[28]. We will not continue in this direction which will be further discussed in other lectures.

We would, however, like to end this talk with a remark on the observability of superlight axions. Several authors have shown that even after dressing up by other interactions, the superlight axion would keep an extremely small coupling with ordinary matter, thus forbidding direct observation[29]. We would like to mention however that, while it seems impossible to disprove the existence of this particle, it might be conceivable that some hints to its presence can be obtained by exhibiting the doubling of the Higgs structure required by the Peccei-Quinn symmetry.

That the vacuum expected value of S is large (> 10^9 GeV) does not imply anything on the value of λ in Eq. (25). Two situations may then arise:

i) λ is comparable to the other constants in the problem. The mass matrix of ϕ_1 and ϕ_2 is then diagonalized to give two systems of Higgses, namely the usual ones (physical H^0, longitudinal part of Z and $W^\pm$), and a superheavy set ($H^{0\prime}$, $H^{0\prime\prime}$, $H^\pm$). The latter set only is coupled with normal strength to S and consequently to the superlight axion. No hope of observing the doubling of the Higgs structure in this case exists.

ii) For some unknown reason λ is very small. In that case, the rest of the Higgs may stay light, and we have a rich structure (H^0, $H^{0\prime}$, $H^{0\prime\prime}$, $H^\pm$) at an accessible energy.

Such a doubling of the Higgs structure would either be observable directly, or via "renormalization effects", since they would affect the value of $\rho = m^2/m_Z^2 \cos\theta^2$, and the location of the physical poles of the Z and W's. Work in this direction is in progress in collaboration with J. Vermaseren. (Furthermore, in grand unified theories the ratio m_b/m_τ might also be affected.)

REFERENCES

1. G. 't Hooft, Phys. Rev. Lett. 37:8 (1976); [also Phys. Rev. D14:3432 (1976)].
2. S. Adler, Phys. Rev. 177:2426 (1969);
   J.S. Bell and R. Jackiw, Nuovo Cim. 60A:49 (1969);
3. V. Baluni, Phys. Rev. D19:2227 (1974).
   R. Crewther, P. Di Vecchia, G. Veneziano and E. Witten, Phys. Lett. 88B:123 (1979).
4. R.D. Peccei and H.R. Quinn, Phys. Rev. D16:1791 (1977).
5. A.I. Vainshtein, V.I. Zakharov and M.A. Shifman, Usp. Fiz. Nauk. 131:537 (1980), english translation Sov. Phys. Usp. 23:429 (1980).
6. S. Weinberg, Phys. Rev. Lett. 40:223 (1978);
   F. Wilczek, Phys. Rev. Lett. 40:279 (1978).
7. a) W. Bardeen and S.-H. Tye, Phys. Lett. 74B:229 (1978);
   b) J. Kandaswamy, P. Salomonson and J. Schechter, Phys. Lett. 74B:377 (1978), Phys. Rev. D19:2757 (1979).
8. T.W. Donnelly, S.J. Freedman, R.S. Lytel, R.D. Peccei and M. Schwartz, Phys. Rev. D18:1607 (1978).
9. A. Zehnder, Phys. Lett. 104B:494 (1981).
10. P. Coteus et al., Phys. Rev. Lett. 42:1438 (1979);
    A. Soukas et al., Phys. Rev. Lett. 44:564 (1980);
    P. Alibran et al., Phys. Lett. 74B:134 (1978);
    P.C. Bossetti et al., Phys. Lett. 74B:143 (1978);
    T. Hansl et al., Phys. Lett. 79B:139 (1978).
11. J. Ellis and M.K. Gaillard, Phys. Lett. 74B:374 (1978).
12. W. Bardeen, S.-H. Tye and J.A.M. Vermaseren, Phys. Lett. 76B:580 (1978).
13. See Ref. 8 and also
    A.F. Rothenberg, SLAC report 147 (1972).
14. D.J. Bechis et al., Phys. Rev. Lett. 42:1511 (1979).
15. J.H. Klems, R.H. Hildebrand and R. Stiening, Phys. Rev. D4:66 (1971);
    T. Goldman and C.M. Hoffman, Phys. Rev. Lett. 40:220 (1978).
16. M.B. Wise, Phys. Lett. 103B:121 (1981).
17. J.-M. Frère, M.B. Gavela and J.A.M. Vermaseren, Phys. Lett. 103B:129 (1981);
    L.J. Hall and M.B. Wise, Harvard preprint HUTP 81/A010 (1981).
18. F.C. Porter, Invited talk at the EPS Conf. on High Energy Physics, Lisbon, July 1981, SLAC preprint SLAC-PUB 2785, CALT 68-852 (1981).
19. H. Faissner, "Evidence for axions - or something like that ?", Invited talk at the International Neutrino Conference 1981, Mani, Hawaii, Aachen preprint PITHA 81/32 (1981).
20. J. Ellis and M.K. Gaillard, Nucl. Phys. B150:141 (1979).
21. R.N. Mohapatra and G. Senjanovic, Phys. Lett. 79B:283 (1978);
    M.A.B. Beg and H.S. Tsao, Phys. Rev. Lett. 41:278 (1978).

22. K.T. Mahantappa and M.A. Sher, Phys. Rev. D18:4354 (1978).
23. E. Eichten and K.D. Lane, Phys. Lett. 90B:125 (1980);
S. Dimopoulos and L. Susskind, Nucl. Phys. B155:237 (1979).
24. S.-H. Tye, "Axion", Cornell preprint CLNS-81/479 (1981).
25. J.E. Kim, Phys. Rev. Lett. 43:103 (1979);
M.A. Shifman, A.I. Vainshtein and V.I. Zakharov, Nucl. Phys. B166:493 (1980).
26. M. Dine, W. Fischler, M. Srednicki, Phys. Lett. 104B:199 (1981).
27. K. Sato and H. Sato, Progr. Theor. Phys. 54:912,1564 (1975);
D.A. Dicus, E.W. Kolb, V.L. Teplitz and R.V. Wagoner, Phys. Rev. D18:1829 (1978), Phys. Rev. D22:839 (1980).
28. M. Wise, H. Georgi and S.L. Glashow, Phys. Rev. Lett. 47:402 (1981).
29. R. Barbieri, R.N. Mohapatra, D.V. Nanopoulos and D. Wyler, Phys. Lett. 107B:80 (1981);
J. Ellis, M.K. Gaillard, D.V. Nanopoulos and S. Rudaz, Phys. Lett. 106B:303 (1981)

CP VIOLATION, COSMOLOGICAL BARYON ASYMMETRY AND NEUTRINO MASSES - THE EFFECT OF INTERMEDIATE MASS SCALES

Antonio Masiero

Max-Planck -Institut für Physik and Astrophysik
Föhringer Ring 6, 8000 Munich 40, Fed. Rep. Germany
and Associato INFN, Sezione di Padova, Italy

The question whether the 10^2 - 10^{15} GeV big desert is completely wild or contains some oases at intermediate mass scales (IMS) is still open. As we are denied any direct verification (at least for IMS some orders of magnitude away from M_W), it behooves us to sharpen any possible motivation and/or consequences of such IMS. Apart from the "classical" emotional (aesthetical) repulsion for a desert which does not present any analog with previous physics, some interest in IMS has been renewed by new trends in particle physics: i) technicolour (including supercolour or technicolour/supersymmetric technicolour scenarios) and the composite models introduce a new fundamental strong interaction, whose scale must be in the TeV region to reproduce the phenomenological weak interaction scale M_W (no matter how poor, they represent the only candidates for a dynamical understanding of M_W); ii) some standard problems of the fruitful particle physics-cosmology marriage could require a prolonged first order phase transition, thus introducing some physical scale between the grand unification mass (M_{GU}) and M_W.

Certainly more important than all this is the fact that IMS models lead to remarkable <u>phenomenological consequences</u> testable in the near future. Limiting myself to those models in which the IMS is connected with parity restoration at low energy[1], I recall that neutrinoless double β-decay and $\mu^- \to e^-\gamma$ are predicted within experimentally accessible ranges if such restoration occurs just near M_W. In this case, also high energy experiments, such as the measurement of the polarization asymmetry in $e_{L,R}p$ scattering, $(\sigma(e_L^-) - \sigma(e_R^-))/(\sigma(e_L^-) + \sigma(e_R^-))$, and the production of right handed neutrinos, i.e., $\sigma(e_R + p \to N_R + X)/\sigma(e_L^- + p \to \nu_L + X)$, would play a decisive role[2]. Dramatic changes have also to be expected in the baryon (B) number violating processes, in particular if parity breaks down in the 10 - 100 TeV region. As the B non-conserving operators of lowest dimension that respect SU(3) × SU(2) × U(1) also conserve B-L, the detection of processes such as $n \to e^- \pi^+$, $n \leftrightarrow \bar{n}$,..., at a level of strength comparable to that of $\Delta B = \Delta L$ processes would indisputably signal the presence of IMS[3].

In order to continue this line of probing the feasibility of IMS models, my paper deals with three new topics which have recently been introduced into the IMS debate: can CP violation, cosmological baryon (B) asymmetry and neutrino masses provide new insights into grand unified theories as signals of where IMS oases blossom in the "big desert"?

CP VIOLATION AND IMS

The standard model allows, in its six-quark version, for the appearance of a phase which can be used to parametrize the amount of CP violation in the kaon system[4]. In this model CP violation is necessarily hard (i.e., it appears in operators of dimension four), so that the θ parameter, measuring the amount of CP violation in strong interactions[5], gets an infinite renormalization. On the other hand, the neutron electric dipole moment (d_n^e) cruelly dictates $\theta < 10^{-9}$ [6]. The only remedy appears to be the imposition of an extra global chiral symmetry (the Peccei-Quinn (PQ) symmetry) by means of which it is possible to rotate θ away[7]. However, the original version of the PQ mechanism implies the existence of a light pseudoscalar particle, the axion[8], whose presence seems to be phenomenologically discarded. Things are not so hopeless, as one can envisage grand unified models where either the axion is very light and does not interact appreciably with matter[9], or the axion does not exist altogether due to a soft breaking of the PQ symmetry[10].

The perspectives change radically once one starts with a Lagrangian which is CP conserving and only the vacuum expectation value (v.e.v.) of the Higgs scalars break CP (spontaneous CP violation)[11]. In this case, <u>θ is finite and calculable</u>[12]. More explicitly, denoting the quark mass matrix by M, the amount of strong CP violation is parametrized by an effective $\bar{\theta}$ parameter given by $\bar{\theta}_{eff} = \theta + \mathrm{Arg}\det M$, with θ being the purely QCD contribution arising from an, in principle allowed, $F\tilde{F}$ term in the QCD Lagrangian. If one demands that the theory be P and/or CP conserving <u>before</u> symmetry breakdown, then $\theta = 0$ by assumption. The experimental bound on d_n^e then implies:

$$\arg\det M < 10^{-9} \tag{1}$$

All previous solutions of the θ-problem which made use of soft CP violation tried to satisfy (1) by imposing some ad hoc discrete symmetry in order to guarantee Arg det M = 0, at least up to one-loop level.

Abandoning this kind of strategy, Mohapatra, Peccei and I proposed a model with spontaneous CP violation, where there exists a tight connection between $\bar{\theta}$ and the scale at which P, B-L and the weak CP are simultaneously broken: the smallness of $\bar{\theta}$ is guaranteed by the mass hierarchy present in the model[13]. The gauge group is $SU(2)_L \times SU(2)_L \times U(1)_{B-L} \times SU(3)_C$ with the quarks $q_{L,R}$ and leptons $\ell_{L,R}$ of each generation assigned to left-right symmetric doublets[1]. Allowing for the possibility that the Higgs fields may eventually (and hopefully!) be displaced by a more dynamical mechanism, we assume that the only Higgses are those that correspond to bilinears in fermion-fermion or fermion-antifermion fields. Thus their transformation properties under $SU(2)_L \times SU(2)_R \times U(1)_{B-L}$ are:

$$\Phi = \left(\tfrac{1}{2}, \tfrac{1}{2}, 0\right) \qquad \text{fermion-antifermion}$$

$$\left.\begin{array}{l} \Delta_{qq_L} = (1, 0, 2/3);\ \Delta_{qq_R} = (0, 1, \tfrac{2}{3}) \\ \Delta_{\ell\ell_L} = (1, 0, 2);\ \Delta_{\ell\ell_R} = (0, 1, 2) \end{array}\right\} \text{fermion-fermion} \tag{2}$$

Charge conservation ensures that only Φ, $\Delta_{\ell\ell_L}$ and $\Delta_{\ell\ell_R}$ can develop a v.e.v. There exist two possible classes of minima in the theory, for two distinct domains of the parameters in the Higgs potential:

i) $$v_L = v_R \quad ; \quad k \neq 0 \tag{3a}$$

ii) $$v_L \ll k \ll v_R \quad ; \quad v_L v_R = \gamma k^2 \tag{3b}$$

where $v_L = \langle\Delta_L\rangle$, $v_R = \langle\Delta_R\rangle$ and $\langle\Phi\rangle = \begin{pmatrix} k' & 0 \\ 0 & k\end{pmatrix}$, but $k' \ll k$ will be taken throughout this paper. Here γ is a function of dimensionless parameters of the potential. Obviously, solution (3a) is experimentally discarded. As for the possibility of spontaneous CP violation in the case (3b), one can prove:

Theorem I: if only one set of $\Delta_{\ell\ell}$ is present, then no spontaneous CP violation is possible in the model at all;

Theorem II: if at least two sets of $\Delta_{\ell\ell}$'s are present, then unremovable CP violating phases appear in the v.e.v.'s.

Remarkably enough, they obey specific relations which reveal the intimate connection between the phases and the mass hierarchy:

$$\varphi \simeq (v_L/v_R)\,\delta_L \quad ; \quad \delta_R \simeq (v_L/v_R)^2\,\delta'_L \tag{4}$$

where φ, δ_L, δ_L' and δ_R indicate the unremovable phase of $\langle\Phi\rangle$ $\langle\Delta_{1L}\rangle$ $\langle\Delta_{2L}\rangle$ and one of the two $\langle\Delta_R\rangle$ (notice that, indeed, the overall symmetry properties allow to choose one of the $\langle\Delta_R\rangle$ real and to rotate away one of the two phases of $\langle\Phi\rangle$). As $M_{W_L} \sim gk$ and $M_{W_R} \sim gv_R$, (4) can be rewritten as follows:

$$\varphi \simeq \gamma \left(M_{W_L}/M_{W_R}\right)^2 \delta_L \; ; \; \delta_R \simeq \gamma \left(M_{W_L}/M_{W_R}\right)^4 \delta_L' \tag{5}$$

Equations (4) or (5) may be seen as a consequence of the Appelquist-Carazzone theorem[14]: the fields whose v.e.v. is the largest must have rather negligible phases.

Quarks get mass from $\langle\Phi\rangle$; since φ is non-vanishing, Arg det M is different from zero <u>already</u> at tree level:

$$\bar{\Theta}_{tree} = \mathrm{Arg\,det}\, M_{tree} \approx \varphi \simeq \gamma \left(M_{W_L}/M_{W_R}\right)^2 \delta_L \tag{6}$$

Imposing the bound (1), we see that

$$M_{W_R} \gtrsim \gamma \times 10^6 \,\mathrm{GeV} \tag{7}$$

appears sufficient. For $\gamma \sim O(1)$, the bound (7) is in fair consistency with the value proposed for the scale of the left-right (L-R) symmetry breakdown in order to get detectable n-$\bar{n}$ oscillations and, as we shall see, for a consistent picture of cosmological baryon asymmetry in the model [I shall comment on the possibility $\gamma \ll O(1)$ later on].

One oculd still fear the one-loop corrections, which involve only $\langle\Delta_R\rangle$ and $\langle\Delta_L\rangle$. A careful analysis shows, however, that propagator effects compensate for the appearance of $\langle\Delta_R\rangle$ and of the unrestricted phase of $\langle\Delta_L\rangle$ in such a way that Arg det $M_{1\text{-loop}} \sim \bar{\theta}_{tree}$. Furthermore, one can easily check that no additional contribution to d_n^e, besides those due to $\bar{\theta}$, are large: <u>d_n^e is really only determined by $\bar{\theta}$</u>.

The model presents another interesting experimental signature in the weak CP violation sector: differently from the Kobayashi-Maskawa model[4], ε'/ε is here expected to be 10^{-5} or so, so that a non-zero measurement of ε'/ε would rule this model out.

The source of CP violation at IMS I discussed is one of the key points in our approach to the particularly serious problem of the survival and generation of any cosmological B asymmetry once an IMS is present.

COSMOLOGICAL BARYON ASYMMETRY VERSUS IMS ?

In the standard GUT-Big Bang scenario, cosmological B generation takes place at a scale near M_{GU}, when the lightest superheavy Higgs bosons can freely decay[15]. This asymmetry ΔB then survives as no subsequent B violating interactions can be rapid in comparison with the expansion rate of the Universe. The picture may change dramatically once IMS are introduced. As observed by Weinberg[16], if B violating forces are associated with particles whose masses M are well below 10^{13} GeV, then their B violating decays will be in equilibrium at $T \gtrsim M$, thus erasing any pre-existing B number, unless the new kind of B violating interactions conserve a quantum number B + aL, $a \neq 0, -1$. This latter possibility does not arise in the L-R model we are considering, since the following two selection rules in B and L non-conservation hold[3]:

i) $\Delta B = 0$, $L = 2$ leading to Majorana neutrinos, $(\beta\beta)_0$ decay ...

ii) $\Delta L = 0$, $\Delta B = 2$ leading to n-$\bar{n}$ oscillations, $n + p \rightarrow \pi$'s,...

Indeed, as $Q = I_{3L} + I_{3R} + (B-L)/2$ and $SU(2)_L \times U(1)_Y$ is expected to be a good symmetry above $\sim 10^2$ GeV, we get $\Delta I_{3R} = \Delta(B-L)/2$. Parity breakdown is due to $\langle\Delta_R\rangle$, i.e. $I_{3R} = 1$, so that $\Delta(B-L) = 2$ in our model.

Explicitly, B violation is introduced into the theory by the $SU(2)_L \times SU(2)_R \times U(1)_{B-L}$ allowed scalar self-coupling:

$$\lambda \epsilon_{i_1 j_1 k_1} \epsilon_{i_2 j_2 k_2} \Delta^a_{\ell\ell} \Delta^a_{q_{i_1} q_{i_2}} \Delta^b_{q_{j_1} q_{j_2}} \Delta^b_{q_{k_1} q_{k_2}} + h.c. \tag{8}$$

where we have antisymmetrized with respect to colour indices and L, R indices have been omitted for simplicity. Thus, the dilepton $\Delta_{\ell\ell}$ does not possess only the decay channel $\Delta_{\ell\ell} \rightarrow \ell\ell$ (B = 0 final state), but also the channel: $\Delta_{\ell\ell} \rightarrow \Delta_{ii}\Delta_{jj}\Delta_{kk}$, carrying B = 2 in the final state as the diquark Δ_{ii} will eventually decay into quark pairs. For $m_{\Delta_{\ell\ell}} \sim$ IMS << M_{GU} (as required for sizable Higgs mediated n-$\bar{n}$ oscillations), Weinberg's warning then would become a real threat for our baryons! Things would not improve even if one tried desparately to push $m_{\Delta_{\ell\ell}}$ to M_{GU}. Indeed, Kuzmin and Shaposhnikov, and indirectly Yanagida and Yoshimura[17], have proved that in a L-R symmetric model the amount of cosmological B asymmetry generated in the B violating decays of superheavy particles is proportional to the ratio M_{LR}/M_{GU}, denoting by M_{LR} the scale at which L-R symmetry is broken[17]. As in our case M_{LR} = IMS $\sim$ $\sim 10^6$ GeV, no chance for a dynamical B generation at M_{GU} exists.

Fortunately, this conundrum is not without any hope. It is true that with an IMS any pre-existing B asymmetry is irremediably wiped out (or, even worse, no B can dynamically arise, if the IMS is to be identified with the scale at which L-R symmetry breakdown arises), however, there still exists the logical possibility of generating a B number anew at $T \lesssim$ IMS. I propose here an example worked out in detail with Mohapatra[18], in the context of the aforementioned $SU(2)_L \times SU(2)_R \times U(1)_{B-L}$ model.

The breaking pattern (3b) entails that left- (ν) and right-handed (N) neutrino fields become two separate Majorana spinors, with $m_N \simeq h_5 V_R$ (I shall return later to m_N), h_5 being defined by the coupling:

$$h_5 \left(\ell_R^T \tau_2 \Delta_{\ell\ell_R} C \ell_R + R \leftrightarrow L \right) + h.c. \tag{9}$$

N violates B in its decay, as it possesses two decay channels with different B number in the final states: $N \to \ell_L \varphi$ ($B = 0$), where φ is the usual Weinberg-Salam Higgs doublet and $N \to e^+ \Delta_{ii,R} \Delta_{jj,R} \Delta_{kk,R}$ ($B = 2$), through the decay mode depicted in Fig. 1. Clearly this second decay mode is allowed if $m_N > m_{ij}$; for definiteness, suppose the following mass spectrum arises

$$M_{W_R} > m_{\Delta_{\ell\ell}} \sim 10^5\,\text{GeV} > m_N \sim 10^4\,\text{GeV} > m_{\Delta_{ij}} \sim 10^3\,\text{GeV} \tag{10}$$

Let us consider the N belonging to the electron-generation, N_e. Since the width of the decay $N \to e^+ \varphi$ is of the order of the expansion rate of the Universe at $T \sim m_N$, and for reasons given in Ref. 18, we shall neglect this decay mode as long as we are around $T \sim m_N$.

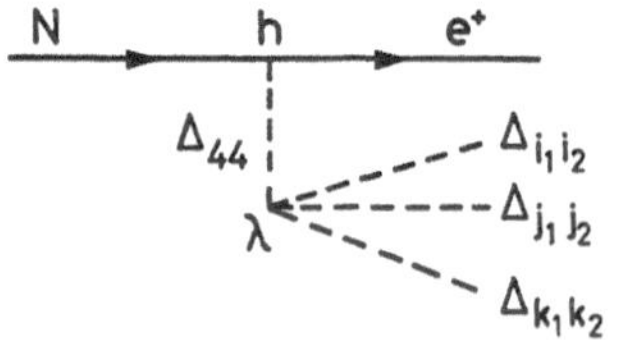

Fig. 1 B violating N decay diagram responsible for the generation of baryon asymmetry at IMS.

Let us, then, follow the sequence of events. The $\Delta_{\ell\ell}$'s decouple at $T_* \sim m_{\Delta_{\ell\ell}}/20$ when their relative number density is $(n_{\Delta_{\ell\ell}}/n_\gamma) \sim \sim 10^{-7}$; these few remaining $\Delta_{\ell\ell}$'s disappear immediately, mainly via the $\Delta_{ij}\Delta_{ij}$ decay mode. After the disappearance of the $\Delta_{\ell\ell}$'s the only operative processes concern the N's. Their effective annihilation mode is $NN \to e^+e^-_R$ via W_R exchange. This process is slow (out of equilibrium) compared to the expansion rate at $T \lesssim m_N$ for $m_{W_R} \sim 5 \cdot 10^6$ GeV. Thus, putting out of equilibrium the N annihilation at $T \sim m_N$ implies a lower bound on m_{W_R}, which roughly concides with the one coming from d_n^e for $\gamma \sim O(1)$ in Eq. (7). For $T \sim m_N$, the decay rate for the process in Fig. 1 is $(\lambda h)^5 m_N^4/m_{\Delta_{\ell\ell}}^4$ and it exceeds the expansion rate $\sim T^2/M_P$ (M_P = Planck mass) for $T \sim 1$ TeV. When the Universe cools down to this temperature all the three ingredients required to generate a net B number, B, C and CP violation and departure from equilibrium, are met. I do not want to enter here into the numerical details of this model: apart from the problem of slowing down sufficiently the $N \to \ell\varphi$ decay channel, one must choose the couplings rather carefully in order to avoid substantial B dilution from B violating processes taking place after N decays. On the other hand, these same couplings are also involved in n-$\bar{n}$ oscillations, so that only a narrow range is allowed in order to guarantee the observability of this last process and prevent any excessive erasure of B. The major point of the model is the clear strategy it envisages for B generation at IMS: in order to produce a net B number at IMS, B asymmetry must be given by effective non-renormalizable interactions so that their reaction rates scale with temperature T^n with $n \gtrsim 5$, thus ensuring the possibility of decoupling for $T < M_{GU}$. Particles violating B via renormalizable interactions (the $\Delta_{\ell\ell}$'s in the above example) must be heavy enough so that they are highly non-relativistic at the moment of B generation.

I close this section devoted to the study of B generation at IMS with two remarks:

i) a valid alternative to the above recipe is provided by the possibility of B violating decays at IMS of particles very weakly coupled to ordinary matter, in order to implement the out of equilibrium condition. A specific example has been given by Yanagida and me[19] in the context of a SU(5) model. In particular. Barbieri, Nanopoulos and I[20] observed that even superheavy fermions whose mass $\lesssim M_{GU}$ could decay extremely late, $T \sim 10$ TeV, producing a consistent amount of ΔB if they cannot couple directly to ordinary matter[20].

ii) in a variant scheme L-R symmetry is broken at $T \sim 2$-$3\ M_W$, the Kuzmin-Shaposhnikov argument[17] may be avoided by considering a scheme where L-R symmetry is not restored at high temperature[21]. In this case, as $(M_{LR}/M_{GU})|_{T\sim M_{GU}} \sim O(1)$ no suppression appears in B. The details of this strategy have been worked out by Senjanovic and me[22] in the context of the $SU(2)_L \times SU(2)_R \times U(1)_{B-L}$ embedding in SO(10) à la Rizzo-Senjanovic[23].

NEUTRINO MASSES AND LEFT-RIGHT SYMMETRIC MODELS AT IMS

Let us come back to the $SU(2)_L \times SU(2)_R \times U(1)_{B-L}$ model with B-L breaking at an IMS $\sim 10^6$ GeV. The Higgs structure I have introduced gives rise to the following mass terms in the neutrino sector:

$$\mathcal{L}^{\nu}_{Y} = \sum_{i,j=1}^{3} \left\{ h_{5_{ij}} \left[v_L \left(\nu^T_{L_i} C \nu_{L_j} + h.c. \right) + L \leftrightarrow R \right] + h_{1_{ij}} k \left[\bar{\nu}_{L_i} \nu_{R_j} + \bar{\nu}_{R_j} \nu_{L_i} \right] \right\} \tag{11}$$

where the notation of Ref. 24 is adopted. In the approximation of small intergenerational mixings, the eigenvalues of the neutrino mass matrix read[24]:

$$m_{\nu_i} \approx \left(h_{5_i} \gamma + h^2_{1_i} / h_{5_i} \right) k^2 / v_R \tag{12}$$

$$m_{N_i} \simeq h_{5_i} v_R \tag{13}$$

Since h_{1i}, h_{5i} and γ are free parameters, no prediction is possible. Roncadelli and I[25] tried to make a systematic study of Eqs. (12) and (13) in the framework of just the two following assumptions: i) intergenerational mixings are small so that Eqs. (12) and (13) are valid to a good approximation; ii) the Yukawa couplins respons- h_{1i} are taken of the same order as the Yukawa couplings responsible for the charged lepton masses. This sounds rather reasonable if one thinks that charged and neutral leptons sit in the same $SU(2)_L \times SU(2)_R \times U(1)_{B-L}$ multiplets and the h_{1i}'s are connected with the neutrino's Dirac mass term. No hypothesis is made on the h_{5i}'s which are involved in the neutrino Majorana mass terms. As for the constraints we impose on the neutrino sector, we supplement the usual requirements, i.e., experimental upper bounds on m_ν and cosmological upper bounds on the sum of the masses of long-lived neutrinos, whose masses are <2 GeV [26], with the new request of a cosmological B asymmetry generated in the decays of N's[18]. This last constraints seems really necessary in view of the B erasure taking place at IMS $\sim T \ll M_{GU}$, as I have previously explained. A nice result arises: the above assumptions and requests single out a neutrino mass spectrum where m_{ν_e} and m_{ν_μ} are below 100 eV and 100 KeV $< m_{\nu_\tau} <$ 1 MeV, so that ν_τ can decay into lighter neutrinos in a time abundantly sufficient for red-shifting its released energy (i.e., $\tau_{\nu_\tau} < 10^{10}$s)[27]. The parameter γ, defined in Eq. (3b), has the upper bound $\gamma < 6 \cdot 10^{-5}$, so that, taking this analysis at face value, one could only infer that no θ problem is present in the model, but one could not directly connect d_n^e and M_{W_R} [see Eq. (7)]. The most important result consists, in any case, of showing that this class of left-right models allows for a rather precise neutrino mass spectrum <u>compatible</u> with <u>all</u> the cosmological constraints, thus constrasting some wide-spread feeling of uneasiness about respecting the bound $\Sigma_i m_{\nu_i} < 100$ eV for sufficiently long-lived neutrinos[28].

In conclusion, I hope to have convinced big desert fans that indeed IMS models provide alternatives also to some old and new theoretical probes, such as the weak and strong CP problem, neutrino masses and cosmological baryon asymmetry. Very soon a wide class of IMS models is going to face some "crucis experimentum": some selection among IMS theories is the first step for a direct confrontation of big desert-IMS grand unified theories.

ACKNOWLEDGEMENTS

It is a great pleasure to thank R. Barbieri, R.N. Mohapatra, D.V. Nanopoulos, R.D. Peccei, G. Senjanovic, M. Roncadelli and T. Yanagida for having shared with me some of their insights. I am grateful to the organizers J. Ellis and S. Ferrara for a very nicely run meeting.

REFERENCES

1. J.C. Pati and A. Salam, <u>Phys. Rev.</u> D10:275 (1974);
   R.N. Mohapatra and J.C. Pati, <u>Phys. Rev.</u> D11:566 and 2558 (1975);
   G. Senjanovic and R.N. Mohapatra, <u>Phys. Rev.</u> D12:1502 (1975);
   R.N. Mohapatra and G. Senjanovic, <u>Phys. Rev. Lett.</u> 44:912 (1980), <u>Phys. Rev.</u> D23:165 (1981).
2. T. Rizzo, <u>Phys. Rev.</u> D21:1214 (1980);
   T. Rizzo and D.P. Sidhu, <u>Phys. Rev.</u> D21:1209 (1980), D22:787E (1980);
   L.M. Sehgal, Aache preprint PITHA-80/17 (1980).
3. R.N. Mohapatra and R.E. Marshak, <u>Phys. Lett.</u> 91B:222 (1980), 94B:183 (1980), <u>Phys. Rev. Lett.</u> 44:1316 (1980);
   R.N. Mohapatra, Proc. of 1st Workshop on Grand Unification, ed. by P. Frampton, S. Glashow and A. Yildiz, Durham, New Hampshire (1980);
   R.E. Marshak, report VPI-HEP 81/3 (1981).
4. M. Kobayashi and T. Maskawa, <u>Progr. Theor. Phys.</u> 49:654 (1973).
5. G. 't Hooft, <u>Phys. Rev.</u> D14:3432 (1976);
   C. Callan, R. Dashen and D. Gross, <u>Phys. Lett.</u> 63B:334 (1976);
   R. Jackiw and C. Rebbi, <u>Phys. Rev. Lett.</u> 37:172 (1976).

6. V. Baluni, Phys. Rev. D19:2227 (1979);
R. Crewther, P. di Vecchia, G. Veneziano and E. Witten, Phys. Lett. 88B:123 (1979).
7. R.D. Peccei and H. Quinn, Phys. Rev. Lett. 38:1440 (1977), Phys. Rev. D16:1791 (1977).
8. S. Weinberg, Phys. Rev. Lett. 40:223 (1978);
F. Wilczek, Phys. Rev. Lett. 40:279 (1978).
9. M. Dine, W. Fischler and M. Srednicki, Phys. Lett. B104:199 (1981).
10. A. Masiero and D. Wyler, preprint MPI-PAW/PTh 74/81 (1981).
11. T.D. Lee, Phys. Rev. D8:1226 (1973), Phys. Rep. C9:148 (1979);
P. Sikivie, Phys. Lett. 65B:141 (1976); for a review, see
G. Senjanovic, Proc. XX International Conference on High Energy Physics, Madison (1980).
12. H. Georgi, Hadronic Journal 1:155 (1978);
M.A.B. Beg and H.S. Tsao, Phys. Rev. Lett. 41:278 (1978);
R.N. Mohapatra and G. Senjanovic, Phys. Lett. 79B:283 (1978).
13. A. Masiero, R.N. Mohapatra and R.D. Peccei, Nucl. Phys. B192: 66 (1981).
14. T. Appelquist and J. Carazzone, Phys. Rev. C11:2856 (1975).
15. For a review on Big Bang baryosynthesis, see M.S. Turner, Lectures given at the 1981 Les Houches Summer School on Gauge Theories, preprint NSF-81-85 (1981) and references therein.
16. S. Weinberg, Phys. Rev. D22:1694 (1980).
17. V.A. Kuzmin and M.E. Shaposhnikov, Phys. Lett. 92B:115 (1980);
T. Yanagida and M. Yoshimura, Phys. Rev. Lett. 45:71 (1980).
18. A. Masiero and R.N. Mohapatra, Phys. Lett. 103B:343 (1981).
19. A. Masiero and T. Yanagida, preprint MPI-PAE/PTh 53/81 (1981).
20. R. Barbieri, D.V. Nanopoulos and A. Masiero, Phys. Lett. 98B:191 (1981); for the role played by superheavy fermions in cosmological baryon generation, see also J. Harvey, E. Kolb, D. Reiss and S. Wolfram, preprint CALT-68-784 (1980).
21. R.N. Mohapatra and G. Senjanovic, Phys. Rev. Lett. 42:1657 (1978), Phys. Rev. D20:3390 (1979), Phys. Lett. 89B:57 (1979).
22. A. Masiero and G. Senjanovic, preprint MPI-PAE/PTh 13/81 (1981), to appear in Phys. Lett. B.
23. T. Rizzo and G. Senjanovic, Phys. Rev. Lett. 46:1315 (1981), Phys. Rev. D24:704 (1981) and BNL preprint (1981), to appear in Phys. Rev. D.
24. R.N. Mohapatra and G. Senjanovic, Phys. Rev. D23:165 (1981).
25. A. Masiero and M. Roncadelli, preprint MPI-PAE/PTh 60/81 (1981).
26. For a review see, for example, G. Steigman, Proc. of Neutrino '81, Maui, Hawaii (July 1981), and references therein.
27. Y. Hosotani, Nucl. Phys. 191B:411 (1981);
M. Roncadelli and G. Senjanovic, preprint MPI-PAE/PTh 16/81 (1981), to appear in Phys. Lett. B.
28. See, for example, the third paper of Ref. 3, pages 81-82.

SOME ASPECTS OF SUPERSYMMETRY BREAKING

M. T. Grisaru

California Institute of Technology, Pasadena, CA 91125

I wish to describe some recent results on breaking of supersymmetry. The work in two dimensions was done with L. Alvarez-Gaumé and D. Z. Freedman[1], the work on soft breaking in four dimensions was done with L. Girardello[2].

There are several possibilities for breaking of supersymmetry:

1) Spontaneous breaking à la O'Raifeartaigh or Fayet - Iliopoulos. I have nothing new to add to this subject[3].
2) Breaking by radiative corrections: this is believed to be impossible in four dimensions, at least in perturbation theory. As I will show, it can be achieved in two dimensions.
3) Breaking by nonperturbative effects, e.g., instantons. We looked at this a number of years ago[4] and Witten has discussed it recently[5].
4) Explicit breaking: this can be done softly, so as to keep some of the good features of supersymmetry (no quadratic divergences induced) and seems just as good a mechanism as spontaneous breaking. Besides Ref. [2] there is earlier work by Capper[6] on this subject.

What are the signals of spontaneous supersymmetry breaking? For example, in the Wess-Zumino model, with transformation laws

$$\delta(A+iB) = \bar{\varepsilon}\psi \qquad \delta(F-iG) = \bar{\varepsilon}\not{\partial}\psi$$

$$\delta\psi = \not{\partial}(A+iB)\varepsilon + (F-iG)\varepsilon \tag{1}$$

if F acquires a vacuum expectation value, supersymmetry is broken and $\delta\psi \sim \langle F\rangle\varepsilon$ = a constant shows that ψ has become a Goldstone fermion. Alternatively, from the supersymmetry algebra $\{Q,Q\} = \gamma \cdot P$ it follows that in a suitable basis $H = Q_i^\dagger Q_i$. Spontaneous supersymmetry breaking means $Q|vac\rangle \neq 0$ which implies that the vacuum energy is no longer zero. In the Wess-Zumino model, for example, the classical potential $V = F^2 + G^2$, and $\langle 0|H|0\rangle = V_{min} > 0$ is equivalent to $\langle F\rangle \neq 0$ (or $\langle G\rangle \neq 0$ if parity is broken); the potential no longer has zero minimum value.

One question that comes up is whether, when supersymmetry is spontaneously broken, one still has $H = Q_i^\dagger Q_i$ or at least one can assert $H \geq 0$. This question is motivated in part by a recent paper by Zanon[7] who shows that in the large-N limit of a supersymmetric O(N) model supersymmetry is broken and the potential has a local minimum where it is negative. I believe that this is an artifact of the approximation and that the large-N limit has to be handled with more care. A formal argument for positivity goes as follows (as described to us by B. Zumino):

Let J^μ_α be the supercurrent, $T^{\mu\nu}$ the energy momentum tensor. Then, in standard current algebra fashion

$$\lim_{q_\mu \to 0} q_\mu \int d^4x\, e^{-iq.x} \langle 0|TJ^\mu_\alpha(x)\bar{J}^\nu_\beta(0)|0\rangle$$

$$= -i\langle 0|\{Q_\alpha, \bar{J}^\nu_\beta(0)\}|0\rangle$$

$$= -2i\langle 0|T^{\mu\nu}|0\rangle(\gamma_\mu)_{\alpha\beta} \tag{2}$$

using the supersymmetry algebra. On the other hand, the limit $q_\mu \to 0$ must be dominated by zero mass states (Goldstone fermions) with $\langle 0|J^\mu_\alpha|p\rangle = if\gamma^\mu u_\alpha(p)$ from which one concludes

$$\langle 0|T_{\mu\nu}|0\rangle = \frac{1}{2} f^2 g_{\mu\nu} \tag{3}$$

and

$$\langle 0|T_{00}|0\rangle \geq 0 .$$

Let me briefly discuss now why in perturbation theory it seems impossible to break supersymmetry by radiative corrections in four dimensions. It will then also become clear why two dimensions is different. The easiest way to see the result is by imagining computing the effective potential using superfield Feynman rules[8]. Typical renormalizable Lagrangians for chiral (ϕ) and real (V) superfields are

$$\mathcal{L} = \int d^4\theta \; V D^\alpha \bar{D}^2 D_\alpha V + \int d^4\theta \; \bar{\phi} e^V \phi$$

$$+ \int d^2\theta [m\phi^2 + \lambda\phi^3 + \zeta\phi] + \text{h.c.}$$

$$+ \int d^4\theta [V D^\alpha V \bar{D}^2 D_\alpha V + V V D^\alpha V \bar{D}^2 D_\alpha V + \ldots] \; . \tag{4}$$

Here (in two-component spinor notation)

$$D^\alpha = \frac{\partial}{\partial\theta_\alpha} - i\,(\sigma^\mu\bar{\theta})^\alpha \partial_\mu \; , \quad \bar{D}^{\dot{\alpha}} = \frac{\partial}{\partial\bar{\theta}_{\dot{\alpha}}} - i(\theta\sigma^\mu)^{\dot{\alpha}} \partial_\mu \tag{5}$$

$$\phi = e^{i\theta\sigma.\partial\bar{\theta}} [(A+iB) + \theta^\alpha \psi_\alpha + (F-iG)\theta^2] \tag{6}$$

$$\bar{\phi} = e^{-i\theta\sigma.\partial\bar{\theta}} [(A-iB) + \bar{\theta}^{\dot{\alpha}} \bar{\psi}_{\dot{\alpha}} + (F+iG)\bar{\theta}^2] \tag{7}$$

$$V = C + \ldots + \theta\sigma^\mu\bar{\theta} A_\mu + \bar{\theta}^2 \theta^\alpha \lambda_\alpha + \theta^2 \bar{\theta}^{\dot{\alpha}} \bar{\lambda}_{\dot{\alpha}} + \theta^2 \bar{\theta}^2 D \; . \tag{8}$$

One can now prove the following result[8]: to any order of perturbation theory, the effective action has the form

$$\Gamma(\phi,\bar{\phi},V) = \sum_n \int d^4\theta \; d^4x_1 \ldots d^4x_n \; G(x_1 \ldots x_n)$$

$$\times \text{ Polynomial in } \phi(x_i,\theta),\; D_\alpha\phi(x_j,\theta) \;\ldots\; V(x_k,\theta) \;\ldots \tag{9}$$

local in θ and integrated with $d^4\theta$ not $d^2\theta$. This has the well-known consequence that the chiral superfield terms $\int d^2\theta(m\phi^2+\lambda\phi^3+\zeta\phi)$ do not receive any radiative corrections and that supersymmetry cannot be broken radiatively. This can be shown as follows:

For simplicity let us consider the case of just a Wess-Zumino multiplet described by ϕ. (Even classically one cannot break supersymmetry for such a simple model - one needs several multiplets - but the argument is the same.) We obtain the effective potential from the effective action by setting spinor fields and all external momenta to zero which means that in eq. (9) $D_\alpha = \partial/\partial\theta^\alpha$, $\bar{D}_{\dot{\alpha}} = \partial/\partial\bar{\theta}^{\dot{\alpha}}$. But then the only sources of $\theta,\bar{\theta}$ factors needed to make the θ-integral nonzero ($\int d^4\theta\theta^2\bar{\theta}^2 = 1$, all other integrands give zero) are the θ^2 and $\bar{\theta}^2$ factors in ϕ, which bring with them at least one factor of F each. We can conclude that the four-dimensional effective potential has the form to any order of perturbation theory

$$V_{eff} = -\frac{1}{2} F^2 + F\mathcal{F}(A) + F^2 \mathcal{G}(F,A) \tag{10}$$

where the first two terms are the standard classical potential and we have assumed parity conservation (B = G = 0).

We extremize the effective potential with respect to F and A, i.e., we look for solutions of

$$\frac{\partial V}{\partial F} = -F + \mathcal{F} + 2F\mathcal{G} + F^2\mathcal{G}'_F = 0$$

$$\frac{\partial V}{\partial A} = F\mathcal{F}' + F^2\mathcal{G}'_A = 0 \ . \tag{11}$$

If supersymmetry is not broken at the classical level (ignore $\mathcal{G}$), it means that there exists a solution F = 0, $\mathcal{F}(A_o) = 0$ for some A_o. But then clearly this is still a solution when $F^2\mathcal{G}$ is included (and it leads still to V = 0) so that quantum corrections do not break supersymmetry. This argument can be extended to more interesting models with the same conclusion. The fact that radiative corrections to the effective potential are quadratic in F-terms is the essential feature. (There can also be terms linear in D from the vector multiplet, but this does not lead to any interesting possibilities.) In the argument we must also assume that the extremum is still an absolute minimum which will automatically be the case if we believe that the vacuum energy is non-negative (c.f. our earlier remarks).

In two dimensions the situation is quite different. Here with $\phi = A + \theta^\alpha\psi_\alpha + \theta^2 F$ the effective action involves only a $d^2\theta$ integration and, in general

$$V_{eff} = -\frac{1}{2}F^2 + F\mathcal{F} + F\mathcal{G}(F,A) \tag{12}$$

so that

$$\frac{\partial V}{\partial F} = -F + \mathcal{F}(A) + \mathcal{G}(F,A) + F\mathcal{G}'_F$$

$$\frac{\partial V}{\partial A} = F\mathcal{F}' + F\mathcal{G}'_A \ . \tag{13}$$

This time the classical minimum at F = 0, $\mathcal{F}(A_o) = 0$ is not, in general, a solution since now we require if F = 0

$$\mathcal{F}(A) + \mathcal{G}(0,A) = 0 \ . \tag{14}$$

Thus if we can arrange that $\mathcal{F}(A_o) = 0$ but $\mathcal{F}(A) + \mathcal{G}(0,A) \neq 0$ for all A, there will be no classical breaking but there will be radiative breaking of supersymmetry. This can be done easily at the one-loop level as follows[1].

In two dimensions a general (one superfield) renormalizable supersymmetric action is, in components

$$I = \frac{1}{2} \int d^2x[(\partial_\mu A)^2+i\bar{\psi}\partial\!\!\!/\psi+F^2-2mFS(A)-m\ S'(A)\bar{\psi}\psi] \tag{15}$$

where S(A) is an arbitrary function. The one-loop renormalization group invariant effective potential is (rescaling so that m=1)

$$V_{eff} = -\frac{1}{2} F^2 + F\ S(A) + \frac{\hbar}{8\pi} \{(S')^2 \ell n\ c(S')^2-[(S')^2+S''F]\ell n\ c[(S')^2+S''F]+FS''\} \tag{16}$$

where $c = e^\gamma/4\pi\mu^2$ and μ is a renormalization point. Then

$$\frac{\partial V}{\partial F} = 0 = -F + S - \frac{\hbar S''}{8\pi} \ell n\ c[(S')^2+S''F] \tag{17}$$

and to break supersymmetry radiatively we can choose S so that this equation has no solution for F = 0 (we can also show that at the minimum $V_{eff} \geq 1/2[F(A)]^2 \geq 0$).

The easiest way to achieve this is to choose S(A) to have a zero and at the same time

$$S - \frac{\hbar}{8\pi} S''\ell n\ c(S')^2 > 0 \tag{18}$$

always, which is not difficult to do with, for example, $S = A^2 - b$ (leading to quartic scalar interactions) or $S = a^2(\ell n \cosh A - b)$, $ca < 1$ and b small.

Finally we observe that while the breaking of bosonic global symmetries is not allowed in two-dimensions (Coleman's theorem) fermionic symmetries may be broken. This is essentially because of different growth properties of bosonic and fermionic two-point functions.

Returning to four-dimensions, I will discuss now soft explicit breaking[2] which seems to provide a phenomenologically viable alternative[9] to spontaneous breaking, whether actually done by hand, or perhaps as the low energy description of a high energy spontaneous breaking of local supersymmetry. The idea here is to break the symmetry in such a way as to keep control of the divergences, or renormalization point dependences of Green's functions. Specifically it means starting with a supersymmetric model where the divergences are rigidly controlled by supersymmetry: aside from a quadratically divergent D-term, only logarithmically divergent wave function and gauge coupling constant renormalizations, and no chiral multiplet mass or coupling constant renormalization, then adding breaking terms in such a way that nothing worse than new logarithmic divergences are induced.

The easiest way to achieve this is to couple to a supersymmetric system external (spurion) superfields in a supersymmetric manner consistent with the renormalizability requirements of superfield power counting rules. New conventional divergent terms will involve these fields and the quantum fields, in a manner which is completely determined by power counting. We break supersymmetry (= translational invariance in superspace) by giving the spurion fields fixed x-independent, θ-dependent values, and read off from the original couplings the form of the breaking terms and from the divergences the form of the new induced divergences.

The power counting rules[8,10], reduce to the following simple statements: the effective action, and hence its divergent part, contains an integral $\int d^4\theta$ which has dimension 2($\mathrm{Dim}\theta=-1/2$). The fields enter polynomially with $\mathrm{Dim}\phi = 1$, $\mathrm{Dim}V = 0$ and $\mathrm{Dim}D_\alpha = 1/2$. The resulting expression must have dimension four. Furthermore, if a diagram contains $<\phi\phi>$ or $<\bar{\phi}\bar{\phi}>$ propagators which carry mass factors m in the numerator, such mass factors contribute to the dimensionality and reduce the degree of divergence, as does gauge invariance which may require that a gauge field V enter through its field strength $W_\alpha = \bar{D}^2 D_\alpha V$. Also, a term which contains only chiral fields ϕ (or only $\bar{\phi}$) must have the form $\phi^{n-1}D^2\phi$ (since $\int d^4\theta\phi^n = 0$), and again the explicit D's reduce the degree of divergence.

Finally, the <u>renormalizability criterion</u> (no dimensional coupling constants) requires that the interactions be such that no more than four factors D_α, D_β, $\bar{D}_{\dot{\alpha}}$, $\bar{D}_{\dot{\beta}}$ appear at any vertex when the Feynman rules are worked out ($\int d^4\theta_i$ at each vertex, propagators proportional to $\delta^4(\theta_i-\theta_j)$, and factors $\bar{D}^2$ or D^2 for each chiral or antichiral field entering a vertex except for one such factor which is used up in a purely chiral interaction $\int d^2\theta$ for converting it to $\int d^4\theta$ form).

We can compile now a list of allowed $\Delta\mathcal{L}$ breaking terms. They must have the form of those appearing in eq. (4). Although listed singly, they can appear in combinations and, in fact, one of them may induce the others. The component results are obtained by doing the θ-integration in the superfield terms. The divergences induced are all logarithmic:

1) $$\Delta\mathcal{L} \sim \int d^4\theta\bar{\phi}U\phi = \mu^2(A^2+B^2) \tag{19}$$

where $U = \mu^2\theta^2\bar{\theta}^2$ is a dimension zero "gauge" superfield. Then

$$\Delta\Gamma_\infty \sim \int d^4\theta[U\bar{\phi}\phi+MU(\phi+\bar{\phi})+U(D^2\phi+\bar{D}^2\bar{\phi})] = \mu^2(A^2+B^2) + \mu^2 mA + \mu^2 F \tag{20}$$

2) $\Delta\mathcal{L} \sim \int d^2\theta\chi\phi^2 + \text{h.c.} = \mu^2(A^2-B^2)$ (21)

where $\chi = \mu^2\theta^2$ is a dimension one chiral superfield

$$\Delta\Gamma_\infty \sim \int d^4\theta(\bar{\chi}\phi+\chi\bar{\phi}) = \mu^2 F \tag{22}$$

3) $\Delta\mathcal{L} \sim \int d^2\theta\eta(\bar{D}^2D^\alpha V)(\bar{D}^2D_\alpha V) = \mu\lambda^\alpha\lambda_\alpha$ (23)

where η is a dimension zero chiral superfield, giving a mass to the fermion of a vector multiplet

$$\Delta\Gamma_\infty \sim \int d^4\theta[\bar{\eta}\phi+\eta\bar{\phi}\phi+\eta\bar{\eta}\phi+\eta\bar{\eta}\bar{\phi}\phi] = F + (FA+GB) + A + (A^2+B^2) \tag{24}$$

4) $\Delta\mathcal{L} \sim \int d^2\theta\eta\phi^3 + \text{h.c.} = \mu(A^3-3A^2B)$ (25)

has the same effect as above.

This is the complete list of soft breaking possibilities. Among breakings which are not soft (although the component theory is still renormalizable), two examples stand out: one is

$$\Delta\mathcal{L} \sim \int d^4\theta UD^\alpha\phi D_\alpha\phi = \mu\psi^\alpha\psi_\alpha \tag{26}$$

which changes the mass of the fermion in a Wess-Zumino multiplet. This interaction has six $D,\bar{D}$ at the vertex and leads to quadratic divergences. The other example is

$$\Delta\mathcal{L} \sim \int d^4\theta U(\phi+\bar{\phi})^3 \sim A^3$$

modifying the interaction, again leads to quadratic divergences.

To conclude, note that the above analysis has not taken into account special circumstances, such as global or gauge invariance which might eliminate some of the possible $\Delta\Gamma_\infty$ terms. Finally, it is worth remarking that while the possibilities of soft breaking are quite limited, they are sufficient to handle most phenomenological needs. Since they do not really begin with more parameters than needed for spontaneous breaking, still keep the theory renormalizable but allow for far more flexibility, they seem to provide a viable alternative to other breaking mechanisms.

REFERENCES

1. L. Alvarez-Gaumé, D. Z. Freedman and M. T. Grisaru (to be published).

2. L. Girardello and M. T. Grisaru, Nucl. Phys. (to be published).
3. For a review see P. Fayet and S. Ferrara, Phys. Reports, 32:249 (1977).
4. L. Abbott, M. T. Grisaru and H. J. Schnitzer, Phys. Rev., D16:3002 (1977).
5. E. Witten, Nucl. Phys., B188:513 (1981).
6. D. Capper, J. Phys., G3:731 (1977).
7. D. Zanon, Phys. Letters, 104B:127 (1981).
8. M. T. Grisaru, M. Roček and W. Siegel, Nucl. Phys., B159:429 (1979).
9. S. Dimopoulos and H. Georgi, Harvard preprint HUTP-81/A022.
10. M. T. Grisaru, Four lectures on supergraphs, in: "An Introduction to Supergravity," edited by S. Ferrara and J.G. Taylor (Cambridge University Press, to be published).

FERMION MASSES, GLOBAL SYMMETRIES AND THE STRONG CP PROBLEM *

R. Barbieri
Scuola Normale Superiore, Pisa
INFN, Sezione di Pisa, Italy
D. Wyler
Theoretische Physik ETH, 8093 Zürich, Switzerland

ABSTRACT

We briefly describe how a general unified theory can account for the observed hierarchical pattern of fermion masses. An explicit SU(5) model predicts at the unification mass $m_b = m_\tau$ and, necessarily, $m_s = m_\mu/3$. It also incorporates a Peccei-Quinn symmetry broken at superlarge energies.

1.

From the phenomenological point of view family replication is described by the standard model in a completely satisfactory way. Even quite subtle aspects, like for example the suppression of the flavour-changing neutral currents, are consistently accounted for. One may, however, express reservations on the deepness of this description at a basic level. As a consequence, in extending the standard model to an unified theory, attempts have been made to improve this situation. A solution to this problem might indeed come from a suitable unified model, even though none of the present proposals - not even the interesting SU(8) - approach[1] suggested by supergravity - can be considered successful.

A striking aspect of the family replication is the hierarchical structure of the fermion masses. Waiting for a deep understanding of the origin of families, we point out in this paper that unified

* Based on a talk given by R.B. at the second Workshop on Unification of the Fundamental Interactions, Erice, October 1981.

theories - or in general theories with two different mass scales M_W and M - are potentially capable of accounting for this hierarchical pattern. We actually describe the main features of an explicit SU(5)-model when the mass matrix in generation space for a fermion of given charge has the structure

$$M = M^{(0)} + \delta M^{(1)} + \delta M^{(2)} \tag{1}$$

Here $M^{(0)}$ has only 1 nonvanishing eigenvalue - the mass of the third family fermion - and the perturbations $\delta M^{(1)}$, $\delta M^{(2)}$ of relative order

$$\begin{aligned} \delta M^{(1)} &\sim \frac{\lambda^2}{4\pi^2} M^{(0)} \\ \delta M^{(2)} &\sim \frac{\lambda^2}{4\pi^2} \delta M^{(1)} \quad , \end{aligned} \tag{2}$$

introduced by radiative corrections, give masses to the fermions of the lighter families. λ is a typical Yukawa coupling perhaps of the same order as the unified gauge coupling g.

To implement the above picture two things are needed: i) a symmetry that naturally makes vanish all the eigenvalues of $M^{(0)}$ but one; ii) a mechanism[2] giving rise to the hierarchical perturbations $\delta M^{(1)}$, $\delta M^{(2)}$. As regards i), we invoke global chiral symmetries either broken explicitly by soft terms or spontaneously by the tree level potential, at the large energy scale M. In the actual explicit model the imposition of these symmetries will be rather ad hoc. This may be attributed to the lack of a fundamental theory of families. Concerning ii) the general mechanism that we exploit is the following. The gauge model has two scales M and M_W with $G \equiv SU(3) \times SU(2) \times U(1)$ unbroken at M. The generic scalar singlet under G is denoted by χ , whereas ϕ indicates a scalar, (one or more) transforming under G as the standard Higgs doublet. We expect χ to get a vacuum expectation value $\langle\chi\rangle \sim M$ whereas $\langle\phi\rangle \sim M_W$. The chiral global symmetries act on ϕ,χ and on the fermion multiplet f in such a way that:

- for some components of f, essentially those corresponding to the lighter families, a direct renormalizable coupling

$$\lambda \; f^T C^{-1} f \, \phi \tag{3}$$

is not allowed;

- on the contrary, for the same f-components, some nonrenormalizable couplings are allowed of the form

$$\sim \lambda \left(\frac{\lambda^2}{4\pi^2}\right)^n \frac{\chi^m}{M^m} f^T C^{-1} f \phi \; ; \; n,m \text{ integers} > 0 \; . \qquad (4)$$

In (4) we have made explicit the dependence on the Yukawa coupling constants due to the radiative origin of the non-renormalizable couplings, as well as the dependence on the mass scale M which gives to (4) the correct dimensionality. After χ and ϕ get a v.e.v., the induced coupling (4) gives rise to mass terms which are naturally suppressed relative to those originated by the direct coupling (3). The problem is to find in the explicit model suitable chiral symmetries which allow the heaviest third generation to get a direct coupling (M^o) and let the radiative masses be generated at the appropriate order for the first and the second family. The chiral global symmetries which are spontaneously broken at the large energy scale $M \gg M_W$ may play the role of the Peccei-Quinn symmetry in connection with the strong CP problem; in such a case the smallness of the effective θ-parameter should not be attributed to the vanishing of a quark mass - phenomenologically unlikely - but rather to the smallness of some of them.

2.

Let us now briefly describe the essential features of an explicit SU(5)-model which realizes the above general framework. The model has been previously developed in collaboration with D.V. Nanopoulos(3).

In an SU(5)-model with no other fermion than the standard ones - transforming as $(\bar{5} + 10)_i$, i = 1, 2,3 - there is some evidence for the coupling of the heaviest third generation to a scalar Higgs ϕ_5 as the only direct coupling to a Higgs that can acquire a v.e.v.. As is well known one gets in this way the successful relation $m_\tau = m_b$ at the unification mass $M \sim 10^{15}$ Gev. On the contrary no direct coupling of the e- and the μ-families should be allowed to ϕ_5 or to ϕ_{45} , which can also acquire a v.e.v. breaking $SU(2) \times U(1) \to U(1)_{em}$. To break the chiralities of these lighter families one may couple them, as well as the τ-family, to scalars which cannot get a v.e.v. consistently with $SU(3)_c \times U(1)_{em}$ remaining exact. The possibly relevant scalars are ϕ_{10}, ϕ_{50} with couplings

$$f_5^T C^{-1} f_5 \phi_{10} \; , \; f_{10}^T C^{-1} f_{10} \phi_{50} \qquad (5)$$

These Yukawa terms will give rise, through radiative corrections, to the smaller induced couplings (4). A closer inspection of the actual diagrams shows[3] that there is only one relevant one-loop graph, which is given in Fig. 1.

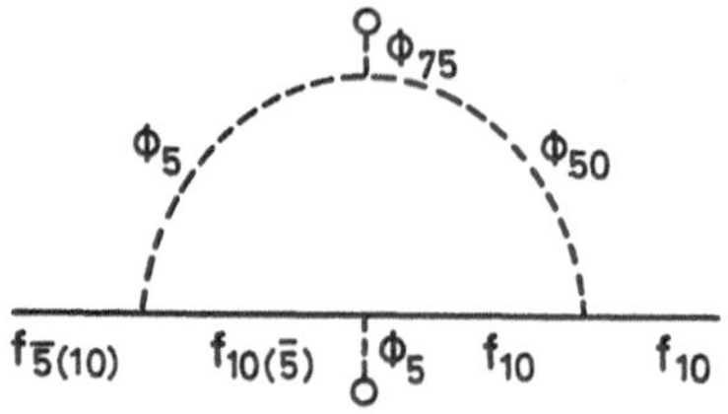

Fig. 1 One-loop diagram giving mass to the μ-generation.

Notice the occurence of the interaction $f^T_{10}\ C^{-1}\ f_{10}\ \phi_{50}$ as well as the presence of ϕ_{75}, whose SU(5)-tensor structure is $T^{\{\gamma\delta\}}_{\{\alpha\beta\}}$. The induced coupling is of the form

$$\sim \lambda \frac{\lambda^2}{4\pi^2} \frac{\phi_{75}}{M} \phi_5^* f^T_{\bar{5}(10)} C^{-1} f_{10} \tag{6}$$

if the trilinear Higgs interaction $\phi_5\phi_{50}\phi_{75}$ has a coupling constant of order M as the masses of the virtual scalars. This in turn, after spontaneous symmetry breaking, will give masses to the μ-generation of order $\lambda^2/4\pi^2$ relative to the one of the τ-generation. Since $\phi_{75}\phi_5^*$ couples to $f^T_{\bar{5}}\ C^{-1}\ f_{10}$ through an effective $\underline{45}^*$, the general mechanism implies the relation

$$m_s = \frac{m_\mu}{3} \tag{7}$$

at the unification mass, as opposed to $m_s = m_\mu$ which one would get from a direct coupling to ϕ_5.

If the family structure of the various Yukawa terms is properly chosen - again a matter of global symmetries - the lightest generation will stay massless even after the inclusion of (6). It is only at the two loop level that such a mass will arise. A consistent Yukawa Lagrangian which impliments the above effects is[3]

* The antisymmetry of the coupling $f_{10}f_{10}\phi_{45}$ forces also the c quark to get a direct mass, since the induced mass would be too small.

$$L_y = a\ \bar{5}_3 10_3 \bar{5} + b\ 10_3 10_3 5 + c\ \bar{5}_2 10_3 \bar{5}' + d\ 10_2 10_2 5' + e\ 10_2 10_3 \overline{50} + f\ 10_3 10_1 \overline{50}' + g\ 5_1 5_3 10 \quad (8)$$

where the subscripts 1,2,3 denote the family index associated with the fermions and the third number in each term stands for the SU(5) transformation property of the scalar fields. This Lagrangian has a $U(1)^4$ global symmetry which forbids any other direct coupling. These symmetries may be broken by soft Higgs self-interactions. The interaction $\phi_5\phi_{50}\phi_{75}$ in Fig. 1 is an example. Alternatively some of them may be broken spontaneously, in which case they can play the role of a Peccei-Quinn symmetry[4] in solving the strong CP-problem. Being broken at M[5], they give rise to very light ($m_A \sim 10^{-8}$eV), unobservable Axions[6]. In general terms, it has been realized in the past year that Goldstone bosons - or quasi-Goldstone bosons - associated with global symmetries broken at a very large energy scale M, are essentially decoupled from fields which are light relative to M since these couplings will go to zero as $M \to \infty$. This could be the fate of the global symmetries which distinguish among the various fermion families.

REFERENCES

1. J. Ellis, M.K. Gaillard, L. Maiani and B. Zumino, in "Unification of the Fundamental Interactions", eds. S. Ferrara, J. Ellis and P. van Nieuwenhuizen (Plenum Press, N.Y., 1980), p. 69.
2. R. Barbieri and D.V. Nanopoulos, Phys. Lett. 91B:369 (1980); 95B:43 (1980);
   S. Barr, Phys. Rev. D21:1421 (1980);
   P. Ramond, University of Florida preprint UFTP 81-8 (1981);
   L. Ibáñez, Oxford preprints OUTP-25/81 and OUTP-40/81 (1981).
3. R. Barbieri, D.V. Nanopoulos and D. Wyler, Phys. Lett. 103B:433 (1981).
4. R. Peccei and H. Quinn, Phys. Rev. D16:1791 (1977).
5. J. Kim, Phys. Rev. Lett. 43:103 (1979);
   M.A. Shifman, A.I. Vainstein and V.I. Zakharov, Nucl. Phys. B166:493 (1980).
6. M. Dine, W. Fischler and M. Sredniki, Phys. Lett. 104B:199 (1981);
   M.B. Wise, H. Georgi and S.L. Glashow, Phys. Rev. Lett. 47:403 (1981);
   R. Barbieri, R.N. Mohapatra, D.V. Nanopoulos and D. Wyler, Phys. Lett. 107B:80 (1981).

SUPERSYMMETRY BREAKING THROUGH A COSMOLOGICAL CONSTANT

Julius Wess

Institut für Theoretische Physik
Universität Karlsruhe
D-7500 Karlsruhe 1, F.R.G.

In this lecture I would like to discuss the assumption that supersymmetry is broken by adding a cosmological constant (not supersymmetric) to a Lagrangian which is invariant under N = 1 gauged supersymmetry transformations. Expanding this Lagrangian around flat space leads to an explicit breaking of supersymmetry. This breaking scheme depends on one parameter only and it turns out to be renormalizable.

Consequences of this breaking for various models of weak and electromagnetic interactions have been investigated by Dan Freedman and myself and radiative corrections have been calculated by Burt Ovrut. These results will be published in a forthcoming paper.

Here I will be mainly concerned with the formulation of the Lagrangian. We shall use the superfield formalism in curved space, as discussed in Ref. 1.

The gravitational part of the Lagrangian is:

$$\mathcal{L}_{Gr} \sim \frac{1}{K^2} \int d^4x\, d^2\theta\{\mathcal{E}\,[R+\Delta]\} + \frac{c}{K^2} \int d^4x e + h.c. \tag{1}$$

The superfields $\mathcal{E}$ and $\mathcal{R}$ are defined in Ref. 1. The constant Δ will parametrize supersymmetry breaking in flat space and the constant c has to be chosen so as to compensate the over-all cosmological constant and to allow an expansion of $e_m{}^a$ around flat space, $e_m{}^a = \delta_m{}^a$.

The part of the Lagrangian (1) which depends on the gravitational field $e_m{}^a$ and on the subsidiary field M has the form:

$$\mathcal{L}_{Gr} \sim -\frac{1}{k^2} \int d^4x\, e\{\mathcal{E} + \frac{5}{2}(MM^* + \Delta M^* + \Delta^* M) + c\} \qquad (2)$$

where $\mathcal{R}$ is the usual Riemannian curvature scalar.

The Euler equation of (2) allows the solution $e_m{}^a = \delta_m{}^a$, $M = -\Delta$ if c is chosen properly.

When coupled to matter fields the equation for M takes the form:

$$\frac{5}{2}\frac{1}{k^2}(M+\Delta) = \frac{\partial}{\partial M^*}\mathcal{L}_{mat.}(M,\chi) \qquad (3)$$

where χ stands for all the matter fields. We expand around flat space $e_m{}^a = \delta_m{}^a$ and we keep only terms of order zero in k^2. Substituting the solution of (3) into the Lagrangian $\mathcal{L} = \mathcal{L}_{Gr.} + \mathcal{L}_{mat.}$ yields

$$\mathcal{L} = \mathcal{L}_{mat.}(-\Delta,\chi) + \text{order}(k^2) \qquad (4)$$

To evaluate the Lagrangian (4) we have to find the M dependence of the matter Lagrangian.

Let me first discuss a Lagrangian with chiral superfields only. In flat space, without symmetry breaking (Δ=0), the Lagrangian is supposed to be:

$$\mathcal{L}_{sym.} = \int d^4x\, d^4\theta\, \phi_i^+\phi_i - \int d^4x\{\int d^4x\{\int d^2\theta\,[\rho^i\phi_i + \frac{1}{2}\mu_{ik}\phi_i\phi_k + \frac{1}{3}\lambda_{ijk}\phi_i\phi_j\phi_k] + \text{h.c.}\} \qquad (5)$$

We have to generalize this Lagrangian to curved superspace; this means that we have to find a covariant Lagrangian in curved space which reduces to (5) in flat space.

The chirality condition $\bar{D}_{\dot\alpha}\phi = 0$ has the covariant generalization $\mathcal{D}_{\dot\alpha}\phi = 0$, with $\mathcal{D}_{\dot\alpha}$ the full covariant derivative[1]. We also have to generalize $\bar{D}\bar{D}\phi^+$ to the kinetic superfield Ξ

$$\Xi = (\mathcal{D}_{\dot\alpha}\mathcal{D}^{\dot\alpha} - 8R)\phi^+ \qquad (6)$$

such that $\mathcal{D}_{\dot\alpha}\,\Xi = 0$ as well.

A possible generalization of (5) to curved space is then:

$$\mathcal{L} \sim -\frac{1}{4}\int d^4x\, d^2\theta\, 2\mathcal{E}\phi\,\Xi - \int d^4x\{d^2\theta\, 2\mathcal{E}\,[\rho_i\phi_i + \frac{1}{2}\mu_{ik}\phi_i\phi_k + \frac{1}{3}\lambda_{ijk}\phi_i\phi_j\phi_k] + \text{h.c.}\} \qquad (7)$$

We could have added a term:

$$w^* \int d^4x\, d^2\theta\, 2\mathcal{E}\,\Xi + \text{h.c.} \tag{8}$$

to (7). In flat space $e_m{}^a = \delta_m{}^a$, $M = 0$ (8) reduces to

$$\int d^4x\, d^2\theta\, \bar{D}\bar{D}\phi = \int d^4x\, d^2\theta\, d^2\bar{\theta}\phi = 0 \tag{9}$$

A term like (8) will, however, contribute if $M = -\Delta$. We shall see that this contribution will be necessary to render a renormalizable Lagrangian.

The M dependence of the Lagrangian (7) and (8) is:

$$\begin{aligned}\mathcal{L} = &\int d^4x\{A_i^+ \Box A_i + i\, \partial_m\bar{\psi}_i\bar{\sigma}^m\psi_i + (F_i-\tfrac{1}{3}A_iM^*)(F_i^*-\tfrac{1}{3}A_i^*M) - \\ &- \int d^4x\{[\rho_i(F_i-M^*A_i) + \mu_{ik}(A_iF_k-\tfrac{1}{2}\psi_i\psi_k-\tfrac{1}{2}M^*A_iA_k + \\ &+ \lambda_{ikj}(A_iA_jF_k-A_i\psi_j\psi_k-\tfrac{1}{3}M^*A_iA_jA_k)] - \\ &- \tfrac{1}{3}w_iM(F_i-\tfrac{1}{3}M^*A_i) + \text{h.c.}\}\end{aligned} \tag{10}$$

This Lagrangian suggests that we should shift the auxiliary field F:

$$\mathcal{F} = F - \tfrac{1}{3}AM^* \tag{11}$$

If we now substitute $M = -\Delta$ in (10) we obtain the result:

$$\mathcal{L} = \mathcal{L}_{\text{sym.}} + \tfrac{1}{3}(\Delta w_i\mathcal{F}_i+\text{h.c.}) - \{\Delta(\tfrac{2}{3}\rho_iA_i+\tfrac{1}{6}\mu_{ik}A_iA_k) + \text{h.c.}\} \tag{12}$$

The first part $\mathcal{L}_{\text{sym.}}$ is the Lagrangian (5) with F replaced by $\mathcal{F}$. Note, however, that the coefficient of $\mathcal{F}_i$ is $-\tilde{\rho}_i = -\rho_i + \frac{1}{3}w_i\Delta$ and can therefore be chosen to be independent of the coefficient of A_i which is $-\frac{2}{3}\Delta\rho_i$.

The final Lagrangian (12) will have the same form for supersymmetric gauge theories as well. $\mathcal{L}_{\text{sym}}$ has to be replaced by the respective gauge invariant Lagrangian, the Δ dependent terms and, therefore, the symmetry breaking part of the Lagrangian will be unchanged. This can be understood from conformal invariance.

To study the renormalizability of the Lagrangian (12), we observe that it can be written with a directional field $\chi = \theta\theta\Delta$

$$\begin{aligned}\mathcal{L} = &\mathcal{L}_{\text{sym.}}(\tilde{\rho}_i,\mu_{ik},\lambda_{ijk}) - \\ &- \int d^4x\{\textstyle\int d^2\theta(\tfrac{2}{3}\rho_i\phi_i\chi+\tfrac{1}{6}\mu_{ik}\phi_i\phi_k\chi) + \text{hc}\}\end{aligned} \tag{13}$$

If we use superfield propagators we know that the only new divergent superfield diagrams are[2]

Their contributions can be absorbed in the renormalization of $\tilde{\ }_i$. The Lagrangian (12) still has several shortcomings. In the tree approximation, the spinorial partners of the gauge fields remain degenerate in mass with the vector fields, and in particular the photino remains massless. In higher order, the photino will acquire a mass through the contributions of the effective superfield potential of the form $W_\alpha W^\alpha \chi\chi$ and $\bar{W}^{\dot\alpha} \bar{W}_{\dot\alpha} \chi\chi^+$. B. Ovrut has calculated the photino mass in a one loop approximation in supersymmetric electrodynamics[3]. He obtained the result:

$$m_\lambda = \frac{e^2}{8\pi^2} \frac{6m^2}{\Delta} \left[(1+\frac{\Delta}{6m}) \ln(1+\frac{\Delta}{6m}) + (1-\frac{\Delta}{6m}) \ln(1-\frac{\Delta}{6m})\right] \tag{14}$$

The maximal value of this mass would be $\propto m^2 \ln^2$ where m is the mass of the charged chiral multiplet. It should be noted, however, that each charged superfield adds to the mass of the photino.

An interesting feature of supersymmetric electrodynamics is that the parameter Δ is restricted to $|\Delta| < \frac{3}{2}m$. For larger values of Δ the potential becomes unstable.

For a Lagrangian of type (13), the mass square formula[4] for bosonic and fermionic degrees of freedom remains valid. Some scalar fields have to remain light compared to their fermionic partners. Higher order contributions to the effective potential of the form $\phi\phi^+\chi$ and $\phi\phi^+\chi^+$ might change this feature.

Finally, let me discuss the simplest model:

$$\mathcal{L} = \phi^+\phi - (\rho\phi + \frac{1}{2}\mu\phi^2 + \frac{1}{3}\lambda\phi^3 + h.c.) \tag{15}$$

This model, without symmetry breaking, has the potential

$$V = FF^+, \quad F^+ = \rho + \mu A + \lambda A^2 \tag{16}$$

This potential has two minima

$$A = \frac{1}{2\lambda} \{-\mu \pm \sqrt{\mu^2 - 4\lambda^2\rho}\} \tag{17}$$

For both values of A, the potential vanishes. This is very characteristic of supersymmetric theories. A potential will have several minima, some might break an internal symmetry. These minima will, however, in general be degenerate. A breaking term of the form (12) will favour one of these minima and it will depend on the strength of supersymmetry breaking (the value of Δ) whether the true minimum will spontaneously break an internal symmetry or not.

If we add the breaking term to (16) we obtain

$$V = (\tilde{\rho}+\mu A+\lambda A^2)(\tilde{\rho}^*+\mu^* A^*+\lambda^* A^{*2}) + + (\frac{2}{3}\Delta\rho A+\frac{1}{6}\Delta\mu A^2 + h.c.) \tag{18}$$

For simplicity, we shall assume that the parameters are real, that $\mu \geq 0$ and that $\mu\tilde{f} + \frac{2}{3}\Delta f = 0$. This guarantees that one of the minima will be at $A \equiv 0$. This minimum will be the absolute minimum if

$$-6\tilde{\rho}\,\frac{\lambda}{\mu} < \Delta < 3\mu - 6\tilde{\rho}\,\frac{\lambda}{\mu}.$$

REFERENCES

1. J. Wess and B. Zumino, Phys. Lett. 79B:394 (1978).
2. L. Girardello and M. Grisaru, Harvard University preprint (1981).
3. J. Wess and B. Zumino, Nucl. Phys. B78:1 (1974).
4. S. Ferrara, L. Girardello and F. Palumbo, Phys. Rev. D20:403 (1979).

SYMMETRY BREAKING IN SUPERSYMMETRIC GUTs*

F. Buccella [1], J.-P. Derendinger, and C.A. Savoy

Département de Physique Théorique
Université de Genève

and

S. Ferrara

CERN - Geneva

INTRODUCTION

Supersymmetry [1] is one of the most attractive and puzzling symmetries of relativistic quantum field theory. It is attractive because of several reasons : among them we mention the enlargement of the space-time symmetry algebra giving rise to supermultiplets of fields with different spins and the softening of quantum divergences [2]. Ultraviolet improvement of radiative corrections is especially dramatic in the supersymmetric extensions of Einstein theory, the so-called supergravities [3], which provide us with theories of interacting particles with spin ranging from $J = 0$ up to $J = 2$ and possessing amazing convergence properties [4]. This is related to the previously mentioned softening of quantum divergences in supersymmetric field theories. More specifically in a globally supersymmetric renormalizable theory with or without gauge interactions there is a set of non-renormalization theorems [5] which ensure to all orders of perturbation theory the absence of radiative corrections to masses and coupling constants; this would certainly not be true in an ordinary gauge theory. As an important consequence of these theorems, it becomes technically natural to impose numerical relations between the bare parameters

* Work partly supported by the Swiss National Science Foundation.
1) Permanent address : Istituto di Fisica, Università di Napoli, Italy.

of the superpotential which describes the self-interactions of the $J = 0,1/2$ chiral multiplets : such relations will survive renormalization. Moreover, there will be no quadratically divergent corrections to masses of scalar particles.

On the other hand, supersymmetry is puzzling since it has to be broken in order to have any relevance to Physics. And there is no indication at what scale should this breaking occur. In some attempts at superunification [6], it was assumed that supersymmetry is broken at the Planck scale. In this case the low energy implications of supersymmetry are very remote and one can only hope to get indirect predictions like constraints on the gauge group unifying the non gravitational interactions and some limitations on the fermion representations. Recently many authors [7-14] have pointed out that a supersymmetric world could emerge if one supersymmetry remains unbroken all the way down to the weak interaction scale. The possible relevance of a supersymmetry broken at low energy, $M_W \ll M_{Planck}$, is suggested by the so-called hierarchy problem [15] one encounters in attempts to unify electroweak and strong interactions. In conventional unified theories (GUTs), the scale of symmetry breaking are strongly related to the masses of the Higgs scalars which acquire v.e.v.'s and are responsible for the breaking of the gauge group. However there is no symmetry to ensure masslessness of scalar fields, as chiral symmetry does for fermions. Thus there is no natural mechanism to provide us with the huge ratio between the scale of baryon number violating processes ($\gtrsim 10^{15}$GeV) and the scale of weak interactions (~ 100GeV). Supersymmetry however potentially possesses the desirable features which allow one to solve the hierarchy problem : if the bare parameters of the Lagrangian allow for two different scales, the higher order corrections will not be able to mix these scales as long as supersymmetry is unbroken (or softly broken at the lower scale). Then a Weinberg-Salam Higgs doublet with bare mass of order 250GeV will not receive mass "corrections" of order 10^{15}GeV as it is usually the case in unified theories.

In the present work, we analyze the first step of symmetry breaking in $N = 1$ supersymmetric unified theories. As usual, these theories have (at least) two scales of breaking : at superheavy energies ($\sim 10^{15}$GeV) the gauge group G is broken down into $G' \supseteq SU(3) \times SU(2) \times U(1)$ and baryon number nonconserving gauge interactions decouple. Supersymmetry however remains unbroken : its breaking will occur only at the weak interaction scale, the necessary hierarchy being technically natural due to supersymmetry. Our aim is to characterize the possible patterns of gauge symmetry breaking consistent with supersymmetry.

Let us first review some well-known properties of the scalar potential in supersymmetric gauge theories [16-18]. The scalar fields are complex : $z = A + iB$, transforming according to some representation $\underline{R}$ of the gauge group G, which is in general reducible. We assume G to be simple, but our results apply as well to semisimple gauge groups. The scalar potential is then given by the following non-negative G-invariant function of z and z* :

$$V(z,z^*) = \sum_a |F_a(z)|^2 + \sum_\alpha |D^\alpha(z)|^2 \geq 0 \tag{1}$$

where the "gauge term", $D^\alpha(z)$, $(\alpha = 1,2,\ldots,\dim G)$ is

$$D^\alpha(z) = g \sum_{a,b} z_a^*(T^\alpha)^a_b z^b \tag{2}$$

the matrices T^α being the generators of G in the representation R, in the basis where z has components $z_a (a = 1,2,\ldots,\dim R)$. We refer to the second term in $V(z,z^*)$ as the D^2-term. The first term, or F^2-term, is defined in terms of the gradient :

$$F_a = \frac{\partial F(z)}{\partial z^a} \tag{3}$$

where F(z) is a G-invariant polynomial in the z-fields (but not in the z*-fields) which is at most cubic in a renormalizable theory.

Supersymmetry survives the spontaneous symmetry breaking of G if $V(\langle z\rangle,\langle z^*\rangle) = 0$, where $\langle z\rangle$ is the vacuum expectation value (v.e.v.) of z; the residual gauge group is the little group of $\langle z\rangle$. Since $V(z,z^*)$ is non-negative, if a supersymmetry preserving solution exists, all the solutions are supersymmetric. In this case the v.e.v.'s have to satisfy the following conditions

$$F_a(\langle z\rangle) = 0, \quad \forall a \tag{4}$$

$$D^\alpha(\langle z\rangle) = 0, \quad \forall \alpha \tag{5}$$

If F(z) has no linear term in some component of z (which has to be a singlet under G) then all possible sets of v.e.v.'s preserve supersymmetry since the trivial solution, without spontaneous gauge symmetry breaking, $\langle z\rangle = 0$ gives the minimum value $V(0,0) = 0$. In this case all the v.e.v.'s are solutions of Eqs (4) and (5). Since we are interested in supersymmetric solutions we will first assume that there are no G-invariant components in the (reducible) multiplet of the scalar fields z, and we will discuss later the possible effects when there are G-singlets among the scalar fields.

In this paper, we introduce some simple but rather powerful methods to discover which v.e.v.'s of a given multiplet of scalar fields are consistent with Eqs (4) and (5). Since we assume supersymmetry to be preserved, we say that all v.e.v.'s which are excluded by these conditions, Eqs (4) and (5), are protected. In general, besides the trivial solution, $\langle z \rangle = 0$, there are several non-trivial ones which break the gauge group G. In practice, we would like to have $SU(3) \times SU(2) \times U(1)$ (perhaps a larger group, consistent with a breaking at low energies) as the residual little group and protect all other solutions (but $\langle z \rangle = 0$). This, in general, is not possible because there are degenerate sets of supersymmetric v.e.v.'s. In these cases we restrict ourselves to ensure that a solution exists with $SU(3) \times SU(2) \times U(1)$ as little group and that the v.e.v.'s which subsequently break $SU(3) \times SU(2) \times U(1)$ are protected. In this sense, we will say that the supersymmetric gauge theory has a natural hierarchy. The breaking of both supersymmetry and the Weinberg-Salam gauge group at much lower energies should need including new terms in the Lagrangian with parameters of the order of 1TeV. In the models proposed so far these new terms are also used to resolve the degeneracy of the supersymmetric solutions in favour of the $SU(3) \times SU(2) \times U(1)$ residual group at high energies.

It must be stressed, however, that in this paper we do not address ourselves to the construction of realistic models. Even if we illustrate our discussion with examples which mimic some well-known GUTs, the emphasis is on the protection of v.e.v.'s and on the methods to select the allowed supersymmetric gauge breaking patterns.

The paper is organized as follows. The two contributions to the scalar potential are analyzed separately. In Section 2, we discuss the vanishing of the D^2-term, and we derive simple but powerful lemmas; we illustrate their application with several examples, leading to a fairly simple selection of the unprotected v.e.v.'s. Section 3 is devoted to the study of the vanishing of the F^2-term. We find a sort of complementarity between the D^2-term and the F^2-term in the sense that v.e.v.'s which are allowed by Eq. (5) are in general forbidden by Eq. (4) and vice-versa. Simple group theoretical arguments can be used to observe this property. In Section 4, we apply the results of the preceding sections to some possible candidates to a supersymmetric gauge theory, based on the gauge groups SU(5), O(10) and E_6. Even if the models involve several irreducible representations (irreps) of chiral supermultiplets, by exploiting the properties of the potential in Eq. (1), one is able to find possible patterns of symmetry breaking consistently with supersymmetry. Section 5 presents a short discussion of the next step in the construction of supersymmetric unified theories, namely the mechanism of supersymmetry breaking.

VANISHING OF GAUGE POTENTIAL TERM

Let us start by discussing the condition $D^\alpha(<z>) = 0$. We introduce the notation

$$D^\alpha(z) = g\langle z|T^\alpha|z\rangle = \sum_i g\langle z_i|T^\alpha|z_i\rangle,$$

where the sum is over the irreps appearing in the representation of the chiral multiplets, the last identity following from the characteristic property of the generators T^α.

We characterize the possible v.e.v.'s, z, by their little group, g_z. For the generators of g_z, by definition $T^\alpha|z\rangle = 0$, so that the corresponding $D^\alpha(z)$ trivially vanish. The remaining generators of G, associated to the coset G/g_z, are such that $T^\alpha|z\rangle \neq 0$. However, for these T^α of G/g_z which do not transform as singlets under g_z, one has $\langle z|T^\alpha|z\rangle = 0$, by the Wigner-Eckart theorem as z is a g_z-singlet. Therefore, one has the sufficient condition :

Lemma 1 : If G/g_z does not contain any generator transforming as a singlet under g_z, then all $D^\alpha(z) = 0$ (i.e., the v.e.v. z is allowed by the D^2-term).

In general this condition is not necessary, since if a T^α of G/g_z is a g_z-singlet, in principle, $T^\alpha|z\rangle$ can be orthogonal to $|z\rangle$. If z is on a critical orbit (which means that there is no element in the representation with a little group containing g_z), then $T^\alpha|z\rangle \propto |z\rangle$, so that the above condition is also necessary :

Lemma 2 : For z on a critical orbit, if G/g_z has a generator T^α transforming as a singlet under g_z, then the corresponding $D^\alpha(z) \neq 0$ (the critical v.e.v. z is then protected by the D^2-term).

For $z = A + iB$ transforming as a real representation of G, the vanishing of the D^2-term imposes particularly simple conditions. Indeed, the generators T^α can be represented by antisymmetric matrices $T^\alpha_{ab} = -T^\alpha_{ba}$, so that the only contribution to D^α comes from $iT^\alpha_{ab}(A_aB_b - B_aA_b)$. For instance

$A \propto B$ is sufficient to have all $D^{\alpha}(z) = 0$. But, in general, the $D^{\alpha}(z)$ vanish in more general circumstances : if, e.g., z transforms as the adjoint representation, A and B can be written as Hermitean matrices and the condition $[A,B] = 0$ ensures the vanishing of the D^2-term. This is an example of Lemma 1 since, if $[A,B] = 0$, z can be rotated into a diagonal form and has a little group which contains the Cartan subgroup of G, so that G/g_z does not contain any g_z-singlet.

These results apply as well to the contribution $D^{\alpha}(z_i)$ of each irrep appearing in the decomposition of z. We first analyze according to Lemmas 1 and 2, some complex irreps of various (simple) groups that could be relevant for grand unification.

The spinorial representation of O(10), $\underline{16}$, has two critical orbits, with little groups SU(5) and O(7), respectively. But both of them correspond to a non-vanishing D^2-term, since O(10)/SU(5) has a SU(5)-singlet generator, and O(10)/O(7) has an O(7)-singlet generator. Also the $\underline{126}$ of O(10) has an SU(5)-invariant critical orbit which gives a non-vanishing D^2-term. However, there are directions of the $\underline{126}$ for which the D^2-term vanishes, like that with $O(5) \times O(5)$ invariance, for instance. In O(14), the next group in the $O(4n+2)$ series, the spinorial representation, $\underline{64}$, has a $(G_2 \times G_2)$-invariant direction for which the D^2-term vanishes.

In the case of the E_6 group, for the $\underline{27}$ fundamental representation, the direction with F_4-invariance gives a vanishing D^2-term, while the direction with the little group O(10) contributes to the D^2-term, since $E_6 \supset O(10) \times U(1)$. For the $\underline{351}_s$ representation, contained in the symmetric product $(27 \times 27)_s$, one finds that among all $[SU(3) \times SU(2) \times U(1)]$-invariant directions, the D^2-term vanishes only on one (non-critical) orbit with little group $O(6) \times O(4)$. Indeed, $E_6/O(6) \times O(4)$ contains just one $[O(6) \times O(4)]$-singlet generator, T, and this orbit is precisely characterized by the fact that $T|z\rangle$ is orthogonal to $|z\rangle$.

We now study some complex irreps of SU(N) : i) any element of the fundamental representation has little group $SU(N-1)$ and so contributes to the D^2-term; ii) the D^2-term vanishes for an antisymmetric tensor, $z^{[ab]}$ of rank two, only for even N, on the critical orbit with little group Sp(N); iii) the symmetric tensor of rank two, $z^{\{ab\}}$ has the same property (for any N) in the critical direction with little group O(N); for the N-fold symmetric representation, $z^{\{a_1 a_2 \dots a_n\}}$, the critical orbit with the little group $[U(1)]^{N-1}$ (Cartan subgroup) does not contribute to the D^2-term.

The vanishing of the D^2-term for a given direction of an irrep means that a v.e.v. in this direction is not protected by the D^2-term. The converse is not true, in general. Indeed, in physical applications one always needs a reducible representation of chiral supermultiplets. The non-vanishing of the D^2-term on a given orbit of an irrep does not ensure the protection of the associated v.e.v.'s, since a cancellation may occur between different contributions. In this case, however, the scales and the directions of the different v.e.v.'s are strictly related. This implies a much more restricted pattern of possible symmetry breakings that in the case of non-supersymmetric theories with the same set of irreps for the scalars. As a trivial example, take a real reducible representation, $\underline{R} + \underline{\bar{R}}$: the D^2-term obviously vanishes if the v.e.v.'s in $\underline{R}$ and $\underline{\bar{R}}$ have the same little group and the same strength.

Let us present some examples where two complex irreps may conspire to give a vanishing overall contribution to the D^2-term :
a) In SU(N), the fundamental representation, z^a, and the symmetric tensor of rank r, $Z_{\{a_1 \ldots a_r\}}$: if their v.e.v.'s are related by :

$$Z_{\{a_1 a_2 \ldots a_r\}} = \frac{1}{\sqrt{r}} \delta^{a_1 b} \ldots \delta^{a_r b} z^b , \quad z^a = \delta^a_b z^b$$

they both have SU(N - 1) as little group and the U(1) in the coset SU(N)/SU(N - 1) has a generator with opposite signs for the two irreps.
b) In SU(2n + 1), the fundamental representation, z^a, and the two-fold antisymmetric representation, $Z^{[ab]}$: if $Z^{[ab]}$ has the little group Sp(2n) which is contained in the little group SU(2n) of z^a, the D^2-term will vanish for $\mathrm{Tr}ZZ^* = n|z|^2$.
c) In O(10), a $\underline{\overline{126}}$ and a $\underline{16}$: if the v.e.v.'s are SU(5)-invariant and if the strengths of the $\underline{16}$ and the $\underline{\overline{126}}$ are in the ratio $\sqrt{2}$, the D^2-term vanishes. The same occurs in E_6 , with a $\underline{\overline{351}}_s$ and a $\underline{27}$ if the v.e.v.'s have O(10) as little group.
d) In O(10), the complex v.e.v. of a $\underline{10}$ with little group O(8) and the v.e.v. of a $\underline{16}$ with little group O(7) can cancel to give a vanishing D^2-term.

In all these examples, the cosets G/g_Z of the v.e.v.'s in each irrep have the same g_Z-invariant generator which takes opposite sign for the two v.e.v.'s. We now construct simple examples where a cancellation occurs between contributions of v.e.v.'s in several irreps :
e) In SU(N), take the r-fold antisymmetric and r fundamental representations, $Z_{[a_1 \ldots a_r]}$ and $z^a_{(i)}$, (i = 1,...,r) : if their

v.e.v.'s are such that $z^a_{(i)} = \mu\delta^{ai}$ and that $Z_{[a_1...a_r]}$ has only one non-vanishing component, $Z_{[12...r]} = \mu$, then the D^2-term vanishes and the little group of the whole set of v.e.v.'s is SU(N - r).
f) In SU(N), with N fundamental representations $z^a_{(i)}$, (i = 1,...,N) the D^2-term vanishes when the v.e.v.'s are all orthogonal, e.g., $z^a_{(i)} = \mu\delta^{ai}$; the SU(N) symmetry is completely broken.

From the examples above it should be clear how to build sets of irreps such that the D^2-term vanishes when the v.e.v.'s are properly chosen. Notice that in the cases above the strengths of all v.e.v.'s are given in terms of only one arbitrary scale. Obviously, more scales may be introduced if one uses a larger set of v.e.v.'s to obtain a vanishing D^2-term.

A remarkable property in these examples is that a G-invariant exists for the set of irreps. This is shown in Table I. If one substitutes in the G-invariant the v.e.v.'s which cancel in the D^2-term, the result does not vanish. Furthermore, the strength of each v.e.v. is proportional to $\sqrt{n}$ where n is the number of times the corresponding irrep appears in the G-invariant.

VANISHING OF THE SUPERPOTENTIAL TERM

We now turn to the discussion of the F^2-term in the potential. It is determined by the superpotential, which may have terms mixing different irreps with several arbitrary parameters. Therefore, for a given (in general, reducible) representation of chiral multiplets many cases are possible depending on the structure of the superpotential. In practice, the analysis simplifies a lot as only directions which are unprotected by the D^2-term are to be considered.

The F^2-term vanishes at the stationary points of the superpotential. We have noticed that a direction with vanishing D^2-term corresponds to a non-vanishing contribution to some invariant under the gauge group which may be present in the superpotential. Examples are the quadratic invariant for self-conjugate representations and the invariants in Table I for complex representations. The gradient of these invariants in the directions selected by the D^2-term will not vanish so that these v.e.v.'s could be protected by the F^2-term. There is a sort of complementarity between the protection by the D^2-term and the F^2-term. The v.e.v.'s which are not protected by the D^2-term will be protected by the F^2-term unless one of the following conditions is satisfied :
I) The gradients of two (or more) invariants are such as to exactly

Table I Examples of solutions of Eq. (5). The little groups of the corresponding orbits are given. In the last column, the non-vanishing invariants associated to these solutions are indicated (see text).

Group Representations	Little group	Invariant
SU(N) $z^{[ij]}$ (N even)	Sp(N)	$\varepsilon_{i_1\ldots i_N} z^{[i_1 i_2]}\ldots z^{[i_{N-1} i_N]}$
$z^{\{ij\}}$	O(N)	$\varepsilon_{i_1\ldots i_N}\varepsilon_{j_1\ldots j_N} z^{\{i_1 j_1\}}\ldots z^{\{i_N j_N\}}$
$z^{\{i_1\ldots i_N\}}$	$[U(1)]^{N-1}$	$\varepsilon_{i_1\ldots i_N}\varepsilon_{j_1\ldots j_N}\ldots\varepsilon_{r_1\ldots r_N}$ $z^{\{i_1 j_1\ldots r_1\}}\ldots z^{\{i_N j_N\ldots r_N\}}$
z^{i} $z_{\{i_1\ldots i_r\}}$	SU(N-1)	$z^{i_1}\ldots z^{i_r} z_{\{i_1\ldots i_r\}}$
z^{i} $z^{[ij]}$ (N odd)	SU(N-1) ∪ Sp(N-1)	$\varepsilon_{i_1\ldots i_N} z^{i_1} z^{[i_2 i_3]}\ldots z^{[i_{N-1} i_N]}$
$z^{i}_{(1)}\ldots z^{i}_{(r)}$ $z_{[i_1\ldots i_r]}$	SU(N-r) SU(r) × SU(N-r)	$z^{i_1}_{(1)}\ldots z^{i_r}_{(r)} z_{[i_1\ldots i_r]}$
$z^{i_1}_{(1)}\ldots z^{i_N}_{(N)}$	Identity	$\varepsilon_{i_1\ldots i_N} z^{i_1}_{(1)}\ldots z^{i_N}_{(N)}$
O(10) $\underline{126}$	e.g. O(5) × O(5)	$(\underline{126}\times\underline{126}\times\underline{126}\times\underline{126})_{\underline{1}}$
$\overline{\underline{126}}$ $\underline{16}$	e.g. SU(5) SU(5)	$[\overline{\underline{126}}\times(\underline{16}\times\underline{16})_s]_{\underline{1}}$
$\underline{16}$ $\underline{10}$ (complex)	e.g. O(7) O(8)	$[(\underline{16}\times\underline{16})_s\times\underline{10}]_{\underline{1}}$
E_6 $\underline{27}$	F4	$(\underline{27}\times\underline{27}\times\underline{27})_{\underline{1}}$
$\underline{351}_s$	e.g. O(6) × O(4)	$(\underline{315}_s\times\underline{351}_s\times\underline{351}_s)_{\underline{1}}$
$\overline{\underline{351}}_s$ $\underline{27}$	O(10) O(10)	$[(\underline{27}\times\underline{27})_s\times\overline{\underline{351}}_s]_{\underline{1}}$

compensate each other, and the relations which are needed between v.e.v.'s are consistent with the vanishing of the D^2-term.
II) The G-invariants one is able to construct for the irreps in correspondence with $|D^2| = 0$ are all of degree ≥ 3 (several examples are available in Table I).
III) Invariants that imply the protection of some particular set of v.e.v.'s can be excluded from the superpotential. As already noticed, the procedure of fixing the parameters in the superpotential, although unaesthetic, is (technically) natural in supersymmetric theories. However, the pattern of fermionic masses is also related to the superpotential and some G-invariants are needed for this purpose in the superpotential. Also, the elimination of some invariants may introduce additional global symmetries in the superpotential, whose breaking might lead to the existence of Goldstone bosons.

We shall illustrate these properties with some simple examples. If there is only one irrep of chiral multiplets, a quadratic invariant may appear in the superpotential if the irrep is real. If there is no cubic invariant, the v.e.v.'s of these fields are protected by the F^2-term. Examples are the adjoint representations of $O(N)$, $(N \neq 6)$ and of the exceptional groups. (This remains true, in general, with two adjoint representations.)

Therefore, it is interesting to consider self-conjugate irreps which admit a cubic invariant. Take for instance, the adjoint of $SU(N)$, $(N \geq 3)$ and the rank-two symmetric traceless tensor of $O(N)$, $(N \geq 5)$, which can be written as a matrix ϕ. If both the invariants, $Tr\phi^2$ and $Tr\phi^3$ are present in the superpotential, the vanishing of the F^2-term requires ϕ to be in a critical orbit, with little groups $SU(M) \times SU(N-M) \times U(1)$ and $O(M) \times O(N-M)$, respectively $(2M \neq N$, in both cases). With the cubic invariant $Tr\phi^3$ alone, only the directions with little groups $SU(N/2) \times SU(N/2) \times$ $\times\, U(1)$ and $O(N/2) \times O(N/2)$ are allowed, provided that N be even.

More generally, two self-conjugate irreps can conspire to give a vanishing F^2-term (if they share some bilinear invariant), in which case the little groups of their v.e.v.'s are closely related. Consider, for instance, the antisymmetric tensor (adjoint representation) and the symmetric tensor representations of $O(2N)$, represented by the traceless matrices : $A^{ab} = -A^{ba}$, $S^{ab} = S^{ba}$, $(a,b = 1,\ldots,2N)$. The possible $O(2N)$ invariants are : TrA^2, TrS^2, TrS^3, TrA^2S. The F^2-term vanishes when the v.e.v. of S has little group $O(2p) \times O(2k-2p) \times O(2N-2k)$, $(p < k/2,\ k \neq N/2)$ while the v.e.v. of A has little group $O(2k) \times U(N-k)$, the residual gauge symmetry being $O(2p) \times O(2k-2p) \times U(N-k)$. This property is easy to understand by noticing, for instance, that at a stationary point of the superpotential the v.e.v.'s of A and of the adjoint part of (AS) have the same invariances.

Also in the case of two conjugate representations, R and $\bar{R}$, of chiral multiplets, the invariant $(R\bar{R})$ in the superpotential would tend to protect their v.e.v.'s. Again, if there are bilinear couplings in the superpotential which involve R and (or) $\bar{R}$ a conspiration may occur allowing for v.e.v.'s of $\bar{R}$ and (or) R. As an example, take an O(10) gauge model with chiral multiplets in the $(\underline{16}+\overline{\underline{16}}+\underline{45})$ representation. Take the v.e.v.'s of the $\underline{16}$ and $\overline{\underline{16}}$ to be parallel and of equal strength to get a vanishing D^2-term. These v.e.v.'s have a little group $\supset SU(4)$ [19]. The possible invariants in the superpotential are : $(\overline{\underline{16}}\times\underline{16})$, $(\underline{45})^2$, $(\overline{\underline{16}}\times\underline{16}\times\underline{45})$. Hence, the v.e.v. of the $\underline{45}$ has to have little group $\supset SU(4)\times U(1)\times U(1)$ in order to compensate for the v.e.v. of the $\underline{45}$ components of $(\overline{\underline{16}}\times\underline{16})$. But since the v.e.v. of the $\underline{16}$-part of $(\underline{16}\times\underline{45})$ has to be proportional to the v.e.v. of the $\underline{16}$, the only possible solution must have an overall SU(5) invariance.

The property that the little groups of the v.e.v.'s for the various irreps have to be strongly related is general. It explains why supersymmetry allows for, and often requires, a richer pattern of Higgs representations than in non-supersymmetric GUTs in order to achieve a certain pattern of symmetry breaking.

It is interesting to consider complex irreps with a cubic invariant. Take for instance the $\underline{27}$ of E_6. The F_4-invariant v.e.v. is allowed by the D^2-term and protected by the F^2-term, while the O(10)-invariant v.e.v. is allowed by the F^2-term and protected by the D^2-term. The complementarity between these two terms in the potential can be proved here in a very illustrative way [20]. The D^2-term has the form $(\underline{27}\times\overline{\underline{27}})_{\underline{78}}\times(\underline{27}\times\overline{\underline{27}})_{\underline{78}}$, while the F^2-term is of the form $(\underline{27}\times\underline{27})_{\overline{\underline{27}}}\times(\overline{\underline{27}}\times\overline{\underline{27}})_{\underline{27}}$. Since $(\underline{27}\times\underline{27})_s = \underline{351} + \overline{\underline{27}}$, there are only two independent quartic invariants built with $\overline{\underline{27}}\times 27$, so that the F^2-term can be written as a linear combination of the D^2-term and a term proportional to $|(\underline{27}\times\overline{\underline{27}})_{\underline{1}}|^2$. When the D^2-term vanishes the potential is proportional to the square of the norm of the v.e.v. so that this v.e.v. is protected. Another example is the antisymmetric tensor of rank two $(\underline{15})$ of SU(6), for which v.e.v.'s in the Sp(6) invariant critical orbit are allowed by the D^2-term, while the F^2-term vanishes for v.e.v.'s with the little group $SU(4)\times SU(2)$.

Up to now, gauge singlet superfields were omitted, since

supersymmetry was left unbroken. However singlet superfields may have interesting implications, if they are used with some real multiplets, even without any breaking of supersymmetry. It is clear that the analysis of the D^2-term remains unchanged in presence of singlet fields : they do not contribute to it. In general, the introduction of a singlet X will modify the superpotential by a polynomial $\mu^2 X + mX^2 + \ell X^3$ and by couplings to conjugate representations R_i and $\bar{R}_j$ of the form $\lambda_{ij} R_i \bar{R}_j X$. The superpotential in this case contains already terms like $M_{ij} R_i \bar{R}_j$. Thus in general the study of the F^2-term remains qualitatively unchanged, with M_{ij} replaced by $M_{ij} + \lambda_{ij} \langle X \rangle$. The v.e.v. of X has to satisfy the constraint

$$0 = \mu^2 + 2m\langle X\rangle + 3\ell\langle X\rangle^2 + \sum_{i,j} \lambda_{ij} \langle R_i \rangle \langle \bar{R}_j \rangle$$

The main difference appears when some v.e.v.'s $\langle R_i \rangle$, $\langle \bar{R}_j \rangle$ are protected by the $M_{ij} R_i \bar{R}_j$ term. It is then possible to choose parameters such that $M_{ij} + \lambda_{ij}\langle X \rangle = 0$, and thus to avoid protection of $\langle R_i \rangle$ and $\langle \bar{R}_j \rangle$. However in this case the directions of the v.e.v.'s are not determined.

APPLICATIONS TO GUTS

We have constructed a method to determine if a given pattern of gauge symmetry breaking is consistent with supersymmetry. From the little groups of the v.e.v.'s in the various irreps one can quickly select the generators T^α of the gauge group for which the corresponding $D^\alpha(z)$ might receive contributions from the v.e.v.'s and so check if $D^\alpha(z) = 0$. If this is the case, one has to consider the superpotential, take the v.e.v. of its gradient, $F^a(z)$, and see if a cancellation may occur between terms arising from quadratic and cubic invariants in the superpotential. Since this obviously requires these terms to have the same invariances, most of the analysis of the F^2-term is performed in terms of the little groups of the v.e.v.'s.

In this section we study the possibility of obtaining the phenomenological $SU(3) \times SU(2) \times U(1)$ gauge group as the residual symmetry in some supersymmetric models, with spontaneously broken gauge groups : $SU(5)$, $O(10)$ and E_6.

a) SU(5). - A minimal supersymmetric model with $SU(5)$ gauge group has been constructed by Dimopoulos and Georgi [9] and independently by Sakai [10]. One needs at least one adjoint representation $(\underline{24})$ and one $(\underline{5} + \underline{\bar{5}})$ of "Higgs" chiral supermultiplets, and $n(\underline{\bar{5}} + \underline{10})$ of "matter" chiral supermultiplets (n is the number of fermion families). The v.e.v.'s of the $\underline{5},\underline{\bar{5}}$'s and $\underline{10}$'s have to be protected in order to preserve $SU(3)\times SU(2)\times U(1)$ down to low energies (where Susy is broken). According to the discussion in Section 2, there are four basic ways in which these complex representations can conspire to give a vanishing D^2-term, and the associated v.e.v.'s must be protected by the F^2-term. To these four possibilities there correspond the four invariants : $(\underline{\bar{5}}\times\underline{5})$, $(\underline{5}\times\underline{10}\times\underline{10})$, $(\underline{\bar{5}}\times\underline{\bar{5}}\times\underline{10})$ and $(\underline{\bar{5}}\times\underline{\bar{5}}\times\underline{\bar{5}}\times\underline{\bar{5}}\times\underline{\bar{5}})_A$. The last invariant, of degree five, cannot appear in the superpotential and the corresponding v.e.v.'s are not protected. Since it requires at least five $\bar{5}$'s, the hierarchy is not naturally ensured by supersymmetry if there are more than four families of light fermions.

The superpotential of Dimopoulos, Georgi and Sakai includes all the possible invariants except those which are odd in the "matter" fields and so are forbidden by the stability of the proton. The presence in the superpotential of both the quadratic and the cubic invariants of the $\underline{24}$ allows for these fields to take v.e.v.'s on the critical orbits with little group $SU(3)\times SU(2)\times U(1)$ or $SU(4)\times U(1)$ as discussed in Section 3. There are also colour and/or e.m. breaking solutions with non-vanishing v.e.v.'s for the scalars in the "matter" multiplets (if $n > 2$).

b) $\underline{O(10)}$. - As we have advocated on more general grounds at the end of Section 3, a minimal supersymmetric model based on $O(10)$ also requires more irreps of Higgs scalars than the non-supersymmetric versions [21-25]. We will discuss the case with the (Higgs) chiral multiplets in the reducible representation : $\underline{54}+\underline{45}+\underline{16}+\underline{\overline{16}}+\underline{10}$. The D^2-term vanishes if one takes the solutions discussed in Section 2 for the real irreps and v.e.v.'s for the $\underline{16}$ and $\underline{\overline{16}}$ with the same little group and equal strengths The superpotential has the following types of invariants : $(\underline{54})^2$, $(\underline{54})^3$, $(\underline{45})^2$, $(\underline{54}\times\underline{45}\times\underline{45})$, $(\underline{16}\times\underline{\overline{16}})$, $(\underline{16}\times\underline{\overline{16}}\times\underline{45})$, $(\underline{10})^2$, $(\underline{10}\times\underline{16}\times\underline{16})$, $(\underline{10}\times\underline{\overline{16}}\times\underline{\overline{16}})$. There is a supersymmetric solution where the v.e.v. of the $\underline{10}$ is indeed protected, the $\underline{16}$ and $\underline{\overline{16}}$ have SU(5)-invariant v.e.v.'s, the $\underline{54}$ takes v.e.v.'s with little group $O(2k)\times O(10-2k)$ and the v.e.v.'s of the $\underline{45}$ have little group $U(k)\times U(5-k)$, where $k = 1,2$. For $k = 2$, the residual symmetry is then $SU(3)\times SU(2)\times U(1)$, as a consequence of the conspiration between all the irreps. The $\underline{54}$ is needed, otherwise the $\underline{45}$ would choose the SU(5)-invariant direction. The $\underline{16}$ and $\underline{\overline{16}}$ are included to avoid a larger residual symmetry, $O(2k)\times U(5-k)$, $k = 1,\dots,4$. The $\underline{45}$ is also needed in order to link the little groups of the $\underline{54}$ and the $\underline{16}+\underline{\overline{16}}$. The SU(5) invariance of the $\underline{16}+\underline{\overline{16}}$ helps in protecting the $\underline{10}$, which contains the Higgs doublets that break the Weinberg-Salam

$SU(2) \times U(1)$ and give masses to the light fermions once supersymmetry is broken at low energy (the matter supermultiplets are in additional $\underline{16}$'s and appear in the superpotential through invariants of the type $(\underline{10} \times \underline{16} \times \underline{16})$).

c) E_6. - As an illustration, let us consider the supersymmetric version of an E_6 GUT [24,26,27] with the matter supermultiplets in $\underline{27}$'s and the Higgs chiral multiplets only in the representations that couple to matter in the superpotential, i.e., the $\underline{27}$ and the $\underline{351}_S$. The most general Higgs superpotential with these irreps contains the following invariants : $(\underline{27})^3$, $(\underline{351}_S)^3$, $(\underline{351}_S \times \underline{27} \times \underline{27})$. We have already shown that there is no non-trivial solution if only the $\underline{27}$ is present. We find that there is no solution with little group containing $SU(3) \times SU(2) \times U(1)$ if all the invariants appear in the superpotential. On the other hand, if the invariant $(\underline{351}_S)^3$ is absent, the $\underline{27}$ is still protected, but the $\underline{351}_S$ can take the v.e.v. allowed by the D^2-term, with little group $O(6) \times O(4)$, as discussed in Section 2.

It seems that a feature of these superGUTs is that more irreps of chiral supermultiplets are needed than the irreps of Higgs scalars in usual GUTs without supersymmetry, and also that complex irreps require the addition of their conjugates, in order to allow for physically interesting v.e.v.'s. In this case all the particles in the Higgs supermultiplets tend to acquire a large mass.

THE PROBLEM OF SUSY BREAKING

In the previous sections, we have considered the first step of symmetry breaking at superheavy energies. In order to protect the hierarchy, supersymmetry was left unbroken. The second step of symmetry breaking will occur at the weak interaction scale; at this stage, supersymmetry and $SU(2) \times U(1)$ must break.

Spontaneous breaking of supersymmetry has well-known features which render its use in realistic models difficult. It will happen when there is no solution to the vanishing of the scalar potential, Eqs (4) and (5). However D^α is quadratic in the scalar fields, except if the gauge group has a $U(1)$ factor which allows a Fayet-Iliopoulos ξ term [28]. In particular if there is no $U(1)$ factor, the D^2-term will vanish for zero value of the scalar fields. Correspondingly the F^2-term will be zero for vanishing scalar fields if therc is no linear term in the gauge invariant superpotential. Spontaneous supersymmetry breaking might thus occur only if one has either a $U(1)$ factor in the gauge group [28] or gauge singlet chiral multiplets in the theory [29]. We recall that if a semi-simple gauge group is broken at superheavy energies

($\sim 10^{15}$GeV), the effective supersymmetric theory at intermediate scales (far from 1TeV or 10^{15}GeV) will not contain a Fayet-Iliopoulos ξ term : the D^{α}'s in the effective theory correspond to generators of unbroken symmetries and these generators annihilate all superheavy v.e.v.'s [8].

Additional U(1) factors must be chosen anomaly-free and traceless in order to avoid quadratic divergences spoiling the naturalness of the hierarchy. Gauge singlet superfields which break supersymmetry are also useful in some other cases, as was mentioned in Section 3, to obtain a correct pattern of superheavy gauge symmetry breaking.

The main problem with spontaneous supersymmetry breaking is to reconcile the mass spectrum with the known light states [30]. Unwanted light colored scalars or massless fermions are in general difficult to avoid.

Breaking of supersymmetry might also be introduced explicitly, with non-supersymmetric gauge invariant terms in the Lagrangian [9,10]. Restrictions on the possible breaking terms follow from the necessity that the non-renormalizability properties of the theory are not spoiled [31]. In particular, these "soft" terms have to be at most of dimension three : they all carry a mass scale which should be of the TeV order. Apart from supersymmetry breaking it is possible in this way to lift the masses of some unwanted light scalars or fermions up to the TeV scale.

Since the breaking of supersymmetry occurs at the TeV scale, the energy of the non supersymmetric ground state will be of the same order. In our analysis of the gauge symmetry breaking in the supersymmetric limit the unprotected v.e.v.'s were of the order of 10^{15}GeV. The value of the potential in the protected directions was of the same order. The breaking of supersymmetry will then not be able to alter the pattern of superheavy gauge breaking; however, if there are degenerate supersymmetric ground states, the true vacuum might be selected with supersymmetry breaking. For instance, we have shown that the adjoint representation of SU(N) and the second rank symmetric traceless tensor of O(N) give a vanishing potential for v.e.v.'s with little group $SU(k) \times SU(N-k) \times U(1)$ and $O(k) \times O(N-k)$ respectively and with $k \neq N/2$. Adding to the Lagrangian a non-supersymmetric negative term quadratic in these representations will select unambiguously $k = [\frac{N}{2}] + 1$. Then for instance SU(5) will break into $SU(3) \times SU(2) \times U(1)$ with the $\underline{24}$, SU(8) into $SU(5) \times SU(3) \times U(1)$ with the $\underline{63}$ and O(10) into $O(6) \times O(4)$ with the $\underline{54}$. The addition of these non-supersymmetric terms in the Lagrangian allows us to select the most desirable one among the degenerate supersymmetric ground states.

REFERENCES

1. J. Wess and B. Zumino, Nucl. Phys. B70:79 (1974).
   Yu.A. Gol'fand and E.P. Likhtman, JETP Lett. 13:323 (1971).
   D.V. Volkov and V.P. Akulov, Phys. Lett. 46B:109 (1973).
2. For reviews see : P. Fayet and S. Ferrara, Phys. Rep.32C:251 (1977).
   A. Salam and J. Strathdee, Fortsch. Phys. 26:57 (1978).
3. D.Z. Freedman, P. van Nieuwenhuizen and S. Ferrara, Phys. Rev. D13:3214 (1976).
   S. Deser and B. Zumino, Phys. Lett. 62B:335 (1976).
4. For a review see : P. van Nieuwenhuizen, Phys. Rep. 68:189 (1981).
5. J. Wess and B. Zumino, Phys. Lett. 49B:52 (1974).
   J. Iliopoulos and B. Zumino, Nucl. Phys. B76:310 (1974).
   S. Ferrara, J. Iliopoulos and B. Zumino, Nucl. Phys. B77:413 (1974).
   J. Wess and B. Zumino, Nucl. Phys. B78:1 (1974).
   S. Ferrara and O. Piguet, Nucl. Phys. B93:261 (1975).
6. J. Ellis, M.K. Gaillard, L. Maiani and B. Zumino, in : Unification of the Fundamental Particle Interactions", (eds. S. Ferrara, J. Ellis, P. van Nieuwenhuizen), Plenum Press, New York 1980, p. 69.
   J. Ellis, M.K. Gaillard and B. Zumino, Phys. Lett. 94B:343 (1980); LAPP-TH-44/CERN-TH 3152, to be published in Acta Physica Polonica.
   J.-P. Derendinger, S. Ferrara and C.A. Savoy, Nucl. Phys. B188:77 (1981).
7. L. Maiani, Proc. of the 1979 Summer School at Gif-sur-Yvette, p. 3.
8. E. Witten, Nucl. Phys. B188:513 (1981).
9. S. Dimopoulos and H. Georgi, Nucl. Phys. B193:150 (1981).
10. N. Sakai, Zeit. für Phys. C11:153 (1982).
11. S. Dimopoulos and S. Raby, Nucl. Phys. B192:353 (1981).
12. H.P. Nilles and S. Raby, Stanford preprint SLAC-PUB-2743 (1981).
13. M. Dine, W. Fischler and M. Srednicki, Inst. Adv. Study Princeton preprint (1981).
14. S. Weinberg, Harvard preprint HUTP-81/A047 (1981).
15. E. Gildener and S. Weinberg, Phys. Rev. D13:333 (1976).
    S. Weinberg, Phys. Lett. 82B:387 (1979).

16. S. Ferrara and B. Zumino, Nucl. Phys. B79:413 (1974).
A. Salam and J. Strathdee, Phys. Lett. 51B:353 (1974).
17. B. de Wit and D.Z. Freedman, Phys. Rev. D12:2286 (1975).
18. S. Ferrara, L. Girardello and F. Palumbo, Phys. Rev. D20:403 (1979).
19. F. Buccella, H. Ruegg and C.A. Savoy, Phys. Lett. 94B:491 (1980).
20. F. Buccella and H. Ruegg, University of Geneva preprint UGVA-DPT 1981/06-300.
21. H. Georgi and D.V. Nanopoulos, Phys. Lett. 82B:392 (1979); Nucl. Phys. B155:52 (1979); ibid. B159:6 (1979).
22. M.S. Chanowitz, J. Ellis and M.K. Gaillard, Nucl. Phys. B128: 506 (1977).
23. M. Gell-Mann, P. Ramond and R. Slansky, in : "Supergravity" (eds P. van Nieuwenhuizen and D.Z. Freedman) North Holland 1979, p. 315.
24. P. Ramond, Sanibel Symposium, Florida 1979, Caltech preprint 68-709 (unpublished).
25. Q. Shafi, M. Sondermann and C. Wetterich, Phys. Lett. 92B:304 (1980).
26. R. Barbieri and D.V. Nanopoulos, Phys. Lett. 91B:369 (1980).
27. B. Stech, in : "Unification of the Fundamental Particle Interactions" (eds S. Ferrara, J. Ellis and P. van Nieuwenhuizen) Plenum Press 1980, p. 17.
28. P. Fayet and J. Iliopoulos, Phys. Lett. 51B:461 (1974).
29. L. O'Raifeartaigh, Nucl. Phys. B96:331 (1975).
30. P. Fayet, Nucl. Phys. B90:104 (1975); Phys. Lett. 69B:489 (1977).
31. L. Girardello and M.T. Grisaru, Brandeis University preprint (July 1981).

QUANTIZATION AND AUXILIARY FIELDS IN 11 DIMENSIONS

A. Van Proeyen*

CERN, Gèneva, Switzerland

INTRODUCTION

At present, supergravity places its hope on the fact that the N = 8 theory can perhaps unify all fundamental interactions. However, it is difficult to verify this expectation, as the structure of the N = 8 theory is only partially known. We have an on-shell version which has been obtained by reducing the N = 1 theory in 11 dimensions[1]. The structure of a theory can be clarified by looking to the complete off-shell version. An easier way of establishing the off-shell version of the N = 8, d = 4 theory is probably by means of the more simply structured N = 1, d = 11 theory. Moreover, as we heard from M. Duff at this conference[2], the N = 1, d = 11 theory could also be used itself for studying the divergence structure, the other important issue in supergravity.

Using the N = 1 theory in 11 dimensions we could obtain the result that the theory can be quantized without sixth order ghost terms. This result is important in view of the existence of auxiliary fields. If a closed algebra is possible with auxiliary fields, M^i occurring at most quadratic in the Lagrangian ($L \sim M^iM^i$ as usual) then the theory cannot have sixth order ghost terms. Indeed, suppose we have a closed set of auxiliary fields, then the usual quantization procedure applies, and the Lagrangian contains only C^*C terms. Also, the field equations can then be at most quadratic in ghosts ($M \sim C^*C$). Eliminating the auxiliary fields will thus produce at most C^*C^*CC terms. The theories with and without auxiliary fields must be equivalent. Therefore the quantization of the theory without auxiliary fields should have no sixth order ghost terms if we hope for a set of auxiliary fields. So we can get information on the auxiliary field structure from the quantization of the

* Aangesteld navorser, N.F.W.O., Belgium.

on-shell theory. This method will be used for actually obtaining the auxiliaries which produce, after their elimination, the four ghost terms in the Lagrangian - which could be a starting point for finding a complete set of auxiliary fields for the N = 1 theory in 11 dimensions.

First (Section 2) we will give some general results on rigid algebras in 11 dimensions. These have been obtained in collaboration with J.W. van Holten. Section 3 will treat the quantization of the on-shell theory, which covers work done in collaboration with B. de Wit and P. van Nieuwenhuizen[3]. Next we will obtain the auxiliary fields which enter in ghost actions[4] (Section 4). We will conclude with some considerations on other possible auxiliary fields. This last section covers work in progress in collaboration with H. Nicolai.

Our notations are the same as in Ref. 5. Antisymmetrization ($[\ ]$) is always done such that the final weight is 1.

$$\Gamma_{ab} = \gamma_{[a}\gamma_{b]} = \tfrac{1}{2}(\gamma_a\gamma_b - \gamma_b\gamma_a)$$

A,B,... denote general symmetry indices (G,L,Q,...).
$\alpha,\beta,\ldots$ denote spinor indices in 11 dimensions, running from 1 to 32.
i,j,... are indices denoting general fields.
a,b,... are Lorentz indices (1 → 11).
$\mu,\nu,\ldots$ are world indices (1 → 11).

When discussing auxiliary fields, we treat only the linearized theory, and therefore there is no real distinction between the last two types of indices.

RIGID N = 1 ALGEBRAS

We start from the de Sitter or Poincaré (x = 0) algebra.

$$[P_a, P_b] = x^2 M_{ab} \qquad [M_{ab}, M_{cd}] = 4\, M_{[b}{}^{[c}\,\delta^{d]}{}_{a]}$$
$$[M_{ab}, P_c] = 2\, P_{[a}\,\delta_{b]c} \tag{2.1}$$

We then introduce one anticommuting spinor Q_α

$$[M_{ab}, Q] = -\tfrac{1}{2}\,\Gamma_{ab}\, Q \tag{2.2}$$

If we allow no other spinors, then Jacobi identities teach us that

$$[P_a, Q] = \tfrac{1}{2}\, x\, \gamma_a\, Q \tag{2.3}$$

For the $\{Q_\alpha, Q_\beta\}$ anticommutator we allow the most general expression. By definition, this must be symmetric in $\alpha\beta$. In dimensions d = 2, 3, 4 mod 8 a charge conjugation matrix $\mathcal{C}$ can be found such that the symmetric matrices are $(\Gamma^{(i)}_{a_1\ldots a_i}\,\mathcal{C}^{-1})_{\alpha\beta}$ with i (indicating the number of Lorentz indices of the gamma matrix) = 1, 2 mod 4. If d (dimension) is odd, then i = 0 to (d-1)/2 forms already a complete set of gamma matrices (the others are the duals of these). So for d = 11 we can write

$$\{Q_\alpha, Q_\beta\} = (\gamma_a\,\mathcal{C}^{-1})_{\alpha\beta}\, Z^{(1)}_a + \tfrac{1}{2}(\Gamma^{(2)}_{ab}\,\mathcal{C}^{-1})_{\alpha\beta}\, Z^{(2)}_{ab} + \tfrac{1}{5!}(\Gamma^{(5)}_{abcde}\,\mathcal{C}^{-1})_{\alpha\beta}\, Z^{(5)}_{abcde} \tag{2.4}$$

Then we only use Jacobi identities to prove that

$$\begin{aligned} \left[Q, Z^{(i)}_{a\ldots}\right] &= y(-)^i\ \Gamma^{(i)}_{a\ldots}\, Q \\ \left[P, Z^{(i)}\right] &= x\, \{{}^{1}{}_{k}{}^{i}\}\, Z^{(k)} \\ \left[Z^{(i)}, Z^{(j)}\right] &= 2y\, \{{}^{i}{}_{k}{}^{j}\}\, Z^{(k)} \end{aligned} \tag{2.5}$$

where the symbol { } (working also on Lorentz indices) is defined by

$$\Gamma^{(i)}\,\Gamma^{(j)} = \{{}^{i}{}_{k}{}^{j}\}\,\Gamma^{(k)} \tag{2.6}$$

With (2.1) - (2.6) all Jacobi identities are satisfied for arbitrary values of x and y. To arrive at a supergravity theory we want P_a to appear in {Q, Q}. This fixes y as a function of x. The normalization is arbitrary, say

$$Z^{(1)}_a = \tfrac{1}{2} P_a \qquad \text{then} \qquad y = x/4 \tag{2.7}$$

If $x \neq 0$ (de Sitter supergravity) then one has

$$Z^{(2)}_{ab} = \tfrac{1}{2}x\, M_{ab} \tag{2.8}$$

and because $[Q, Z^{(5)}] \neq 0$, $Z^{(5)}$ has to exist.

If x = 0 (Poincaré), then $Z^{(2)}$ and $Z^{(5)}$ are "central charges", (they commute with everything) except for Lorentz transformations, for which they transform as indicated by their indices. In this case they are optional. They can be omitted such that all Jacobi identities remain satisfied. We can still remark that the formulae derived in this section are valid for any dimension d = 2, 3, 4 mod 8. This restriction was necessary to allow the formula (2.4). We included here the de Sitter algebra, because it is very similar to the conformal case. One can start with the conformal algebra, include one Q and one S supersymmetry and continue with only Jacobi identities. One arrives at similar results, although all identities have not yet been checked.

QUANTIZATION OF THE ON-SHELL THEORY

If we take the Poincaré algebra of the previous section (x=0) and allow a $Z^{(2)}$, one can represent it as[1]

$$\delta e_\mu{}^a = \tfrac{1}{2}\bar{\varepsilon}\gamma^a\psi_\mu$$

$$\delta\psi_\mu = D_\mu(\hat{\omega})\varepsilon + \sqrt{2}/288\,(\gamma_\mu\Gamma^{abcd} - 12\,e_\mu{}^{[a}\Gamma^{bcd]})\varepsilon\hat{F}_{abcd} \qquad (3.1)$$

$$\delta A_{\mu\nu\rho} = -\sqrt{2}/8\,\bar{\varepsilon}\Gamma_{[\mu\nu}\psi_{\rho]} + 3\,\partial_{[\mu}Z_{\nu\rho]}$$

ω is as usual the gauge field of the Lorentz rotations (^ denotes covariantization) and is determined by the constraint which converts translations into general co-ordinate transformations

$$R_{\mu\nu}(P) = 0 \qquad (3.2)$$

$Z_{\mu\nu}$ is the parameter of $Z^{(2)}$. One could also use $Z^{(5)}$ instead of of $Z^{(2)}$. Then (3.1) would involve a six index field. The linear theory is then equivalent, but one cannot obtain the non-linear part of the on-shell Lagrangian[6]. By going to the local algebra the commutators have changed. As already mentioned, translations are replaced by general co-ordinate transformations. The corresponding commutators have derivative terms in their right-hand side. Also the gauge fields occur in the right-hand side of the {Q, Q} anticommutator

$$[\delta_G(\xi), \delta_G(\eta)] = \delta_G(\eta\cdot\partial\xi - \xi\cdot\partial\eta)$$

$$[\delta_G(\xi), \delta_L(\lambda)] = \delta_L(\xi\cdot\partial\lambda)$$

$$[\delta_L(\lambda), \delta_L(\varepsilon)] = \delta_L(-4\,\varepsilon^{[m}{}_k\lambda^{n]k})$$

$$[\delta_Z(Z), \delta_G(\xi)] = \delta_Z(\xi\cdot\partial Z^{\mu\nu} - \partial^{[\mu}\xi_\lambda Z^{\nu]\lambda})$$

$$[\delta_Q(\varepsilon), \delta_G(\xi)] = \delta_Q(\xi\cdot\partial\varepsilon)$$

$$[\delta_Q(\varepsilon), \delta_L(\lambda)] = \delta_Q(\tfrac{1}{2}\lambda\cdot\Gamma^{(2)}\varepsilon)$$

$$[\delta_Q(\varepsilon_1), \delta_Q(\varepsilon_2)] = \delta_G(\tfrac{1}{2}\bar{\varepsilon}_2\gamma^\mu\varepsilon_1) + \delta_Q(-\tfrac{1}{2}\bar{\varepsilon}_2\gamma^\nu\varepsilon_1\psi_\nu)$$

$$+ \delta_L \left[\tfrac{1}{2}\, \bar{\varepsilon}_2\, \gamma^{\nu}\, \varepsilon_1\, \hat{\omega}_{\nu}{}^{mn} + \sqrt{2}/288\, \bar{\varepsilon}_2\, (\Gamma^{mnabcd} + 24\delta^{ma}\delta^{nb}\,\Gamma^{cd})\, \varepsilon_1\, \hat{F}_{abcd}\right]$$

$$+ \delta_Z\, (-\tfrac{1}{2}\, \bar{\varepsilon}_2\, \gamma^{\sigma}\, \varepsilon_1\, A_{\sigma\mu\nu} - \sqrt{2}/24\, \bar{\varepsilon}_2 \Gamma_{\mu\nu}\, \varepsilon_1) \qquad (3.3)\ \text{cont.}$$

The last equation is, however, not satisfied on ψ_μ

$$\begin{aligned}
&[\delta_Q(\varepsilon_1),\ \delta_Q(\varepsilon_2)]\ \psi_\mu = \text{as in (3.3)} \\
&+ 1/1152\ (\bar{\varepsilon}_2\gamma^a\varepsilon_1)\ [-29\ \Gamma_{\mu a\nu} + 32\ \gamma_{(\mu}e_{\nu)a} - 187\ \gamma_a\ g_{\mu\nu}]\ R^{\nu} \\
&+ 1/(1152\times 2)(\bar{\varepsilon}_2\ \Gamma^{ab}\varepsilon_1)\ [-7\ \Gamma_{\mu ab\nu} - 68\,\Gamma_{a(\mu}e_{\nu)b} + 29\,\Gamma_{ab}\ g_{\mu\nu} - \\
&\qquad - 422\ e_{\mu a}\ e_{\nu b}]\ R^{\nu} \qquad (3.4) \\
&+ 1/(1152\times 24)(\bar{\varepsilon}_2\ \Gamma^{abcde}\varepsilon_1)\ [-\Gamma_{\mu abcde\nu} + g_{\mu\nu}\ \Gamma_{abcde} + 14\ e_{a(\mu}\Gamma_{\nu)bcde} \\
&\qquad + 68\ e_{\mu a}\ e_{\nu b}\ \Gamma_{cde}]\ R^{\nu}
\end{aligned}$$

with the ψ_μ field equation $R^{\mu} = \Gamma^{\mu\rho\sigma}\ D_\rho\psi_\sigma$.

We will write (3.3) and (3.4) symbolically as

$$[\delta(\xi_1),\ \delta(\xi_2)]\ \phi^i = \delta_C\ (f^C_{AB}\ \xi^B_1\ \xi^A_2) \qquad (3.5)$$

$$[\delta_Q(\varepsilon_1),\ \delta_Q(\varepsilon_2)]\ \psi_{\mu\alpha} = \text{closure} + (\eta_{\mu\nu}^{J})_{\alpha\beta}\ I^{c\ell}{}_{,(\nu\beta)}\ \bar{\varepsilon}_1\ O^J\ \varepsilon_2$$

O^J stands for γ, $\Gamma^{(2)}$, $\Gamma^{(5)}$ (such that the right-hand side is antisymmetric in $1 \leftrightarrow 2$). By comparing with (3.4) we see that

$$(\eta_{\mu\nu}^{J})_{\alpha\beta} = (\eta_{\nu\mu}^{J})_{\beta\alpha}\ , \qquad (3.6)$$

a property which can also be derived quite generally[5,7].

We will now follow the usual procedures to derive the quantum action. All these steps are generally proven. One obtains fourth order ghost terms, but then also sixth order, eighth order,... and in principle the series does not stop[7]. We will immediately narrow down to the special case of N = 1, d = 11, and show how the series stops after the fourth order ghost terms. To be able to define propagators, one adds a gauge fixing term to the Lagrangian

$$I = I^{c\ell} + \tfrac{1}{2}\ F_A\ \gamma^{AB}\ F_B \qquad (3.7)$$

The new action is of course no longer invariant under the gauge transformation. To be able to prove unitarity, Slavnov-Taylor identities,..., one implements the theory with a global invariance

(BRST-invariance) with an anticommuting parameter Λ

$$\delta\phi^i = \delta_A\ (C^A\Lambda)\ \phi^i \qquad (3.8)$$

C^A is a ghost field corresponding to the transformation A. Eq. (3.8) is a gauge transformation, so it leaves $I^{c\ell}$ invariant. However, because of the second term in (3.7) we have

$$\delta I = -\Lambda\, F_A \gamma^{AB}\ \delta_C(C^C)\ F_B \qquad (3.9)$$

To cancel this transformation one adds a ghost action to the theory

$$I\ (2\text{-gh}) = C^{*B}\delta_C(C^C)\ F_B \qquad (3.10)$$

and defines BRST transformations for the antighosts C^* as

$$\delta C^{*B} = \Lambda F_A \gamma^{AB} \qquad (3.11)$$

However, because we added (3.10) to the theory, we also have to consider the effect of $\delta\phi^i$ (3.8) on (3.10).

$$\delta I = C^{*B}\ \delta_A(C^A\Lambda)\ \delta_C(C^C)\ F_B \qquad (3.12)$$

Because of the commutation properties of ghosts (opposite of the corresponding symmetry), (3.12) contains $\frac{1}{2}$ × a commutator on F_B. If we had a closed set of fields, then the first equation of (3.5) would imply

$$\delta I = \tfrac{1}{2}\ C^{*B}\delta_D(f^D_{CA}\ C^A\Lambda C^C)\ F_B \qquad (3.13)$$

Comparison with (3.10) then means that one can obtain complete invariance by defining

$$\delta C^D = -\tfrac{1}{2}\ f^D_{CA}\ C^A\Lambda C^C \qquad (3.14)$$

This is the usual procedure for a closed algebra. Only terms quadratic in ghosts are needed. The non-closure part of (3.5) now generates an extra

$$\delta I = \tfrac{1}{2}\ C^{*B}F_{B,(\mu\alpha)}(\eta^J_{\mu\nu})\ I^{c\ell}_{,\nu\beta}\ \bar{C}^Q\ O^J C^Q\Lambda \qquad (3.15)$$

Because this is proportional to $I^{c\ell}_{,\nu\beta}$ it can be eliminated by an extra ψ transformation in $I^{c\ell}$

$$\delta^*\psi_{\nu\beta} = -\tfrac{1}{2}C^{*B}F_{B,(\mu\alpha)}(\eta^J_{\mu\nu})_{\alpha\beta}\ \bar{C}^Q O^J C^Q\Lambda \qquad (3.16)$$

This $\delta^*\psi_{\nu\beta}$ then also acts on the gauge fixing term and on the ghost action. For the gauge fixing, the general procedure of (3.8) - (3.10) applies. Any transformation of the gauge fixing can be eliminated by new ghost terms: $C^{*A}\delta F_A$. So (3.16) implies a new ghost action

$$I(4\text{-gh}) = \frac{1}{2} C^{*A}F_{A,(\nu\beta)}(-\tfrac{1}{2}) C^{*B}F_{B,(\mu\alpha)}(\eta_{\mu\nu}{}^{g})_{\alpha\beta}\, \bar{C}^{Q}{}_{O}{}^{J}C^{Q} \tag{3.17}$$

The compensation by (3.11) occurs twice in the four-ghost action (3.17). Therefore we included an extra factor $\frac{1}{2}$.

Before proceeding we give the example where ψ_μ appears only in F_Q, and F_Q has its usual form

$$F_Q = \gamma \cdot \psi \qquad \text{or} \qquad F_{Q,\mu} = \gamma_\mu \tag{3.18}$$

In this case the four-ghost Lagrangian is explicitly

$$L(4\text{-gh}) = -1/4608 \left[245\, \bar{C}^*\gamma^a C^*\, \bar{C}\gamma_a C + 43/2\, \bar{C}^*\Gamma^{ab}\, C^*\, \bar{C}\,\Gamma_{ab}C + \right.$$
$$\left. + 1/24\, \bar{C}^*\Gamma^{abcde}C^*\, \bar{C}\,\Gamma_{abcde}C\right] \tag{3.19}$$

All ghosts in (3.19) are supersymmetry ghosts.

Which transformations have not yet been included?

a) $\delta^*\psi$ in $I(2\text{-gh})$

b) $\delta^*\psi$ in $I(4\text{-gh})$

c) $\delta\phi^i$ in $I(4\text{-gh})$

d) δC in $I(4\text{-gh})$

Transformations b) seem unnecessary in (3.19) as this does not contain ψ. This is a general fact if the gauge fixings are not quadratic in ψ. So if we restrict our gauge choices as such, then we are left with a), c) and d). All of these are of the form

$$\delta I = -\frac{1}{4} C^{*A}F_{A,i}\, C^{*B}\, F_{B,j}\, B^{ij} \tag{3.20}$$

B is proportional to the non-closure function. Contribution c) includes $\delta\eta$. Comparison with (3.5) shows that this is a part of a Jacobi identity. Indeed, one can show that in case of structure functions and non-closure the Jacobi identities read

$$\delta_E\left[(-f^E_{CD}\, f^D_{BA} + \delta_A f^E_{CB})(C^A \Lambda C^B C^C(-)^C)\right]\, \phi^i + I^{c\ell}{}_{,j}B^{ji} = 0 \tag{3.21}$$

The multiplication with the ghosts provides the cyclicity. B is the same expression as in (3.20). If we denote the quantity in [ ] in (3.21) by A^E then the on-shell Jacobi identities read

$$\delta_E(A^E)\, \phi^i = 0 \tag{3.22}$$

Until now the definition of summation over all symmetries has not been clearly fixed. We will now state that this sum includes all linear independent on-shell symmetries. The inclusion of all on-shell symmetries means:

$$\text{If } I^{c\ell}{}_{,j}\, f^j = 0 \text{ and } f^j \text{ does not vanish on-shell then} \quad f^j = \delta_A(\xi^A)\phi^j \tag{3.23}$$

The independence of symmetries implies:

$$\text{If for all } i\ \delta_A(\xi^A)\ \phi^i = 0 \text{ then } \xi^A = 0 \tag{3.24}$$

Because of (3.24), the on-shell equation (3.22) implies $A^E = 0$ on-shell, so generally

$$A^E = I^{c\ell}{}_{,j}\ A^{jE} \tag{3.25}$$

To find out which E appears in (3.25) we scan all structure functions. The only place where a field equation occurs is in $\delta_Q\ f^L_{QQ} \simeq \delta_Q\ (\hat{\omega}+\tilde{F}) \simeq I^{c\ell}{}_{,(\mu\alpha)}$. Therefore the E can only denote a Lorentz transformation, and the Jacobi identity (3.21) reads

$$I^{c\ell}{}_{,j}\left[\delta_L(A^{jL})\ \phi^i + B^{ji}\right] = 0 \tag{3.26}$$

So we can use (3.23) to obtain

$$\delta_L(A^{jL})\ \phi^i + B^{ji} = \delta_A\ (X^{iA})\ \phi^j \tag{3.27}$$

To justify this, one has to show that the left-hand side is not proportional to a field equation. In A^{jL} a field equation was already extracted [see (3.25)]. B contains only non-closure functions and transformations of the non-closure function. From the explicit form (3.4) we see that this cannot produce field equations. Because of the symmetry of the non-closure function (3.6) B^{ij} is also symmetric. This can be used to show from (3.27) that X = A. Then we can finally use the expresssion for B in (3.20)

$$\begin{aligned}\delta I &= -\tfrac{1}{2}\ C^{*A}F_{A,i}\ C^{*B}F_{B,j}\ \delta_L\ (A^{jL})\ \phi^i \\ &= -\ C^{*A}\delta_L\ (\tfrac{1}{2}C^{*B}F_{B,j}A^{jL})\ F_A\end{aligned} \tag{3.28}$$

This last form makes it clear that a new transformation of the Lorentz ghost in the two-ghost action (3.10) eliminates δI.

$$\delta C^L = \tfrac{1}{2}\ C^{*B}F_{B,j}\ A^{jL} \tag{3.29}$$

In general one should now consider the new ghost transformation on the four-ghost action (3.17). However, in this case the four-ghost action contains only supersymmetry ghosts, while (3.29) concerns only Lorentz ghosts. This means that the construction is finished. The quantum action contains only four-ghost terms and three-ghost terms are present in $\delta\psi_\mu$. As explained in the introduction, these could arise after elimination of auxiliary fields in a fully closed theory. The extra terms $\delta^*\psi_\mu$ indicate that $\delta\psi_\mu$ in the fully closed theory has auxiliary fields whose field equations are quadratic in ghosts.

AUXILIARY FIELDS IN $\delta\psi_\mu$

In this section we will derive those auxiliary fields which occur in $\delta\psi_\mu$, construct the corresponding Lagrangian and derive the quantum action. We will allow a $\delta\psi_\mu$ such that closure is not disturbed on the elfbein. The structure functions are allowed to change. So we allow extra $\delta\psi_\mu$ terms such that

$$\delta_Q e_\mu{}^a = \tfrac{1}{2}\bar{\varepsilon}\,\gamma^a\,\psi_\mu \tag{4.1}$$

$$[\delta_Q(\varepsilon_1),\ \delta_Q(\varepsilon_2)]\, e_\mu{}^a = \Lambda^{ab}\, e_\mu{}^b, \qquad \Lambda^{ab} = -\Lambda^{ba}$$

This means a change in the f^L_{QQ} structure function. The solution to this is

$$\begin{aligned}\delta\psi_\mu{}^{extra} &= -1/72(\Gamma^{\rho\sigma}\varepsilon\, N^{(3)}_{\mu\rho\sigma} + \Gamma^{\rho\sigma\tau}\varepsilon\, N^{(4)}_{\mu\rho\sigma\tau}) \\ &\quad - 1/72\,\gamma_\mu(M^{(0)} + \Gamma^{abc}\, M^{(3)}_{abc} + \Gamma^{abcd} M^{(4)}_{abcd})\,\varepsilon \equiv \beta^i_\mu\,\varepsilon\, M^i\end{aligned} \tag{4.2}$$

We have also written a symbolic form where M^i stands for the N fields as well as the M fields. The {Q, Q} anticommutator is now changed into

$$\begin{aligned}[\delta_Q(\varepsilon_1),\ \delta_Q(\varepsilon_2)] &= \delta_G^{cov}\,(\tfrac{1}{2}\,\bar{\varepsilon}_2\gamma^\nu\varepsilon_1) + \delta_Z(-\sqrt{2}/24\ \bar{\varepsilon}_2\Gamma^{\mu\nu}\varepsilon_1) + \\ &+ \delta_L\,[\sqrt{2}/288\ \bar{\varepsilon}_2\ (\Gamma^{mnabcd} + 24\,\delta^{ma}\delta^{nb}\Gamma^{cd})\,\varepsilon_1\,\hat{F}_{abcd} + \\ &+ \tfrac{1}{2}\,\bar{\varepsilon}_2\ (\gamma_m\beta^i_n + \tilde{\beta}^i_n\gamma_m)\,\varepsilon_1\, M^i] \\ \tilde{\beta} &= C^{-1}\,\beta^T\end{aligned} \tag{4.3}$$

To obtain closure on ψ_μ one has to introduce transformations of M and N to R_μ, such that $\psi_\mu \overset{\varepsilon_1}{\rightarrow} M^i \overset{\varepsilon_2}{\rightarrow} R_\mu$ cancels (3.4). We take the most general form

$$\delta\, M^i = \bar{\varepsilon}\,\alpha^i_\nu\, I^{c\ell}{}_{,\nu} \tag{4.4}$$

The arbitrary α_ν are parametrized by nine arbitrary variables. Closure on ψ_μ means

$$\eta^J_{\mu\nu}\,(\bar{\varepsilon}_1\, O^J\,\varepsilon_2) + \beta^i_\mu\,\varepsilon_2\,\bar{\varepsilon}_1\,\alpha^i_\nu - \beta^i_\mu\,\varepsilon_1\,\bar{\varepsilon}_2\,\alpha^i_\nu = 0 \tag{4.5}$$

This gives 14 equations for the nine parameters. However, it turns out that there is still a one parameter solution.

$$\begin{aligned}
\delta M^{(0)} &= a\,\bar{\varepsilon}\,\gamma\cdot R \\
\delta M^{(3)}_{abc} &= -5/8\,\bar{\varepsilon}\,\Gamma_{[ab}R_{c]} + (1/18 + a/90)\,\bar{\varepsilon}\,\Gamma_{abc}\,\gamma\cdot R \\
\delta M^{(4)}_{abcd} &= -7/24\,\bar{\varepsilon}\,\Gamma_{[abc}R_{d]} + (1/36 - a/360)\,\bar{\varepsilon}\,\Gamma_{abcd}\,\gamma\cdot R \\
\delta N^{(3)}_{abc} &= 27/4\,\bar{\varepsilon}\,\Gamma_{[ab}R_{c]} - 5/8\,\bar{\varepsilon}\,\Gamma_{abc}\,\gamma\cdot R \\
\delta N^{(4)}_{abcd} &= 3\,\bar{\varepsilon}\,\Gamma_{[abc}R_{d]} - 7/24\,\bar{\varepsilon}\,\Gamma_{abcd}\,\gamma\cdot R
\end{aligned} \tag{4.6}$$

Equations (4.2) and (4.6) then give the closure on ψ_μ. However, there is no closure on the auxiliary fields M^i and the contribution $A_{\mu\nu\rho} \overset{\xi_1}{\to} \psi_\mu \overset{\xi_2}{\to} M^i$ now also implies non-closure on $A_{\mu\nu\rho}$ proportional to the $M^i{}_\mu$.

The linearized part of the Cremmer-Julia[1] Lagrangian was

$$L = -\tfrac{1}{2}R - \tfrac{1}{2}\bar{\psi}_\mu R^\mu - 1/48\, F_{abcd}^{\;2} \tag{4.7}$$

The transformations (4.2) undo the invariance of this Lagrangian

$$\delta L = -\bar{R}^\mu \beta^i_\mu\, \varepsilon\, M^i \tag{4.8}$$

Therefore we add a new term quadratic in auxiliary fields

$$L^* = M^i\, x^{ij} M^j \tag{4.9}$$

with arbitrary coefficients x^{ij}. The transformations (4.4) then give

$$\delta L^* = 2M^i\, x^{ij} \bar{R}_\mu \alpha^{jT}_\mu\, \mathcal{C}\, \varepsilon \tag{4.10}$$

The cancellation of (4.8) and (4.10) gives equations for the x^{ij} which turn out to be solvable for the x^{ij}

$$\begin{aligned}
L^* = 1/144\,\Big[\tfrac{1}{a}\, M^{(0)^2} &+ 1/((-1/64)+(3a/40))\,\big((1/18+a/90)N^{(3)^2} \\
&+ (5/4)N^{(3)}\cdot M^{(3)} + 27/4M^{(3)^2}\big) \\
&- 1/(1/576+a/120)\,\big((1/36-a/360)N^{(4)^2} + \\
&+ 7/12N^{(4)}\cdot M^{(4)} + 3M^{(4)^2}\big)\Big]
\end{aligned} \tag{4.11}$$

If we add (4.11) to (4.7) we have again a Lagrangian invariant under (3.1), (4.2) and (4.6).

Is quantization now easier? Let us repeat that we have now closure on $e_\mu{}^a$ and ψ_μ, but not on M^i, and also non-closure on $A_{\mu\nu\rho}$ proportional to M^i. First, we do the usual steps as for a closed system. Equation (3.7) gets the ghost terms (3.10). For example, with the usual gauge fixing (3.18) we get, for the supersymmetry ghosts,

$$L_Q\,(2\text{-gh}) = \bar{C}^* \not{D} C - \sqrt{2}/288\ \bar{C}^*\ \Gamma \cdot \hat{F} C + \bar{C}^*\ \gamma \cdot \beta^i\ C\ M^i \qquad (4.12)$$

But because of non-closure (3.12) had extra contributions. In fact, we see that only the non-closure on the gauge fixing function is important. Furthermore, we will only take gauge fixing functions which are independent of auxiliary fields. This is of course always possible. However, one cannot avoid $A_{\mu\nu\rho}$ in gauge fixings (to fix the Z gauge). Thus the non-closure on F_B is proportional to the auxiliaries, and hence also on their field equations. Let us define

$$[\delta_Q(\epsilon_1),\ \delta_Q(\epsilon_2)]\ F_B = \text{closure terms} + (\bar{\epsilon}_1 O^J \epsilon_2)\ \eta_{Bi}^{\ J}\ x^{ij} M^j \qquad (4.13)$$

Then we get

$$\delta I = \tfrac{1}{2} C^{*B} \eta_{Bi}^{\ J}\ x^{ij} M^j (\bar{C}^Q O^J C^Q)\Lambda \qquad (4.14)$$

So this is again proportional to a field equation and can be cancelled by a $\delta^* M^i$ in $I^{c\ell}$

$$\delta^* M^i = -1/4\ C^{*B} \eta_{Bi}^{\ J}\ (\bar{C}^Q O^J C^Q)\Lambda \qquad (4.15)$$

In Section 3 the corresponding $\delta^*\psi$ was applied on the gauge fixing, which gave rise to the four-ghost action (3.17). However, here we already mentioned that M^i does not occur in gauge fixings. Therefore, no four-ghost terms are generated. We still have to apply $\delta^* M^i$ on the two-ghost terms

$$\delta I = -1/4\ C^{*A}\ F_{A,\mu} \beta^i_\mu\ C^Q\ C^{*B} \eta_{Bi}^{\ J}\ \bar{C}^Q O^J C^Q \Lambda \qquad (4.16)$$

Now we are at the point where we applied Jacobi identities in Section 3. Equation (4.16) is contribution a). Contributions b), c) and d) do not occur here. Because (4.16) is again not proportional to a field equation, we can apply the same arguments. However, this time we do not have to scan the structure functions to find which symmetry indices occur in A^E. In Section 3 this was important to see that no symmetry occurs for which the ghost appears in the four-ghost action. Now we have no four-ghost action so we need only that δI can be cancelled by some transformations δC.

To show consistency of this section with Section 3, one should prove that elimination of the auxiliary fields produces again the same four-ghost Lagrangian and three-ghost $\delta^*\psi_\mu$.

The complete field equations read

$$2\,x^{ij}M^j + C^{*A}F_{A,\mu}\,\beta^i_\mu\,C^Q = 0 \tag{4.17}$$

So, elimination of the M^i in

$$\delta\psi_\mu = \beta^i_\mu\,C^Q\Lambda\,M^i \tag{4.18}$$

produces $C^{*A}C^QC^Q$ terms. In the action the terms

$$I \sim MM + C^{*A}F_{A,\mu}MC^Q$$

produce $C^{*A}C^{*B}C^QC^Q$ terms. In fact one can prove generally that if i) there is a solution to the closure equations (4.5) and ii) we can find an invariant Lagrangian, so there is a solution for the x^{ij} such that (4.8) and (4.10) cancel

$$\beta^i_\mu = 2\,x^{ij}\alpha^{jT}_\mu\mathcal{C}\ , \tag{4.19}$$

then we get again the same four ghost action[4]. We will prove here explicitly that $\delta^*\psi_\mu$ (3.16) is also reobtained. We first insert (4.19) in (4.18). Then we can eliminate the auxiliaries with (4.17) to get

$$\delta\psi_{\nu\beta} = -C^{*A}F_{A,\mu\alpha}(\beta^i_\mu)_{\alpha\varepsilon}\,C^\varepsilon_Q\,(\alpha^i_\nu)_{\delta\beta}\,\mathcal{C}_{\delta\gamma}\,C_{Q\gamma}\,\Lambda \tag{4.20}$$

The closure equation (4.5) can be written as

$$(\beta^i_\mu)_{\alpha\varepsilon}\,C_{Q\varepsilon}\,C_{Q\gamma}\,\mathcal{C}_{\gamma\delta}\,\alpha_{\nu\delta\beta} = \tfrac{1}{2}(\eta^J_{\mu\nu})_{\alpha\beta}\,\bar{C}_Q O^J C_Q \tag{4.21}$$

(We used the commuting properties of the ghosts and $\bar{\varepsilon}_\alpha = \varepsilon_\beta\,\mathcal{C}_{\beta\alpha}$.) Finally, using the antisymmetry of the charge conjugation $\mathcal{C}$, (4.21) allows us to write (4.20) in the form (3.16).

OTHER AUXILIARY FIELDS?

In looking for other auxiliary fields, we can use the same method as for obtaining the M^i. We now want to preserve the closure on ψ_μ. So we can only allow transformations of M^i which do not contribute to the commutator on ψ_μ, or give a ∂_μ derivative:

$$[\delta_Q(\varepsilon_1),\ \delta_Q(\varepsilon_2)]\ \psi_\mu = \delta_\mu(f^Q_{QQ}\varepsilon_1\varepsilon_2) \tag{5.1}$$

Terms which give a ∂_μ derivative change the f^Q_{QQ} structure function. This can indeed occur. The M^i transform in the derivative of a scalar-spinor η. This η contributes to the structure functions. We will now generally investigate into which spinors the M^i can transform without affecting the commutator on ψ_μ. Antisymmetric bosons can only transform in antisymmetric fermions. In principle each boson transforms into different fermions. For example, $M^{(4)}_{abcd}$

transforms into a $\xi^{(0)}$, $\xi^{(1)}_a$, $\xi^{(2)}_{ab}$, $\xi^{(3)}_{abc}$, $\xi^{(4)}_{abcd}$ where each $\xi^{(i)}$ is an irreducible representation of the Lorentz group.

$$\gamma^a \xi^{(i)}_{a\ldots} = 0 \tag{5.2}$$

The closure on ψ_μ relates the $\xi^{(i)}$ appearing in the different M^i, and it turns out that only one $\xi^{(0)}$, one $\xi^{(2)}_{ab}$ and one $\xi^{(3)}_{abc}$ are possible. [The $\xi^{(0)}$ was expected because it is the same degree of freedom as the parameter a in (4.6).] Finally, if $A_{\mu\nu\rho}$ transforms in a spinor $\lambda_{\mu\nu\rho}$ (which is necessary to allow closure on $A_{\mu\nu\rho}$), then this again disturbs the closure on ψ_μ. So this must be compensated by M^i transformations into

$$\Lambda_{\mu\nu\rho\sigma} = \partial_{[\mu}\lambda_{\nu\rho\sigma]} \tag{5.3}$$

All these transformations can be visualized in the following scheme

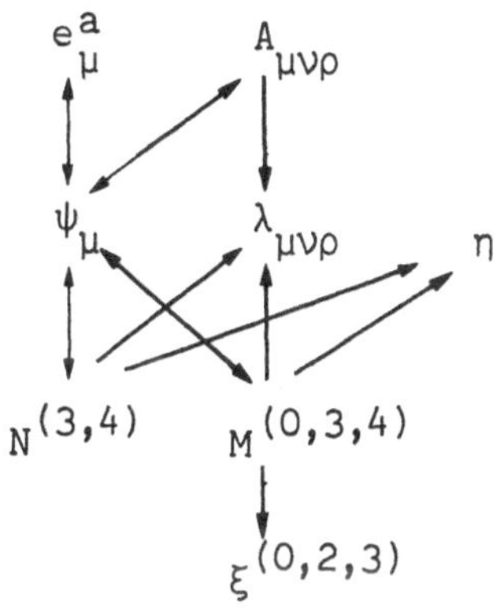

(5.4)

The fields are placed in rows according to their dimension. This implies that an arrow upwards involves a derivative in the corresponding transformation. The explicit form of the new transformations is

$$\begin{aligned}
\delta A_{\mu\nu\rho} &= -\sqrt{2}/8\ \bar{\varepsilon}\ \lambda_{\mu\nu\rho} \\
\delta N^{(3)}_{abc} &= -81\ \bar{\varepsilon}\ \Gamma_{[ab}\partial_{c]}\ \eta + 6\ \bar{\varepsilon}\ \Gamma_{abc}\ \partial\!\!\!/\eta \\
\delta N^{(4)}_{abcd} &= -12\ \bar{\varepsilon}\ \Gamma_{[abc}\partial_{d]}\ \eta + \bar{\varepsilon}\ \Gamma_{abcd}\ \partial\!\!\!/\eta + 18\ \bar{\varepsilon}\ \Lambda_{abcd} \\
\delta M^{(0)} &= \bar{\varepsilon}\ \xi^{(0)} \\
\delta M^{(3)}_{abc} &= 6\ \bar{\varepsilon}\ \Gamma_{[ab}\partial_{c]}\ \eta - 3/5\ \bar{\varepsilon}\ \Gamma_{abc}\ \partial\!\!\!/\eta + \\
&\quad + 1/90\ \bar{\varepsilon}\ \Gamma_{abc}\xi^{(0)} + 6\ \bar{\varepsilon}\ \gamma_{[a}\xi^{(2)}_{bc]} - \bar{\varepsilon}\ \xi^{(3)}_{abc}
\end{aligned} \tag{5.5}$$

$$\delta M^{(4)}_{abcd} = \bar{\varepsilon}\,\Gamma_{[abc}\partial_{d]}\,\eta - 1/10\,\bar{\varepsilon}\,\Gamma_{abcd}\,\partial\!\!\!/\eta - 3/2\,\bar{\varepsilon}\,\Lambda_{abcd} - $$

$$- 1/360\,\bar{\varepsilon}\,\Gamma_{abcd}\xi^{(0)} + \bar{\varepsilon}\,\Gamma_{[ab}\xi^{(2)}_{cd]} + \bar{\varepsilon}\,\gamma_{[a}\xi^{(3)}_{bcd]} \qquad (5.5)\ \text{cont.}$$

The next step could then be the determination of the fermionic transformations $\delta\lambda$, $\delta\eta$, $\delta\xi$ from closure on the bosonic fields A, N and M.

We now address the question of the uniqueness of the transformations. First of all one can of course always redefine fields in (5.5). This also includes the possibility that, for example, η could be proportional to $\lambda^{(0)}$ (the scalar-spinor irreducible part of $\lambda_{\mu\nu\rho}$). Furthermore, not all fields are absolutely necessary. This corresponds to the constraints in the superspace approach. For example, as far as we have calculated up to now, $M^{(0)}$ could be omitted. The "constraint" $M^{(0)} = 0$ also eliminates $\xi^{(0)}$ and puts a = 0. The fermionic fields λ, η, ξ are all the possible fields at this level of the calculations, but up to now we do not know which are necessary. More important is our restriction that we did not change the elfbein transformation (4.1). If we run into problems at later stages of the calculations, this is the point we have to return to. This transformation is the only one where we have not been general. In fact, if necessary, we could add more terms to the elfbein transformation, e.g.,

$$\begin{aligned} \delta e^a_\mu &= \bar{\varepsilon}\,\chi^{ab}\,e_{\mu b} \\ \chi^{ab} &= \chi^{ba};\ \gamma_a\chi^{ab} = \chi^a{}_a = 0 \end{aligned} \qquad (5.6)$$

this would allow a whole new class of auxiliary fields. The other non-uniqueness of our approach is the possible addition of extra symmetries. If these enter in the {Q, Q} commutation relation they can allow possibilities which are presently forbidden. This could in the end result in smaller multiplets, in the same way as central charges accomplish this in higher N theories. These extra symmetries can eventually exist only locally and not in the rigid algebra. This occurs if the corresponding structure functions are proportional to fields which vanish on-shell. This situation has been found in conformal N = 2 supergravity, where central charges are only possible in the local algebra[8].

We finally remark that conformal supergravity can be important here too. Even if no action exists for the conformal theory in d = = 11 (it would probably involve spins higher than two), it can still simplify the transformations. Comparing (4.2) to the gravitino transformations of N = 1 and N = 2 in four dimensions, it is tempting to write the second line as an S supersymmetry transformation

$$\delta\psi_\mu = -\gamma_\mu\,\varepsilon_S \qquad (5.7)$$

Breaking of conformal symmetry with a compensating multiplet[9] (involving higher spins which are now auxiliary fields) could result in a "decomposition law"

$$\varepsilon_s = 1/72\ (M^{(0)} + \Gamma^{abc} M^{(3)}_{abc} + \Gamma^{abcd} M^{(4)}_{abcd}\ )\ \varepsilon \tag{5.8}$$

This presumption is supported by the fact that we already obtained the λ and η transformation in the M^i

$$\delta\lambda_{abc} = \Gamma_{abc}\ \varepsilon_s;\ \delta\eta = 3/4\ \varepsilon_s \tag{5.9}$$

Also, only the $M^{(0,3,4)}$ transform in $\dot{\xi}^{(0,2,3)}$, such that these last fields could belong to the compensating multiplet.

All this indicates that we have just discovered a small section of a very rich structure which the 11 dimensional off-shell theory possesses.

REFERENCES

1. E. Cremmer, B. Julia and J. Scherk, Phys. Lett. 76B:409 (1978); E. Cremmer and B. Julia, Phys. Lett. 80B:48 (1978); Nucl. Phys. B159:141 (1979).
2. M. Duff, contribution to these proceedings.
3. B. de Wit, P. van Nieuwenhuizen and A. Van Proeyen, Phys. Lett. 104B:27 (1981).
4. A. Van Proeyen, Ghost interactions from auxiliary fields in d = 11 supergravity, CERN preprint TH-3151 (1981) to be published in Nucl. Phys. B.
5. P. van Nieuwnehuizen, Phys. Reports 68:189 (1981).
6. H. Nicolai, P.K. Townsend and P. van Nieuwenhuizen, Lettere Nuovo Cimento 30:315 (1981).
7. E.S. Fradkin and T.E. Fradkin, Phys. Lett.72B:343 (1978); B. de Wit and J.W. van Holten, Phys. Lett. 79B:389 (1978); J.W. van Holten, University of Leiden Ph.D. Thesis (May 1980).
8. B. de Wit, J.W. van Holten and A. Van Proeyen, Phys. Lett. 95B:51 (1980); Nucl. Phys. B184:77 (1981).
9. B. de Wit, contribution to these proceedings.

GEOMETRICAL CONSTRAINTS ON SUPERGRAVITY COUPLING

K.S. Stelle

Laboratoire de Physique Théorique
de l'Ecole Normale Supérieure
24 rue Lhomond, 75231 Paris Cedex 05 (France)

INTRODUCTION

The various geometrical formalisms for supergravity underline the similarities and differences with general relativity. Unlike the straightforward differential geometry which is appropriate for the description of Einstein's theory, the description of supergravity requires the imposition of constraints upon the connections and vielbeins in superspace. Ultimately, when these constraints are solved, a new geometry emerges in terms of the unconstrained "prepotentials", which is more akin to the view of general relativity as a non-linear realization of GL(4, R) than to the standard differential geometry. The constraints that must be imposed are not unique, and the various choices correspond to the various possible sets of auxiliary fields when the theory is expressed in components. Aside from the mathematical aspects, this choice is important also for the applications of supergravity, for the possibilities of coupling to matter fields are strongly influenced by the choice of auxiliary fields.

In the following, we shall see that the minimal[1] set of auxiliary fields is the most versatile in that although they imply conformally invariant couplings in the kinetic terms of scalar multiplet matter fields, they permit arbitrary scalar multiplet self-interactions. The non-minimal[2] set and the new minimal set,[3] while offering different possibilities for the coefficients of improvement terms in the kinetic part of the matter action, share the requirement that scalar multiplet self-interactions possess a U(1) symmetry that in the flat space limit corresponds to an R-symmetry.[4] The coefficients of the improvement terms in the kinetic action for a given scalar multiplet depend upon its charge under the R-symmetry in that case.

In the new minimal case, there is actually a local U(1) symmetry. The construction of the supergravity Lagrangian in this case is complicated by the fact that the usual volume of superspace $\int d^4x d^4\theta E$ vanishes. At the same time, in this case alone is there the possibility of coupling the Fayet-Iliopoulos term of a U(1) vectormultiplet to supergravity in a direct generalization of the flat-space expression $\int d^4x d^4\theta V$. This leads to the correct way to write the supergravity action in this case, which takes the form of the Fayet-Iliopoulos term, but for the supergravity theory's own local U(1). The work presented in this article was done in collaboration with P.S. Howe and P.K. Townsend.

1. LINEARIZED SUPERGRAVITY-MATTER COUPLINGS

In superspace, the differential geometric formulation[5] of supergravity starts with inverse vielbeins $E_A{}^M$, with tangent space indices A acted upon by the local Lorentz group and world indices M acted upon by the group of general coordinate transformations in superspace, $Z^M = (x,\theta)$. Adding the connections $\Omega_A{}^{bc}$, we construct the covariant derivatives, torsions and curvatures :

$$\nabla_A = E_A{}^M \partial_M + \Omega_A{}^{bc} M_{bc} \tag{1}$$

$$[\nabla_A, \nabla_B\} = -T_{AB}{}^C \nabla_C - R_{AB}{}^{cd} M_{cd} \tag{2}$$

$A = \alpha, \dot{\alpha}$ (spinor), a(Bose)

where the connections and curvature components are contracted with local Lorentz generators M_{cd} that act on both boson and fermion fields.

If one were to try to work directly with the above formalism, it would not be possible to write the supergravity action, nor would it be possible to couple to chiral superfields. Moreover, there is no necessity to keep the connections as independent variables, for, as in general relativity, they may be determined by the vielbeins through conventional constraints on the torsions. In addition, the vielbein components $E_a{}^M$ can be determined in terms of $E_\alpha{}^M$, $E_{\dot{\alpha}}{}^M$. Collectively, the integrability conditions that allow for coupling to chiral matter fields taken together with the conventional constraints are[6,7]

$$\begin{aligned} T_{\alpha\beta}{}^a &= T_{\alpha\beta}{}^{\dot{\gamma}} = T_{\alpha\beta}{}^{\gamma} = T_{ab}{}^c = 0 \\ T_{\alpha\dot{\beta}}{}^a &= -i\,\sigma_{\alpha\dot{\beta}}^a \\ T_{\alpha\dot{\beta}}{}^{\dot{\gamma}} &= -\delta_{\dot{\beta}}{}^{\dot{\gamma}}\, T_\alpha \\ T_{\alpha b}{}^c &= \delta_b{}^c\, T_\alpha \end{aligned} \tag{3}$$

The above constraints actually suffice for the description of conformal supergravity, which is the generalization of the Weyl invariant theory $\int d^4x \sqrt{-g}(C_{\mu\nu\alpha\beta})^2$. For Poincaré supergravity, there can only be an action giving non-trivial field equations if we impose a further constraint. It is at this stage that a choice enters. The minimal choice is

$$T_\alpha = 0 . \tag{4}$$

In this case, the theory will turn out to have the minimal set of six Bose auxiliary fields A_m,M,N. The general non-minimal choice[8] is

$$\begin{aligned} &R^* - \zeta\nabla^\alpha T_\alpha - 2\zeta^2 T^\alpha T_\alpha = 0 \\ &R^* := \tfrac{1}{6} R_{\alpha\beta}{}^{\alpha\beta} . \end{aligned} \tag{5}$$

In this case, there are 14 Bose auxiliary fields A_m,M,N,V_m,U_m and 8 Fermi auxiliary fields λ_α, χ_α. The special value $\zeta=1$ describes the new minimal case, where 8 Bose and 8 Fermi fields drop out of the theory, leaving just A_m and V_m, subject to the constraint $\partial^m V_m = 0$ and with A_m becoming a gauge field for a U(1) local symmetry. The constraints (3) taken with either (4) or (5) (but not both) do not imply any field equations.

In order to see the consequences of the various choices for matter coupling, we consider the theory in its linearized limit. The vielbeins can be expanded about their flat-superspace values $\overset{(0)}{E}{}_A{}^M$:

$$\begin{aligned} &\overset{(0)}{E}{}_a{}^m = \delta_a{}^m \\ &\overset{(0)}{E}{}_\alpha{}^\mu = \delta_\alpha{}^\mu \qquad \overset{(0)}{E}{}_\alpha{}^{\dot\mu} = 0 \\ &\overset{(0)}{E}{}_\alpha{}^m = i\,\sigma^m_{\alpha\dot\beta}\,\bar\theta^{\dot\beta} . \end{aligned} \tag{6}$$

The flat space supersymmetric covariant derivative is

$$D_\alpha := \overset{(0)}{E}{}_\alpha{}^M \partial_M . \tag{7}$$

The linearized supergravity fields then describe the deviations from flat space

$$E_A{}^M = \overset{(0)}{E}{}_B{}^M(\delta_A{}^B - \kappa H_A{}^B). \tag{8}$$

After some gauge choices, the constraints (3) can be solved by

$$\begin{aligned} &H_\beta{}^a = i D_\beta h^a \qquad h^a = \sigma^a_{\alpha\dot\beta} h^{\alpha\dot\beta} \\ &H_\alpha{}^\beta = \delta_\alpha{}^\beta A \\ &H_\alpha{}^{\dot\beta} = 0 , \end{aligned} \tag{9}$$

where h^a is a real vector superfield and A is a complex scalar superfield. Then some important quantities are

$$T_\alpha = D_\alpha\{(2A+A^*) - \tfrac{1}{4}[D_\beta,\bar{D}_{\dot\beta}]h^{\beta\dot\beta} - \tfrac{i}{2}\partial_{\beta\dot\beta}h^{\beta\dot\beta}\}$$

$$R^* = D^2 A \tag{10}$$

$$E \equiv \mathrm{Ber}(E_M{}^A) = 1 + 2(A+A^*) - \tfrac{1}{2}[D_\alpha,\bar{D}_{\dot\alpha}]h^{\alpha\dot\alpha} ,$$

where E is the superdeterminant of the vielbeins, the density for full superspace.

The constraints (4) or (5) impose further restrictions on the complex scalar superfield A. In the minimal case, from (4) we find

$$A = \tfrac{1}{12}[D_\alpha,\bar{D}_{\dot\beta}]h^{\alpha\dot\beta} + \tfrac{i}{2}\partial_{\alpha\dot\alpha}h^{\alpha\dot\alpha} + C + 2C^* \tag{11}$$

where C is an arbitrary chiral superfield, $\bar{D}_{\dot\alpha}C = 0$. The chiral superfield C has the role of realizing local superconformal invariance even in the Poincaré theory, by compensation, just as in the form of general relativity with a compensating scalar field $R\phi^2 - 6\partial_\mu\phi\partial^\mu\phi$. C may thus be set to zero by a conformal gauge choice. The only remaining superfield is then $h^{\alpha\beta}$, which contains the graviton field $h_m{}^a$, the gravitino field $\psi_m{}^\alpha$ and the auxiliary fields.

In the non-minimal case, we obtain a constraint on A and A^*, since (5) implies

$$D^2[(1-2\zeta)A - \zeta A^*] = 0 . \tag{12}$$

If $\zeta \neq 1$, this may be solved by

$$A = (1-4\zeta+3\zeta^2)^{-1}[(1-2\zeta)L + \zeta L^*] \tag{13}$$

with $D^2 L = \bar{D}^2 L^* = 0$.
On the other hand, if $\zeta=1$, $A-A^*$ is unconstrained while the real superfield $A+A^*$ satisfies

$$D^2(A+A^*) = 0. \tag{14}$$

In this case, there is a local U(1) symmetry in superspace allowing $(A-A^*)$ to be gauged away, so it must not be present in any supergravity-matter couplings.

To investigate the matter couplings, we must promote the flat superspace action for the Wess-Zumino multiplet to superspace :

$$\int d^4x\, d^4\theta\, E\, \phi^*\phi . \tag{15}$$

The chirality constraint that ϕ satisfies must also be covariantized, and in the general case there is some freedom in how this may be done :

$$(\nabla_\alpha + 2\rho T_\alpha)\,\phi = 0\,. \tag{16}$$

The parameter ρ is the chiral weight of the superfield ϕ. In terms of the linearized solutions to the vielbein constraints (3) given in (9,10), the chiral constraint (16) may be solved in terms of an ordinary flat-space chiral field ϕ_0 :

$$D_\alpha\,\phi_0^* = 0 \tag{17}$$

$$\phi^* = \left[1 + i h^{\alpha\dot\alpha}\partial_{\alpha\dot\alpha} + i\tfrac{\rho}{2}\partial_{\alpha\dot\alpha}h^{\alpha\dot\alpha} - 2\rho(2A + A^*) + \tfrac{\rho}{2}[D_\alpha, \bar D_{\dot\alpha}]h^{\alpha\dot\alpha}\right]\phi_0^* + \cdots\,.$$

Using the solutions (10) and (17), the supergravity coupling to the kinetic term for a chiral superfield is

$$\int d^4x\, d^4\theta\, E\,\phi^*\phi =$$

$$\int d^4x\, d^4\theta\left[\phi_0^*\phi_0 + h^{\alpha\dot\alpha}\left\{(1-2\rho)\bar D_{\dot\alpha}\phi_0^* D_\alpha\phi_0 + i\rho\,\phi_0\overleftrightarrow{\partial}_{\alpha\dot\alpha}\phi_0^*\right\} + 2(1-3\rho)(A+A^*)\,\phi_0^*\phi_0\right] + \cdots \tag{18}$$

In deriving eq. (18), we have not yet used the extra constraint (4) or (5).

In the minimal case, the constraint (4) allows us to express A in terms of $h^{\alpha\dot\beta}$ and C, and the latter can be gauged to zero. Thus we obtain

$$\int d^4x\, d^4\theta\, E\,\phi^*\phi = \int d^4x\, d^4\theta\left[\phi_0^*\phi_0 + \tfrac{1}{3}h^{\alpha\dot\alpha}V_{\alpha\dot\alpha}\right]$$

$$V_{\alpha\dot\alpha} = \bar D_{\dot\alpha}\phi_0^* D_\alpha\phi_0 + i\;\phi_0\overleftrightarrow{\partial}_{\alpha\dot\alpha}\phi_0^*\;, \tag{19}$$

ie., coupling of $h^{\alpha\dot\alpha}$ to the well-known improved supercurrent of the Wess-Zumino model. This supercurrent $V_{\alpha\dot\alpha}$ contains the fully improved energy-momentum tensor $T_{\mu\nu}$ for the scalar and spinor fields, which is traceless if the free field equations are used, since we have not yet put in any self-interaction. It also contains the supersymmetry current satisfying $\gamma^\mu S_\mu = 0$ and a conserved axial U(1) current. Note that the chiral weight ρ does not appear in (19). This is because for the minimal case, $T_\alpha = 0$, so ρ in (16) is irrelevant.

In the non-minimal case generally, the chiral weight of ϕ appears in the coupling, as can be seen from (18), noting that A is not related to $h^{\alpha\dot\alpha}$ by the constraint (12). In the general case, it is not possible to represent the supercurrent by a single superfield $V_{\alpha\dot\alpha}$, since A remains in the formalism (containing further auxiliary and gauge fields). In the new-minimal case, however, for $\zeta = 1$, we have the constraint (14) on $A+A^*$, limiting it to a real, "linear" superfield. There is actually sufficient gauge freedom in the theory to gauge $A+A^*$ to zero, since the gravitational gauge invariances are contained in a spinor superfield, and

$$\delta(A+A^*) = D^\alpha \bar{D}^2 \Lambda_\alpha + h.c. \quad (20)$$

Gauging $A+A^*$ to zero in this way restricts the residual transformations of $h^{\alpha\dot{\alpha}}$, but we may use this gauge to identify the improvement coefficients on the scalar fields, for then $h^{\alpha\dot{\alpha}}$ couples to

$$V'_{\alpha\dot{\alpha}} = (1-2\rho)\bar{D}_{\dot{\alpha}}\Phi_0^* D_\alpha \Phi_0 + \frac{i\rho}{2}\Phi_0 \overleftrightarrow{\partial}_{\alpha\dot{\alpha}} \Phi_0^* . \quad (21)$$

Expanding this supercurrent in components,[9] we find that there are improvement terms $f(\Box\eta_{\mu\nu} - \partial_\mu\partial_\nu) A^2$ in the stress tensors for the scalar and pseudoscalar fields of the multiplet, but f does not take its conformal value $-1/6$. Instead, the result from ref. 9, is

$$f = \frac{-(4-11\rho)}{6(3-8\rho)} . \quad (22)$$

Only for $\rho = 1/3$ does this take the conformal value $f = -1/6$. As we shall see, this corresponds to a conformally invariant self-interaction $\int d^4x d^2\theta \phi^3$ in the flat space limit.

The difference $(A-A^*)$ does not appear in (18) and as one can see from (14), this superfield satisfies no constraint at all. In fact, $(A-A^*)$ will not appear in the gravitational action either, and corresponds to a further gauge invariance that is present in the new-minimal case only.[3] In considering matter couplings, we must continue to require that $(A-A^*)$ can be gauged to zero, thus requiring a U(1) invariance of the matter system that becomes gauged when it is coupled to supergravity. In this case, the only auxiliary fields that cannot be gauged away are A_m and V_m with A_m a component U(1) gauge auxiliary field, and V_m being identically conserved, since it is the field strength for a gauge auxiliary antisymmetric tensor,

$$V_m = \epsilon_{mnrs}\partial^n A^{rs} . \quad (23)$$

2. SELF-INTERACTIONS OF MATTER FIELDS

In flat superspace, self-interactions of chiral superfields must be written using integrals over $d^2\bar{\theta}$:

$$\int d^4x\, d^2\bar{\theta}\, \phi^{*n} + h.c. \quad (24)$$

Another way to view this is to solve the constraint $D_\alpha(\phi^{*n}) = 0$ by $\phi^{*n} = D^2 X$ (since $D^3 \equiv 0$), and take $\int d^4x d^2\theta d^2\bar{\theta} X$ + c.c. In curved space, we must use the following two results :

$$\int d^4x\, d^4\theta\, E(\nabla^\alpha + 2T^\alpha)V_\alpha = 0 , \quad (25)$$

allowing us to integrate by parts, and the identity

$$(\nabla_\alpha + 2kT_\alpha)\nabla^2(k) \equiv 0 \tag{26}$$

$$\nabla^2(k) := \nabla^\beta\nabla_\beta - 2R^* + 2k\nabla^\alpha T_\alpha + 4k^2 T^\alpha T_\alpha + 4kT^\alpha \nabla_\alpha .$$

Thus, in order to have an invariant written locally in terms of some integrand Σ , we must have

$$(\nabla_\alpha + 2T_\alpha)\Sigma = 0 , \tag{27}$$

in which case,

$$\Sigma = \nabla^2(k=1)X = (\nabla^\alpha + 2T^\alpha)(\nabla_\alpha + 2T_\alpha)X - 2R^*X \tag{28}$$

so

$$X = -\frac{\Sigma}{2R^*} - \frac{(\nabla^\alpha + 2T^\alpha)(\nabla_\alpha + 2T_\alpha)X}{R^*} . \tag{29}$$

Using the constraints on the torsions (3), it can be proved[7] that $\nabla^\alpha R^* = 0$, so

$$\int d^4x\, d^4\theta\, EX = \int d^4x\, d^4\theta\, E\left\{-\frac{\Sigma}{R^*} - (\nabla^\alpha + 2T^\alpha)(\nabla_\alpha + 2T_\alpha)\left(\frac{X}{R^*}\right)\right\}$$

$$= -\int d^4x\, d^4\theta\, E\,\frac{\Sigma}{R^*} \tag{30}$$

since the second term vanishes using (24). This result, derived in refs. 8,10 gives the same formula as was known in the minimal case[5,8], but now requiring the new term in T_α in the constraint (26) on Σ. According to the definition of chiral weight given in (16), this means that the integrand Σ must have chiral weight one if we use non-minimal or new minimal auxiliary fields.

The requirement of a definite chiral weight for the integrand of the self-interaction terms is equivalent to requiring that these self-interaction terms have an R-symmetry.[4] In flat space, a superpotential $V(\phi_i)$ has an R-symmetry if there exist a set of charges ρ_i such that if

$$\phi_i \rightarrow e^{i\alpha\rho_i}\phi_i \qquad i = 1 \cdots N$$

then

$$V(\phi_i) \rightarrow e^{i\alpha}V(\phi_i) . \tag{31}$$

In this case, the phase acquired by $V(\phi_i)$ can be cancelled by transforming $\theta^\beta \rightarrow e^{i\alpha/2}\theta^\beta$, so $d^2\theta \rightarrow e^{-i\alpha}d^2\theta$.

In curved space, the R-symmetry is linked to chiral transformations of the supergravity fields rather than to transformations of θ ; in this case, the phase of $V(\phi)$ is cancelled by a phase coming from the chiral measure E/R^*. In the new minimal case, this U(1) symmetry is promoted to a local symmetry, as we have seen earlier in our discussion of the linearized theory, requiring $(A-A^*)$ to be able to be gauged to zero. In the case of minimal auxiliary fields, the constraint required of the integrand Σ is just ordinary chirality, $\nabla_\alpha \Sigma = 0$, without any notion of chiral weight being defined. Thus, if minimal auxiliary fields are used, supergravity may be coupled to arbitrary self-interaction terms for chiral superfields, and R-symmetry is not required.

3 THE VANISHING VOLUME OF SUPERSPACE AND THE ACTION FOR NEW-MINIMAL SUPERGRAVITY

In general, the supergravity action is the only generally covariant object in superspace of the right dimension that can be written in terms of the differential geometric building blocks of the theory, i.e., the volume of superspace[5], $\int d^4x d^4\theta\, E$.

This construction breaks down in the case of the new-minimal formulation of supergravity, however, for in this case the volume of superspace vanishes[10] due to the constraint (5) with $\zeta = 1$. This is easy to see if we consider the full superspace integral of a chiral field of chiral weight zero, $\nabla_\alpha \phi^* = 0$, of which a particular example is a constant chiral field, $\phi^* = 1$. We may solve the chiral constraint on ϕ^* as we did in section 2, but now with the chiral weight $\rho = k = 0$. Thus, we have

$$\phi^* = (\nabla^\alpha \nabla_\alpha - 2R^*)\, X \; . \tag{32}$$

If we also use the constraint (5), we can re-write this as

$$\phi^* = (\nabla^\alpha + 2T^\alpha)(\nabla_\alpha - 2T_\alpha) X - 2(\zeta-1)(\nabla^\alpha T_\alpha + (\zeta+1) T^\alpha T_\alpha) X \tag{33}$$

and taking into account the result (25), this implies

$$\int d^4x\, d^4\theta\, E\, \phi^* = -2(\zeta-1) \int d^4x\, d^4\theta (\nabla^\alpha T_\alpha + (\zeta+1) T^\alpha T_\alpha) X \; , \tag{34}$$

which vanishes for $\zeta = 1$, implying for $\phi^* = 1$ that

$$\int d^4x\, d^4\theta\, E = 0 \; . \tag{35}$$

This problem provides the seed of its own resolution, however. The vanishing of the full superspace integral of a chiral field of weight zero allows one to write the Fayet-Iliopoulos term[4] for a U(1) gauge multiplet in interaction with supergravity in a direct generalization of its flat space expression, as was found originally in a tensor calculus formulation of the new-minimal supergravity. Thus we have simply

$$\int d^4x\, d^4\theta\, E\, V \tag{36}$$

rather than an exponentiated expression requiring compensating conformal transformations, as is the case in the minimal formulation of supergravity.[12] The expression (36) is gauge invariant due to (34) under the gauge transformation

$$V \to V + i(\Lambda - \Lambda^*)$$
$$\nabla_\alpha \Lambda^* = 0 \ . \tag{37}$$

The possibility of writing the Fayet-Iliopoulos term in the form (36) also allows us to construct the superspace action for supergravity.[10] In the new-minimal formulation, the torsion component T_α acts as a connection for the extra U(1) gauge invariance of the theory,

$$\delta T_\alpha = i \nabla_\alpha S \ , \tag{38}$$

where S is a real general scalar superfield.

It is in fact for this reason that the parameter ρ in the definition (16) of a chiral field is called the chiral weight of that field, since T_α acts as a gauge field under the chiral U(1) transformations. Moreover, T_α satisfies the condition, following from the constraints (3) alone[7],

$$\nabla_\alpha T_\beta + \nabla_\beta T_\alpha = 0 \ . \tag{39}$$

This result implies that T_α is itself the superspace gradient of a complex scalar superfield,

$$T_\alpha = \nabla_\alpha T \ , \tag{40}$$

which is confirmed by the linearized result given in (10), where the gauge transformation behaviour of T_α follows from $\delta(A - A^*) \sim i\ S$.

It is the object T rather than one of the geometrical objects of the theory (vielbeins, torsions, etc.) that must be used in constructing the action for supergravity in the new minimal case. Since this construction departs from the standard differential geometric formalism, we must specifically verify that all the required gauge invariances are respected. In particular, we must ensure invariance under the U(1) transformation (38). Since $\delta\ T = i\ S$, with S real, this requires that we construct the supergravity Lagrangian from $(T+T^*)$. However, we must also take into account the fact that T is not completely determined by (40), for an arbitrary chiral superfield may be added to T without changing T_α :

$$T' = T + C^*$$
$$\nabla_\alpha C^* = 0 \ . \tag{41}$$

The arbitrariness (41) in the definition of T amounts to an extra "pregauge" transformation for which invariance must be respected. But this is just the type of invariance that is ensured for the Fayet-Iliopoulos term (36) in the new minimal case. Thus we may take for the supergravity action the expression

$$I = \int d^4x\, d^4\theta\, E(T+T^*)\,. \tag{42}$$

The field equations that follow from this action may be derived using standard techniques ; for details, see ref. 10. The result is the expected field equation for new minimal supergravity,

$$G_{\alpha\dot{\beta}} = 0 \tag{43}$$

where the constraints (3) restrict $T_{\alpha\dot{\alpha},\beta,}{}^{\gamma}$ to $G_{\beta\dot{\alpha}}$:

$$T_{\alpha\dot{\alpha},\beta,\gamma} = -2i\,\epsilon_{\alpha\gamma}\,G_{\beta\dot{\alpha}}\,. \tag{44}$$

Various aspects of the new minimal supergravity have been studied in other recent papers.[11,13-15] Reference 11 presents the tensor calculus formalism, which may equally be used for the above discussion. The linearized superspace action is given in Ref. 13, agreeing with the linearization of (42) upon explicit solution of the constraints (3,5). The solution of the constraints to obtain an unconstrained "prepotential" formulation of the non-linear theory is given in Ref. 14, while the geometrical implications of the new-minimal theory are treated in Ref. 15.

Constraints upon matter couplings that are needed for consistent coupling to supergravity are not restricted to the non-minimal and new minimal formulations. The requirement of an R-symmetry in chiral matter self-interactions also applies when coupling to minimal supergravity if one wishes to include a Fayet-Iliopoulos term to trigger symmetry breaking. Details of these very general constraints are given in Ref. 16.

REFERENCES

1. K.S. Stelle and P.C. West, Phys. Lett. 74B:330 (1978) ; S. Ferrara and P. van Nieuwenhuizen, Phys. Lett. 74B:333 (1978).
2. P. Breitenlohner, Phys. Lett. 67B:49 (1977) ; B. de Wit and M.T. Grisaru, Nucl. Phys. B139:531 (1978).
3. M.F. Sohnius and P.C. West, Phys. Lett. 105B:353 (1981).
4. For a review, see P. Fayet and S. Ferrara, Phys. Rep. 32C:249 (1977).
5. For a review of supergravity in superspace, see B. Zumino in: "Recent Developments in Gravitation," M. Levy and S. Deser, eds., (Plenum Press, N.Y., 1978), p. 405.
6. S.J. Gates, Jr, and W. Siegel, Nucl. Phys. B163:519 (1980).
7. S.J. Gates, Jr, K.S. Stelle and P.C. West, Nucl. Phys. B169:347 (1980).

8. W. Siegel and S.J. Gates, Jr, Nucl. Phys. B147:77 (1979).
9. M. Duff and P.K. Townsend, in: Proc. Nuffield Workshop on Quantum Gravity, C.U.P., (to be published).
10. P.S. Howe, K.S. Stelle and P.K. Townsend, Phys. Lett. 107B:420 (1981).
11. M.F. Sohnius and P.C. West, in: Proc. Nuffield Workshop on Quantum Gravity, C.U.P., (to be published).
12. K.S. Stelle and P.C. West, Nucl. Phys. B145:175 (1978).
13. S. Bedding and W. Lang, Max-Planck-Institute preprint (1981).
14. S.J. Gates, Jr, M. Rocek and W. Siegel, Caltech preprint CALT-68-868 (1981).
15. A. Galperin, V. Ogevetsky and E. Sokatchev, in: Proc. II Intern. Seminar on Quantum Gravity, Moscow (1981).
16. R. Barbieri, S. Ferrara, D.V. Nanopoulos and K.S. Stelle, CERN preprint TH-3243 (1982).

ON KALUZA-KLEIN THEORIES *

J. Strathdee

International Centre for Theoretical Physics
P.O.B. 586, Miramare
34100 Trieste, Italy

ABSTRACT

Some technical aspects are considered for the case in which the internal space of the Kaluza-Klein ground state is a quotient space.

INTRODUCTION

The notion that spacetime may have more than four dimensions, and that the extra dimensions may be associated with internal symmetries in particle physics, goes back to the work of Kaluza[2] and Klein[3]. The original proposal of Kaluza was to interpret electric charge as the generator of motions in a fifth dimension. This was generalized by De Witt[4] to the case of Yang-Mills symmetry by assigning to the extra dimensions the geometry of a compact non-Abelian group. The motivation for studying generally covariant field theories in spacetime of more than four dimensions is to obtain a geometrical interpretation of internal quantum numbers such as electric charge, i.e. to place them in the same context as energy and momentum. The latter observables are associated with translational symmetry in 4-dimensional Minkowski space; the internal observables would be associated with symmetry motions in the extra dimensions.

---

* The work reported in this talk was developed in collaboration with Abdus Salam. A detailed account will be submitted for publication elsewhere[1].

In theories of the Kaluza-Klein type one starts with the hypothesis that spacetime has 4+K dimensions. One assumes general covariance and adopts the 4+K-dimensional curvature scalar as one component of the Lagrangian. There may be additional terms associated with matter fields. Next, one supposes that, owing to some dynamical mechanism, there is a spontaneous compactification[5] of the extra K dimensions in the ground state of this system. The origin of this mechanism is obscure, and its operation may be dependent upon the presence of appropriate matter fields to act as sources to generate the necessary curvature in the extra dimensions. Such matter field sources are not very welcome in the theory since they represent non-geometrical features and so are somewhat contrary to the spirit of the enterprise. However, an interesting exception[6] could be the extended supergravities where certain matter-like fields are justified by an underlying supersymmetry and, possibly, the related geometry of superspace. In any case, one assumes that the ground state has the geometry of $M^4 \times B^K$ where M^4 is 4-dimensional Minkowski space and B^K is a compact K-dimensional manifold. The size of B^K must be sufficiently small to make it unobservable at the currently available energies (i.e. $< 10^{-17}$ cm). The main interest is in the possible symmetries of B^K. These could be expected to show themselves in the patterns of observed particles.

If the extra dimensions are more than a book-keeping device then the excitations associated with them should be considered. Since the excitations correspond to propagation in a compact space, their spectrum of masses is necessarily discrete. Perhaps unfortunately, the scale of these masses tends to be comparable to the Planck mass. The reason for this can be illustrated most simply in the K = 1 theory where B^K is a circle of radius R. The 5-dimensional line element may be assumed to have the form

$$ds^2 = (dx^m\, e_m{}^a(x))^2 - (dy - dx^m\, \kappa\, A_m(x))^2 ,$$

where m,n,a,b,... take the values 0,1,2,3 and $e_m{}^a(x)$ represents the vierbein field on 4-dimensional spacetime. The 4-vector $A_m(x)$ is to be interpreted as the electromagnetic potential. Indeed, the 5-dimensional curvature scalar reduces here to the form

$$\frac{1}{\kappa^2}\, R_4(e) - \frac{1}{4} F^2 ,$$

where $R_4(e)$ is the 4-dimensional curvature scalar and $F_{mn} = \partial_m A_n - \partial_n A_m$. Gauge transformations take the form of co-ordinate transformations,

$$x^m \to x^m \quad , \quad y \to y + \Lambda(x)$$

with

$$e_m{}^a \to e_m{}^a \quad , \quad A_m \to A_m + \frac{1}{\kappa} \partial_m \Lambda$$

which leave invariant the 1-forms $dx^m e_m^a$ and $dy - dx^m \kappa A_m$. The parameter κ which appears in these formulae is of course the Planck length ($\sim 10^{-33}$ cm.). The invariant coupling to a scalar field $\phi(x,y)$ would be described by the Lagrangian

$$\frac{1}{2} (e_a{}^m(\partial_m\phi + A_m\partial_y\phi))^2 - \frac{1}{2} (\partial_y\phi)^2 + \ldots$$

On comparing this to the usual minimal coupling theories in 4 dimensions where the gauge covariant derivative is

$$\partial_m\phi - i\, q\, A_m\, \phi$$

one sees that the electric charge operator, q, must be given by

$$q = i\, \kappa\, \partial_y \, ,$$

the generator of displacements in the new dimension. Since $\partial_y \sim O(1/R)$ it follows that electric charges are proportional to κ/R. To have realistic charges it is necessary to assume that R is comparable to $\kappa \sim 10^{-33}$ cm. Notice further that, because of the term, $(\partial_y\phi)^2$, in the above Lagrangian, the scalar particles will receive mass contributions proportional to their charges,

$$\delta m = i\partial_y = \frac{q}{\kappa} \sim 10^{19} \text{ GeV} \quad .$$

Because of this one is naturally interested to discover - in the non-Abelian generalizations - any exceptional cases where, for reasons of symmetry, the mass contribution is suppressed.

Although the excitations are generally very heavy, they are not without interest. First, it is necessary to examine them in order to test the classical stability of any proposed ground state. Secondly, the short range forces generated by the exchange of these objects are comparable to graviton exchange. They could be important in any discussion of ultraviolet properties of Kaluza-Klein theories.

The purpose of this note is to consider some of the features of Kaluza-Klein theories, particularly, the structure of the normal mode expansions on internal spaces of the form G/H, where G is a compact Lie groups and H is one of its subgroups. A more detailed treatment is in preparation[1].

The generalization to 4+K dimensions of the Kaluza-Klein ansatz for the line element is

$$ds^2 = (dx^m\, e_m{}^a(x))^2 - ((dy^\mu - dx^m\, \kappa A_m{}^{\hat\alpha}(x)\, K_{\hat\alpha}{}^\mu(y))\, e_\mu{}^\beta(y))^2,$$

where the extra co-ordinates are denoted by y^μ, $\mu = 1,2,\ldots,K$. The covariant vectors $e_\mu{}^\beta(y)$, $\beta = 1,2,\ldots,K$ constitute an orthonormal basis on the manifold B^K. The Yang-Mills vectors $A_m{}^{\hat\alpha}(x)$, $\hat\alpha = 1,2,\ldots,\dim G$ are associated with Killing vectors, $K_{\hat\alpha}{}^\mu$, corresponding to the symmetry group, G, of the manifold. This Ansatz[7], which incorporates at least some of the massless excitations of the presumed ground state, can be looked upon as the leading term of a systematic normal mode expansion on B^K. In the following we discuss some properties of the symmetric spaces, G/H, which are useful for constructing such an expansion.

SOME PROPERTIES OF G/H

Let y^μ, $\mu = 1,2,\ldots,K$ label the cosets of G with respect to its subgroup, H. It is convenient to choose a representative element from each coset, the "boost" $L_y \in G$. Multiplication from the left by an arbitrary element, $g \in G$, generally carries L_y into some other coset whose representative element would be $L_{y'}$. Hence there must exist an element $h \in H$ such that

$$g\, L_y = L_{y'}\, h$$

This defines the left translation of G/H, the mapping $y \to y'$ corresponding to g. It also defines the little group element $h = h(y,g)$.

To make geometry on G/H, a covariant basis is needed. For this consider the 1-form

$$e(y) = L_y^{-1}\, d\, L_y$$

which belongs to the Lie algebra of G.

$$\begin{aligned} e(y) &= e^{\hat\alpha}(y)\, Q_{\hat\alpha} \\ &= dy^\mu\, e_\mu{}^{\hat\alpha}(y)\, Q_{\hat\alpha} \end{aligned}$$

where $Q_{\hat\alpha}$, $\hat\alpha = 1,\ldots,\dim G$, are the generators. Under left translation

$$\begin{aligned} e(y) \to e(y') &= h\, L_y^{-1}\, g^{-1}\, d(g\, L_y\, h^{-1}) \\ &= h\, e(y)\, h^{-1} + h\, d\, h^{-1} + h\, L_y^{-1}\, g^{-1}\, dg\, L_y h^{-1}. \end{aligned}$$

In component form this reads

$$e^{\hat{\alpha}}(y') = e^{\hat{\beta}}(y)\, D_{\hat{\beta}}{}^{\hat{\alpha}}(h^{-1}) + (h\, d\, h^{-1})^{\hat{\alpha}} + (g^{-1}\, dg)^{\hat{\beta}}\, D_{\hat{\beta}}{}^{\hat{\alpha}}(L_y\, h^{-1}),$$

where the matrices $D_{\hat{\beta}}{}^{\hat{\alpha}}$ define the adjoint representation of G, i.e.

$$g^{-1}\, Q_{\hat{\beta}}\, g \;=\; D_{\hat{\beta}}{}^{\hat{\alpha}}(g)\, Q_{\hat{\alpha}} \;.$$

It is useful to separate those generators, $Q_{\bar{\alpha}}$ which span the Lie algebra of H from the remainder, Q_α , $\alpha = 1,\ldots,K$ which span the orthogonal complement. With this notation, the coefficients $e_\mu{}^{\hat{\alpha}}$ separate into $e_\mu{}^\alpha$ and $e_\mu{}^{\bar{\alpha}}$. The K × K matrix $e_\mu{}^\alpha$ defines a covariant basis on G/H. Under left translation,

$$e_\mu{}^\alpha(y') \;=\; \frac{\partial y^\nu}{\partial y'^\mu}\; e_\nu{}^\beta(y)\; D_\beta{}^\alpha(h^{-1}) \;.$$

The left invariant metric on G/H is given by

$$g_{\mu\nu}(y) = e_\mu{}^\alpha\; e_\nu{}^\alpha$$

(and is to be used as the ground state expectation value of $g_{\mu\nu}(x,y)$ in the 4+K-dimensional Einstein theory).

It is important to note the tangent space rotation D(h) associated with left translations of the co-ordinates, y^μ . Since $D_\alpha{}^\beta(h)$ is a K × K orthogonal matrix, this defines an embedding of H in SO(K).

EINSTEIN THEORY IN 4 + K DIMENSIONS

The basic variables of a 4+K-dimensional gravitation theory are the 4+K-bein

$$\begin{aligned} E^A &= dz^M\; E_M{}^A(z) \\ &= dx^m\; E_m{}^A + dy^\mu\; E_\mu{}^A \end{aligned}$$

and the spin connection

$$B_A{}^B = dz^M\; B_{MA}{}^B = \ldots$$

The action should be invariant under both the 4+K-dimensional general co-ordinate transformations and the group of frame rotations, SO(1,3+K). To make contact with ordinary physics one assumes that spontaneous compactification yields a ground state with

$$< E_M{}^A(x,y) > = \begin{pmatrix} \delta_m{}^a & 0 \\ 0 & e_\mu{}^\alpha(y) \end{pmatrix} .$$

This implies the ground state invariance:

4-dimensional Poincaré × Yang-Mills group .

The Yang-Mills group consists of the x-dependent left translations of y^μ combined with tangent space rotations $h \in H \subseteq SO(K)$. The embedding of H in SO(K) is given by $D_\alpha{}^\beta(h)$. Infinitesimally,

$$\begin{aligned} D_{\alpha\beta}(h) &= \delta_{\alpha\beta} + \delta h^{\bar{\gamma}} c_{\alpha\gamma\beta} \\ &= \delta_{\alpha\beta} + \omega_{\alpha\beta} , \end{aligned}$$

where $\omega_{\alpha\beta} = -\omega_{\beta\alpha}$ and $c_{\hat{\alpha}\hat{\beta}\hat{\gamma}}$ are structure constants of G. Viewed as an infinitesimal SO(K) transformation this implies

$$\frac{1}{2} \omega_{\alpha\beta} \Sigma^{\alpha\beta} = \delta h^{\bar{\gamma}} Q_{\bar{\gamma}}$$

with SO(K) generators $\Sigma^{\alpha\beta}$, i.e.

$$Q_{\bar{\gamma}} = \frac{1}{2} c_{\alpha\bar{\gamma}\beta} \Sigma^{\alpha\beta} .$$

This implies, for example, that the K-vector of SO(K) has the H-content of G/H.

The ground state values of the spin connection, B, are generally model dependent. If zero torsion is assumed then

$$< B_a{}^b > = < B_a{}^\beta > = 0$$

$$< B_\alpha{}^\beta > = e^\gamma \frac{1}{2} c_{\gamma\alpha}{}^\beta + e^{\bar{\gamma}} c_{\bar{\gamma}\alpha}{}^\beta .$$

The internal space generally has a non-vanishing curvature.

EXPANSIONS

To obtain an effective Lagrangian in 4 dimensions one could expand the fields in harmonics of G/H and integrate over y^μ using their orthogonality properties. For a function defined on a compact manifold, G, the expansion would include all matrix elements of all irreducible unitary representations of G,

$$\phi(g) = \sum_n \sum_{pq} \sqrt{d_n} \, D_{qp}^n(g) \, \phi_{pq}^n ,$$

where d_n = dimensions of D^n. For functions on G/H, the expansion is subject to some restrictions. Suppose that

$$\phi_i(hg) = \mathbb{D}_{ij}(h)\,\phi_j(g)\ ,$$

where $\mathbb{D}(h)$ is a particular representation of H. Such functions are essentially defined on G/H since

$$\phi_i(g_1) = \mathbb{D}_{ij}(g_1 g_2^{-1})\,\phi_j(g_2)$$

for g_1 and g_2 in the same coset. The representations $\mathbb{D}$, appropriate to a given case are determined by the reduction of the tangent space group SO(1,3+K) with respect to SO(1,3) × H - making reference to the embedding of H in SO(K). Thus, the Yang-Mills transformations which leave the ground state invariant act on the fields $\psi_i(x,y)$ such that

$$\psi_i(x,y) \to \psi_i'(x,y') = \mathbb{D}_{ij}(h)\,\psi_j(x,y).$$

For each irreducible $\mathbb{D}$ the appropriate expansion is

$$\psi_i(x,y) = \sum_n \sqrt{\frac{d_n}{d_{\mathbb{D}}}}\ \sum_{\zeta,p} D^n_{i\zeta,p}(L_y{}^{-1})\,\psi^{\,n}_{p\zeta}(x)\ ,$$

where the sum includes all those representations, D^n, of G which contain $\mathbb{D}$ on restriction to H. (If $\mathbb{D}$ is contained more than once in D^n then a supplementary label, ζ , is needed to distinguish the states. The rows of D^n are here labelled so as to emphasize its H-content. $d_{\mathbb{D}}$ = dimension of $\mathbb{D}$.) The coefficient fields are irreducible with respect to G,

$$\psi^n_{p\zeta}(x) \to \psi'^n_{p\zeta}(x) = D^n_{pq}(g)\,\psi^n_{q\zeta}(x)\ .$$

To integrate out the y-dependence of the 4+K-dimensional action, use orthogonality properties such as

$$\int_{G/H} d^Ky \det e(y) \sum_i D_{p,i\zeta}(L_y)\,D_{i\zeta',p'}(L_y^{-1}) =$$

$$= \delta_{nn'}\,\delta_{pp'}\,\delta_{\zeta\zeta'}\,\frac{d_{\mathbb{D}}}{d_n}\,V_K\ ,$$

where V_K = volume of G/H.

In the expansion of the 4+K-bein $E_M{}^A(x,y)$, one may choose to retain only the leading terms

$$E_m{}^a(x,y) = E_m{}^a(x) + \ldots$$

$$E_m{}^\alpha(x,y) = A_m{}^{\hat\beta}(x)\, D_{\hat\beta}{}^\alpha(L_y) + \ldots$$

$$E_\mu{}^\alpha(x,y) = e_\mu{}^\alpha(y) + \ldots$$

The 4+K-dimensional curvature scalar then reduces to

$$R_{4+K} \sim R_4 - \frac{1}{4} F_{ab}{}^{\hat\alpha} F_{ab}{}^{\hat\beta} D_{\hat\alpha}{}^\gamma(L_y)\, D_{\hat\beta}{}^\gamma(L_y) + R_K \quad ,$$

where $R_4 = R_4(E)$, the usual 4-dimensional scalar

R_K = constant curvature scalar of G/H

$F_{ab}{}^{\hat\alpha}$ = Yang-Mills field strength.

Integration over y^μ gives

$$\frac{1}{V_K} \int d^K y \det e(g)\, D_{\hat\alpha}{}^\gamma(L_y)\, D_{\hat\beta}{}^\gamma(L_y) = \delta_{\hat\alpha\hat\beta} \frac{\dim G/H}{\dim G}$$

and so the usual Yang-Mills action is produced. One can show that the G-covariant couplings of the Yang-Mills vector to matter fields is implied by the 4+K-dimensional covariance. However, since the Yang-Mills group involves both co-ordinate and tangent space transformations, the minimal coupling of A_m must arise partly in E and partly in B. There will generally be non-minimal couplings as well. Details are given elsewhere[1].

REFERENCES

1. Abdus Salam and J. Strathdee, ICTP Trieste preprint IC/81/211 (198
2. Th. Kaluza, Sitzungsber. Preuss. Akad. Wiss. Berlin, Math. Phys. Kl, 966 (1921).
3. O. Klein, Z. Phys. 37, 895 (1926).
4. B. De Witt, Dynamical Theory of Groups and Fields (Gordon and Breach, N.Y.,1965).
5. J. Scherk and J.H. Schwarz, Phys. Letters 57B, 463 (1975).
6. P.G.O. Freund and M.A. Rubin, Chicago preprint EFI 80/35 (1980).
7. J.F. Luciani, Nucl. Phys. 135B, 111 (1978).

GRAVITY AS A DYNAMICAL CONSEQUENCE OF THE STRONG, WEAK, AND ELECTROMAGNETIC INTERACTIONS

A. Zee

Department of Physics, FM-15

University of Washington, Seattle, WA 98195

ABSTRACT

La lumière fut, donc la pomme a chu.[1]

Is Newton's gravitational constant G a fundamental constant of Nature or it is calculable in terms of the other fundamental constants?

We would like to argue here that G is indeed calculable by physicists. After outlining the general philosophy and motivation behind this statement, we present a specific calculation of G, unfortunately not in the real world but for a class of gauge theories which presumably does not describe the real world.

That G is positive is certainly one of the experimental facts in physics least open to doubt. As physicists we should try to understand why.

In gauge theories, we insist that the coefficient of $F_{\mu\nu}F^{\mu\nu}$ is positive (and write it as $1/g^2$) so that the Euclidean action is positive definite and hence bounded below. This also insures that the gauge bosons propagate normally and not as ghosts. The situation for gravity is somewhat murkier. The scalar curvature R, even in Euclidean space, can be either positive or negative. We thus can no longer rely on the condition that the Euclidean path integral be defined to yield an "a priori" argument for the sign of G. It is still true the normal propagation of the graviton requires G to be positive. However, a theory of gravity with a Lagrangian of the form $\alpha R^2 + \beta R_{\mu\nu}^2 + G^{-1}R$ would have, with suitable choices of the signs of α and β, a corresponding Euclidean action bounded from below even if G is taken to be negative. Although it assuredly does not describe the real world it may well be sensible.

The positivity of G insures the normal propagation of the graviton. As long as the graviton propagates normally, the attractiveness of gravity follows from the fact that the spin of the graviton is even, according to a fundamental theorem of quantum field theory.

We begin with the simple but perhaps profound observation that in Yang-Mills theory the non-Abelian structure of the gauge field $F_{\mu\nu} = \partial_\mu A_\nu - \partial_\nu A_\mu - i[A_\mu, A_\nu]$ forces A_μ to have the dimension of ∂_μ. Thus, the Yang-Mills Lagrangian $F_{\mu\nu}^2$ has dimension four regardless of the dimension of space-time. In contrast, the Einstein-Hilbert Lagrangian R involves the scalar curvature and has dimension two in space-time of any dimension. Thus, in this sense, Yang-Mills theory is perfectly matched to the four dimensional space-time in which we find ourselves. It is tempting to speculate that this fact may be intimately connected to the observation that the world is, indeed, four dimensional.

The fact that the dimensions of the gauge field and the metric field do not depend on the dimension of space-time is of course due to the intimate connection of these fields to geometry. In contrast, the requirement that the Dirac Lagrangian be scale-invariant will assign to fermion fields a dimension dependent on the space-time dimension.

Why is our theory of gravity matched to two, rather than four, dimensional space-time? We would like to attach some significance to this apparent mismatch. More precisely, scale invariance forbids Einstein's Lagrangian for gravity. As was first fully appreciated by Coleman and E. Weinberg,[2] it is extremely attractive to impose scale invariance on the fundamental Lagrangian of the world. In a scale invariant theory with n dimensionless coupling constants, all dimensionless ratios of physical quantities are calculable in terms of (n-1) dimensionless couplings. This is because the renormalization procedure introduces a mass scale μ whose choice is arbitrary and for which one dimensionless coupling may be traded through "dimensional transmutation". If the fundamental Lagrangian is indeed scale invariant then the ratio of G to some other dimensional physical quantity such as the proton mass should be calculable.

The importance of scale invariance was realized as long ago as 1938 by Heisenberg[3] who remarked that theories with dimensionless coupling constants are interestingly located on the boundary between theories with "nasty" and theories with "nice" short-distance behavior. Of the four fundamental interactions, Fermi's theory of weak interaction and Einstein's theory of gravity are both characterized by couplings of dimension of inverse mass squared. A great triumph of physics over the last twenty years is the discovery that the weak interaction is actually secretly characterized by a dimensionless coupling and that Fermi's constant is given by the vacuum expectation value of some scalar field: $G_F = \langle\phi\rangle^{-2}$. It is natural to ask if gravity is also secretly dimensionless and if Newton's constant is also determined by the vacuum expectation value of some scalar field. The physics of the weak interaction and of gravity is of course quite different. Gravitation is long-ranged and so the vacuum expectation value cannot be associated with the mass of a mediating particle. Indeed, 1/G appears in the Einstein-Hilbert action multiplying what is essentially the kinetic energy term for the graviton. G and the Yang-Mills coupling g^2 measure the "stiffness" of the graviton and the gluon field respectively against excitation.

It is trivially easy to realize the idea that G^{-1} is given by the vacuum expectation of a scalar field. One merely has to replace the Einstein-Hilbert action by the action

$$\int d^4x\sqrt{-g}\,[\tfrac{1}{2}\,\varepsilon\phi^2 R + \tfrac{1}{2}\,\partial_\mu\phi\partial_\nu\phi g^{\mu\nu} - V(\phi)].$$

This was done independently by a large number of people.[4-9] Here we follow the treatment given in Ref. 4. The potential $V(\phi)$ is assumed to attain its minimum value when $\phi = V$. Then

$$G_N = \frac{1}{16\pi}\,(\tfrac{1}{2}\,\varepsilon V^2)^{-1}.$$

The introduction of scalar fields into gravity has a long history. Here, the crucial feature is the incorporation of spontaneous symmetry breaking. As a consequence the scalar field is "anchored" in a deep potential well $V(\phi)$ and thus the physical consequences of the present theory are indistinguishable from Einstein's theory except under extreme conditions of space-time curvature. This is in sharp contrast to earlier work such as that of Brans and Dicke. The coupling ε is dimensionless and has to be taken to be positive. Thus, this theory does not shed any light on the sign of G.

One is tempted to identify ϕ as the Higgs field responsible for the breaking of grand unified theory into strong, weak and electromagnetic interactions. In that case, gravity is weak because the other three coupling constants move so slowly (logarithmically) under the renormalization group. This "modern" view of why gravity is weak was in fact known to Landau. Unfortunately, the relevant symmetry scale of the SU(5) grand unified theory is only about 10^{-4} $M_{P\ell}$ ($M_{P\ell}$ denotes the Planck mass $\sim 10^{19}$ GeV). The SU(5) theory is however incomplete in a number of ways and we may hope that eventually the present theories will be extended to a theory set at the Planck mass.

As has already been remarked, the present theory is indistinguishable from Einstein's theory except under extreme conditions such as may exist in the early Universe. Newton's gravitational "constant" may then vary with temperature. It is not inconceivable that this phenomenon of a varying gravitational "constant" may be relevant for the horizon problem.[10] Any serious discussion is necessarily highly speculative,[11,12] however.

This theory is not only both trivial to construct and dull in its consequences but also rather unattractive. Our motivating philosophy is based on scale invariance. It was suggested in Ref. 10 that the theory of the world including gravitation should contain no dimensional parameter. We may take $V(\phi) = -\lambda\phi^4$ and, following Ref. 2, generate the symmetry breaking by radiative one-loop effects.

The next logical step was taken by Adler[13] who asked what would happen if there are no elementary scalar fields at all? After all, elementary scalar fields are generally regarded with some repugnance by physicists. Remarkably enough, in the absence of elementary scalar fields, gauge invariance and scale invariance combine to forbid terms proportional to the scalar curvature R in the Lagrangian. Thus, the term $\bar{\psi}\psi R$ has dimension five and is forbidden by scale invariance. On the other hand, the term $A_\mu A^\mu R$ does have dimension four but is not gauge invariant. We find it rather satisfying that terms proportional to the scalar curvature are excluded by scale invariance and gauge invariance, two fundamental symmetries with deep geometrical significance.

Since scale invariance is "automatically" broken by the renormalization procedure, or if one prefers, by quantum fluctuations, terms proportional to R will be induced[14]. Thus, we have the possibility that gravity is generated by the dynamical breaking of some grand unified gauge theory near the Planck mass.

It may well turn out that gravity is actually, in some sense, a consequence of the other three interactions. "La lumière fut, donc la pomme a chu". We find this philosophy rather appealing in that it obviates the need for a marriage of gravity with the other three interactions. Gravity, mediated by a spin-two field, does look quite different from the other three interactions, which have now been revealed to be all mediated by spin-one fields. It is true that various ingenious individuals have invented clever symmetries relating particles of different spins. However, the ensuing marriage arranged by local supersymmetry is attended by unwanted, or at least so far unobserved, spin -3/2 particles. This is not to deny the great beauty of supergravity. The philosophy behind supergravity is in some sense the exact opposite of the one advocated here in that it seeks to determine the other three interactions starting with Einstein's theory of gravity. Regrettably, the philosophy of induced gravity also appears to be incompatible with the beautiful idea of Kaluza that gauge invariance is merely the manifestation of general coordinate invariance in higher dimensions, an idea which is in turn not unrelated to supergravity[15]. Perhaps some way could be found to reconcile induced gravity and the Kaluza-Klein theory[16].

Strictly speaking, the remarks above must be amended somewhat if R^2 and $R_{\mu\nu}^2$ terms are included in the Lagrangian. It is then a matter of semantics whether one regards gravity as just as fundamental as the other three interactions. We still prefer to say, perhaps somewhat too picturesquely, that gravity is generated by the other three interactions in the sense that the known propagation of the graviton at long distance is a consequence of the other three interactions.

We remark parenthetically that in certain supergravity theories the sign of G is correlated[17] to the normal propagation of the gauge bosons. This should count as a plus for supergravity.

Before proceeding further, we would like to remark briefly on the question of quantizing gravity. There are two possible points of view. (I) Perhaps gravity should not be quantized at all. Some people feel that gravity does have some mysterious connection with space-time geometry[18] and the graviton should not be treated as just any particle, as particle physicists are wont to do. We are also so pitifully ignorant of physics at the Planck scale and beyond. (Strictly speaking, this argument actually suggests that if gravity is to be quantized one should not worry too much about renormalizability of quantum gravity.) The price one has to pay for this view is that the action principle which leads to Einstein's equation is then ad hoc. Unfortunately, this view, that $g_{\mu\nu}$ describes a classical arena in which quantized matter fields play, is probably inconsistent[19]. For instance, by measuring the classical gravitational field we could in principle determine our distance from a massive object to arbitrary accuracy, in contradiction with the uncertainty principle. (II) Gravity should be quantized. In path integral lingo, the metric $g_{\mu\nu}(x)$ is to be integrated over like any other fields. In this case, we are obliged to put in the R^2 and $R^2_{\mu\nu}$ terms, while, with view (I), it is arguably optional.

For simplicity and as a first try, we adopt view (I) in what follows, keeping in mind that it is most likely inconsistent. Many of our remarks apply equally well to both views but some of the formulas below have to be modified. To the extent that semi-classical radiation theory has a limited but well-defined domain of validity, we expect view (I) to represent an approximation to the full theory with gravity itself fully quantized.

In some sense, the roots of gravity lie in Lorentz invariance. To write down Lorentz-invariant interactions between fields we have to introduce the Minkowski metric $\eta_{\mu\nu}$. Once we admit the possibility of $\eta_{\mu\nu}$ depending on space-time, we promote the metric to a field $g_{\mu\nu}(x) = \eta_{\mu\nu} + h_{\mu\nu}(x)$. In field-theory language, $h_{\mu\nu}$ is then a field without a proper kinetic energy term. In the view discussed here, the appropriate kinetic energy term arises as a consequence of dynamical scale-symmetry breaking. It is amusing to ask whether there are other hitherto unobserved interactions which are generated in the same way[20].

Scale invariance also forbids the appearance of a cosmological constant term $\int d^4x\sqrt{g}\,\Lambda$ in the action which would in general appear upon symmetry breaking. At the moment, no one knows how to avoid generating this undesirable term. This is perhaps the weakest point in the program to generate gravity spontaneously, and, indeed, this problem afflicts all current theories in particle physics which utilize the notion of spontaneous symmetry breaking. We imagine that the ultimate correct theory of dynamical symmetry breaking will not produce a cosmological term.

The discussion here represents, in some sense, the modern realization of ideas of Sakharov[21] who identified gravity as the "elasticity of space-time". Very similar ideas have been discussed under the name of "pre-geometry"[22]. The main distinction of the discussion here and in Refs. 21 and 22 is our insistence that the other three interactions be described by a renormalizable and scale invariant theory so that G is finite and calculable and hence is not cut-off dependent.

After all this discussion, we now give an actual formula for G derived independently by Adler[13] and by the present author[23]. We follow here the derivation given in Ref. 23. Starting with the Lagrangian $\mathcal{L} = \int d^4x(-1/2\, T^{\mu\nu})h_{\mu\nu} + \cdots$ (which defines $T_{\mu\nu}$)

we expand the generating functional $\langle 0|T^* e^{+i\int \mathcal{L} d^4x}|0\rangle$ to extract the term quadratic in h. We treat $h_{\mu\nu}$ as a c-number classical field and thus the effective order $-h^2$ Lagrangian is given by

$$i\int d^4x \mathcal{L}_{eff}(x) = \frac{1}{2!}\left(\frac{-i}{2}\right)^2 \int d^4x d^4y h_{\mu\nu}(x) h_{\lambda\rho}(y) \langle 0|T^* T^{\mu\nu}(x) T^{\lambda\rho}(y)|0\rangle . \tag{1}$$

The T* product is understood to be the connected part. We specialize to the form $h_{\mu\nu} = 1/4\ \eta_{\mu\nu} h$ and choose h(x) to be a slowly varying function so that we can expand

$$h(x) = h(y) + z^\mu \partial_\mu h(y) + \frac{1}{2!} z^\mu z^\nu\ \partial_\mu \partial_\nu h(y) + \cdots \tag{2}$$

$$z = x - y .$$

Defining $T(x) = \eta_{\mu\nu} T^{\mu\nu}(x)$ we find

$$i\,\mathcal{L}_{eff}(x) = \frac{-1}{2^7} h^2(x) \int d^4z \langle 0|T^* T(z) T(0)|0\rangle + \frac{1}{2^{10}} [\partial h(x)]^2 \int d^4z\ z^2 \langle 0|T^* T(z) T(0)|0\rangle + \cdots .$$

A simple computation shows the order $-h^2$ term in $\sqrt{-g}\,R$ to be $-3/32(\partial h)^2$. Thus we find the following representation for Newton's coupling constant:

$$\frac{1}{16\pi G} = \frac{i}{96} \int d^4x\ x^2 \langle 0|T^* T(x) T(0)|0\rangle . \tag{3}$$

It is worth emphasizing that the right-hand side of Eq. (3) is a purely flat-space quantity determined completely by the other three interactions. If we understand the other three interactions thoroughly, we could in principle calculate $\langle 0|T^* T(x) T(0)|0\rangle$ precisely, evaluate the integral in Eq. (3), and thus obtain G.

If the R^2 and $R^2_{\mu\nu}$ terms are included and if gravity is quantized, there will be an extra term on the right-hand side of Eq. (3) due to the contribution of virtual gravitons. This extra term has recently been worked out by Adler[24].

If the strong, electromagnetic, and weak interactions are described by a grand unified gauge theory with massless fermions, the operator T(x) is determined via the trace anomaly[25] to be

$$T = [2\beta(g)/g]\ \frac{1}{4}\ F^{\alpha}_{\mu\nu} F^{\mu\nu\alpha}.$$

The formula for Newton's constant in Eq. (3) expresses in precise terms our philosophy that gravity is induced as a result of quantum fluctuations. A heuristic understanding of this is not difficult to find and is readily suggested by[26] the Feynman diagram representing $\langle 0|T^*T(x)T(0)|0\rangle$ which is shown here:

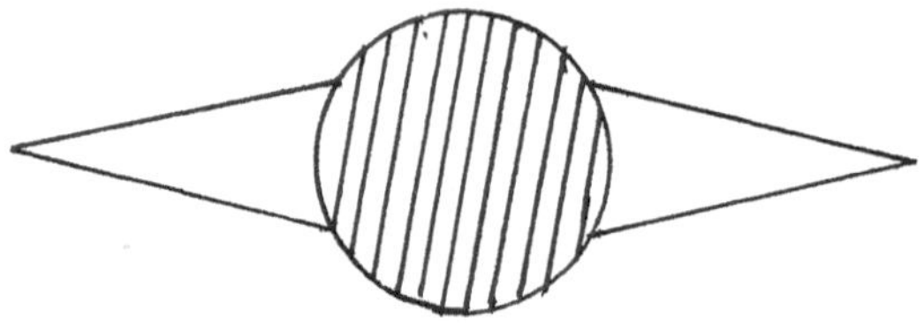

(The trace operator $T(x)$ couples to a pair of particles which interact in the shaded blob.) In the absence of $(\partial h)^2$ term in the action, a spatially varying gravitational field contains no energy. While this may be true classically it cannot be true quantum mechanically. With quantum fluctuation a pair, say e^-e^+ pair (actually a gauge boson pair), could always be created. With $\vec{\nabla} h_{\mu\nu} \neq 0$ this pair will then fall, gaining kinetic energy. By bringing the e^-e^+ pair together and annihilating them we can always extract energy out of the gravitational field. Another version given by Adler[14] involves the legendary Elevator of Einstein. The usual statement is that for a small enough elevator one cannot say whether the elevator is uniformly accelerating or whether it is falling in a gravitational field (say that of the earth). But with quantum fluctuations an e^+e^- pair may be created and the e^- could tunnel out of the elevator to make a grand tour sampling the curvature tensor before coming back to annihilate the positron. Clearly, these two arguments represent out intuitive understanding of the Feynman diagram displayed above.

An evaluation of Eq. (3) will give G in terms of the scale mass μ. We envisage that a calculation of the proton mass m_N, say, in terms of μ is done independently so that eliminating μ one then determines the dimensionless ratio $G\, m_N^2$. As was remarked in the introduction, the sign of G is of great interest. If in a realistic calculation we should obtain a negative G this would clearly indicate that either the idea of induced gravity or the theory we are calculating with is wrong.

Define

$$\psi(-x^2) = \langle 0|T^*T(x)T(0)|0\rangle_{\text{connected}}\ .$$

Then the formula for G may be rotated to Euclidean space to read

$$\frac{1}{16\pi G} = - \frac{1}{96} \int_E d^4x \, x^2 \, \psi(x^2). \tag{3'}$$

A crude calculation of G_{ind} taking into account the short-distance ultraviolet region was given in Ref. 23 and motivated the derivation of Eq. (3). Unfortunately, we were led to conclude that the sign of G_{ind} depends on the long-distance infrared region of which we are totally ignorant. There was also a calculation[27] of G using a dilute instanton gas approximation, but again, due to our ignorance of the long-distance physics, the infrared region was excluded by an artificial cut-off on the instanton size. Thus, neither of these calculations is conclusive as regards the sign of G_{ind}. In order to include the infrared region, Adler has outlined[24] a program based on numerical lattice calculations.

In a recent paper[28] we remarked that there is a class of gauge theories in which the infrared region is completely known -- in fact, the function ψ may be computed explicitly. These are gauge theories with the property that in the expansion of the renormalization group function $\beta(g) = - 1/2\, g^3(b_0 + b_1 g^2 + \cdots)$ the coefficient b_0 is positive and small (so that the theory is barely asymptotically free) while the coefficient b_1 is negative. (Such theories are well-known to exist; quantum chromodynamics with sixteen quark triplets provides an example[29]. There is then an infrared stable fixed point given by $g_*^2 = (-b_0/b_1)$. By choosing appropriately the gauge group and the fermion representations, we can make g_*^2 arbitrarily small. The calculation described below is "exact" to the extent that g_*^2 is small.

We refer the interested reader to Ref. 28 for details of the calculation. Suffice it to say here that after some manipulations we obtain

$$\frac{1}{16\pi G\mu^2} = - \frac{\pi^2}{96} \left(\frac{c b_0^2 g_*^2}{16}\right) \int_0^\infty \frac{d\tau}{\tau^2} F(\tau). \tag{4}$$

Here we have introduced a dimensionless distance variable $\tau = \mu^2 x^2$. c is a known numerical constant. The scale mass μ has been fixed by the condition $g^2(\mu^2) = 1/2\, g_*^2$ and thus has physical meaning.

It turns out the integrand $F(\tau)$ is positive. Naively, this would imply that G will be negative. The fact that $F(\tau)$ is positive simply confirms a general spectral function analysis. One may naively write a Källen-Lehmann representation for the two-point function $\psi(x^2)$ and conclude from the positivity of the spectral function that G will always come out negative. The fallacy in the argument is that the Källen-Lehmann representation actually does not exist[13] due to the fact that T(x) is a dimension four operator and so $\psi(x^2)$ behaves at short distances like x^{-8}. There is thus no general theorem about the sign of G in induced gravity. The short-distance behavior of $\psi(x^2)$ is reflected here by the behavior of $F(\tau)$ as $\tau \rightarrow 0$. One verifies easily by a renormalization group analysis that

$$F(\tau) \underset{\tau\rightarrow 0}{\rightarrow} \left(\frac{1}{\log \tau} + \cdots\right)^2.$$

Thus, the integral diverges in the ultraviolet region and must be regulated.

In Ref. 28 we dimensionally regulated the integral by continuing to space-time with dimension = 2ω and taking the limit $\omega \rightarrow 2$. The continuation produces a function analytic in the complex ω plane with a cut along the positive real axis. We approach the cut symmetrically. A justification for this prescription has been given by Adler[24].

It is well known that within dimensional regularization, to any finite perturbative order, one only encounters poles in the complex ω plane. Here, however, the trace anomaly incorporates effects to all orders in perturbation theory. Also, we have not continued the β function and the logarithmic behavior of $F(\tau)$. Had we done that, our integral would have a cut on the real axis to the right of $\omega = 2$ and an infinite number of poles to the left of $\omega = 2$. These poles coalesce to form a cut in the suitable limit.

We extract in this way a finite value for G. The sign of G turns out to be given by

$$\text{sign } G = (-1)^{[\gamma]}$$

where [x] = largest integer less than or equal to x. The parameter γ is defined as $2/(b_0 g_*^2)$. As is well known, dimensional regularization often reverses the naive sign of integrals. Precisely this phenomenon occurs here. The ultraviolet region contributes positively to G and may or may not overwhelm the infrared region, depending on the parameter γ. This toy calculation suggests that a general argument on the sign of G may not exist.

We claim that this calculation verifies general arguments given by Adler that G should be finite and calculable. We do not have to subtract off any poles. This is because quadratic divergences do not occur in dimensional regularization and the potential logarithmic divergence is softened logarithmically by asymptotic freedom.

One might feel that the regularization procedure is a bit delicate. Certainly, the calculation should be performed with another regularization scheme, namely the higher derivative regularization[30], as a check. However, if Adler's rigorous argument[13] that G is finite and calculable in the context of induced gravity is correct, then the value of G obtained should be regularization independent. We see no reason to doubt Adler's argument.

Eventually, one would like to do a calculation for the real world. Realistically ψ is not known. Adler[24] has proposed that the integral K in Eq. (4) be divided into two integrals, one from 0 to τ_0, the other from τ_0 to ∞. Let us refer to these two pieces as K_{UV} and K_{IR} so that $K = K_{UV}(\tau_0) + K_{IR}(\tau_0)$. Adler envisages the calculation of the integrand in K_{IR} by Monte-Carlo methods on the lattice. No regularization is needed to evaluate K_{IR}. In contrast, we must dimensionally regulate the ultraviolet piece K_{UV} so we need to have an analytic form for the integrand. For τ_0 small enough, we can use the asymptotic freedom form for the integrand. We define $K^n_{UV}(\tau_0)$ as the value of the ultraviolet integral in Eq. (4) with $F(\tau)$ replaced by the first n terms of an asymptotic freedom expansion of $F(\tau)$ (with the integral dimensionally regulated). The plan is to match this into a numerical calculation of $K_{IR}(\tau_0)$. This matching needs to be done with care, however. The reason is that in the expansion of $F(\tau)$, as $\tau \to 0$ each succeeding term is only logarithmically less singular. Thus, for any finite n

$$\lim_{\tau_0 \to 0} [K^n_{UV}(\tau_0) + K_{IR}(\tau_0)] = \infty .$$

The correct procedure[31] is to evaluate the limit

$$\lim_{n \to \infty} [K^n_{UV}(\tau_0) + K_{IR}(\tau_0)]$$

for a fixed but small enough τ_0.

Without actually doing a realistic calculation, we could anticipate what the value of G will turn out to be. Consider a scale invariant gauge theory which does not have a small infrared fixed point but which has a growing coupling constant in the infrared limit. Then G should come out to be of order of the square of the mass scale at which the theory becomes strongly coupling. This is essentially the same conclusion as reached in the version of induced gravity involving Higgs fields. Thus, if the standard SU(5) theory minus its Higgs fields actually manages to generate its own breaking at a scale of M_X, then the value of G induced in such a theory will be of order M_X^2.

In the derivation of the formula for G, we made the simplification of taking $h_{\mu\nu} = 1/4\ \eta_{\mu\nu}h$. (See Eqs. (1,2).) This is of course not a necessary assumption[32]. Consider Eq. (1). Introducing the Fourier transform

$$h_{\mu\nu}(x) = \int \frac{d^4k}{(2\pi)^4} e^{ikx} h_{\mu\nu}(k)$$

we find

$$i\int d^4x\, \mathcal{L}_{eff} = \frac{1}{2}\left(\frac{-i}{2}\right)^2 \int \frac{d^4k}{(2\pi)^4} h_{\mu\nu}(k) h_{\lambda\rho}(-k) \int d^4y\, e^{iky} \quad (5)$$
$$\langle 0|T^*T^{\mu\nu}(y)T^{\lambda\rho}(0)|0\rangle .$$

To determine the general form of $\langle 0|T^*T_{\mu\nu}(x)T_{\lambda\rho}(0)|0\rangle$ we must derive the appropriate divergence condition it satisfies. To do this, one notes that[33] the stress-energy density tensor $\mathfrak{T}_{\mu\nu} = \sqrt{-g}\, T^{\mu\nu}$ in curved space-time is covariantly conserved

$$\partial^\mu \langle \mathfrak{T}_{\mu\nu} \rangle + \Gamma^\nu_{\sigma\tau} \langle \mathfrak{T}^{\sigma\tau} \rangle = 0. \quad (6)$$

The expectation value is taken in the curved manifold. Varying this equation with respect to $g_{\lambda\rho}$ and taking the flat space limit we obtain

$$i\partial^\mu \langle 0|T^*T_{\mu\nu}(x)T_{\lambda\rho}(0)|0\rangle = \Lambda(\eta_{\nu\rho}\partial_\lambda + \eta_{\nu\lambda}\partial_\rho - \eta_{\lambda\rho}\partial_\nu)\delta^{(4)}(x). \quad (7)$$

The expectation value here is taken in flat space. We can now derive[32] the general form

$$\int d^4x\, e^{ikx}\, i\langle 0|T^*T_{\mu\nu}(x)T_{\lambda\rho}(0)|0\rangle$$
$$= L_{\mu\nu\lambda\rho} \int_0^\infty ds\left[\rho_2(s)\frac{k^2}{k^2 - s} + \frac{3}{4}\rho_0(s)\right] \quad (8)$$
$$+ \Pi_{\mu\nu}\Pi_{\lambda\rho} \int_0^\infty ds \frac{\left(\frac{4}{3}\rho_2(s) + \rho_0(s)\right)}{k^2 - s} - \Lambda(\eta_{\nu\rho}\eta_{\lambda\mu} + \eta_{\nu\lambda}\eta_{\rho\mu} - \eta_{\lambda\rho}\eta_{\mu\nu}).$$

We define the projection operator

$$\Pi_{\mu\nu} = \eta_{\mu\nu} k^2 - k_\mu k_\nu$$

$$L_{\mu\nu\lambda\rho} = k^{-2}(\Pi_{\mu\lambda}\Pi_{\nu\rho} + \Pi_{\mu\rho}\Pi_{\nu\lambda} - 2\Pi_{\mu\nu}\Pi_{\lambda\rho}) .$$

The operator $L_{\mu\nu\lambda\rho}$ contracted with $h_{\mu\nu}h_{\lambda\rho}$ gives an expression proportional to the quadratic part of Einstein's Lagrangian. The presence of the last term in Eq. (8) is required by the divergence condition, Eq. (7). The spectral functions ρ_2 and ρ_0 are the same as those defined in Ref. 33. The representation is formal in the sence that the integration over s may not converge but it indicates the correct tensor structure for each contribution to the spectral function.

Using this general representation one could readily[32] derive the formula for G (Eq. (3)) without the simplifying assumption $h_{\mu\nu} = 1/4\ \eta_{\mu\nu}h$.

Next, we make a number of miscellaneous remarks.

Suppose the theory is exactly scale invariant (that is to say, the theory is at a fixed point $g = g^*$, $\beta(g^*) = 0$, so $T = 0$). In this case, we have $1/G = 0$ or $G = \infty$. This is exactly what one would expect: The Einstein-Hilbert term is not generated if scale invariance holds. N = 4 supersymmetric Yang-Mills theory may have $\beta(g) = 0$.

For the induced gravity idea to work we have to assume that we know the other three interactions over all distance scales. One possible view may be that one should cut off the integral in Eq. (3') at the Planck length. This is unacceptable in that, as is explained above, G will always come out negative.

Creatures living in six-dimensional space-time will find the Yang-Mills action rather peculiar and note that it is forbidden by scale invariance. A philosophy of induced Yang-Mills would be proposed by physicists in this six-dimensional world[23] and a discussion similar to ours could be given.

The induced gravity idea is also reminiscent of the situation in the Gross-Neveu[34] model and the CP^N model. In these two dimensional models, kinetic energy terms for boson fields, forbidden by scale invariance and renormalizability, are dynamically induced. However, it is also instructive to note that the generation of these terms are infrared in the sense that the terms in question have dimensions four, greater than the dimension two of space-time. In contrast, the Einstein-Hilbert term has dimension less than the space-time dimension. Incidentally, formulas analogous to Eq. (3) can be readily written down. For instance, in the Gross-Neveu model

$$\mathcal{L} = \bar{\psi}\, i\partial\!\!\!/\psi - \sigma\bar{\psi}\psi - \frac{1}{2g^2}\sigma^2$$

the missing kinetic energy term for the σ field $1/2\,(\partial_\mu\sigma)^2$ will be induced. The coefficient of this term is given by

$$\frac{1}{M^2} = -\frac{1}{4}\, i \int d^2x\, x^2 \langle 0|T^*\, \bar{\psi}\psi(x)\, \bar{\psi}\psi(0)|0\rangle \; .$$

Since $\bar{\psi}\psi$ has dimension one this integral is not singular in the short-distance limit. It is perfectly finite. The dispersion argument which did not go through for induced gravity goes through here. We find

$$\frac{1}{M^2} = \int_0^\infty d\sigma^2\, \rho(\sigma^2)/\sigma^4 .$$

The spectral function ρ is positive-definite and so $1/M^2$ is positive as it should be. Alas, things are not so simple in the real world.

We remark in passing that even in the presence of scalar fields the term ϕ^2R may be forbidden if the theory has a global supersymmetry[24,25].

Induced gravity may have amusing implications for cosmology. Presumably, when the temperature is high enough, the induced term will disappear. The effect of temperature on the induced gravitational constant deserves to be investigated.

Finally, we mention that in the induced gravity framework the Weyl-Eddington action

$$\int d^4x\sqrt{g}\; (-)(\rho R^2 + \gamma\, C^2_{\mu\nu\sigma\tau})$$

will also be induced. Here $C_{\mu\nu\sigma\tau}$ denotes the Weyl tensor. The representation in Eq. (8) allows us to derive formulas[32] for ρ and γ.

In general, one would expect that ρ and γ would be logarithmically divergent. A more careful analysis reveals that if the other three interactions are described by an asymptotically free theory the induced ρ will in fact be finite[36]. Also its sign is determined by a general argument and comes out to be just the right sign so a tachyon does not appear in the theory.[36]

In conclusion, we feel that a coherent and reasonable account of gravitational physics may be given along the following line. The three non-gravitational interactions are described by a scale and conformal invariant and asymptotically free Yang-Mills theory with massless fermions. We insist on conformal invariance so that the gravitational sector of theory is given by the Weyl action $\int d^4x\, C^2_{\mu\nu\sigma\tau}$. The theory is renormalizable[37] and, thanks to the Lee-Wick mechanism[38], has a unitary S-matrix. Possible breakdown of causality is observable only at the Planck length. In this theory, Einstein's theory of gravity is induced as an effective long-distance theory. An R^2 term is also induced with a finite and physically desired sign.

ACKNOWLEDGMENTS

We have benefited over the years from comments by many colleagues, including S. Adler, R. Barbieri, S. Barr, J. Barrow, S. Bludman, D. Boulware, L. Brown, N. Cabibbo, E. Cremmer, R. Dashen, J. Ellis, F. Englert, J. Iliopolous, A. Linde, L. McLerran, L. Maiani, A. Neveu, P. Ramond, D. Reiss, P. Rossi, L. Smolin, H. Terazawa, M. Testa, M. Tonin, H.A. Weldon.

We are especially indebted to S. Adler and L. Brown for many useful discussions.

This work supported in part by the U.S. Department of Energy.

REFERENCES

1. E. Cremmer and H. Lubatti, private communication.
2. S. Coleman and E. Weinberg, Phys. Rev. D7: 1888 (1973).
3. W. Heisenberg, Z. Physik 110: 251 (1938); S. Sakata, H. Umezawa, and S. Kamefuchi, Prog. Theo. Phys. 7: 327 (1952).
4. A. Zee, Phys. Rev. Lett. 42: 417 (1979).
5. L. Smolin, Nucl. Phys. B160: 253 (1979).
6. Y. Fujii, Phys. Rev. D9: 874 (1974).
7. P. Minkowski, Phys. Lett. 71B: 419 (1977).
8. T. Matsuki, Prog. Theor. Phys. 59: 235 (1978).
9. A.D. Linde, Pis'ma Zh. Eksp. Teor. Fiz. 30: 479 (1979)(JETP Lett. 30: 447 (1979)].
10. A. Zee, Phys. Rev. Lett. 44: 703 (1980).
11. A. Linde, Phys. Lett. 93B: 394 (1980); H. Sato, Prog. Theo. Phys. 64: 1498 (1980); M.D. Pollock, Padova preprint IFPD 40/81 (1981).

12. H. Fleming, Sao Paulo preprint (1981); B. Meyer, preprint (1981).
13. S. Adler, Phys. Rev. Lett. 44: 1567 (1980); Phys. Lett. 95B: 241 (1980).
14. For further references and for a review of the physics and philosophy behind induced gravity see S.L. Adler, Rev. of Mod. Phys. (to be published) and in The High Energy Limit, ed. by A. Zichichi; A. Zee, "Grand Unification and Gravity", in the Proceedings of the 4th Kyoto Summer School (to be published) and in Proceedings of the Erice Conference (October 1981, to be published).
15. E. Cremmer, Lectures at the Trieste and Seattle Schools, (1981).
16. H.Y. Guo, Beijing preprint ASITP-81-001 (1981).
17. E. Cremmer, K. Stelle, and P. Townsend, private communications.
18. See, for example, E. Wigner's remark cited in the second paper in Ref. 14.
19. We thank T. Banks, F. Englert, and S. Mandelstam for conversations on this point.
20. A. Zee, to be published.
21. A. Sakharov, Dokl. Akad. Nauk. 177: 70 (1967); O. Klein, Phys. Scr. 9: 69 (1974); C.W. Misner, K.S. Thorne, and J.A. Wheeler, "Gravitation" (Freeman, San Francisco, 1973), p. 426.
22. K. Akama, Y. Chikashige, T. Matsuki, and H. Terazawa, Prog. Theo. Phys. 60: 1900 (1980); see also recent work by G. Veneziano and collaborators (these Proceedings).
23. A. Zee, Phys. Rev. D23: 858 (1981).
24. S.L. Adler, Rev. of Mod. Phys. (to be published) and in The High Energy Limit, ed. A. Zichichi.
25. R.J. Crewther, Phys. Rev. Lett. 28: 1421 (1972).
26. Historically, the diagram suggested the formula.
27. B. Hasslacher and E. Mottola, Phys. Lett. 95B: 237 (1980).
28. A. Zee, to be published.
29. W. Caswell, Phys. Rev. Lett. 33: 244 (1974).
30. B. Lee and J. Zinn-Justin, Phys. Rev. D5: 3121 (1972); A. Slavnov, Nucl. Phys. B31: 301 (1971).
31. S. Adler, private communication.
32. L. Brown and A. Zee, to be published.
33. D. Boulware and S. Deser, J. of Math. Phys. 8: 1468 (1967), Eq. (13).
34. D. Gross and A. Neveu, Phys. Rev. D10: 3235 (1974).
35. M. Kaku and P. Townsend, Phys. Lett. 76B: 54 (1978).
36. A. Zee, preprint.
37. K. Stelle, Phys. Rev. D16: 953 (1977) and references therein.
38. T.D. Lee and G. Wick, Phys. Rev. D2: 1033 (1970) and references therein.

UNIFICATION OF GAUGE AND GRAVITY INTERACTIONS FROM COMPOSITENESS

G. Veneziano

CERN, Geneva, Switzerland

Within our present understanding, it is conceivable that all elementary interactions can be described by gauge theories. This common structure opened the way towards unification, first with the successful electroweak theory of Glashow-Weinberg-Salam and then by its more speculative merging with the SU(3) theory of strong interactions in the so-called grand unified theories[1] (GUTs). The energy scale at which this GUT unification takes place, M_{GUT}, is very large, ranging between 10^{14} and 10^{17} GeV depending on the details of the actual theoretical scheme.

In spite of their doubtless appeal and partial success, GUTs are still facing hard, unresolved problems (origin of families, hierarchy problem, etc.). Even more frustrating has been the impossibility so far of including gravity in a unified scheme. Not only is gravity a gauge theory (of space-time rather than of an internal symmetry), it is also expected to become strong and hence non-negligible at a scale (of order 10^{19} GeV) which is embarrassingly close to M_{GUT} itself.

On the other hand, whereas in usual GUTs electroweak and strong interactions are renormalizable interactions becoming equally weak at M_{GUT}, the gravitational interaction shows such a steady increase of strength with energy that its meaning as a quantum theory is itself doubtful. This difference is of course an obvious obstacle to unifying gravity with the other interactions. Another more formal difference is that, in the conventional approach, the gauge connection is considered as a fundamental field whereas in gravity it is given in terms of the fundamental metric of vierbein fields.

In this talk I shall propose and discuss a theoretical framework in which all gauge symmetries are introduced in the same way, i.e., in terms of fundamental fermionic fields only. Gauge bosons, as well as gravitons, will be composite and their interactions will be automatically unified (and strong) at the same scale, the inverse radius of compositeness. If we exclude gravity from the picture, this approach reduces to the one of Ref. 2. My presentation here will follow closely the papers written by Amati and myself on the subject[3].

This unconventional unification scheme leads to the usual framework in a "low energy" regime to be carefully defined later. Indeed, as we move down from the unification mass, the effective low energy interactions will be the ones dictated by gauge invariance and the principle of equivalence, by the presence of the corresponding dynamically generated gauge bosons and by dimensional arguments. Conventional gauge theories and gravity appear then as phenomenological low energy approximations. Away from that regime, our theory will differ from the conventional ones, not even allowing, for instance, an unambiguous definition of a space-time metric. This last point clearly indicates the difference between our approach and those which try to generate quantum gravity from matter fluctuations in a background metric field[4]. Let us remark that, as discussed in Ref. 2, the vector gauge theories appearing in the low energy regime have the property of going towards unification at higher energies in a crescendo. This sort of asymptotically non-free GUT has been proposed and shown to be satisfactory on pure phenomenological grounds in Refs. 5 and 6. For us this sort of unification is an automatic consequence of gauge field compositeness.

In the preceding discussion I have taken the attitude that the gauge symmetries observed at low energies ($\lesssim$ 100 GeV) are those relevant for our unification procedure. One could take the alternative attitude that the gauge symmetry that unified with gravity at M_{Planck} is the GUT interaction itself. In this case GUT would appear as an effective "low energy" theory valid at $E \approx M_{GUT} << M_{P\ell}$ which could further break at lower energies down to SU(3) $\otimes$ SU(2) $\otimes$ U(1) following one of the conventional GUT schemes. This scheme would necessitate several GUT families above M_{GUT} in order to reach a small unified gauge coupling at M_{GUT}. This idea is perhaps less appealing in our view, but shows the possibility of the co-existence of two very different unification mechanisms.

The outline of the talk is as follows. I shall start by constructing actions invariant under a set of gauge transformations on the basis of only fundamental fermions. To avoid internal indices (colour, flavour, etc.) I shall limit the treatment to Lorentz and U(1) gauge invariances (i.e., gravitation and QED), (see Ref. 2 for inclusion of other gauge symmetries). After having written the invariant action, I shall discuss a strategy to analyze it through the introduction of a suitable set of auxiliary fields. Next I shall discuss the vacuum properties, associated with homogeneous classical solutions. They are shown to generate a scale Λ through a spontaneous breaking of local Lorentz and general co-ordinate transformations. This scale, which appears as a vacuum property, represents the only dimensional quantity of the theory.

Subsequently, I shall outline the computation of the quadratic fluctuations around the previous stationary field configuration up to second order derivatives. This will allow us to recognize induced mass and kinetic terms of the eigenmodes. A light sector (massless bosons and arbitrarily light fermions) is identified, together with a heavy sector (masses of order Λ). If the heavy eigenmodes are not excited - as expected when all energies are much smaller than Λ - the remaining light modes are shown to represent a spinor, a photon and a graviton described by an effective action which is the usual Einstein-Dirac-Maxwell action with the Newton constant and the electric charge given in terms of Λ. This relation can be described by saying that Λ is to be identified with both the Planck mass and the Landau pole position. At the end, I shall comment briefly on how this theory differs from the conventional one in the short distance domain where one can hope it solves some of the well-known difficulties of quantum gravity.

2. We begin by writing, in terms of a single spinor field ψ, an action invariant under local U(1) and Lorentz transformations. In their infinitesimal form these read*):

$$\psi(x) \to (1 - i\alpha(x))\psi(x) \; ; \quad U(1) \tag{1a}$$

$$\psi(x) \to (1 - \frac{i}{4}\sigma_{\beta\gamma}\omega^{\beta\gamma}(x))\psi(x) \; ; \quad O(3,1) \tag{1b}$$

Besides, we shall write the action as a space-time integral of a scalar density thus enforcing invariance under general co-ordinate transformations $x_\mu \to x'_\mu(x)$. Our first task will be to construct a covariant derivative. Defining

$$D_\mu = \partial_\mu + A_\mu + \sigma^{\beta\gamma}A_{\mu,\beta\gamma} \, , \tag{2}$$

$$A_\mu = \frac{1}{8}\,\mathrm{Tr}(H^{-1}\overleftarrow{H}_\mu - \overrightarrow{H}_\mu H^{-1})$$

$$A_{\mu,\beta\gamma} = \frac{1}{16}\,\mathrm{Tr}(\sigma_{\beta\gamma}H^{-1}\overleftarrow{H}_\mu - \sigma_{\beta\gamma}\overrightarrow{H}_\mu H^{-1}) \tag{3}$$

*) We adopt the normalizations and conventions of J.D. Bjorken and S.D. Drell, Relativistic Quantum Fields, McGraw-Hill (1965). The usual contraction of Lorentz indices through $\eta_{\alpha\beta} = (1,-1,-1,-1)$ is tacitly understood.

and the Dirac index matrices

$$H = \psi\bar{\psi}, \quad \overleftrightarrow{H}_\mu = \psi \overleftrightarrow{\partial}_\mu \bar{\psi} \tag{4}$$

it is easy to check that D_μ is a good covariant derivative under the transformations (1). This means that $D_\mu\, \psi(x)$ transforms as $\psi(x)$ in (1).

As a consequence, the composite operator

$$W_{\alpha\mu}(x) = \bar{\psi}(x)\frac{i}{2}\left[\gamma_\alpha, D_\mu\right]\psi(x) \equiv \frac{i}{2}\,\bar{\psi}(x)\gamma_\alpha\overrightarrow{D}_\mu\psi(x) - \frac{i}{2}\,\bar{\psi}(x)\overleftarrow{D}_\mu\gamma_\alpha\psi(x) \tag{5}$$

transforms as a usual vierbein, i.e., as a set of four four-vectors (one for each index α) under general co-ordinate transformations and, for each μ, as a vector under local Lorentz transformations. Therefore $\det W_{\alpha\mu}$, where the determinant refers to the 4×4 matrix of indices $\alpha\mu$, is a scalar density. This can still be multiplied by a scalar invariant which, due to the lack of a space-time metric tensor to saturate space-time indices, can only be an arbitrary scalar function $V(H)$. We are thus led to define

$$A_f = \int d^4x (\det W_{\alpha\mu}) V(H) \tag{6}$$

as the most general relativistic and U(1) invariant action of our single spinor system.

Let us notice that even if non-polynomial in ψ and $\bar{\psi}$ [due to H^{-1} in Eqs. (3)] A_f is polynomial of fourth order in the derivatives and only first order in the derivative with respect to any space-time co-ordinate. The four derivatives compensate the dimension of d^4x so that ψ is dimensionless and so is H. Thus $V(H)$ and therefore A_f are free of any dimensional parameters. We shall see that the arbitrariness of $V(H)$ will not jeopardize the possibility of analyzing the consequences of (6). $V(H)$ will indeed enter into the condition for the minimum of A_f in terms of H. As we shall see, the existence of a minimum will be important while its location (arbitrary for arbitrary V) will be irrelevant. Further details of V will influence the spectrum of the theory but in a way which will not affect its basic physical aspects.

We shall be interested in extending the single spinor case discussed up to now by introducing possible replicas, i.e., different fields ψ_i transforming analogously under the local transformations (1). In so doing we may assign a different weight K_i to the contribution of the ith replica to the definition (4) of H that enters into the definition (2), (3) of the covariant derivative. This introduces therefore a set of dimensionless parameters K_i into the invariant action (6) where now

$$W_{\alpha\mu} = \frac{i}{2} \sum_i \bar{\psi}_i [\gamma_\alpha, D_\mu] \psi_i \tag{7}$$

D_μ being given by (3) in terms of

$$H = \sum_i K_i \psi_i \bar{\psi}_i \ , \quad \overleftarrow{\vec{H}}_\mu = \sum_i K_i \psi_i \overleftarrow{\vec{\partial}}_\mu \bar{\psi}_i \tag{8}$$

We shall actually use this extension in two ways. One will be to use N equal weights (K_i=1, i=1, ..., N) in order to define a 1/N expansion. The other will be to enforce a Pauli-Villars type of regularization.

The quantum theory we are considering implies the functional integration over the fundamental fields ψ_i, $\bar{\psi}_j$ of the exponential of the action A_f of Eq. (6) to which source terms $\bar{j}_i\psi_i + j_i\bar{\psi}_i$ are tacitly added. This is obviously a complex task because of the high non-linearity of A_f. Our strategy is to introduce as many auxiliary bosonic fields as needed in order to render A bilinear in ψ_i, $\bar{\psi}_i$ and to perform next the fermion functional integral. This leads us to a new action which, though equivalent to the original one, is written only in terms of the auxiliary fields.

For this purpose we introduce four functional Lagrange multipliers $\xi^{\mu\alpha}$, $\lambda^{\mu,\beta\gamma}$, λ^μ and ϕ needed to enforce the definitions of $W_{\alpha\mu}$, $A_{\mu,\beta\gamma}$, A_μ and H as given in Eqs. (3), (7) and (8)*). The action written in terms of the original fermions and of the auxiliary bosonic fields, reads:

$$A_{f,b} = \int d^4x \det W\{[V(H) - \xi^{\mu\alpha}W_{\alpha\mu} - \lambda^{\mu,\beta\gamma}A_{\mu,\beta\gamma} - \lambda^\mu A_\mu + \mathrm{Tr}(H\phi)] +$$

$$+ \sum_i \bar{\psi}_i (\xi^{\mu\alpha} \tfrac{i}{2}[\gamma_\alpha, (\partial_\mu + A_\mu + \sigma_{\beta\gamma}A_{\mu,\beta\gamma}] - K_i\phi + \tag{9}$$

$$+ \frac{K_i}{4} H^{-1}\overleftrightarrow{\partial}_\mu \lambda^\mu + \frac{K_i}{16} (\sigma_{\beta\gamma} H^{-1}\overleftarrow{\partial}_\mu - H^{-1}\sigma_{\beta\gamma}\overrightarrow{\partial}_\mu)\lambda^{\mu,\beta\gamma})\psi_i\}$$

The integration over ψ_i, $\bar{\psi}_i$ is now formally straightforward. However, in order to give meaning to the tr log obtained in that way, an ultra-violet regularization procedure must be introduced. It is crucial to do this while respecting all the invariances of the theory. We shall adopt a Pauli-Villars (PV) prescription by choosing replicas characterized by parameters K_i in Eq. (8) and by PV parameters C_i satisfying

$$\sum_i C_i K_i^{2n} = 0, \quad n = 0, 1, \ldots \tag{10}$$

*) Notice that ϕ and H are dimensionless matrices in Dirac space, that $W_{\alpha\beta}$, $A_{\mu\ \beta\gamma}$, A_μ have dimensions of a mass and that $\xi^{\mu\alpha}$, $\lambda^{\mu,\beta\gamma}$ and λ^μ have dimension m^{-1}.

extending the usual PV conditions. Recall that, as usual, negative C_i represent PV ghosts. We shall discuss later the physical relevance of the regularization parameters.

Let me stress that in order to regularize the integration over the fermion fields we had to introduce into the theory a set of dimensionless parameters satisfying a set of conditions. The important point, nevertheless, is that all invariances of the theory have been preserved. In particular, our regularization has not broken conformal invariance unlike what inevitably happens when mass and kinetic terms in the action have the same degree of homogeneity in ψ, $\bar{\psi}$.

The condition (10) allows one to give a meaning to the formal expression arising from integration over the fermion fields and hence to the effective bosonic action A_b defined by

$$\exp A_b = \int \prod_i \mathcal{D}\psi_i \mathcal{D}\bar{\psi}_i \exp A_{f,b} \tag{11}$$

Using (9) for $A_{f,b}$ we immediately obtain

$$\begin{aligned} A_b = \int d^4x \det W \big[V(H) - \dot{\xi}^{\mu\alpha} W_{\alpha\mu} - \lambda^{\mu,\beta\gamma} A_{\mu,\beta\gamma} - \lambda^{\mu} A_{\mu} + \mathrm{Tr}(H\phi)\big] + \\ + \sum_i C_i \operatorname{tr} \log\{\dot{\xi}^{\mu\alpha} \frac{i}{2} \big[\gamma_\alpha, (\partial_\mu + A_\mu + \sigma_{\beta\gamma} A_{\mu,\beta\gamma}\big] - K_i\phi + \\ + \frac{K_i}{4} H^{-1} \overleftrightarrow{\partial}_\mu \lambda^\mu + \frac{K_i}{16}(\sigma_{\beta\gamma} H^{-1} \overleftarrow{\partial}_\mu - H^{-1} \sigma_{\beta\gamma} \overrightarrow{\partial}_\mu) \lambda^{\mu,\beta\gamma}\} \end{aligned} \tag{12}$$

We have therefore translated our initial action, written in terms of only fermions, into the action (12) written in terms of only bosons. This language will be better suited for analyzing the spectrum of the theory. Before doing that, let us stress that among the variety of boson fields we are left with, none has the properties of a metric field. We have two fields $W_{\alpha\mu}$ and $\xi^{\mu\alpha}$ which could be associated with lower and upper index vierbeins and could be used to define space-time metric tensors such as $W_{\alpha\mu}W_{\alpha\nu}$ and $\xi^{\mu\alpha}\xi^{\nu\alpha}$. However, these two tensors cannot be identified with $g_{\mu\nu}$ and $g^{\mu\nu}$ unless one is the inverse of the other. We thus see that the possibility of defining a metric needs the identification of W with ξ^{-1} up to a proportionality constant. This is not of course the case in our theory since W and ξ are independent fields. We shall be able nevertheless to determine an asymptotic kinematical regime in which that identification will be correct, thus allowing the definition of a space-time metric as an approximate concept.

3. We shall attempt the analysis of our action (12) written in terms of auxiliary bosonic fields by a semi-classical treatment, which consists of finding stationary points of the action and expanding around them. We note that this procedure becomes justified as a 1/N expansion if N identical replicas of each fermion plus regulator families are considered. A factor N then multiplies the tr log term of (12) playing the role of the $1/\hbar$ factor of the other terms in the action.

Let us now discuss the stationarity condition which implies the vanishing of the first derivatives of A_b with respect to all auxiliary fields. The only first order functional derivative not involving the tr log in (11) is

$$\frac{\delta A_b}{\delta W_{\alpha\mu}} = \det W\{\dot{\xi}^{\mu\alpha} - (W^{-1})^{\mu\alpha}(V(H) - \dot{\xi}^{\mu\alpha}W_{\alpha\mu} - \lambda^{\mu}A_{\mu} - \lambda^{\mu,\beta\gamma}A_{\mu,\beta\gamma} + \mathrm{Tr}(\phi H))\} \tag{13}$$

Its vanishing gives a proportionality between $<\dot{\xi}>$ and $<W>^{-1}$ so that neither $<\xi>$ nor $<W>$ can be zero, implying a spontaneous breaking of local Lorentz invariance. In order to discuss further the stationarity conditions we encounter the difficulty of having to evaluate derivatives of the tr log operator. This operator is a functional of the classical field configurations. Therefore the classical equations will themselves depend on the class of functions among which one looks for solutions. We will be interested in a vacuum that preserves global Poincaré invariance. We therefore will look for classical homogeneous solutions (i.e., space-time independent) in the absence of external sources. In this case the evaluation of the tr log expression and its derivatives is simple.

Let us first discuss the stationarity condition on H. The presence of H^{-1} in A_b implies that $H \sim 0$ configurations should be strongly damped. Indeed, the vanishing of $\delta A_b/\delta H$ equates $V'(H)$ with an expression that, depending on the fields, becomes singular for $H \to 0$. We assume that this equation, which has a well-defined meaning for $H \neq 0$, may be satisfied by some value of H. This may possibly imply some condition on V(H) which we assume to be met. By taking into account Lorentz invariance we thus write

$$<H> = v\,\mathbf{1} \tag{14}$$

and as we shall see, the actual value of v will be immaterial as long as it is non-zero.

Continuing our analysis, it is easy to see that the four equations

$$\frac{\delta A_b}{\delta\lambda^\mu} = \frac{\delta A_b}{\delta A_\mu} = \frac{\delta A_b}{\delta\lambda^{\mu,\beta\gamma}} = \frac{A_b}{\delta A_{\mu,\beta\gamma}} = 0 \qquad (15)$$

are solved by

$$\langle\lambda^\mu\rangle = \langle A_\mu\rangle = \langle\lambda^{\mu,\alpha\beta}\rangle = \langle A_{\mu,\alpha\beta}\rangle = 0 \qquad (16)$$

in accordance with global Lorentz invariance. The remaining two stationary conditions give

$$\frac{\delta A_b}{\delta\phi} = 0 \Rightarrow v \det\langle W\rangle = -\frac{1}{4}\sum_i C_i \int\frac{d^4p}{(2\pi)^4}\,\mathrm{Tr}\,\frac{K_i}{\langle\xi\rangle^{\mu\alpha}\gamma_\alpha p_\mu - K_i\langle\phi\rangle} = \qquad (17)$$

$$= \frac{1}{\det\langle\xi\rangle}\,\frac{1}{u}\sum_i C_i \int\frac{d^4r}{(2\pi)^4}\,\frac{K_i^2u^2}{r^2-K_i^2u^2}$$

$$\frac{\delta A_b}{\delta\xi^{\mu\alpha}} = 0 \Rightarrow \langle W_{\alpha\mu}\rangle \det\langle W\rangle = -\sum_i C_i \int\frac{d^4p}{(2\pi)^4}\,\mathrm{Tr}\,\frac{\gamma_\alpha p_\mu}{\langle\xi\rangle^{\nu\beta}\gamma_\beta p_\nu - K_i\langle\phi\rangle} = \qquad (18)$$

$$= \frac{(-1)}{\det\langle\xi\rangle}\,\langle\xi^{-1}\rangle_{\alpha\mu}\sum_i C_i \int\frac{d^4r}{(2\pi)^4}\,\frac{r^2}{r^2-K_i^2u^2}$$

where

$$r^\alpha \equiv \langle\xi\rangle^{\mu\alpha}p_\mu \qquad (19)$$

and

$$\langle\phi\rangle = u\,\mathbb{1} \qquad (20)$$

as required by global Lorentz invariance.

By using the condition $\Sigma_i C_i = 0$ we recognize the equality of the last two integrals of Eqs. (17) and (18). Introducing now the dimensional proportionality constant η by

$$\langle W\rangle_{\alpha\mu} = \eta\langle\xi^{-1}\rangle_{\alpha\mu} \qquad (21)$$

equations (14), (17) and (18) imply

$$uv = \eta = V(v)$$

$$I_4 \equiv \sum_i C_i \int\frac{d^4r}{(2\pi)^4}\,\frac{r^2}{r^2-h_i^2} = \eta^5, \quad h_i \equiv K_i u \qquad (22)$$

where, we remark, the h_i satisfy the same PV conditions (10) as the K_i.

Since W and ξ have dimensions, their non-zero constant v.e.v. implies a vacuum generated scale Λ which, taking into account Eq. (21), can be defined without loss of generality by

$$<\xi>^{\mu\alpha} = \Lambda^{-1}\,\eta_{\mu\alpha},\ <W>_{\alpha\mu} = \eta\Lambda\cdot\eta_{\mu\alpha}$$

$$\eta_{\mu\alpha} \equiv \mathrm{diag.}\ (1,-1,-1,-1) \tag{23}$$

Equations (23) show explicitly that our vacuum breaks (spontaneously) local Lorentz as well as general co-ordinate transformation invariances leaving a diagonal global Lorentz invariance. The spontaneous breaking of scale invariance generates the scale Λ as a vacuum property[7]. This is the only dimensional scale of the theory, so that all masses (as well as other dimensional quantities) will be proportional to it. In the case of fermions, the masses which appear as poles in the propagator are given by

$$m_i = h_i\Lambda \tag{24}$$

We may, of course, choose $h_0 << h_i$, $i \neq 0$ in order to have the original fermion arbitrarily light. This mass is protected by a chiral invariance for the light fermion in the limit $K_0 \to 0$ ($h_0 \to 0$). Indeed, as is evident from the form (9) of the action, K_i measures the strength of the γ_5 non-invariant terms. Furthermore, for $K_0 \to 0$, the light fermion is excluded from the condensate (14).

Let us call h the smallest h_i ($i\neq 0$) so that $M = h\Lambda$ represents the mass of the lightest regulator (or set of regulators). M acts as a cut-off in our theory and will appear explicitly in Green's functions. M cannot be sent to infinity since an evaluation of I gives $I_4 \sim h^4$, hence:

$$V(v)^5 = I_4 \sim h^4 = (\frac{M}{\Lambda})^4 \tag{25}$$

Let us discuss for a moment the meaning of Eq. (25). In order for the homogeneous flat-space solution*) to exist, the parameters h_i and V(v) have to be related by Eq. (22). This can be seen either as a condition on the regulator mass M in terms of a given V(v) and Λ or, once M/Λ is arbitrarily chosen, as a fine tuning of an additional constant contribution to V(H). This is not surprising if one realizes that both V and the PV parameters are part of the regularization needed to give meaning to the theory. In this connection let us remark that we will be able to identify some heavy fields (as, for instance, $\lambda^{\mu,\alpha\beta}$), the integration over which would induce an additional contribution to the effective potential

*) This will also imply the absence of an induced cosmological term as discussed later.

V(H) which depends on the PV parameters and is singular for $H \to 0$. One may thus consider V(H) of Eq. (24) as a driving term. We notice, however, that had we considered the regularization mass M as the basic scale of the theory, Eq. (25) would determine, without fine tuning, Λ and hence the vacuum scale of Eq. (23). This could appear to follow from having explicitly broken scale invariance through the introduction of M. Nevertheless, the just stated equivalence with the spontaneous breaking of scale invariance shows that our regularization procedure represents a soft breaking.

The fact that our breaking is really of a spontaneous nature will be substantiated later when we will show a typical generation of massive O(3,1) gauge bosons through the eating up of Goldstone particles associated with broken generators.

4. We shall now compute the quadratic part of the action (12) in the shifted bosonic fields. Furthermore, we shall expand this quadratic action up to second order in the field momenta q_i. The q independent terms, once diagonalized, will allow us to determine masses and, in particular, to identify the massless modes. Terms quadratic in the momenta will provide induced kinetic terms for the massless fields and will lead to the identification of the induced Newton and fine structure constants.

We shall denote by a tilded field its fluctuation around its v.e.v., lowering for convenience all its upper indices by the flat η tensor

a) <u>q Independent Terms</u>: since the field $W_{\alpha\mu}$ does not appear inside the tr log there is no dependence of A_b on the derivatives of W. One then finds, at all momenta

$$\frac{\delta A_b}{\delta\tilde{W}_{\alpha\mu}\delta\tilde{W}_{\beta\nu}} = -\eta^3\Lambda^2(\eta_{\mu\alpha}\eta_{\nu\beta} + \eta_{\mu\beta}\eta_{\nu\alpha}) \tag{26}$$

$$\frac{\delta A_b}{\delta\tilde{W}_{\alpha\mu}\delta\tilde{\xi}_{\nu\beta}} = -\eta^4\Lambda^4(\eta_{\mu\alpha}\eta_{\nu\beta} + \eta_{\alpha\beta}\eta_{\mu\nu}) \tag{27}$$

the derivatives being of course evaluated for vanishing fluctuations.

The dependence of A_b on ξ is more complicated. One finds:

$$\frac{\delta A_b}{\delta\tilde{\xi}_{\mu\alpha}\delta\tilde{\xi}_{\nu\beta}} = -\Lambda^6\, T_{\alpha\beta,\mu\nu}(q) \tag{28}$$

where $T_{\alpha\beta,\mu\nu}(q)$ is given by a one-loop Feynman diagram, i.e.:

$$T_{\alpha\beta,\mu\nu}(q) = \sum_i C_i \int \frac{d^4p}{(2\pi)^4} \frac{p_\mu p_\nu}{D_q D_{-q}} \mathrm{Tr}\left[\gamma_\alpha(\not{p}+\frac{\not{q}}{2}+m_i)\gamma_\beta(\not{p}-\frac{\not{q}}{2}+m_i)\right] \tag{29}$$

$$D_{\pm q} = (p \pm \frac{q}{2})^2 - m_i^2$$

with m_i given in Eq. (24). At q = 0 one finds

$$T_{\alpha\beta,\mu\nu}(0) = \sum_i C_i \int \frac{d^4p}{(2\pi)^4} \frac{4p_\mu p_\nu \eta_{\alpha\beta}(m_i^2-p^2) + 8p_\mu p_\nu p_\alpha p_\beta}{(p^2-m_i^2)^2} = \tag{30}$$

$$= -\eta_{\alpha\beta}\eta_{\mu\nu}I_4 + \frac{1}{3}(\eta_{\mu\nu}\eta_{\alpha\beta}+\eta_{\mu\alpha}\eta_{\nu\beta}+\eta_{\mu\beta}\eta_{\nu\alpha})\cdot\sum_i C_i\int\frac{d^4p}{(2\pi)^4}\frac{p^4}{(p^2-m_i^2)}$$

with I_4 defined in Eq. (22). With our PV regularization the last integral in (30) is equal to $3I_4$ so that:

$$\left.\frac{\delta A_b}{\delta\tilde{\xi}_{\mu\alpha}\delta\tilde{\xi}_{\nu\beta}}\right|_{q=0} = -I_4\Lambda^6(\eta_{\mu\alpha}\eta_{\nu\beta} + \eta_{\alpha\nu}\eta_{\beta\mu}) = -\eta^6\Lambda^6(\eta_{\mu\alpha}\eta_{\nu\beta} + \eta_{\alpha\nu}\eta_{\beta\mu}) \tag{31}$$

where Eq. (22) has been used.

The mass matrix formed by the quadratic terms (26), (27) and (31) can be written as

$$-\Lambda^4\eta^5(\Lambda\tilde{\xi}_{\alpha\mu} + \frac{1}{\Lambda\eta}\tilde{W}_{\alpha\mu})(\Lambda\tilde{\xi}_{\beta\nu} + \frac{1}{\Lambda\eta}\tilde{W}_{\alpha\mu})(\delta_{\alpha\mu}\delta_{\beta\nu} + \delta_{\alpha\nu}\delta_{\beta\mu}) \tag{32}$$

from which one can recognize the two eigenmodes

$$\tilde{V}_{\alpha\mu} = \frac{1}{2}(-\Lambda\tilde{\xi}_{\alpha\mu} + \frac{1}{\Lambda\eta}\tilde{W}_{\alpha\mu}) \;;\quad \tilde{U}_{\alpha\mu} = \frac{1}{2}(\Lambda\tilde{\xi}_{\alpha\mu} + \frac{1}{\Lambda\eta}\tilde{W}_{\alpha\mu}) \tag{33}$$

the first of which is massless. The mode $\tilde{U}$ is a heavy mode, i.e., a field with mass of $O(\Lambda)$. Later on we shall be able to see that, in a suitable low energy approximation, $\tilde{V}$ of Eq. (33) is the fluctuation of a vierbein field V satisfying the equations of the usual Einstein theory.

Let us now turn to the q independent quadratic terms in the connections $A_{\mu,\beta\gamma}$, A_μ and in their corresponding Lagrange multipliers $\lambda^{\mu,\beta\gamma}$, λ^μ.

Straightforward calculations give

$$\frac{\delta A_b}{\delta A_{\mu,\alpha\beta}\delta A_{\nu,\gamma\beta}} = 8\Lambda^2 I_2 O^{\alpha\beta,\gamma\delta}_{\mu\nu} \tag{34}$$

$$\frac{\delta A_b}{\delta\lambda_{\mu,\alpha\beta}\delta A_{\nu,\gamma\delta}} = \frac{1}{2}\Lambda^4\eta^4\,\tilde{O}^{\alpha\beta,\gamma\delta}_{\mu\nu} \tag{35}$$

$$\frac{\delta A_b}{\delta\lambda_{\mu,\alpha\beta}\delta\lambda_{\nu,\gamma\delta}} = \Lambda^6\hat{O}^{\alpha\beta,\gamma\delta}_{\mu\nu} \tag{36}$$

where I_2 is a regularized (and otherwise quadratically divergent) integral whose evaluation gives:

$$I_2 = c\eta^{5/2} = c'(M^2/\Lambda^2) \tag{37}$$

with c and c' numerical constants (see Ref. 3). The tensors appearing in Eqs. (34)-(36) consist of products of η's and are antisymmetric in $\alpha \leftrightarrow \beta$, $\gamma \leftrightarrow \delta$.

Turning now to the A_μ, λ_μ sector, the situation is the same as the one discussed in Ref. 2. The two-by-two mass matrix has only one non-zero entry corresponding to a mass of order Λ for the λ^μ field. The absence of a mass term for A_μ or of a $\lambda^\mu A_\mu$ mixing is of course a consequence of U(1) gauge invariance (the case of $A_{\mu,\beta\gamma}$ is more subtle in this respect, as discussed below).

Finally, the two scalar fields ϕ and M can be shown to be also heavy. We do not give here the explicit expression but just remark that the mass of M depends on the details of V(M) [e.g., on V"(v)].

b) q Dependent Terms: we start with terms linear in the momenta. These mix tensors differing by one index, in particular $A_{\mu,\beta\gamma}$ and $\lambda^{\mu,\beta\gamma}$ with $\xi^{\nu\alpha}$. We find

$$\frac{\delta A_b}{\delta\lambda_{\mu,\alpha\beta}\delta\tilde{\xi}_{\gamma\delta}} = \frac{1}{8}\eta^4\Lambda^5 q_\nu\tilde{O}^{\alpha\beta,\gamma\delta}_{\mu\nu} \tag{38}$$

$$\frac{\delta A_b}{\delta A_{\mu,\alpha\beta}\delta\tilde{\xi}_{\gamma\delta}} = 2\Lambda^3 I_2 q_\nu O^{\alpha\beta,\gamma\delta}_{\mu\nu} \tag{39}$$

where the tensors O and $\tilde{O}$ of Eqs. (34) and (35), being antisymmetric in $\gamma \leftrightarrow \delta$, pick up the antisymmetric part of $\tilde{\xi}$. We shall come back to the above results after discussion of quadratic terms in q for the W,ξ sector which, as mentioned above, contain only ξ. Defining

$$\xi^{\mu\alpha}_{s,a} = \frac{1}{2}(\xi^{\mu\alpha} \pm \xi^{\alpha\mu}) \tag{40}$$

we find up to second order in q:

$$\frac{\delta A_b}{\delta\xi_s\delta\xi_a} = 0 \; ; \quad \frac{\delta A_b}{\delta\tilde{\xi}^s_{\mu\alpha}\delta\tilde{\xi}^s_{\nu\beta}} = O^{(+)}_{\mu\alpha,\nu\beta} \; ; \quad \frac{\delta A_b}{\delta\tilde{\xi}^a_{\mu\alpha}\delta\tilde{\xi}^a_{\nu\beta}} = O^{(-)}_{\mu\alpha,\nu\beta} \tag{41}$$

where

$$\begin{aligned} O^{(+)}_{\mu\alpha,\nu\beta} = \tfrac{1}{6}\Lambda^4 I_2 \big[& q^2(\eta_{\alpha\beta}\eta_{\mu\nu} + \eta_{\alpha\nu}\eta_{\beta\mu} - 2\eta_{\alpha\mu}\eta_{\beta\nu}) - \\ & - q_\alpha q_\beta \eta_{\mu\nu} - q_\mu q_\beta \eta_{\alpha\nu} - q_\mu q_\nu \eta_{\alpha\beta} - q_\alpha q_\nu \eta_{\beta\mu} + \\ & + 2q_\beta q_\nu \eta_{\alpha\mu} + 2q_\alpha q_\mu \eta_{\beta\nu}\big] \end{aligned} \tag{42}$$

$$O^{(-)}_{\mu\alpha,\nu\beta} = \tfrac{1}{2}\Lambda^4 I_2 \left[q_\gamma q_\delta O^{\mu\alpha,\nu\beta}_{\gamma\delta}\right] \tag{43}$$

Recalling that $\tilde{\xi}_s$ does not appear in Eqs. (38) and (39) we see that the full kinetic term in $\tilde{\xi}_s$ is given by $O^{(+)}$, implying a contribution to A_b of the form:

$$\begin{aligned} -\tfrac{1}{6}\Lambda^4 I_2 \big[& \Box(\tilde{\xi}^{\mu\alpha}_s \tilde{\xi}^s_{\mu\alpha} - \tilde{\xi}^{\mu}_{s,\mu}\tilde{\xi}^{\alpha}_{s,\alpha}) - 2\partial_\alpha\partial_\beta\tilde{\xi}^{\mu\alpha}_s\tilde{\xi}^{\beta}_{s,\mu} + \\ & + 2\partial^\alpha\partial^\mu\tilde{\xi}^s_{\mu\alpha}\tilde{\xi}^{\nu}_{s,\nu}\big] \end{aligned} \tag{44}$$

The quadratic pieces that contain ξ_a [Eqs. (38), (39), (43)], can be combined with those containing $A_{\mu,\gamma\delta}$ and $\lambda^{\mu,\gamma\delta}$ to give the following contributions to A_b:

$$4\Lambda^2 I_2 B_{\mu,\alpha\beta}B_{\nu,\gamma\delta}O^{\alpha\beta,\gamma\delta}_{\mu\nu} + \tfrac{1}{2}\Lambda^4\eta^4\lambda_{\mu,\alpha\beta}B_{\nu,\gamma\delta}\tilde{O}^{\alpha\beta,\gamma\delta}_{\mu\nu} \tag{45}$$

where

$$B_{\mu,\alpha\beta} = A_{\mu,\alpha\beta} + \frac{\Lambda}{4}q_\mu\tilde{\xi}^a_{\alpha\beta} \tag{46}$$

The fact that $A_{\mu,\alpha\beta}$ and $\tilde{\xi}^a_{\mu\alpha}$ appear only in the combination $B_{\mu,\alpha\beta}$ of Eq. (46) is due to the O(3,1) gauge invariance of the theory. $A_{\mu,\alpha\beta}$, being a connection, transforms inhomogeneously under the transformation (2.16):

$$A_{\mu,\alpha\beta} \rightarrow A'_{\mu,\alpha\beta} + \frac{1}{4}(\sigma^{\gamma\delta})_{\alpha\beta}\partial_\mu\omega_{\gamma\delta}(x) \tag{47}$$

Naïvely this would seem to forbid terms in the action containing $A_{\mu,\alpha\beta}$ without derivatives as is the case for the electromagnetic field A_μ. On the other hand, in our case we can construct another field transforming as a connection by taking derivatives of $\tilde{\xi}^a_{\alpha\beta}$.

Thus the combination $B_{\mu,\alpha\beta}$ of Eq. (46) is again a good tensor. The first term in Eq. (46) shows that $B_{\mu,\alpha\beta}$ is a massive field which has "eaten up" à la Higgs the six would-be Goldstone bosons $\tilde{\xi}^a_{\alpha\beta}$ of the spontaneously broken O(3,1) invariance. This fact confirms the validity of our statement that the flat-metric expectation values of W and ξ represent a spontaneous rather than explicit breaking of the symmetries of the original action, dilatation symmetry included.

Turning finally to the q dependent terms involving the λ^μ and A_μ electromagnetic fields we find again, as in Ref. 3, that they involve A_μ only through the field strenght tensor $F_{\mu\nu}$. This ensures that A_μ is an exactly massless field contributing to the action with a term

$$-\frac{1}{4} I_0 F^2_{\mu\nu} \; ; \quad I_0 \Big|_{q=0} \cong \frac{1}{12\pi^2} \log(M^2/m^2) + O(1) \tag{48}$$

I_0 here is the usual QED vacuum polarization integral whose dependence on q is well known.

Summarizing the results obtained so far, we have identified in our spectrum the following light particles:

i) one or several light fermions (according to the number of parameters K_i set to be much smaller than one) protected by chiral invariance from obtaining masses of $O(\Lambda)$;

ii) one massless spin one particle associated with the exact U(1) gauge symmetry, i.e., a photon. In a similar way, one would obtain eight gluons from requiring local SU(3) invariance;

iii) a massless excitation with the structure of the vierbein, associated with general relativistic invariance.

Besides these, we have obtained a number of "heavy" excitations, i.e., fields with mass of $O(\Lambda)$. As long as these heavy degrees of freedom carry energies $E_i \ll \Lambda$, they will not be excited away from their v.e.v.'s. In this limit we have, to a very good approximation, the following relations

$$\eta \Lambda \tilde{\xi} = -\frac{1}{\Lambda}\tilde{W} \Rightarrow \dot{\xi} = \eta W^{-1} \tag{49}$$

$$H = v \, \mathbb{1} \; , \quad \phi = u \, \mathbb{1} \tag{50}$$

$$B^{\alpha\beta}_\mu = 0 \Rightarrow A^{\alpha\beta}_\mu = \frac{1}{4} \dot{\xi}^{-1}_{\alpha\nu} \partial_\mu \dot{\xi}^{\nu\beta} \tag{51}$$

$$\lambda^{\mu} = \lambda^{\mu,\alpha\beta} = 0 \tag{52}$$

In this regime Eq. (49) allows, as discussed before, the definition of a metric after identification of $\Lambda\xi$ $(W/\eta\Lambda)$ with the usual vierbein fields with upper (lower) indices. In that case we recognize in Eq. (51) the usual definition of the Lorentz connection in terms of the vierbein. Looking at the momentum-dependent part of the fluctuations, we then see that Eq. (44) is nothing but the quadratic expansion around the flat vierbein solution of

$$\frac{1}{6}\,\Lambda^2 I_2 \sqrt{-g}\;R \tag{53}$$

with g and R the usual metric determinant and scalar curvature expressed in terms of the vierbein.

At fourth order in the derivatives, a term[8] (log M/m) $(R^2-3R_{\mu\nu}R^{\mu\nu})$ appears. Higher order terms compatible with general relativistic invariance will also be generated, their dimensions being compensated by appropriate powers of M^{-1}. All these R^2 and higher order terms are therefore negligible in the low energy regime under discussion.

We notice at this point the absence of an induced $\sqrt{-g}$ cosmological term that, if present, would have appeared with an M^4 scale factor. This is an expected consequence of the fact that in our approach the flat metric is a classical solution and that a cosmological term would contain linear vierbein fluctuations.

Hence, in the low energy regime under discussion, we recover the Einstein-Dirac action for gravity plus spin 1/2 matter, with a vanishing cosmological constant and an induced Newton constant given by

$$(16\pi G_N)^{-1} = \frac{1}{6}\,\Lambda^2 I_2 = \frac{1}{6}\,c' M^2 \;\; ; \; c' = \text{numerical constant} \tag{54}$$

Equation (54) shows that the Newton constant is related to the vacuum generated scale Λ by precisely the coefficient that makes it equal to the momentum cut-off M.

As already said, the quadratic terms in λ^{μ} and A_{μ} provide the usual kinetic term $F^2_{\mu\nu}$. As in Ref. 3, in the low energy regime where we can set $\lambda^{\mu} = 0$, we find nothing but the ordinary QED action with an effective fine structure constant given by the inverse of the quantity I_0 of Eq. (48). For $m^2 << q^2 << M^2$ one finds:

$$\alpha(q^2) = 3\pi\left[\log(M^2/q^2) + O(1)\right]^{-1} \tag{55}$$

In the case of N light fermions of U(1) charges e_i we would have obtained for the ith fermion the effective electromagnetic coupling

$$\alpha_i(q^2) = 3\pi\, e_i^2 \Big[\sum_{j=1}^{N} e_j^2 \log(M^2/q^2) + O(1)\Big]^{-1} = \tag{56}$$
$$= 3\pi e_i^2 \Big[\sum_{j=1}^{N} e_j^2 \log\Big(\frac{1}{G_N N q^2}\Big) + O(1)\Big]^{-1}$$

where we have used Eq. (54) with the appropriate factor coming from the multiplicity of fermions. Equation (56) expresses the fact that, up to some factors O(N), the Landau pole of QED and the Planck mass have to coincide in our approach. A similar statement can be made for other unbroken gauge symmetries (as for instance QCD) provided that enough fermionic thresholds make them eventually asymptotically non-free[2].

5. The regime in which we have seen ordinary gauge theories arise as effective interactions is the one in which all fields carry, by decree, small momenta. This regime, however, does not necessarily coincide with that of the full quantum theory at low energy. It is only so at the tree level or if high momentum quantum fluctuations (as appearing through ultra-violet behaviours of loops) are strongly damped. We do not know yet if our approach will lead to such behaviour.

We can only give a few qualitative hints of why such behaviour could be milder than in the conventional theory, by underlining the modifications implied by our theory for a conventional calculation of gravity effects at short distances.

Sticking to the bosonic language that proved useful in recognizing the low-lying spectrum, we find differences with the usual treatment of gauge and gravity theories at various levels.

i) The rich heavy sector (masses of order M, the Planck mass) we have found will contribute as much as the light one as soon as the momenta they transfer approach M. Cancellations may of course occur. We recall nevertheless that among these heavy particles with well-defined masses and couplings, we find also PV ghosts. These represent a nuisance even if the meaning of tree-unitarity at a level in which not even a space-time metric can be defined is far from clear. It is possible that a supersymmetric extension of our model could alleviate this problem.

ii) Both the light and heavy sectors have quadratic terms in the fields with higher derivatives rescaled, of course, by powers of 1/M. These higher derivatives may be resummed and lead to propagators which are calculable functions of q^2. Only for

$q^2 \ll M^2$ do they coincide, for the light sector, with the conventional propagators. A calculation at this level to test the short-distance behaviour of the theory is in progress. A behaviour milder than the conventional one would already be a gratifying signal for our alternative approach to gravity.

iii) Besides the aforementioned contributions that count as soon as $q^2 \sim M^2$, we also meet an infinite series of induced many field couplings. They would, of course, contribute to any given process at higher and higher loop levels and therefore should be depressed by higher and higher powers of $1/N$. It is nevertheless very unclear whether or not we have any right to expect an ultra-violet behaviour which is uniform in $1/N$.

All the high frequency modifications to the conventional theory, i.e., extra particles, higher derivative interactions and extra couplings, are there to recall the basic fact that the graviton and gauge bosons were composite objects. At short distances we should indeed see the effects of their structure through some sort of form factor which softens their contribution as one approaches momentum transfers of order M.

These considerations show perhaps that the language of auxiliary fields, so convenient for studying the spectrum and the low energy structure of the theory, is not the correct language for studying the short-distance behaviour. The complexity of this language here is to remind us of its duality with respect to the language of components which, in our theory, is the one of the original fermions. This is perhaps better suited to short distances and our hope for a mild behaviour in this limit can be reinforced by the high derivative structure of our original action which should penalize very high frequencies. In this regime we should also lose the scale Λ, provided by the vacuum condensate, and recover therefore all the local symmetries of the theory. This could suggest a deep ultra-violet behaviour which might not be worse than that of ordinary renormalizable theories. All these arguments are only hints, but they nevertheless support our hope that the pregeometric gravity we propose could be a valid alternative for overcoming the pathologies of the conventional theory.

REFERENCES

1. For a review of GUTs cf. for instance J. Ellis in Scottish Universities Summer School, ed. K.C. Bowler and D.G. Sutherland, (1981), p. 201.
2. D. Amati, R. Barbieri, A.C. Davis and G. Veneziano, Phys. Lett, 102B:408 (1981);
   A.C. Davis, to appear in these proceedings.
3. D. Amati and G. Veneziano, Phys. Lett. 105B:358 (1981) and CERN preprint TH.3197 (1981).

4. A.D. Sakharov, Dokl. Akad. Nauk. SSSR 177:70 (1967); [Soviet Physics Dokl. 12:1040 (1968)];
Ya.B. Zel'dovich, Zh. ETF. Pis. Red. 6:922 (1967); [JETP Lett. 6:345 (1967)];
O. Klein, Phys. Scr. 9:69 (1974);
K. Akama, Y. Chikashige, T. Matsuki and H. Terazawa, Progr. Teor. Phys. 60:868 (1978);
K. Akama, Progr. Theor. Phys. 60:1900 (1978);
S.L. Adler, Phys. Rev. Lett. 44:1567 (1980), Phys. Lett. 95B:241 (1980);
B. Hasslacher and E. Mottola, Phys. Lett. 95B:237 (1980);
A. Zee, Phys. Rev. D23:858 (1981);
In a recent paper, to be published in Rev. Mod. Phys., S.L. Adler reviews these approaches to induced gravity and extends the analysis of renormalizable field theories with dynamical scale invariance breaking to the treatment of a quantized metric field.
5. L. Maiani, G. Parisi and R. Petronzio, Nucl. Phys. B136:115 (1978);
N. Cabibbo, L. Maiani, G. Parisi and R. Petronzio, Nucl. Phys. B158:295 (1979).
6. N. Cabibbo and G. Farrar, to be published.
7. We recognize here an analogy with the approach of V. De Alfaro, S. Fubini and G. Furlan [Phys. Lett. 97B:67 (1980)], G. Furlan, to appear in these proceedings, where a scale invariance breaking mass parameter also appears through the v.e.v. of a vierbein (or metric) field. Their approach differs however from ours since its starting point is the conventional Einstein action with a fundamental metric (or vierbein) field.
8. Cf. K. Akama et al., Ref. 4.

SOME REMARKS ABOUT QUANTUM GRAVITY

V. de Alfaro and S. Fubini

CERN, Geneva and Istituto di Fisica Teorica, Università di Torino and I.N.F.N., Sezione di Torino

G. Furlan

Istituto di Fisica Teorica, Università di Trieste

INTRODUCTION

Gravitation is the oldest of the fundamental interactions known to us. It finds an extraordinarily beautiful and simple formulation in Einstein's General Relativitiy, which embodies the equivalence principle by requiring the invariance of the theory under general reparametrization of the co-ordinates $x^{\mu} \rightarrow x'^{\mu} = f^{\mu}(x)$. Experiments confirm the predictions of the classical theory in the macroscopic, large distance, domain.

The recent developments in the description of the fundamental interactions, from gauge theories and spontaneous symmetry breaking to grand unification, have refocused our attention on gravitation as a quantum field theory and on its eventual influence on particle physics.

Quantum effects are expected to become important for very high energies, of the order of the Planck mass $M_{P\ell} = (G_N)^{-1/2} \approx 10^{19}$ GeV corresponding to distances $L = (4\pi G_N)^{1/2} \approx 10^{-33}$ cm. The dimensional quantity setting the scale of these effects is Newton's constant G_N which at the same time determines the strength of the macroscopic gravitational phenomena. Unfortunately, such a dimensional character of the gravitational coupling constant makes the resulting theory non-renormalizable, namely, higher terms of the conventional perturbation expansion exhibit worse and worse asymptotic behaviour at small distances.

Thus we have to face the problem that a satisfactory understanding of gravity is still lacking at the quantum level, in spite of its certainly unique properties of invariance[1].

Several interesting avenues towards a possible solution have been discussed (some of them in this session) which go from considering gravitation as a spontaneously or dynamically broken theory[2] to the more extreme attempt of a composite model[3]. I shall illustrate our proposal[4] which is, in a way, more conservative and tries to exploit fully what a rich theory, such as General Relativity, can do for us.

CONSIDERATIONS ON DIMENSIONALITY

Let us begin by noticing that the fundamental constant G_N is introduced in the theory in a rather phenomenological way, namely, after all, in order to match the dimensions of the minimal Einstein-Hilbert gravitational action, where $g_{\mu\nu}(x)$ is assumed to be dimensionless. One has

$$A_E = -\frac{1}{16\pi G_N}\int d^4x\sqrt{-g}\,R. \tag{1}$$

In fact, the presence of a dimensional parameter is surprising. Indeed, due to its general symmetry properties, the action A_E is in particular invariant under co-ordinate dilatations, $x'^\mu = \rho v^\mu$ and we are accustomed to associating scale invariant field theories with the absence of any dimensional constant. A familiar example is represented by Yang-Mills theories which are invariant under scale transformations in the limit of vanishing masses for all matter fields.

The point presumably lies in the fact that for the particle physicist, General Relativity has to be considered as the reparametrization invariant field theory of a massless, spin two particle, the graviton. The geometrical aspect ($g_{\mu\nu}(x)$ as the space-time metric tensor), even if highly suggestive at the classical level, is not essential and $g_{\mu\nu}(x)$ has finally to be treated as a (quantum) field[5]. In particular, a closer look at the general transformation properties of the fields suggests that the scale dimension (we may refer to it as "group theoretical") to be assigned to $g_{\mu\nu}$ is actually -2 (in units of length). More precisely, one is led to ascribe the following values to the scale dimension of various field quantities:

$$g_{\mu\nu},\ g^{\mu\nu},\ \sqrt{-g} \quad : \quad \Delta = -2, 2, -4. \tag{2}$$

$$R_{\mu\nu},\ R = g^{\mu\nu}R_{\mu\nu} \quad : \quad \Delta = -2, 0$$
$$A_\mu, \phi, \psi \quad : \quad \Delta = -1, 0, 0. \tag{2}$$

where A_μ, ϕ, ψ are vector, scalar and spinor fields respectively.

The argument is as follows: given the general co-ordinate reparametrization

$$x^\mu \to x'^\mu = f^\mu(x) \simeq x^\mu + \varepsilon^\mu(x) \tag{3}$$

we recall that the corresponding infinitesimal variation of a tensor $t^{\mu_1,\dots,\mu_M}_{\nu_1,\dots,\nu_N}$ is

$$\delta t^{\mu_1\cdots\mu_M}_{\nu_1\cdots\nu_N} = (t'(x) - t(x))^{\mu_1\cdots\mu_M}_{\nu_1\cdots\nu_N} = -\varepsilon^\lambda \partial_\lambda t^{\mu_1\cdots\mu_M}_{\nu_1\cdots\nu_N}$$
$$- t^{\mu_1\cdots\mu_M}_{\lambda\cdots\nu_N}\partial_{\nu_1}\varepsilon^\lambda \cdots + t^{\lambda\cdots\mu_M}_{\nu_1\cdots\nu_N}\partial_\lambda\varepsilon^{\mu_1}\cdots + t^{\mu_1\cdots\lambda}_{\nu_1\cdots\nu_N}\partial_\lambda\varepsilon^{\mu_M} \tag{4}$$

Let us now consider the special case of a scale transformation

$$\varepsilon^\mu(x) = -\varepsilon x^\mu \tag{5}$$

From Eq. (4) one immediately obtains

$$\delta t^{\mu_1\cdots\mu_M}_{\nu_1\cdots\nu_N} = \varepsilon(x\cdot\partial - \Delta)\, t^{\mu_1\cdots\mu_M}_{\nu_1\cdots\nu_N} \tag{6}$$

where $\Delta = M - N$ is the difference between the number of contravariant and covariant components respectively. Given the form of Eq. (6) it seems natural to assert that the tensor $t^{\mu_1,\dots,\mu_M}_{\nu_1,\dots,\nu_N}$ has dimensionality Δ. Consequently we find the values of dimensionality listed in Eq. (2).

In order to avoid any confusion, it is important to mention that the above dilatational transformations are different from the so-called Weyl, local scale, transformations

$$g_{\mu\nu}(x) \rightarrow g'_{\mu\nu}(x) = \Omega^{-2}(x)\, g_{\mu\nu}(x) \qquad (7)$$

(in infinitesimal form $\Omega \simeq 1 + \omega$)

$$\delta g_{\mu\nu} = -2\omega(x)\, g_{\mu\nu})$$

The implementation of this additional symmetry would require a more complicated structure of the action. The connection between the two types of transformations and their relevance for the flat limit is discussed for instance in Ref. 6.

One is then immediately convinced that, as a consequence, the Einstein-Hilbert action can be written in a full scale invariant form where no dimensional parameters appear and no additional fields need be introduced*. We simply have

$$A_E = -\frac{1}{4}\int d^4x \sqrt{-g}\, R\,. \qquad (8)$$

Naturally, on pure invariance grounds, additional pieces are possible, such as a cosmological term $\lambda\sqrt{-g}$, λ dimensionless, or terms proportional to $\sqrt{-g}R^2$, $\sqrt{-g}\, R_{\mu\nu}R^{\mu\nu}$ which can be discarded on the basis of a minimality requirement (to avoid four derivatives and ghost poles in the propagator).

Notice the essential rôle of the non-polynomial field quantity $\sqrt{-g}$ in the matching of dimensions.

THE VACUUM

All this may look a little tricky and the natural question at this point is "where has Newton's constant gone?" Our proposal is that Newton's constant finds its natural place, not as a coupling constant in the action, but in the boundary condition which specifies the behaviour of $g_{\mu\nu}(x)$ at large distances. More precisely, when requiring that $g_{\mu\nu}(x)$ asymptotically behaves like the flat solution* of the equation of motion, its form is taken to be

*Our approach is therefore different from the one advocated by several authors (see the talk given at this meeting by A. Zee), which assumes the existence of a new scalar field whose vacuum expectation value gives rise to Newton's constant.

$$g_{\mu\nu}(x)\Big|_{flat} = \frac{1}{L^2}\eta_{\mu\nu}, \qquad L^2 = 4\pi G_N \tag{9}$$

The constant reproduces at the elementary level the dimension -2 of the field.

The essence of this point of view is that Newton's constant with its dimensionality does not represent a general feature of the gravitational action but rather characterizes the particular and fundamental class of solutions to be used in the description of phenomena at large distances. It is thus quite clear that such a formulation is completely equivalent to the conventional one as long as the "Newtonian" results of general relativity are concerned. However, the underlying framework is much more general and offers the possibility of describing "non-Newtonian" phenomena, not weighted by G_N.

This argument finds a meaningful and appropriate reformulation in the quantum language. The heart of the matter is to assume that Newton's constant appears in the theory via the vacuum expectation value of $g_{\mu\nu}$, namely

$$\langle 0|\, g_{\mu\nu}(x)\,|0\rangle = \frac{1}{L^2}\eta_{\mu\nu}, \tag{10}$$

which is the quantum version of Eq. (9).

The situation is reminiscent of the case of the linear σ model[7] (even if analogies should not be pushed too far!), where the starting Lagrangian has chiral isospin $SU(2) \times SU(2) \sim SO(4)$ as (internal) symmetry group. The vacuum has however a lower symmetry, i.e., $SU(2) \sim SO(3)$ and this is clearly exhibited by the existence of a non-vanishing vacuum expectation value of the bosonic field $\phi_\alpha(\alpha=1,\ldots,4)$, namely

$$\langle 0|\, \phi_\alpha(x)\,|0\rangle = f_\pi\, \delta_{\alpha 4} \tag{11}$$

The dimensional constant f_π is a measure of the spontaneous breaking of the symmetry SO(4). As is well known, f_π determines the low energy behaviour of the pion amplitudes and acts in this domain as a universal coupling of pions to any hadronic system, while the rôle of the usual Lagrangian coupling constants is less important.

In a similar way one can say that while the gravitational action is general invariant the vacuum is not; the flat vacuum solution (10) in particular breaks invariance under dilatations (and thus general invariance) leaving Poincaré as symmetry subgroup. Thus the dimensional constant G_N which rules the large distance behaviour of the transition amplitudes for emission and absorption of gravitations does not appear in the Lagrangian. Lagrangian constants, like a possible cosmological term, do not seem to play any important rôle in classical gravitation.

These considerations also maintain their validity when matter is present, and once more only dimensionless constants have to appear in the total Lagrangian.

It is instructive to examine in a little more detail the case where a matter field is present. This will help us to understand the apparent clash between the above value of zero dimensionality for scalar and spinor fields and the canonical ones (-1 and -3/2 respectively).

Let us consider the simple example of a scalar field: the action is

$$A_M = \int d^4x \sqrt{-g} \left\{ \frac{1}{2} g^{\mu\nu} \partial_\mu \phi \partial_\nu \phi + P(\phi) \right\} \tag{12}$$

and $P(\phi)$ is an arbitrary polynomial

$$P(\phi) = \lambda_2 \phi^2 + \lambda_3 \phi^3 + \lambda_4 \phi^4 + \dots \tag{13}$$

where, given the dimension zero of ϕ, the λ_i's are pure numbers. The fact that all these terms are allowed may seem surprising. This is clarified if one goes to the flat limit which corresponds to $g_{\mu\nu} = 1/L^2 \eta_{\mu\nu}$ and

$$A_M \to \int d^4x \left\{ \frac{1}{L^2} \frac{1}{2} \eta^{\mu\nu} \partial_\mu \phi \partial_\nu \phi + \frac{1}{L^4} P(\phi) \right\} \tag{14}$$

We thus see that in order to recover in the flat limit the familiar kinetic term, independent of L, a new field has to be used

$$\varphi(x) = \frac{1}{L} \phi(x) \tag{15}$$

A more refined discussion shows that the appropriate field to be used for flat space theories is actually

$$\varphi(x) = (-g)^{1/8}\,\phi(x)\,, \tag{16}$$

a density of dimensionality -1.

What about the polynomial part and its flat limit? One has

$$\frac{1}{L^4} P(L\varphi) = \frac{\lambda_2}{L^2}\varphi^2 + \frac{\lambda_3}{L}\varphi^3 + \lambda_4\varphi^4 + \ldots \tag{17}$$

Thus a mass term $(\sqrt{\lambda_2/L^2})$ of the order of the Planck mass appears together with couplings of the super-renormalizable, renormalizable and unrenormalizable types.

The connection with the usual (flat) theory of conformal invariance is subtle and the main motivation of the wise choice (16) is to achieve it. In fact it is possible to show that the energy-momentum tensor for a flat space takes on the form

$$\theta_{\mu\nu} = \partial_\mu\varphi\,\partial_\nu\varphi - \frac{1}{4}\eta_{\mu\nu}(\partial\varphi)^2 + \frac{1}{L^4}\eta_{\mu\nu}\left[P - \frac{1}{4}\varphi\frac{dP}{d\varphi}\right] \tag{18}$$

so that

$$\theta_\mu{}^\mu = \frac{1}{L^4}\left(4P - \varphi\frac{dP}{d\varphi}\right). \tag{19}$$

Conformal invariance requires the trace $\theta_\mu{}^\mu$ to vanish: then only the term $\lambda_4\phi^4$ is allowed, as we are used to.

As a conclusion, all masses are measured in units of the Planck mass and are therefore related to a spontaneous breaking of dilatation. It is amusing to notice that in this spirit the universal gravitation law can be written in a form similar to the Coulomb one, i.e., in terms of dimensionless gravitational "charges"

$$\mu_i = m_i L = m_i\sqrt{4\pi G_N}\,:$$
$$U(r) = -G_N\frac{m_1 m_2}{r} = -\frac{\mu_1\mu_2}{4\pi r}. \tag{20}$$

RENORMALIZABILITY AND SMALL DISTANCE BEHAVIOUR

The fact that the fundamental Lagrangian is naturally scale-invariant is expected to have some important consequences. First of all, since G_N characterizes a solution rather than the full theory, different classical solutions can exist which obey non-Newtonian boundary conditions and which may be used as backgrounds to describe phenomena occurring in different space-time domains.

It is well known in fact that classical solutions exist for gravitation with a cosmological term and/or interacting with matter field (gauge fields, non-linear σ model, etc.). These solutions are non-Newtonian in the way explained above and have a dependence on dimensional constants which are always introduced through the boundary conditions. Such constants can be considered as characterizing different vacua, whose relevance to a study of the hadronic structure is an open matter and we shall not discuss it here*.

Secondly, but more importantly, one can expect that some general quantum features of the Green functions, like the small distance behaviour or the commutation relations, should be almost independent of L and substantially be fixed by the invariance properties of the underlying action. In order to discuss this point it is useful to recall how the presence of Newton's constant leads, in the conventional formulation, to the non-renormalizability of quantum gravity. Applying standard perturbative techniques to gravitation is not immediate due to the non-polynomial character of the Lagrangian, which contains the inverse operator $g^{\mu\nu}$ and comlicated animals like $\sqrt{-\det g_{\mu\nu}}$, etc. The problem is usually tackled by separating the field $g_{\mu\nu}$ into a Newtonian background plus a quantum part:

$$g_{\mu\nu}(x) = \frac{1}{L^2}\left(\eta_{\mu\nu} + L\,\varphi_{\mu\nu}(x)\right). \tag{21}$$

Here $\phi_{\mu\nu}$ is an operator of dimension -1 and the subsequent procedure consists of an expansion in $L\phi_{\mu\nu}$. It thus follows that the interaction term is actually a power series in L: L being a

*Excellent reviews exist of classical solutions and we need not repeat them[8]. It is however interesting that the space-time dependence of some solutions allows one to reproduce the elementary dimension -2 of $g_{\mu\nu}$ without introducing G_N. For instance, in the case of gravitation coupled to a gauge field and with a cosmological term λ we can mention, as the simplest example, the so-called meron solution which reads (in a Euclidean metric)

$$g^{M}_{\mu\nu}(x) = \delta_{\mu\nu}\frac{2}{\lambda^2}\frac{1}{x^2},$$

with $\lambda^2 = e^2$ the gauge charge.

dimensional coupling constant, one faces all the unpleasant peculiarities of a non-renormalizable theory.

In our opinion, this is an unavoidable consequence of using, at small distances, the expansion in L which is only suitable for a description at large distances. Taking into account only a finite number of terms in this expansion clearly violates the general invariance properties of the theory (only Poincaré invariance is respected); these indeed require a non-polynomial Lagrangian and such a character is preserved only after summing the whole series. Therefore it is not surprising that the use of the above expansion beyond its limits of validity does not reproduce the correct small distance behaviour.

This discussion also provides a hint about the relative importance of the effects determined by the vacuum on the one side, and by the general invariance of the theory on the other. The main conclusion is that the vacuum essentially represents an infra-red phenomenon: it depends on L (or on other constants according to the physical situation) and is an extra piece of information added independently to the Lagrangian. On the contrary, the ultra-violet behaviour will mainly be inferred from the general properties of the of the underlying Lagrangian only, and consequently there is no dependence (or a soft one) of these results on L. (The additional problem of smoothly joining these two aspects is completely open and an understanding of it has still to come, as is the case in QCD.)

It is fruitful to illustrate these considerations with some explicit examples. The crux of the matter is represented by the use of inverse operators and we shall first consider the case of a free scalar field, of zero mass and dimension -1. We expand it around a (constant) classical background μ, namely

$$h(x) = \mu + h'(x) , \qquad \langle h'(x) \rangle_0 = 0 \tag{22}$$

and its Green function is of the form (as experts know, it is useful to work in a Euclidean metric)

$$\langle h(x)\, h(y) \rangle_0 = \mu^2 + G(x-y) = \mu^2 + \frac{1}{4\pi^2 (x-y)^2} . \tag{23}$$

This simple result already suggests that at large distances the Green function is given by the properties of the vacuum (the constant part) while, at small distances, the leading term is $[(x-y)^2]^{-1}$, in agreement with an elementary dimensional argument.

The interesting question arises when one wants to evaluate the two-point function of the inverse operator $h^{-1}(x)$. The simplest way to obtain it is through an expansion in $h'(x)/\mu$:

$$I(x-y) \equiv \langle h^{-1}(x) h^{-1}(y) \rangle_0 = \left\langle \frac{1}{\mu + h'(x)} \frac{1}{\mu + h'(y)} \right\rangle_0$$

$$= \frac{1}{\mu^2} \left\langle \Sigma_n \left(-\frac{h'(x)}{\mu}\right)^n \Sigma_m \left(-\frac{h'(y)}{\mu}\right)^m \right\rangle_0 = \frac{1}{\mu^2} \Sigma_n n! \left(\frac{G(x-y)}{\mu^2}\right)^n \tag{24}$$

Using the integral representation

$$n! = \int_0^\infty \alpha^n e^{-\alpha} d\alpha \tag{25}$$

we easily reach the already known result[9]

$$I(x-y) = \int_0^\infty \frac{e^{-\alpha} d\alpha}{\mu^2 - \alpha G(x-y)} \tag{26}$$

The correct definition of I(x-y) actually requires a further prescription for the behaviour at the pole (it turns out that the principal value is the right recipe); however, since we are interested in the small distance behaviour, this point, which would affect the spectrum properties, is not of immediate interest to us.

In particular, as $x - y \to 0$ we obtain from Eq. (26) that

$$I(x-y) \underset{x-y \to 0}{\sim} (x-y)^2 \log [\mu^2 (x-y)^2] \tag{27}$$

Therefore, apart from the logarithmic term which contains a residual dependence on the background, the ultra-violet behaviour is the one we should have expected from simple dimensional counting. Notice however, that this result required summing the complete series in $1/\mu^2$. Taking into account only a finite number of terms would, on the contrary, lead to the completely wrong indication of a theory dramatically divergent at each order of the perturbation expansion. Similar results stand if more complicated cases are considered, for instance

$$\langle h^{-m}(x)\, h^{-m}(y)\rangle_0 \underset{x-y\to 0}{\sim} [(x-y)^2]^m \log[\mu^2(x-y)^2] \tag{28}$$

(The logarithmic factor always appears in the first power.)

In the case of gravity the problem is, of course, far more complicated and we have no complete answer concerning full renormalizability. However, the analysis of some simple examples shows that dimensional arguments still keep their validity.

Let us introduce for convenience the vierbein fields $v_\mu{}^a(x)$ $g_{\mu\nu} = \Sigma_1^4 a\ v_\mu{}^a(x) v_\nu{}^a(x)$ of dimension -1; then the use of known integral representations for a determinant leads to results which confirm the above point of view, for instance

$$\langle (\det V_\mu^a(x))^{-1}\, (\det V_\mu^a(y))^{-1}\rangle_0 \underset{x-y\to 0}{\sim} [(x-y)^2]^4 \log[\mu^2(x-y)^2] \tag{29}$$

and the like.

Notice that results like (27), (28) and (29) are quite general in the sense that they depend very little on dividing the fields into a classical background plus a quantum part. On the other hand, for practical purposes, it seems almost unavoidable to rely on a perturbative-like expansion of some kind. In doing so, much care has to be taken, as the previous example shows, before inferring any conclusion on the full theory from the behaviour of a finite number of terms of the series; for this purpose the choice of the initial separation of the field plays a crucial role.

Indeed the usual representation [the analogue of Eq. (21)]

$$V_\mu^a(x) = \frac{1}{L}\,\delta_{\mu a} + V'^{a}_{\mu}(x) \tag{30}$$

is a good starting point only in the infra-red domain, while at small distances the expansion in L leads to the well-known difficulties.

Considering this, this fact is not that surprising since the separation (30) is Poincaré invariant but not scale invariant (the two pieces transform differently) and cannot, therefore, verify arguments based on dilatations. In order to investigate the ultraviolet regime a conformal and scale invariant decomposition is more suitable. A possibility is to write[10]

$$V_\mu^{\ a}(x) = \delta_{\mu a}\, h(x) + \bar{V}_\mu^{\ a}(x) \tag{31}$$

where h(x), $\bar{v}_\mu^{\ a}(x)$ are both operators of dimension -1, and to expand in $\bar{v}/h$ an object of dimension zero. Its two-point function behaves logarithmically and does not substantially influence the theory at small distances:

$$\left\langle \bar{V}_\mu^{\ a}(x)/h(x) \quad \bar{V}_\nu^{\ b}(y)/h(y) \right\rangle_0 \underset{x-y\to 0}{\sim} \log[\mu^2(x-y)^2]\,. \tag{32}$$

More precisely, the interaction Lagrangian turns out to be of the form

$$\mathcal{L}_1 = \partial_{\mu_1} V_{\nu_1}^{a_1}\, \partial_{\mu_2} V_{\nu_2}^{a_2}\, F^{\mu_1 \nu_1 a_1}_{\mu_2 \nu_2 a_2}\left(\bar{V}/h\right). \tag{33}$$

The function F,...,$(\bar{v}/h)$ can be expressed as a series in $\bar{v}/h$ so that $\mathcal{L}_1$ has dimension -4 in any order and should not behave differently from a renormalizable model, apart from first order logarithmic terms like in (32).

The model is highly heuristic, with some internal difficulties (h(x) is a ghost, just to mention one), however it has the good quality of concretely illustrating a possible hint towards the solution of the renormalizability problem of quantum gravity.

CONCLUSIONS

Let us summarize the main points of our approach. These are a) to ascribe the correct dimensionality to the field $g_{\mu\nu}$ (or $v_\mu^{\ a}$): this looks quite natural, even trivial; b) to interpret Newton's constant as a vacuum effect: this requires distinguishing between boundary conditions and fundamental laws, as we are getting accustomed to in particle physics; c) to treat inverse operators, a problem which has already attracted some attention for effective chiral Lagrangians*.

*It is quite interesting that all these features arise naturally in the model of $g_{\mu\nu}$ as a composite field recently proposed by Amati and Veneziano[3].

As a consequence of all this we expect that at small distances the Green functions of quantum gravity are substantially of a power form* with the exponent determined by purely dimensional arguments: gravitation should not behave differently from a renormalizable theory. Perhaps we may be able to paraphrase Coleman "The divergence structure of a field theory respects the symmetry of the Lagrangian even if the vacuum does not"[11].

The statement by Coleman of course refers to renormalizable polynomial theories with internal symmetries, but it is an extremely interesting problem to ascertain whether similar considerations apply to the non-polynomial case with space-time symmetries.

In this spirit we should like to conclude by saying that the problem of renormalizing quantum gravity is, in our opinion, still open and that the same point can be made about the importance of quantum gravitational effects at very high energies.

REFERENCES

1. For a recent review see "Recent developments in gravitation", Cargèse Summer Institute (1978) (Plenum Press, N.Y., 1979).
2. A.D. Sakharov, Dokl. Akad. Nauk. SSR 177:70 (1967); (Sov. Phys. Dokl. 12:1040 (1968));
   Ya.B. Zel'dovich, Zh.ETF. Pis. Red. 6:922 (1967); (JETP Lett. 6:345 (1967));
   O. Klein, Phys. Scr. 9:69 (1974);
   K. Akama, Y. Chikashige, T. Matsuki and H. Terazawa, Prog. Theor. Phys. 60:868 (1978);
   K. Akama, Prog. Theor. Phys. 60:1900 (1978);
   S.L. Adler, Phys. Rev. Lett. 44 (1980); Phys. Lett. 95B:241 (1980).
   B. Hasslacher and E. Mottola, Phys. Lett. 95B:237 (1980);
   A. Zee, Phys. Rev. D23:858 (1981).
3. D. Amati and G. Veneziano, CERN preprint TH-3126 (1981).
4. V. de Alfaro, S. Fubini and G. Furlan, Nuovo Cimento A50:523 (1979); ibid. B57:227 (1980).
5. See, for instance,
   S. Weinberg, "Gravitation and cosmology", ed. J. Wiley, New New York (1972);
   A. Salam, "Impact of quantum gravity in particle physics" in "Quantum gravity - an Oxford Symposium", eds. C. Isham, R. Penrose and D. Sciama, Oxford University Press (1975).
6. J. Wess, "Conformal invariance and the energy-momentum tensor", Springer Tracts in Modern Physics, vol. 60 (1971).

*Apart from logarithmic terms.

7. See, for instance,
V. de Alfaro, S. Fubini, G. Furlan and C. Rossetti, "Currents in hadron physics", North Holland Pub. Co., Amsterdam (1973).
8. Several aspects of classical solutions are reviewed by V. de Alfaro, S. Fubini and G. Furlan, Acta Physica Austriaca Supp. XXII:51 (1980).
9. A quite complete presentation can be found in "Non-polynomial Lagrangians, renormalization, gravity" in Tracts and mathematics and natural sciences, Vol. I ed. A. Salam (1971).
10. V. de Alfaro, S. Fubini and G. Furlan, Phys. Lett. 97B:67 (1980).
11. S. Coleman in "Laws of hadronic matter", ed. A. Zichichi Academic Press, N.Y. (1975).

FROM QUANTUM COSMOLOGY TO QUANTUM GRAVITY*

F. Englert**

Université Libre de Bruxelles, Belgium
and
CERN, Geneva, Switzerland

SUMMARY[1]

The remarkable success of general relativity is marred by two fundamental difficulties at two apparently different scales. First the acausal character of the hot big bang cosmology is unacceptable because it leads to dynamical instabilities when extrapolated back to the Planck temperature. Second, the quantized theory does not yet exist, not only because general relativity is not renormalizable but mainly because of the lack of positive definiteness of the (Euclidean) action. The theory proposed here solves the first problem and the solution sheds unexpected new light on quantum gravity.

The stabilization of the "initial" thermal state is achieved by constructing a semi-classical solution of the coupled gravitational and matter system with zero cosmological constant. This solution is an exponentially expanding de Sitter space-time in which black holes are created by a quantum process out of the expansion energy. The very existence of such a non-trivial self-consistent solution, in addition to the trivial empty Minkowski space solution, is in fact a consequence of the non-positive definiteness of the Euclidean action[2]. The initial state of the classical expansion is entirely determined by the theory: we obtain

* Supported in part by the Belgium State under the contract A.R.C. 79/83-12.

** Permanent address: Pool de Physique, Campus de la Plaine, C.P. 225, Bd. du Triomphe, 1050 - Bruxelles, Belgium.

$$T = \frac{1}{8\pi M} = [48\pi^{3/2} m_{Planck}]^{-1} = 4.57 \times 10^{16} \text{ GeV} \quad (1)$$

$$H = [24\pi^{1/2} t_{Planck}]^{-1} = 4.26 \times 10^{41} \text{ s}^{-1} \quad (2)$$

where T is the initial temperature, H the initial Hubble constant and M the mass of the created black holes.

Subsequent black hole decay signals the onset of an adiabatic regime. The theory predicts that the total number of quanta produced by the decay within a causally connected region of the Universe is greater than 10^{87}, and hence consistent with present observation, provided the number of distinct massless helicity states $\nu(T)$ [at the temperature (1)] is bounded by

$$\nu(T) \lesssim 2000 \quad (3)$$

If ultimate elementary particle models conform to (3), we have thus explained in causal terms the conventional adiabatic expansion.

We argue that the initial nucleation process, which is seized upon by the self-consistent creation mechanism, has to originate from a quantum metric fluctuation. The requirement of no fine tuning for such fluctuations leads to the conclusion that the classical universe is <u>finite</u> in space and time and that it is spatially <u>flat</u>; thus the present deceleration parameter q is predicted to be

$$q = 0.5 \quad (4)$$

The creation era produces a large time contraction for observers confined within a universe, so that the size of a universe in space-time is about 50 Planck units on the scale of external observers; a universe appears therefore as a typical fluctuation of quantum gravity. External observers (and ourselves) have to include universe-like configurations (and TCP reversed configurations) in the path integral over metrics. Universe-like configurations generate a huge entropy which tends to stabilize the path integral in the neighbourhood of paths of stationary phase. We conjecture that <u>this stabilization does occur</u> so that the path integral over metrics is saturated by a "foam of universes", except for infrared corrections which would cut themselves off at the Planck length on the scale considered. In this foam picture the non-renormalizability of gravity is the bonus that renders interactions between universes small, and renormalizable field theory may be viewed as the phenomenological manifestation of the existence of long range interactions within a universe. However, in order to define a background $g_{\mu\nu}$ for a given observer it is necessary to project out correlations and an infinite number of hidden degrees of freedom.

This projection introduces an irreducible element of indeterminacy in the space-time description of quantum field theory as exemplified by the appearance of the temperature (1). The lack of a description in terms of a pure global quantum state is clearly fundamental and suggests that four-dimensional space-time should be explained in terms of new concepts.

REFERENCES

1. This theory is due to
A. Casher and F. Englert, Phys. Lett. B104:117 (1981);
A detailed version and more complete references can be found in:
F. Englert, "Quantum Field Theory and Cosmology" in Cargèse Summer School of Fundamental Interactions, (1981), to be published; Brussels University preprint (1981).
2. This non-trivial solution (with an elementary matter field instead of black holes) was suggested by
R. Brout, F. Englert and E. Gunzig, C.R.G. 10:1 (1979);
For more recent developments see:
R. Brout, F. Englert, J.-M. Frère, E. Gunzig, P. Nardone, C. Truffin and P. Spindel, Nucl. Phys. B170:228 (FS1) (1980).

MINIMAL HYPOTHESES FOR PARTICLE DEFINITION IN CURVED SPACE-TIME

Mario A. Castagnino

Instituto de Física de Rosario
Av. Pellegrini 250
2000 Rosario, Argentina

Diego D. Harari and Carmen A. Nuñez

Instituto de Astronomía y Física del Espacio
Casilla de Correo 67, Sucursal 28
1428 Buenos Aires, Argentina

INTRODUCTION

The interaction between a quantum field and an unquantized gravitational field can be studied in the frame of a semiclassical theory in which the gravitational field enters, according to the theory of general relativity, as the background geometry, where the covariant generalizations of Klein-Gordon, Dirac, Maxwell, etc. equations are considered[1]. Such a theory is supposed to be a good approximation to a more general, as yet unknown, full quantum theory, when the components of the Riemann curvature tensor are small (in atomic units, $h = c = 1$) as compared to the Planck energy (10^{19} GeV) squared. One of the most striking results of this formulation is the possibility of particle- antiparticle-pairs creation by a dynamical gravitational field, the thermal radiation of black holes predicted by Hawking[2] being its most popular version. Particle creation in asymptotically static universes, where the particle model (the plane waves) is perfectly defined in the far future and past, has also been studied[3]. But the universe has undoubtly not been static in the past; then an "in-out" theory is unrealistic. Moreover, it would be very interesting to know the details of the creation mechanism during the universe evolution, and to study the consequences of reintroducing the energy-momentum tensor of the created matter as a source in Einstein's field equations in a more sophisticated cosmological model.

In Minkowski space-time, the vacuum expectation value of normal products of free field operators vanishes, whence its relevance to the elimination of vacuum energy. However, normal ordering is defined as a consequence of considering the positive- and negative-frequency plane waves as the antiparticle and particle models, which is possible for free fields. Interacting fields, instead, admit the particle language only in asymptotic situations, far in space and time from the interacting region. As we are dealing with a quantum field interacting with a classical gravitational field, would it be possible to interpret the quantum states in terms of particles? And if so, how could the properties of plane waves be generalized to curved space-time?

A program to answer these questions was begun by Lichnerowicz who looked for the generalization of the kernels $\Delta(x,x')$ and $\Delta_1(x,x')$ which, in Minkowski space-time allow a covariant decomposition of a scalar field into its positive and negative-frequency parts[4]. The properties imposed by Lichnerowicz on their analogues in curved space, $G(x,x')$ and $G_1(x,x')$, were enough to define unequivocally the commutator G. However, the expression derived for the $G_1(x,x')$ fitting Lichnerowicz conditions depends upon the orthonormal base of Klein-Gordon equation solutions involved, so reflecting the ambiguity of the particle model in curved space. Therefore, additional physical hypotheses must be added.

Some heuristic arguments help finding these selective criteria. A highly energetic particle won't "feel" the curvature effects if its wavelength is much smaller than $R_{max}^{-1/2}$ (R_{max} being the biggest component of the curvature tensor), and thus it must resemble a flat-space particle. Furthermore, as the high energy behaviour of a field theory is governed by the singular structure of the kernels when $x \to x'$, that structure should be reproduced by $G_1\ (x,x')$. These were the ideas underlying the formulation of the Quantum Equivalence Principle[5,6], which essentially consisted in ascribing a different particle definition to each different spatial hypersurface Σ, the particle model at Σ being derived from the kernel $G_1(x,x')$ selected from all possible kernels by the following Cauchy data

$$G_1(x,x')\Big|_\Sigma = \Delta_1\left[s(x,x')\right]\Big|_\Sigma ; \quad \frac{\partial G_1(x,x')}{\partial n^i}\Bigg|_\Sigma = \frac{\partial \Delta_1\left[s(x,x')\right]}{\partial n^i}\Bigg|_\Sigma \tag{1}$$

where $s(x,x')$ is the geodetic distance between x and x', and n^i is the unitary normal vector to the surface Σ.

In a general situation these data imply $G_1^\Sigma(x,x') \neq G_1^{\Sigma'}(x,x')$, particle production arising between Σ and Σ'. The particle creation rate predicted by the Q.E.P. was evaluated in a spatially flat expanding universe up to the second order in a series of powers of the Hubble coefficient[7] and though the particle density is convergent, its energy diverged logarithmically, discrediting in this way the process of normal ordering with respect to that particle definition. Moreover it has been pointed out[8] that if Cauchy data (1-a, b) are fixed, one of the Lichnerowicz conditions automatically determines the third Cauchy datum, the mixed derivative of the biscalar, which is not the mixed derivative of $\Delta_1(s)$. Finally, the attempts to generalize this model to higher spin particles were successful when evaluated up to first order[9,10] but they ran into inconsistencies when developed at higher orders[11].

These unpleasant facts might be evidence that the identification between flat and curved singular strucutres of $G_1(x,x')$ over all a Cauchy surface is an excessively strong requirement. In fact, the non-existence of a global momentum space in an arbitrary curved space indicates that the singular structure of $\Delta_1(s)$ should be reproduced by $G_1(x,x')$ only as $x \to x_0$, $x' \to x_0$, i.e. a different factor could generalize each kind of divergence (quadratic, logarithmic and also the constant and infinitesimal terms) of $\Delta_1(s)$ whenever it tends to its Minkowskian limit as the curvature globally vanishes.

If certain particular symmetries hold over space-time it can be expected that the functions $G_1^{x_0}(x,x')$ verifying the above hypothesis were the same, for all the points x_0, over certain kind of hypersurfaces (for example over $\Sigma_\tau \equiv \{t=\tau= \text{const.}\}$ in a Robertson-Walker spatially flat expanding universe). A global momentum space can then be associated to those surfaces (i.e. to a comoving reference frame) and a particle model will then exist.

This task will be implemented in sections II and III in a spatially flat expanding universe in a power series development up to terms containing the curvature, but not its derivatives or higher powers. The model will be proved to be completely defined (up to the order considered) and no particle creation is obtained. (The method can be extended to spin ½ particles as we will show elsewhere.) In section IV the model is compared to other ones in the literature. In particular it is show that, up to the order considered, our hypotheses lead to De Witt's kernel[12], which was derived using the proper time representation formalism, and also by Parker and Bunch[13], using normal coordinates with origin at x_o to solve Klein-Gordon equation.

MINIMAL HYPOTHESES

The Klein-Gordon equation in curved space-time is:

$$(\Box - m^2 - \xi R)\Phi(x) = 0$$

where $\Box = -g^{ij}\nabla_i\nabla_j$, g_{ij} is the metric tensor (latin indices run from 0 to 3, signature +, −, −, −); ∇_i indicates covariant derivatives, $R = g^{ik}R_{ik} = g^{ik}\,R^h{}_{ihk}$ is the scalar curvature; and ξ defines the kind of coupling to the gravitational field (minimal coupling for $\xi = 0$ and conformal coupling for $\xi = 1/6$, which makes the equation conformally invariant in the massless case).

The existence of a well defined function $G(x,x')$ in every globally hyperbolic universe, having the necessary properties to be considered the commutator of the scalar field ($|\Phi(x),\Phi(x')| = -iG(x,x')$), was proved in ref. 4. A generalization of $\Delta_1(x,x')$ must satisfy the following conditions in order to define a good projector into orthogonal subspaces of the space of Klein-Gordon equation solutions

$$\text{i)}\quad G_1(x,x') \text{ must be real} \tag{2-a}$$

$$\text{ii)}\quad G_1(x,x') = G_1(x',x) \tag{2-b}$$

$$\text{iii)}\quad (\Box - m^2 - \xi R)G_1(x,x') = 0 \tag{2-c}$$

$$\text{iv)}\quad G(x,x') = -i \langle G_1(x,y), G_1(x',y)\rangle \tag{2-d}$$

where the scalar product $\langle u,v\rangle$ is defined as

$$\langle u,v\rangle = i\int_\Sigma \left[(\partial_j u^*)v - u^*(\partial_j v)\right] d\sigma_j \tag{3}$$

But these conditions are not enough to define uniquely $G_1(x,x')$: its expression derived from (2) in terms of any orthonormal base for Klein-Gordon eq. solutions is[5]

$$G_1(x,x') = \sum_A \{\Phi_A(x)\Phi_A^*(x') + \Phi_A^*(x)\,\Phi_A(x')\} \tag{4}$$

which is not invariant under an orthonormal base transformation (Bogolyubov's transformation).

One of these $G_1(x,x)$ must be singled out in order to have a particle definition. In an arbitrary curved space-time there must exist a "good" G_1 associated to each point x_o, which will be called $G_1^{x_o}(x,x')$ having the same high-energy behaviour than its flat-space analogue. But this function will not necessarily be the same for different points of space-time. In other words, there is no global momentum space. Nevertheless, if certain symmetries still hold, all $G_1^{x_o}(x,x')$ for x_o belonging to a certain kind of surface (as we said, for example, over t=const. hypersurfaces in a Robertson-Walker spatially flat universe) will coincide, and we expect a particle definition relative to those surfaces (i.e. to a reference frame or some kind of observers) to be possible. The following minimal properties should be satisfied by these $G_1^{x_o}(x,x')$: i) they must verify the Minkowskian limit and ii) they must lead to a particle model reproducing flat space-time high energy behaviour, i.e., their singular structure must be that of $\Delta_1[s(x,x')]$.

$\Delta_1[s(x,x')]$ as a function of the geodetic interval $s(x,x')$ is

$$\Delta_1(s) = \frac{m^2}{4\pi} \,\mathrm{Im}\left[\frac{H_1^{(1)}(ms)}{ms}\right] \qquad (5)$$

where $H_1^{(1)}$ is the first order and type Hankel function which has a quadratic term, independent of mass, a logarithmic one, a constant one and terms vanishing as s does. For small values of ms the following expansion is valid[12]

$$G_1(x,x') = \frac{m^2}{2\pi}\Big\{\frac{1}{m^2(x-x')^2} + \Big[\gamma - \log 2 + \log m + \tfrac{1}{2}\log|(x-x')^2|\Big]\times \\ \left[\frac{1}{2}+\frac{m^2(x-x')^2}{2^2.4}+\ldots\right] - \frac{1}{4}\frac{m^2(x-x')^2}{2^2.4}(1+\tfrac{1}{4}) - \\ \frac{m^4(x-x')^4}{2^2.4^2.6}\left(1+\frac{1}{2}+\frac{1}{6}\right) - \ldots\ldots\Big\}$$

Different behaviours with respect to s can be associated to successive derivatives of Δ_1 with respect to m^2, i.e., $\Delta_1(s)$ essentially has a quadratic divergence, $\frac{\partial\Delta_1}{\partial m^2}$ starts with a logarithmic one, $\frac{\partial^2\Delta_1}{(\partial m^2)^2}$ is regular, and so on.

The generalized kernel $G_1^{x_o}(x,x')$ should reproduce the different dependences on s but a different biscalar function could in principle correct each of them provided it satisfies its Minkowskian limit. This statement (for $G_1^{x_o}$ and its derivatives), constitutes, together with set (2), the minimal hypotheses to define a "good" $G_1^{x_o}(x,x')$:

$$\text{v)}\quad \lim_{\substack{x\to x_o\\ x'\to x_o}} G_1^{x_o}(x,x') = \lim_{\substack{x\to x_o\\ x'\to x_o}} \sum_{n=0}^{\infty} F_n^{x_o}(x,x')\,\frac{\partial^n\Delta_1(s)}{(\partial m^2)^n} \qquad (2\text{-e})$$

being
$$\left\{\begin{aligned} &F_o^{x_o}(x,x') = 1\\ &F_\alpha^{x_o}(x,x') = 0\end{aligned}\right. \quad \text{if } R_{ijkl} = 0 \text{ (flat space-time)} \qquad (6)$$

and $n = 0, 1, 2, \ldots$; $\alpha = 1, 2, 3, \ldots.$

The functions $F_n^{x_o}(x,x')$ appearing in (2-e) must be real symmetric biscalars as (2-a, b) require. Moreover, the derivatives of $G_1^{x_o}(x,x')$ with respect to m^2 should also be expected to have the same singular structure as the flat space-time ones, i.e.

$$\text{vi)}\quad \lim_{\substack{x\to x_o\\ x'\to x_o}} \frac{\partial^i G_1(x,x')}{(\partial m^2)^i} = \lim_{\substack{x\to x_o\\ x'\to x_o}} \sum^{\infty} F'^{x_o}_{n_i}(x,x')\,\frac{\partial^n\Delta_1}{(\partial m^2)^n} \qquad (2\text{-f})$$

Therefore, it can be proved that F_0 cannot depend on m^2, $F_1^{x_o}$ can only linearly depend on m^2, $F_2^{x_o}$ quadratically, etc.

These conditions might appear as very weak ones to determine uniquely $G_1^{x_o}(x,x')$, but we shall demonstrate how the formalism works in a power series development of the metric derivatives in a spatially flat Robertson-Walker universe.

CONSTRUCTION OF THE MODEL

The functions F_n and its derivatives will be evaluated in a power series of the metric derivatives. The most general expressions for the coincidence limits can only consist on the curvature and the metric tensors and the mass of the particles. Dimensional considerations and the Minkowskian limit assumed in (6) help selecting the limits constituents. Using atomic units ($\hbar = c = 1$) the mass has frequency dimensions and the scalar curvature R, squared frequency dimensions. Terms involving the curvature but not its derivatives nor higher powers than the first are considered, postponing to a forthcoming paper the evaluation of higher orders.

The most general expressions will then be

$$\lim_{\substack{x \to x_0 \\ x' \to x_0}} F_0^{x_0}(x,x') = 1$$

$$\lim_{\substack{x \to x_0 \\ x' \to x_0}} F_\alpha^{x_0}(x,x') = (m^2)^{\alpha-1} A_\alpha R \qquad \alpha = 1, 2, 3, \ldots\ldots$$

$$\lim_{\substack{x \to x_0 \\ x' \to x_0}} \nabla_i F_n^{x_0}(x,x') = 0 \qquad n = 0, 1, 2, \ldots\ldots$$

$$\lim_{\substack{x \to x_0 \\ x' \to x_0}} \nabla_i \nabla_j F_n^{x_0}(x,x') = (m^2)^n [B_n R_{ij} + C_n R g_{ij}] \tag{7}$$

(The curvature and the metric are of course evaluated at x_0).

All higher derivatives will vanish at the order considered, meaning that a cut at second order in the metric derivatives automatically implies cutting at second order a Taylor development of the corresponding $F_n^{x_0}$ and then of $G_1^{x_0}$. $G_1^{x_0}$ could thus be known in a Taylor development around x_0. Wherefore, the coefficients A_α, B_n, C_n, ..., are to be determined.

In a Robertson-Walker spatially flat universe

$$ds^2 = dt^2 - a(t)^2 \, \delta_{\alpha\beta} \, dx^\alpha \, dx^\beta \tag{8}$$

$G_1^{x_o}(x,x')$ is expected to be the same function of x and x' if x_o belongs to the surface $\Sigma(x_o) \equiv \{x/t(x) = t(x_o)\}$. This function, that we shall call $G_1^{\Sigma(x_o)}(x,x')$, will define a good particle model if it verifies conditions (2-a, b, c, d, e, f). Condition 2-d) implies a relationship among the Cauchy data for all possible G_1 satisfying Klein-Gordon equation. This condition will be applied over all space-time where itself and its first and second mixed derivatives with respect to the normal vector to the surface $\Sigma(x_o)$ are satisfied over a spatial hypersurface, because the scalar product <u, v> is independent of the surface Σ if u, v are Klein-Gordon equation solutions.

It can then be proved[8] that condition (2-d) will be satisfied if the Fourier transforms of the Cauchy data for $G_1^{\Sigma(x_o)}(x,x')$ over $\Sigma(x_o)$ obey

$$F_{\vec{k}}[\partial_{o'o} G_1^\Sigma(x,x')] F_{\vec{k}}[G_1^\Sigma(x,x')] - F_{\vec{k}}[\partial_o G_1^\Sigma(x,x')] F_{\vec{k}}[\partial_{o'} G_1^\Sigma(x,x')] = 1 \tag{9}$$

where

$$G_1^\Sigma(x,x') = \frac{1}{(2\pi a)^3} \int d^3\vec{k} \, e^{-i\vec{k}.\vec{r}} \, F_{\vec{k}}[G_1^\Sigma(x,x')]$$

From (2-e) and (2-f) a Taylor development of the terms involved in (9) reads

$$F_o^{x_o}(x,x')\rfloor_{\Sigma(x_o)} = 1 - \frac{a^2 r^2}{2}\left[\left(\frac{B_o}{6} + C_o\right)R - B_o H^2\right]$$

$$F_\alpha^{x_o}(x,x')\rfloor_{\Sigma(x_o)} = (m^2)^{\alpha-1} R \, A_\alpha - \frac{a^2 r^2}{2} (m^2)^\alpha \left[\left(\frac{B_\alpha}{6} + C_\alpha\right)R - B_\alpha H^2\right]$$

$$\partial_o F_n^{x_o}(x,x')\rfloor_{\Sigma(x_o)} = \partial_{o'} F^{x_o}(x,x')\rfloor_{\Sigma(x_o)} = 0$$

$$\partial_{oo'} F_n^{x_o}(x,x')\rfloor_{\Sigma(x_o)} = -(m^2)^n \left[\left(\frac{B_n}{2} + C_n\right)R + 3B_n H^2\right]$$

(due to the spatial symmetry of the metric, $\lim \nabla_{ij} F_n = -\lim \nabla_{ij'} F$ has been assumed). We made use of the expressions $R_{\alpha\alpha} = -a^2(\frac{R}{6} - H^2)$; $R_{oo} = 3H^2 + \frac{R}{2}$; $R = -6(\dot{H} + 2H^2)$; $H = \dot{a}/a$; R, H^2 and a should be evaluated at $t(x_o)$.

The geodesic distance between x and x' can be obtained from Riemannian normal coordinates[6] y^i as

$$s(x, x')^2 = \eta^{ij} y^i y^j$$

η^{ij} being Minkowski metric, and it turns out to be

$$s(x,\, x')^2\rfloor_\Sigma = -\,a^2 r^2 (1 + \frac{a^2 r^2 H^2}{12})$$

$$\partial_o s(x,x')^2\rfloor_\Sigma = \partial_{o'} s(x,x')^2 = -a^2 r^2 H$$

$$\partial_{o'o} s(x,x')^2\rfloor_\Sigma = -2 + \frac{1}{3} a^2 r^2 (\frac{R}{6} - 2H^2)$$

$$\partial_{oo} s(x,x')^2\rfloor_\Sigma = 2 + \frac{2}{3} a^2 r^2 (H^2 + \frac{R}{6})$$

Making use of these expressions to evaluate $\Delta_1(s)$ and its derivatives we obtain

$$\Delta_1(s)\rfloor_\Sigma = \frac{1}{(2\pi a)^3} \int d^3k \, \frac{e^{-i\vec{k}.\vec{r}}}{\omega_{\vec{k}}} \{ 1 - \frac{5}{8} \frac{H^2}{\omega_{\vec{k}}^2} \frac{m^4}{\omega_{\vec{k}}^4} \}$$

$$\partial_o \Delta_1(s)\rfloor_\Sigma = \partial_{o'} \Delta_1(s)\rfloor_\Sigma = \frac{-H}{(2\pi a)^3} \int d^3\vec{k} \, \frac{e^{-i\vec{k}.\vec{r}}}{\omega_{\vec{k}}} (1 + \frac{m^2}{2\omega_{\vec{k}}^2})$$

$$\partial_{o'o} \Delta_1(s)\rfloor_\Sigma = \frac{1}{(2\pi a)^3} \int d^3\vec{k} \, e^{-i\vec{k}.\vec{r}} \, \omega_{\vec{k}} \{ 1 + \frac{7}{8} \frac{H^2}{\omega_{\vec{k}}^2} \frac{m^4}{\omega_{\vec{k}}^4} - (1 + \frac{m^2}{2\omega_{\vec{k}}^2}) \, x$$

$$(\frac{1}{18} \frac{R}{\omega_{\vec{k}}^2} + \frac{5}{3} \frac{H^2}{\omega_{\vec{k}}^2})\}$$

where all the terms have been completely Fourier analyzed taking into account that

$$\int d^3k \, e^{-i\vec{k}.\vec{r}} \, g(k) \, F(x) = \int d^3k \, e^{-i\vec{k}.\vec{r}} \, (f * g)$$

$$f(h) = \int d^3\vec{r} \, e^{i\vec{h}.\vec{r}} \, F(r), \qquad f * g = \int d^3\vec{h} \, f(k-h) \, g(h)$$

We are now able to construct $G_1^{\Sigma(x_o)}(x,x')$ and its derivatives and replacing them in (9) to determine the constants. But it can be seen from (9) that the second term of the l.h.s. does not contain higher powers than the second in m^2. Therefore it is unnecessary to consider $F_n^{x_o}$ having $n > 2$: their coefficients will vanish or appear in a combination which will vanish in $G_1^{x_o}(x,x')$. The Cauchy data are then

$$G_1^\Sigma(x,x')\rfloor_\Sigma = \frac{1}{(2\pi a)^3} \int d^3k \, \frac{e^{-i\vec{k}.\vec{r}}}{\omega_{\vec{k}}^2} \{ 1 + \frac{H^2}{\omega_{\vec{k}}^2} \left[\frac{m^2}{\omega_{\vec{k}}^2} . \frac{3}{2} (B_0 + B_1) + \frac{m^4}{\omega_{\vec{k}}^4} (-\frac{15}{2}) x \right.$$

$$\left. (\frac{B_1}{2} + B_2 + \frac{1}{12}) + \frac{105}{8} B_2 \frac{m^6}{\omega_{\vec{k}}^6} \right] + \frac{R}{\omega_{\vec{k}}^2} \left[-\frac{A_1}{2} - \frac{3}{2} \frac{m^2}{\omega_{\vec{k}}^2} \, x \right.$$

$$\{(\frac{B_0}{6} + C_0)+(\frac{B_1}{6} + C_1)-\frac{A_2}{2}\}+\frac{15}{2}\frac{m^4}{\omega_{\vec{k}}^4}\{\frac{1}{2}(\frac{B_1}{6} + C_1)+(\frac{B_2}{6} +$$

$$+ C_2)\}- \frac{105}{8}\frac{m^6}{\omega_{\vec{k}}^6}(\frac{B_2}{6} + C_2)\Big]$$

$$\partial_o G_1^{\Sigma}(x,x')\rfloor_{\Sigma} = \frac{-H}{(2\pi a)^3}\int d^3\vec{k}\,\frac{e^{-i\vec{k}.\vec{r}}}{\omega_{\vec{k}}}(1+\frac{m^2}{2\omega_{\vec{k}}^2})$$

$$\partial_{o'o} G_1^{\Sigma}(x,x')\rfloor_{\Sigma} = \frac{1}{(2\pi a)^3}\int d^3\vec{k}\, e^{-i\vec{k}.\vec{r}}\,\omega_{\vec{k}}\{1+\frac{H^2}{\omega_{\vec{k}}^2}\Big[(\frac{5}{3}- 4B_0)+\frac{m^2}{\omega_{\vec{k}}^2}\;x$$

$$(\frac{B_0}{2} + \frac{3}{2}B_1+\frac{5}{6})+\frac{m^4}{\omega_{\vec{k}}^4}(\frac{7}{8}+\frac{3}{4}B_1-\frac{3}{2}B_2)-\frac{15}{8}\frac{m^6}{\omega_{\vec{k}}^6}B$$

$$+\frac{R}{\omega_{\vec{k}}^2}\Big](\frac{1}{18}-\frac{B_0}{3}+\frac{A_1}{2})+\frac{m^2}{\omega_{\vec{k}}^2}(\frac{1}{36}+\frac{B_0}{12}+\frac{C_0}{2} + \frac{B_1}{4}+\frac{C_1}{2}-\frac{A_2}{4})+$$

$$+\frac{m^4}{\omega_{\vec{k}}^4}(-\frac{3}{4}C_1-\frac{1}{8}B_1-\frac{B_2}{2}-\frac{3}{2}C_2)+\frac{m^6}{\omega_{\vec{k}}^6}(\frac{15}{48}B_2+\frac{15}{8}C_2)|\}$$

Introducing these results in (9) an identity is obtained for the terms not having R or H^2 (the flat space-time terms) and all the others must vanish as a polynomial in $m^2/\omega_{\vec{k}}^2$ (i.e. each term independently). Moreover, if the particle model is to be defined for any evolution the universe had experienced the terms involving R and H^2 must also vanish independently. The method leads to $B_0=1/6$, A_1 arbitrary and the remaining coefficients vanish.

$G_1^{\Sigma}(x,x')$ can then be written as

$$G_1^{\Sigma}(x,x') = \frac{1}{(2\pi a)^3}\int d^3k\,\frac{e^{-i\vec{k}.\vec{r}}}{\omega_{\vec{k}}}\{1-\frac{5}{8}\frac{H^2}{\omega_{\vec{k}}^2}\frac{m^4}{\omega_{\vec{k}}^4}+\frac{1}{4}\frac{H^2}{\omega_{\vec{k}}^2}\frac{m^2}{\omega_{\vec{k}}^2}-\frac{1}{2}\frac{R}{\omega_{\vec{k}}^2}(A_1 + \frac{1}{12}\frac{m^2}{\omega_{\vec{k}}^2})\}$$

$$= \frac{1}{(2\pi a)^3}\int\frac{d^3k\; e^{-i\vec{k}.\vec{r}}}{W} \qquad (10)$$

where $W = \omega_{\vec{k}}\{1+\frac{5}{8}\frac{H^2}{\omega_{\vec{k}}^2}\frac{m^4}{\omega_{\vec{k}}^4}-\frac{1}{4}\frac{H^2}{\omega_{\vec{k}}^2}\frac{m^2}{\omega_{\vec{k}}^2}+\frac{1}{2}\frac{R}{\omega_{\vec{k}}^2}(A_1+\frac{1}{12}\frac{m^2}{\omega_{\vec{k}}^2})$

If now G_1^{Σ} is asked to satisfy Klein-Gordon equation, A_1 is determined univocally and it turns out to be $(1/6 - \xi)$ and the particle model is completely defined (leading to no particle creation).

Indeed, let us consider an orthonormal basis for Klein-Gordon equation solutions: the functions[7]

$$\Psi_{\vec{k}}(\vec{x},t) = \frac{e^{-\vec{k}.\vec{r}}}{(2\pi a)^{2/3}} \frac{e^{-i\int^{t'}\Omega\, dt'}}{\sqrt{2\Omega}} \tag{11}$$

and their complex conjugates; where

$$\Omega = \omega_{\vec{k}}\{1+\frac{5}{8}\frac{H^2}{\omega_{\vec{k}}^2}\frac{m^4}{\omega_{\vec{k}}^4}-\frac{1}{4}\frac{H^2}{\omega_{\vec{k}}^2}\frac{m^2}{\omega_{\vec{k}}^2}-\frac{1}{2}(\xi-\frac{1}{6})\frac{R}{\omega_{\vec{k}}^2}+\frac{R}{\omega_{\vec{k}}^2}\frac{m^2}{24\omega_{\vec{k}}^2}\}$$

assures $\Psi_k(x,t)$ is a Klein-Gordon equation solution up to the considered order. In terms of these functions, expression (4) for $G_1(x,x')$ reads

$$G_1(x,x')\rfloor_\Sigma = \frac{1}{(2\pi a)^3}\int d^3k\, \frac{e^{-i\vec{k}.\vec{r}}}{\Omega} \tag{12}$$

Comparing (10) and (12) and taking (11) into account, indeed $W = \Omega$ and $A_1 = \frac{1}{6} - \xi$. It can be proved that we would have the same result if we ask only for a convergent energy of the created particles.

COMPARISON WITH OTHER MODELS

Different attempts have been developed in order to construct a curved space-time quantum field theory. An integral representation of the Feynman propagator $G_F + G + \frac{1}{2} iG_1$ was given by De Witt[12] (chapter 17)

$$G_F(x,x') = \frac{\Delta^{\frac{1}{2}}}{(4\pi)^2}\sum_{n=0}^{\infty} a_n(-\frac{\partial}{\partial m^2})^n\int_o^\infty \frac{1}{z^2}\exp\left[-i(m^2z-\frac{1}{4}\frac{s^2}{z})\right]dz$$

where

$$\Delta^{\frac{1}{2}} = g(x)^{-\frac{1}{4}}\,\mathrm{Det}\left[\partial_{i'}\partial_j(\tfrac{1}{2}s^2(x,x')\right]g(x')^{-\frac{1}{4}}$$

is the Van Vleck determinant and a_n are determined by the recurrence relation

$$g^{ij}\nabla_i(\tfrac{1}{2}s^2)\,\nabla_j(a_{n+1}) + (n+1)\,a_{n+1} = \Delta^{\frac{1}{2}}\Box(\Delta^{\frac{1}{2}}\,a_n)$$

and $a_0 = 1$.

This expression was deduced from the minimally coupled Klein-Gordon equation and it leads to

$$\lim_{x\to x'} a_1 = -\frac{1}{6}R. \tag{13}$$

The factor F_0 in (7) obeys:

$$\lim_{x\to x'} F_0 = 1, \quad \lim_{x\to x'} \nabla_i F_0 = 0, \quad \lim_{x\to x'} \nabla_{ij} F_0 = \frac{1}{6} R_{ij}$$

which corresponds to $F_0 = \Delta^{\frac{1}{2}}$ up to the considered order, and if $\xi = 0$, our corresponding expression obeys (13).

Parker and Bunch[13] have also obtained this same kernel by solving Klein-Gordon equation in normal coordinates for arbitrary ξ values. Their kernel also coincides with ours. (Our spin $\frac{1}{2}$ particle model also coincides with De Witt's one).

However, the works mentioned above do not require $G_1^{x_0}(x,x')$ to define a particle model as we do, but $G_F(x,x')$ is constructed to solve the renormalization of a $\lambda\Phi^4$ interacting theory, so both approaches show an interesting coincidence. On the other hand, as the particle creation predicted by the "in-out" theories would not appear if there were a univocally defined $G_1(x,x')$, we believe something is missing in our formalism. In fact, particle creation is related with terms of the form e^{-k} in W that can't be computed with our expansion in powers of the metric and its derivatives i.e. powers of k^{-1}.

The path integral method has also been proposed by Chitres and Hartle[14] to find $G_1(x,x')$ in a linearly expanding universe and a thermal spectrum was predicted for the created matter. However, it was later proved[15] that such a $G_1(x,x')$ does not satisfy the Minkowskian limit and $G_1(x,x') = \Delta^{\frac{1}{2}} \Delta_1(s)$ was suggested for a linear evolution and $\xi = 1/6$, which also coincides with ours, up to the considered order. In this particular case, exact calculations are possible and no particle creation is obtained. But again the Chitre and Hartle $G_1(x,x')$ can be obtained from ours by adding e^{-k} terms.

Therefore, our future work will be focussed on the study and computation of these terms.

REFERENCES

1. Gibbons, G.W., "Quantum field theory in curved space-time", in "General Relativity" eds. S.W. Hawking and W. Israel (Cambridge Univ. Press, Cambridge, 1979).
2. Hawking, S.W., Commun. Math. Phys., 43:199 (1975).
3. Parker, L., in "Asymptotic Structure of Space-Time", eds. F. Esposito and L. Witten (Plenum Press, N.Y., 1977).
4. Lichnerowicz, A., Inst. Haut. Et. Sci. Publ. Math. No. 10 (1961).
5. Castagnino, M., Weder, R., J. Math. Phys. 22:142 (1981).
6. Castagnino, M., Foussats, A., Laura, R. and Zandrón, O.; Il Nuovo Cimento, 60A:138 (1980).
7. Castagnino, M., Chimento, L. and Harari, D.; Phys. Rev. D. In press, (1981).

8. Ceccato, H., Foussats, A., Giacomini, H., and Zandrón, O.: The kernel G_1 and the Q.E.P., submitted to Il Nuovo Cimento (1981).
9. Castagnino, M., Ann. Inst. H. Poincaré, 25, No. 1 55 (1981).
10. Foussats, A., Laura, R. and Zandrón, O., J. Math. Phys. 22:357 (1981).
11. Ceccatto, H., Foussats, A., Giocomini, H., and Zandrón, O.; On the inconsistency of a photon creation mechanism in an expanding universe, submitted to Il Nuovo Cimento, (1981).
12. De Witt, B.; in "Relativity, Groups and Topology", Les Houches, 1963, eds. C. and B. De Witt (Gordon and Breach, N.Y., 1963).
13. Bunch, T.S. and Parker, L.; Phys. Rev. D20:2499 (1979).
14. Chitre, D.M. and Hartle, J.B.; Phys. Rev. D16:251 (1977).
15. Nairai, H. and Azuma, T.; Progr. of Theoret. Phys. 59:1522 (1978)

FERMIONS IN THE DESERT

P. Ramond

Physics Department
University of Florida
Gainesville, Florida 32611

INTRODUCTION

In the following we consider the possibility of adding to the standard Grand Unified Models such as SU_5, SO_{10} or E_6extra vector-like fermions with anomalously low masses. We present some models and in some cases understand the low masses of these fermions in terms of a global chiral symmetry which solves the strong CP problem and leads to an all but invisible axion.

In the standard picture of Grand Unified Theories[(1)], there is a desert of interactions between the Grand Unified scale and that at which electroweak interactions are broken. If we adopt as the fundamental unit of mass the Planckton: 1 Planck mass = 1 Planckton, the Grand Unified scale breaking occurs typically at 10^{-4} (Planckton), the electroweak breaking at 10^{-17} etc... It is felt by many that this large gap of interactions is unnatural in the sense that it is difficult to realize it in a perturbative field theory where radiative corrections tend to undo it since the coupling constants are not small enough to create such a large hierarchy. This difficulty has been considered sufficiently serious by some to abandon GUTs for, say, technicolor[(2)] theories or else load it with supersymmetry[(3)] which leaves only tree level relations to be understood. In both cases, one predicts a plethora of new phenomena at large scales soon to be reached in the laboratory.

Let us remark that there exists another hierarchy problem: the cosmological constant which is experimentally minute with no theoretical understanding for it.

The theoretical and conceptual successes of standard GUTs, such as $\sin^2\theta_w$, m_b/m, n_B/n_γ as well as the new phenomena it predicts (τ_p, ν-oscillations) makes it very appealing to us, while it is blemished only by the hierarchy problem. Perhaps we should not try to solve the hierarchy problem now, but rather wait until we have understood how to incorporate gravity (i.e. completed the unification). In the following, therefore, we take the point of view that standard GUTs are essentially correct with a desert of interaction between 10^{-4} and 10^{-7} (in Planckton natural units).

We address ourselves to the possibility[4] of finding at the edge of the desert vector-like fermions which are relics of the GUTs breaking at 10^{-4}.

Let us first review the phenomenology of fermions. For that purpose it is convenient to use weak isospin. The usual charged fermions appear as left-handed weak doublets and right-handed singlets, resulting in a mass with the quantum number $\Delta I_w = 1/2$:

$$f_L \sim 1/2 \qquad f_R \sim 0 \qquad \rightarrow \qquad f_L^\dagger f_R \sim 1/2 \ .$$

The natural scale for these fermions is the scale at which weak isospin breaks, since it is known experimentally to break in the $\Delta I_w = 1/2$ channel as well. Thus in that sector the numbers of interest are the mass ratios

$$\frac{m_e}{m_w} \sim 10^{-5} \ , \quad \frac{m_\mu}{m_w} \sim 10^{-3} \quad \frac{m_\tau}{m_w} \sim 10^{-2} \ , \quad \text{etc...}$$

The neutrinos on the other hand have only left-handed doublets, and they can have Majorana masses with the quantum numbers $\Delta I_w = 1$:

$$\nu_L \sim 1/2 \quad \rightarrow \quad \nu_L^T \sigma_2 \nu_L \sim 1 \ .$$

There we are faced with very small limits on the mass ratios

$$\frac{m_{\nu_e}}{m_w} < 10^{-11} \ , \quad \frac{m_{\nu_\mu}}{m_w} < 10^{-8} \ , \quad \frac{m_{\nu_\tau}}{m_w} < 10^{-3} \ .$$

Their smallness can be understood in terms of lepton-number conservation or in Grand Unified Models where the enormous suppression comes from the desert ratio

$$\frac{m_{\nu_e}}{m_e} \sim \frac{m_w}{m_x} \ ,$$

where m_x characterizes GUTs breaking[5]. All known fermions have either $\Delta I_w = 1/2$ or 1 masses. However, theories such as E_6 lead us to envisage other types of fermions which have vector-like masses with the quantum number $\Delta I_w = 0$:

$$f_L \sim 1/2 \; , \quad f_R \sim 1/2 \qquad f_L^{\dagger} f_R \sim 0.$$

Such fermions have not been observed in the laboratory. One likely reason is that their natural scale is at m_x or beyond, i.e. they are allowed to acquire their mass when that breaking occurs. For instance, in the standard SU_5 model, one can add fermions transforming according to the $5_L + \bar{5}_L$ representation. Their mass would therefore transform as

$$5_L \cdot \bar{5}_L \sim 1 + 24 \; ,$$

with the result that, when the 24 of Higgs present in the standard model gets a vacuum expectation value (vev) of order m_x, then new fermions get a mass of that order of magnitude[6].

On the other hand we have no real understanding of the ratio of masses of the usual fermions to their natural scale (e.g. m_e/m_w). In this light it does not seem entirely unreasonable to ask whether these $\Delta I_w = 0$ fermions could be very light with respect to m_x with the hope that they might play a role in phenomenology. In other words, can we sprinkle the desert with these fermions? The main obstacle is to find models where suppression by 10 orders of magnitude can be realized!

One example of such a suppression occurs in the neutrino to W-boson mass ratios. There, we think we can understand it in several possible ways: – the neutrino is strictly massless because of lepton number conservation, – the neutrino mass is very small because it occurs in a different quantum number ($\Delta I_w = 1$) than the enabling breaking mechanism ($\Delta I_w = 1/2$), and using the postulated hierarchy "on its head"[5].

Another possible way to obtain a mass much smaller than its natural scale is to fix it in such a way that it can turn on only in higher order of perturbation theory. In GUTs there are some effective very small Yukawa couplings ($m_e/m_w \sim h/e$) of the order of 10^{-7}, so it is not inconceivable that one can obtain numbers $\sim 10^{-15}$ with very few iterations of the Yukawa couplings. However, in order to prevent masses from appearing at tree or one-loop level, we need one extra symmetry which is necessarily chiral in nature. Later we will find that the postulated Peccei-Quinn[7] symmetry just fits the bill.

To be explicit, take the standard SU_5 model with one extra $5_f + \bar{5}_f$ of left-handed fermions. In the absence of any extra symmetries, the following renormalizable coupling are allowed:

bare mass term   : $5_f \cdot \bar{5}_f m$.

Yukawa coupling : $5_f \cdot \bar{5}_f 24_H$,

where 24_H is the usual Higgs representation. For our
$\Delta I_w = 0$ masses to be small, we need a symmetry that forbids the
above bare mass and Yukawa couplings, but allows non-renormalizable couplings of higher dimension such as

$$5_f \cdot \bar{5}_f \; 24_H 24_H, \; 5_f \bar{5}_f (24_H)^3 \; etc...$$

This can be done by a chiral continuous symmetry of the Peccei-
Quinn type. This global symmetry was introduced to understand
the apparent lack of CP violation in QCD even in the presence of
instanton effects. Since quarks have masses, this phase
symmetry was spontaneously broken, resulting in a pseudo-Nambu-
Goldstone boson called the axion or Higglet (we prefer the name
Pecceiquinno) with a small mass (because of explicit instanton
breaking). Originally it was thought that the Peccei-Quinn (PQ)
symmetry was broken at m_w, resulting a Pecceiquinno with well-
defined properties – it was not found. Recently it was
noticed[8] that if the breaking of the P.Q. symmetry took place
at the Grand Unified Scale m_x, the resulting pseudo Nambu--
Goldstone boson would be all but visible. It is gratifying that
by asking that there exist low mass $\Delta I_w = 0$ fermions we arrive
at the very same scenario!

     As an aside, let us note an amusing model with an automatic
Peccei-Quinn symmetry. As noted earlier[9] E_6 is the simplest
GUTs model which recovers one striking property of the Weinberg-
Salam model, namely that all the symmetry breaking takes place
in the fermion mass matrix. In E_6 the Yukawa term looks like

$$\mathcal{L}_Y \sim 27_f \cdot 27_f \; [27_H + 351'_H] \; ,$$

and it contains a global <u>chiral</u> symmetry with value +1 for the
fermions and –2 for the <u>Higgs</u>. As E_6 is broken down to SO_{10} and
SU_5 this global symmetry survives and is to be identified with
the P.Q. symmetry. Only one assumption in the Higgs potential
needs to be made: no cubic couplings, i.e. classical scale
invariance. This is to be contrasted with the standard SU_5
model which also has a global phase symmetry, but it is <u>vector-</u>
<u>like</u> and cannot serve as a P.Q. symmetry. The essential ingred-
ient in the E_6 model is that the SU_5 24_H is complex since it
resides inside the 351′. It is thus very easy to accommodate
the P.Q. phase symmetry in GUTs[10].

     Let us see in detail how this idea works in our embryonic
SU_5 model. We start with the following fermion content

$$\bar{5}_f + 10_f + (\bar{5}'_f + 5_f).$$

As stated above we need to forbid the bare mass term:

$$\bar{5}_f \cdot 5_f, \quad \bar{5}'_f \cdot 5_f$$

as well as the Yukawa couplings

$$\bar{5}_f 5_f 24_H, \quad \bar{5}'_f \cdot 5_f 24_H \ ,$$

while allowing the non renormalized couplings.

$$\bar{5}_f 5_f (24_H)^n \ , \qquad \bar{5}'_f \cdot 5_f (24_H)^n \ ,$$

where n = 2 or 3, depending on the presence of Higgs cubic couplings. We have presented elsewhere(4) a model with cubic couplings, here we present one without. Note that in order to achieve the above with a continuous symmetry, we need to enlarge the Higgs representation to be complex. Since we need to keep track of the chiral symmetries of the extra 5_f, we need to have it appear in Yukawa couplings, which forces us to add an extra 10_H of Higgs. The Yukawa coupling of our model is

$$\mathcal{L}_Y: \ (a\bar{5}_f + b\bar{5}'_f)10_f\bar{5}_H + 10_f 10_f (c5'_H + d5''_H) + 5_f 10_f 10_H,$$

when a, b, c, d are arbitrary constants, and 5_H, $5'_H$, $5''_H$ are three Higgs quintets. The Higgs potential is ordinary except for the presence of the quartic coupling.

$$\overline{10}_H 24_H 5'_H 5''_H \ , \qquad 24_H 24_H 5_H \bar{5}'_H.$$

As a result, the model has two global symmetries, a chiral one (P.Q.) which is called X and the usual vector-like one which is called Y. Their assignments are given in the following table

	$\bar{5}_f, \bar{5}'_f$	10_f	5_f	5_H	$5'_H, 5''_H$	24_H	10_H
X	-1	0	-1/2	-1	0	1/2	1/2
Y	-3	1	3	-2	-2	0	-4

It is easy to read off the allowed and forbidden couplings:

$\bar{5}_f 5_f$, $\bar{5}'_f 5_f$ have X = -3/2 and are forbidden

$\bar{5}_f 5_f 24_H$, $\bar{5}'_f 5_f 24_H$ have X = -1; they are also forbidden.

The first allowed contribution is, as required, non-renormalizable:

$\bar{5}_f 5_f (24_H)^3$, $\bar{5}'_f 5_f (24_H)^3$ have X = Y = 0 and are therefore allowed.

It is easy to see, for those who like diagrams, that such a contribution occurs first at the two-loop level with only Higgs interactions (no gauge lines!). If x, y, z are Yukawa couplings, and λ_1, λ_2 are Higgs quartic couplings, we find that the diagram is proportional to

$$\frac{\lambda_1 \lambda_2 xyz}{m_H^2} \langle 24 \rangle^3 ,$$

where m_H is the mass of a heavy Higgs.

Take $\lambda_1 \sim \lambda_2 \equiv \lambda$. Then

$$\langle 24 \rangle \sim \frac{m_H}{\lambda^{1/2}} ,$$

leading to the diagram strength

$$xyz \; m_x \frac{\lambda}{e} ,$$

where e is the gauge coupling. Taking Yukawa couplings of $O(10^{-4})$, we see that it is not too hard to get $\Delta I_W = 0$ fermion with masses in the Tev range. One can make similar models with $\Delta I_w = 0$ fermions transforming as $10_f + \overline{10}_f$, etc... .

Since these masses are proportional to the cube of ordinary Yukawa couplings we expect that the $\Delta I_w = 0$ fermion associated with the lightest family (the electronic family) will themselves be the lowest in mass. We find it likely that the fermions associated with the μ- and τ- families will be in the middle of the desert. Note that this model leads to an invisible axion(11).

The next question of interest is the effect of such fermions on phenomenology(12). Again we consider just one extra $5_f + \bar{5}_f$ of fermions, with physical content:

- a charge -1/3 quark D_L, D_R,
- a vector-like lepton doublet $\begin{pmatrix} N \\ E^- \end{pmatrix}_L$, $\begin{pmatrix} N \\ E^- \end{pmatrix}_R$,

with the leptons undergoing vector-like weak interactions in the absence of mixing. When $\Delta I_w = 0$ breaking takes place, the charged and neutral leptons acquire a common mass

$$m_E = m_N ,$$

and the quark acquires a different mass. If it comes about

first from the effect of the 24, we expect in general that it will be lighter than m_E

$$m_D = \left(\frac{2}{3}\right)^n m_E .$$

The effect of such extra fermions on $\sin\theta_w$ is minimal. One finds that

$$\sin^2\theta_w(m_w) = \sin^2\theta_w(m_w)\big|_0 + \frac{\hat{\alpha}(m_w)}{12\pi}\ln\frac{m_D}{m_E} ,$$

from which we see that the value of $\sin'\theta_w$ is essentially unaffected because $\alpha(m_w)$ is so small.

The effect on the proton lifetime is also minimal. First we note that by adding new fermions $\alpha(m_x)$ necessarily increases, but this is not the only effect: if $m_D < m_E$, the SU_3 curve will meet with the SU_2 curve at a larger energy, so that in this case m_x is larger -- this is what happens when the $\Delta I_w = 0$ breaking occurs via the 24_H. These two effects compete with one another. Numerical evaluations shows that under reasonable assumptions τ_p gets reduced by .7 of its standard value. In view of the order of magnitude uncertainties in the calculation of τ_p we do not consider this effect significant.

The addition of vector-like fermions may affect the GIM mechanism since the pattern of charges becomes non standard. This imposes a lower bound on the masses of non-GIM fermions. If we assume that mixing angles between two quarks go as $(M_1/M_2)^{1/2}$, then we find that one is GIM-safe as long as $m_D \gg 10^{4-5} m_d$, which is very easy to satisfy.

These fermion may also play a role in CP-violation[13] -- it could be that all CP violations occurs at m_x (neglecting the ensuing troubling domain problem) and that these fermions by mixing with the ordinary ones induce the observed CP-violation in the K-system. In this case we need $10_f + \overline{10}_f$ to contribute to the K_0-$\bar{K}_0$ mass difference.

Finally let us mention the experimental signature of these fermions. If these extra fermions do not have exotic quantum numbers they will decay by mixing with the ordinary fermions. One case of particular interest arises when $m_E > m_N$. Note that $m_E - m_N$ has $\Delta I_w = 1$ quantum numbers and is therefore expected to be very small. One would find a heavy lepton decaying into a neighboring heavy neutral state, in contradistinction with the usual picture where a heavy lepton decays into its much lighter neutrino.

To conclude, we think that it is very likely that there are

extra $\Delta I_w = 0$ fermions since all models beyond the minimal SU_5 predict them. Under special circumstances it could be that these fermions play a role in phenomenology by having anomalously low masses -- we showed that this could be achieved by the presence of the Peccei-Quinn symmetry. If they exist at low mass, their presence as a relic of the $\Delta I_w = 0$ breaking will shed light on interactions at 10^{-29} cm.

The author wishes to acknowledge the kind invitation of the organizers, J. Ellis and S. Ferrara. This work was supported by DOE contract DSR80136je7.

REFERENCES

(1) For a comprehensive review, see P. Langacker, Phys. Rep. 72C:185 (1981).

(2) For a review see P. Sikivie, University of Florida preprint UFTP-81-26 (1981).

(3) See for instance N. Sakai and T. Yanagida, Max Planck Institute preprint MPI-PAE/PTH 55/81 (1981).

(4) P. Ramond, Lectures on Grand Unification, 4th Kyoto Summer Institute, June 1981, University of Florida preprint UFTP-81-13 (1981).

(5) M. Gell-Mann, P. Ramond and R. Slansky, in "Supergravity" ed. by P. Van Nieuwenhuizen and D. Z. Freedman (North-Holland, 1979, Amsterdam). T. Yanagida, in Proc. of Workshop on the Unified Theory and the Baryon Number of the Universe, ed. by O. Sawada and A. Sugamoto KEK (1979).

(6) H. Georgi, Nucl. Phys. B156:126 (1979).

(7) R. Peccei and H. Quinn, Phys. Rev. Lett. 38:1440 (1977); Phys. Rev. D16:1791 (1977).

(8) J. E. Kim, Phys. Rev. Letters 43:103 (1979). H.P. Nilles and S. Raby, SLAC preprint (1981). M. Dine, W. Fischler, A. Srednicki, Nucl. Phys. B189:575 (1981).

(9) P. Ramond, invited talk at the "Sanibel Symposia", Caltech preprint CALT-68-709 (1979), unpublished.

(10) R. Slansky, private communication.

(11) M. B. Wise, H. Georgi, and S. L. Glashow, Phys. Rev. Lett. 47:402 (1981).

(12) M. Bowick, in preparation.

(13) A. I. Sanda, private communication.

THE WEAK INTERACTION AS AN INDIRECT MANIFESTATION OF HYPERCOLOR

H. Fritzsch+

CERN, Geneva

and

Max-Planck-Institut für Physik und Astrophysik, München

1. INTRODUCTION

Recently a large number of authors has been interested in constructing composite models of leptons and quarks. The idea is that the leptons and quarks are composed of several constituents which are bound together by a superstrong force. There exist various constraints on the sizes of leptons and quarks, e.g. the agreement between theory and experiment of the anomalous magnetic moment of the electron, which imply that those sizes are less than about 10^{-17} cm.

If the leptons and quarks are composite objects, one may expect the same to be true for at least some of the intermediate (gauge) bosons. A possible point of view is to regard the fermions and the intermediate bosons as composite objects at a very high energy scale (of the order of the mass scale entering in the grand unification schemes ($\sim 10^{15}$ GeV) or more (see e.g. ref. (1)). In this case the interactions at relatively low energies (QCD, flavor interactions) can be interpreted as effective gauge theories; essentially no deviations from the standard pattern of the QCD and QFD gauge theory framework are expected.

+ On leave from Sektion Physik, Universität München

Here we should like to explore another possible road. We shall suppose that the W bosons and the fermions are bound states, while the massless bosons (photon, gluons) are elementary[2,3]. The masses of the W bosons are generated dynamically by the binding forces in much the same way the ρ meson mass is generated in QCD. In this approach the weak interactions are indirect manifestations of the strong binding forces inside the W boson. One is reminded of the situation which has evolved during the last 20 years with respect to the strong interactions. Many years ago the strong interactions were interpreted as a gauge theory in which the baryons played the rôle of the elementary fermions while the vector mesons (ρ, ...) were considered to be gauge particles. Today the situation has changed considerably. Both the baryons and the vector mesons are composite objects, consisting of quarks. The strong interactions between the nucleons are regarded as residual effects of the gluonic forces between the quarks.

Perhaps the weak interactions are of a similar nature, and the W bosons consist of constituents which are at the same time the building blocks of the fermions. Of course, we realize that the weak interactions differ in their properties substantially from the strong interactions. First of all, they violate parity. Moreover they are well described by the exchange of vector particles. No exchange forces generated by scalar or tensor particles are observed. Nevertheless we find it useful to explore the point of view outlined above in more detail. The main consequence of the schemes we shall discuss below is that there is no reason to believe that the predictions for the masses of the W and Z bosons made within the minimal SU(2) x U(1) model are valid. We expect these masses to be much ligher than expected within the standard model ($M_W \sim 100 \ldots 160$ GeV, $M_Z \sim 120 \ldots 350$ GeV).

2. COMPOSITE MODELS

We take the point of view that the leptons, quarks and W bosons are composed of constituents, while the photon and the gluons of QCD are elementary[2,3]. In the absence of electromagnetism the global symmetry group of the weak interactions is SU(2) (weak isospin). It will be shown later that the observed structure of the neutral current is obtained if one takes into account the mixing between the photon and the neutral SU(2) boson W_3. This mixing arises dynamically, due to the electromagnetic annihilation of the W_3 constituents. For this reason only models in which the W bosons consist of two constituents seem realistic. This excludes many composite models, including the rishon model[4].

We shall assume that the underlying gauge symmetry is given by the group $SU(3)_c \times G_h \times U(1)_e$ (c: color, e: electric charge). The group G_h is the hypercolor gauge group describing the confining forces responsible for the binding of the hypercolored constituents.

The corresponding gauge theory is called QHD. For simplicity we shall use the hypercolor group SU(n), where n is yet unspecified. The extension to other groups is easily made.

If we regard the weak bosons as bound states, several possibilities are open. For example, the weak bosons could be composed of hypercolored fermions or bosons; these fermions or bosons could be colored or color singlets. The weak bosons could be composed of two or more constituents. For reasons which will become clear afterwards, models in which the weak bosons consist of more than two constituents cannot be accepted, likewise models in which the weak bosons are composed of scalar or pseudoscalar bosons. We shall concentrate on schemes in which the weak bosons are composed of spin 1/2 fermions. Furthermore we shall mainly consider the lightest family of leptons and quarks, consisting of the (u,d)-pair and the (ν_e, e^-)-pair.

Two classes of models are possible:

a) The W-bosons consist of hypercolored fermions, which are color-singlets.

b) The W-bosons consist of hypercolored and colored fermions.

In both classes the W-bosons are, of course, hypercolor and color singlet bound states. For illustration we consider the following schemes, in which the fermions are composed of (pseudo) scalar and fermionic constituents (haplons). The underlying gauge group is

$$U(1)_e \times SU(3)_c \times G_H$$

(G_H: hypercolor gauge group).
The gauge theory QHD based on the group G_H (hypercolor gauge group) describes the dynamics of the haplon constituents bound together by the superstrong hypercolor forces. The QHD confinement parameter Λ_H is supposed to be of the order of a few hundred GeV ($\sim (G_F)^{-1/2}$). The gauge group G_H is not specified, but taken to be SU(n).

Case A (as far as the quantum numbers of the constituents are concerned, this case has also been considered in ref. (5)).

	spin	el. charge	color	hypercolor
α	1/2	1/2	1	n
β	1/2	-1/2	1	n
x	0	1/6	3	$\bar{n}$
y	0	-1/2	1	$\bar{n}$

The simplest QHD singlets one can form are fermions of electric charges (2/3, -1/3) = [(αx), (βx)] and (0, -1)= [(αy),(βy)] identified with (u,d), (ν_e, e^-) respectively.

Scheme B[2)]

	spin	el. charge	color	hypercolor
α	1/2	-1/2	3	n
β	1/2	+1/2	3	n
x	0	-1/6	3	$\bar{n}$
y	0	1/2	$\bar{3}$	$\bar{n}$

This scheme has the special feature that all haplons carry electric charge, color and hypercolor. The simplest QHD singlets are

$$\nu_e = (\bar{\alpha}\ \bar{y})_1 \qquad u = (\bar{\alpha}\ \bar{x})_3$$

$$e^- = (\bar{\beta}\ \bar{y})_1 \qquad d = (\bar{\beta}\ \bar{x})_3$$

(the index denotes the color of the state). Both in scheme A and B there exist vector bosons composed of the fermions ($\bar{\alpha}\beta$, $\bar{\beta}\alpha$, ...). Those are interpreted as the carriers of the weak interactions. The global symmetry group SU(2) generated by the haplon doublet $\binom{\alpha}{\beta}$ is identified with the weak isospin.

The observed parity violation of the weak interaction can be accommodated in two different ways:

1. The bound state structure discussed above is assumed to be valid only for the lefthanded fermions.

2. Both the lefthanded and righthanded fermions are bound states, however the Λ_H parameter of the righthanded fermions is larger than the Λ_H parameter of the lefthanded ones, i.e. two QHD group are needed ($G_H^{(L)}$, $G_H^{(R)}$). In this case the global symmetry groups are weak interactions is $SU(2)_L \times SU(2)_R$; the W-bosons coupling to the righthanded fermions are heavier than those which couple to the lefthanded fermions.

Both in the schemes A and B the weak interaction is an effective interaction of the Van der Waals type, generated by the superstrong QHD force. The universality of the weak interaction between leptons and quarks follows from the global SU(2) symmetry in the α-β-space.

The charge assignments made above are not unique. However we find the assignment chosen in both schemes A and B particularly interesting; note that one of the scalar haplons has charge 1/6. The fermionic haplons α and β have the charges $\pm$ 1/2. This implies in particular that the electric charge is a pure isovector in the weak isotopic space spanned by the (α,β)-doublet.

One could also consider models in which the leptons and quarks are composed of three fermions bound together by a force based on the group $SU(3)_H$. As as illustrative example we consider the following model, based on the group

$$G = SU(3)_H \times SU(3)_c \times U(1)_e .$$

All are fermions, which transform under $SU(3)_h \times SU(3)_c$ as $(3_h, 3_c)$ or $(3_h, \bar{3}_c)$.

We need three different types of constituents:

$$\begin{pmatrix} \alpha \\ \beta \end{pmatrix} = \begin{pmatrix} 3_h, 3_c \\ 3_h, 3_c \end{pmatrix}$$

$$\begin{pmatrix} x \\ y \end{pmatrix} = \begin{pmatrix} 3_h, 3_c \\ 3_h, 3_c \end{pmatrix}$$

$$a = (3_h, 3_c).$$

The leptons and quarks are represented as

$$\begin{pmatrix} \nu_e \\ e^- \end{pmatrix} = \begin{pmatrix} \alpha \; x \; a \\ \beta \; x \; a \end{pmatrix}$$

$$\begin{pmatrix} u \\ d \end{pmatrix} = \begin{pmatrix} \alpha \; y \; a \\ \beta \; y \; a \end{pmatrix}$$

(only hypercolor singlets and color singlets / color triplets are taken into account).
Again the weak isospin is the group spanned by the doublet (α,β). The W-bosons are composed of α and β.

However in this model a serious problem arises - its global symmetry is too large. For example we can easily enlarge the global symmetry from SU(2) to SU(3), generated by the triplet (α,β,a). As a result we obtain too many new leptons and quarks which are degenerate in mass with the observed leptons/quarks. Such objects are not observed.

It can easily be seen that this disease is not specific to the model outlined above, but affects all models in which the leptons and quarks are composed of three spin 1/2 constituents. For this reason we believe that models in which the leptons and quarks are composed of three spin 1/2 hypercolored constituents are not realistic.

3. THE CHARGED WEAK CURRENT

It is assumed that the spectral functions of the weak currents in QHD are qualitatively similar to the ones in QCD. At low energies they are dominated by the lowest lying pole, and at high energies (energies large compared to Λ_H) they can be described by a continuum of "weak quanta" (this term is borrowed from ref. (6)), i.e. a continuum of $\bar{\alpha}\alpha$, $\bar{\alpha}\beta$ $\bar{\beta}\alpha$, or $\bar{\beta}\beta$- pairs. The main difference between our approach and the so-called "standard model" of the weak interaction lies in the strength of the W-lepton or W-quark vertex. In the standard model the coupling constant describing this vertex is given by $e/\sin\theta_W$ (θ_W: SU(2) x U(1) mixing angle, e: electric charge). In our approach the weak interaction is an effective interaction of the van der Waals type, generated by the superstrong QHD force. The universality of the weak interaction follows from the global SU(2) symmetry in the α-β-space. The strength g of the W-quark or W-lepton vertex has nothing to do with the electric charge e, since the electromagnetic interaction is an elementary interaction (like the QCD and QHD interaction). We expect the strength of the W-fermion vertex g to be of similar order as the strength of the ρ-nucleon vertex ($g \approx 1$)(in this respect our approach is similar to the one discussed in ref. (3)). Due to the relation $g^2/8M_W^2 = G_F/\sqrt{2}$ implying $M_W = g \cdot 123$ GeV the mass of the weak boson may well be much larger than expected in the standard SU(2) x U(1) gauge theory (in the case $g = 1$ the W-mass is 123 GeV).

4. THE NEUTRAL CURRENT AND THE SIZE OF THE W-BOSON

In the absence of mixing with the electromagnetic current there exist two neutral current channels, the isovector channel described by $(\bar{\alpha}\alpha-\bar{\beta}\beta)/\sqrt{2}$, and the isoscalar channel described by $(\bar{\alpha}\alpha+\bar{\beta}\beta)/\sqrt{2}$. Especially one expects that the isoscalar spectral function is dominated at low energies by the lowest lying meson with the quantum numbers $(\bar{\alpha}\alpha+\bar{\beta}\beta)/\sqrt{2}$. The observed neutral current is a pure isovector, if we neglect for a moment the mixing with the electromagnetic current. The experimental data allow at most a 10 % isoscalar contribution to the neutral current[7]), i.e. the isoscalar boson W_0 must be much heavier than the isovector boson W_3 (or the Z-boson). The isoscalar boson has the internal quantum numbers of the vacuum, i.e. its mass term may receive contributions from the hypergluon annihilation channel. The situation is quite similar to the one in QCD. Here the η'-meson is about 6 times heavier than the π_0 - meson. It is generally assumed that the source of the mass gap is the gluons of QCD. The isoscalar quark configuration $(\bar{u}u + \bar{d}d) / \sqrt{2}$ can mix with gluonic configurations. This implies that the isoscalar meson is heavier than the isovector meson[8].

Our approach makes sense only if something similar happens in QHD for the weak bosons. The hypercolor annihilation channels must lead to a rise of the mass of the isoscalar weak boson relative to the mass of the isovector boson. For our further discussion we shall assume that this is indeed the case.

The experimental data on the neutral current interaction require a mixing between the photon and the W_3 boson (the neutral, isovector partner of W^+ and W^-), which in the standard SU(2) x U(1) scheme is caused by the spontaneous symmetry breaking.

Within our approach this mixing is due the W_3 - γ transitions, generated dynamically like the ρ-γ transitions in QCD (for an early discussion, based on vector meson dominance, see ref. (9)). The magnitude of $\sin^2\theta_W$ is directly related to the strength of the γ-W_3 transition. The latter is determined by the electric charges of the W-constituents and by the W wave function near the origin.

We suppose that in the absence of electromagnetism the weak interactions are mediated by the triplet (W^+, W^-, W^3), where $M(W^+) = M(W^-) = M(W^3) = O\,(\Lambda_H)$.

After the introduction of the electromagnetic interaction the photon and the W^3 - boson mix. We denote the strength of this mixing by a parameter λ, following ref. (9), which is related to g (W-fermion coupling constant) and the effective value of $\sin^2\theta_W$

$$\sin^2\theta_W = \frac{e}{g} \cdot \lambda$$

Furthermore one has:

$$M_W = g \cdot 123 \text{ GeV}$$

$$M_Z^2 = \frac{M_W^2}{1-\lambda^2}$$

The mixing parameter λ is determined by the decay constant F_W (or f_W) of the W-boson, which we define in analogy to the decay constants of the ρ_0-meson (F_ρ, f_ρ respectively):

$$\langle o|j_\mu^3|W^3\rangle = \langle o|\tfrac{1}{2}(\bar{\alpha}\gamma_\mu\alpha - \bar{\beta}\gamma_\mu\beta)|W^3\rangle = \varepsilon_\mu\, M_W^{\,2}/f_W = \varepsilon_\mu\, M_W\, F_W$$

One finds:

$$\lambda = \frac{e}{f_W} = e \cdot \frac{F_W}{M_W}$$

In the table we have displayed the numerical values for F_W as a function of g. F_W is a rather sensitive function of g; for g = 0.75 one finds F_W = 166 GeV; for g = 1.2 one obtains F_W = 425 GeV[10]).

Table (all masses and energies in GeV; we have used $\sin^2\theta_W = 0.22$ and n = 4)

g	0.75	0.9	1	1.1	1.2
λ	0.54	0.65	0.73	0.80	0.87
M_W	92	111	123	135	148
M_Z	110	146	179	225	300
F_W	166	239	295	357	425
Λ_W	76	103	123	144	166
x	0.82	0.93	1.0	1.06	1.12

In order to study the physics of the bound state structure of the weak bosons in more detail we express the decay constant F_W in terms of the bound state wave function of the weak bosons. We use a non-relativistic wave function, which, of course, cannot be but a very crude approximation.

One has, leaving out irrelevant Lorentz indices in case A:

$$|W^3\rangle = \frac{1}{\sqrt{2}}\,\frac{1}{\sqrt{n}}\sum_{i=1}^{n}(\bar{\alpha}_i\alpha_i - \bar{\beta}_i\beta_i)\phi\,(x)$$

(ϕ (x) coordinate space wave function, i: hypercolor index, n: number of hypercolors)

In case B one finds

$$|W^3\rangle = \frac{1}{\sqrt{2}}\,\frac{1}{\sqrt{n}}\,\frac{1}{\sqrt{3}}\sum_{i=1}^{n}\sum_{j=1}^{3}(\bar{\alpha}_{i,j}\alpha_{i,j} - \bar{\beta}_{i,j}\beta_{i,j})\,\phi\,(x)$$

(i: color index). The current matrix element can be written as

$$<o|j_\mu^{\ 3}|W^3> = \varepsilon_\mu \sqrt{n} \cdot \gamma \sqrt{2M_W} \cdot \phi\ (o) = \varepsilon_\mu \cdot M_W F_W$$

$$F_W = \sqrt{n} \cdot \gamma \cdot \sqrt{2/M_W}\ \phi\ (o)$$

where $\gamma = 1$ in case A and $\gamma = \sqrt{3} = \sqrt{n_c}$ in case B.

$$\sin^2\theta_W = \frac{e^2}{g} \sqrt{n} \cdot \gamma \sqrt{2/M_W^{\ 3}}\ \phi\ (o),$$

$$= e^2/\ g \cdot F_W\ /\ M_W.$$

i.e. $\sin^2\theta_W$ is proportional to the coordinate space wave function of the W-boson at the origin. Taking for example a constant coordinate space wave function, the inverse radius Λ_W of the W-boson and $\phi(o)$ are related: $|\phi(o)|^2 = (4/3\pi)^{-1} \cdot \Lambda_W^3$.

It is useful to introduce the dimensionless quantity $x = (\Lambda_W\ /\ M_W)$, and one obtains

$$\sin^2\theta_W = e^2/g \cdot \sqrt{2n}\ \gamma(\tfrac{4}{3}\pi)^{-1/2}\ x^{3/2}\ .$$

For example for $g = 1$, $\gamma = \sqrt{3}$ (case B) and $n = 4$ one finds $x = 1.0$, i.e. $\Lambda_W \approx M_W$. Unless n is very large ($n > 10$), x is not much smaller than one.

It is instructive to compare the structure of the W-bosons with the one of the ρ-meson. In the ρ-case one has

$$\lambda_\rho = \frac{e}{f_\rho} = 0.054,$$

taking $f_\rho = 5.6$. This leads to

$$x_\rho = \Lambda_\rho\ /\ m_\rho \approx 0{,}28.$$

Thus the structure of the ρ-meson differs substantially from the structure of the W-boson. (If the W were similar to the ρ, one would find $\sin^2\theta_W = 0.016$.)

If the bound state dynamics of the W-bosons were similar to the quark dynamics inside the ρ-meson, the effective SU(2) x U(1) mixing parameter $\sin^2\theta_W$ would be more than an order of magnitude smaller than observed. In order to accommodate the observed value of $\sin^2\theta_W$, the decay constant F_W and the inverse size Λ_W must be of the same order as the mass of the W-boson. This suggests a similarity of the W bound state dynamics to the bound state dynamics of the π-meson in QCD. In the case of the π-meson one has $F_\pi \approx M_\pi$; the inverse radius of the π-meson, defined analogously to Λ_W, and the pion mass are of the same order, while in the case of the ρ-meson the radius is about four times larger than the inverse mass. One may conclude that the mass of the W-boson is anomalously small, compared to its inverse size.

The pion decay constant is about equal to the Λ-parameter of QCD. Thus the corresponding parameter in the hypercolor dynamics Λ_H is expected to be of the same order as the W-mass, i.e. 100 ... 160 GeV.

Thus far we have assumed that the W-bosons consist of spin 1/2 constituents. In some models spin zero constituents are used as W-constituents (see e.g. ref.(3)). We should like to argue that such models can be ruled out, on the basis of the following argument. The effective SU(2) x U(1) mixing parameter $\sin^2\theta_W$ is proportional to the wave function at the origin $\phi(o)$. If the W-bosons consist of spin zero constituents, the coordinate space wave function must be a p-wave, i.e. $\phi(o) = 0$. Thus in a nonrelativistic approach one finds $\sin^2\theta_W = 0$. Even if we allow for fairly large relativistic corrections, it seems impossible to us to obtain the observed large value of $\sin^2\theta_W$ in such models.

5. EXOTIC PARTICLES

If the leptons and quarks are composed of spin zero and spin 1/2 haplons, one expects new particles (exotic quarks, leptons, new types of W bosons) to exist and eventually to show up in the experiments. For example, in case B there exist color sextet quarks as well as color octet analogs of the leptons, whose effective masses should be larger than 50 GeV (see ref. (2) and ref. (11)).

6. THE FLAVOR PROBLEM

Thus far our considerations were based on the first family of leptons and quarks. An interesting way to interpret the other families is to suppose that they are excitations of the first family involving QHD-gluons. For example in case B the muon could be interpreted as the state $(\beta\bar{y}h)$ where h denotes a QHD-gluon. The τ-lepton would be the state $(\beta\bar{y}hh)$, etc. The QHD interaction causes transitions between these states, which lead to the phenomenon of weak-interaction mixing. Orbital excitations of the lepton or quark configurations are expected to have a mass of order Λ_H and are not considered here.

In such an approach the problem of stability of the "heavy" leptons μ,τ against radiative decay arises. Since no new quantum numbers are introduced, a decay like $\mu \to e\gamma$ can occur. The hypergluon turns itself into a photon, and a hypercolor rearrangement takes place. It could well be that such a rearrangement is highly suppressed, and a stability rule similar to the "Zweig rule" in QCD is at work. In any case the interpretation of the higher flavors as hypercolor excitations involving the QHD gluons seems a possible way to resolve the flavor problem.

7. FINAL COMMENTS

Here we have argued that the weak interactions could be a Van der Waals - type interaction, i.e. an indirect manifestation of QHD. If this is true, one expects that the weak interactions look at high energies very unlike what was previously thought. In particular the W boson as well as the Z boson are heavier than expected within the SU(2) x U(1) gauge model. The width of the Z boson is relatively large, e.g. in the case B one expects $\Gamma(Z_0) \approx 10 \ldots 20$ GeV. The leptonic branching ratio $B(Z_0 \to e^+e^-, \mu^+\mu^-, \ldots)$ is much smaller than expected within the SU(2) x U(1) theory[11]. Thus it will be fairly difficult to observe the weak bosons in the p-$\bar{p}$ collider experiments now under way at CERN.

ACKNOWLEDGEMENT:

Most of this work has been done in collaboration with Dr. G. Mandelbaum. I would like to thank Profs. J.J. Sakurai and D. Schildknecht for useful discussions.

REFERENCES

1. See e.g.: J. Ellis, these proceedings, and references therein.
2. H. Fritzsch and G. Mandelbaum, Phys. Lett. 102B:319 (1981).
3. L. Abbott and E. Farhi, Phys. Lett. 101B:69 (1981).
4. H. Harari, Phys. Lett. 86B:83 (1979);
   M.A. Shupe, Phys. Lett. 86B:87 (1979).
5. O.W. Greenberg and C.A. Nelson, Phys. Rev. D10:256 (1974);
   R. Barbieri et al., CERN preprint TH.3089 (1981);
   R. Casalbuoni and R. Gatto, Phys. Lett. 103B:113 (1981).
6. J.D. Bjorken, Phys. Rev. D19:335 (1979).
7. J.J. Sakurai, private communication (1981).
8. See e.g.: H. Fritzsch and P. Minkowski, Nuovo Cimento 30A:395 (1975).
9. P. Hung and J.J. Sakurai, Nucl. Phys. B143:81 (1978).
10. H. Fritzsch and G. Mandelbaum, CERN preprint TH.3203 (1981), to appear in Phys. Lett. B.
11. G. Mandelbaum, University of Munich thesis (1981).

CONSTRAINTS ON COMPOSITE MODELS OF QUARKS AND LEPTONS

R.D. Peccei

Max-Planck-Institute für Physik und Astrophysik
Föhringer Ring 6, Munich, Fed. Rep. Germany

1. INTRODUCTION: GUTS VERSUS COMPOSITENESS

Grand unified theories (GUTs)[1] provide a beautiful extension of the standard model of strong and electroweak interactions. Three features particularly stand out as important consequences of having an underlying grand unified theory:

(1) Charge quantization is a natural consequence of unification to a group where charge is one of the generators.

(2) The strength of weak, strong and electromagnetic forces is the same at the scale where unification occurs.

(3) A plausible scenario for generating a nonvanishing baryon asymmetry in the universe can be constructed in GUTs.

Along with these more general statements in GUTs go specific predictions for the Weinberg angle, the rate of proton decay and the value of the ratio of baryons to photons in the universe, n_B/n_γ .

It will ultimately remain for experiment to decide whether the idea of grand unification is correct. Nevertheless, at this stage, there appear already certain features in particle physics which are not understandable purely from a GUTs point of view and which may require a different approach from the one of unification. I shall mention here three prominent puzzles:

(1) The values of fermion masses and of the various mixing angles (Cabibbo angles) are not predictions of GUTs.

(2) No plausible explanation of family replication exists in GUTs.

(3) The large ratio between the unification mass $M_{GUT} \simeq 10^{15}$ GeV and M_W is not understood. Also, the relative closeness of M_{GUT} to the Planck mass, $M_P \simeq 10^{19}$ GeV, is a mystery.

There have been, of course, attempts to understand these points within the framework of GUTs, but in general one has encountered other difficulties or only partial solutions have been found. For instance, there are some fermion mass predictions in SU(5)[2] - notably $m_b/m_\tau \simeq 3$ - but not all mass ratios and mixing angles can be predicted. Similarly, one can incorporate family replications in GUTs by going to very large groups[3*], but in general one is left with very many other unwanted states. The hierarchy problem also has been studied in GUTs, most recently in connection with some supersymmetric extensions of GUTs[4], but no clear understanding of the reason for the large scale difference is yet available.

Composite models of quarks and leptons[5] could provide a plausible alternative to GUTS. Indeed, in principle, it appears easier to understand some of the open problems of GUTs in this framework. In some sense, the hope underlying composite models of quarks and leptons is that the story concerning QCD and hadronic physics may repeat itself. There is a widespread belief that when the bound state problem in QCD is finally solved, the hadronic spectrum will emerge as the natural output. Similarly, the hope of composite models of quarks and leptons (preon theory) is that the quark and lepton spectrum will also directly follow from these models. Indeed, some authors push the analogy further. From the point of view of QCD, nuclear physics can be thought of as the result of residual "Van der Waal" forces, due ultimately to quark and gluon interactions. In this vein, one can imagine that the weak interactions are just "Van der Waal" manifestations of the fundamental interactions which bind the quark and lepton constituents.

Composite models of quarks and leptons have, however, some important differences from QCD. First of all, one would like to construct these models in a way that the successes of GUTs are retained. In general, there is no problem with relating the charges of the quarks and leptons, since they are made up of the same constituents. However, unless one unifies electromagnetism and QCD, there is no understanding of the relation of α to α_s. There is, in general, also not an electroweak unification, since from the point of view of many models the weak interactions are an induced, residual, effect. Of course, in principle, there is nothing wrong with this, since this is the starting philosophy of these models.

---

* For a different approach see, J. Ellis, these Proceedings.

Furthermore, one should expect that, if the weak interactions are residual, the Weinberg angle should be a calculable parameter. The biggest difficulties that the preon theories have, vis à vis GUTS, is related to the question of proton decay and the establishment of a baryon asymmetry in the universe. We shall devote Section 3 of this paper to discussing these matters.

Besides these clear physical differences between QCD and preon theories, there are also important dynamical differences. At the most basic level, the difference is quite clear: QCD never cares about flavors; preon theory must, on the other hand, in its dynamics have the possibility of generating distinct families.* To my knowledge there is no preon model which can, even semiconvincingly, generate the observed family replications[6]. It seems to me that discovering the mechanism in preon models which produces the replications will be tantamount to finding the Rosetta Stone. Not being in possession of this fabled stone, alas, I will say nothing more on this fundamental problem.

A further dynamical problem which distinguished preon models from QCD concerns the size of the bound states. In QCD, as in all theories which are scale invariant at the classical level, one expects that the radius of the bound states and the mass of the bound states be closely related. In essence in these theories there is only one scale Λ , which can be taken to be the scale at which the running coupling constant goes through unity. Then $\langle r \rangle \sim 1/\Lambda$, while $M \sim \Lambda$. However, we know that quarks and leptons are quite point-like. If the preon theory is again a confining theory - as is to be expected if preons are to be the constituents of confined quarks - one must find a dynamics which produces light, but almost point-like objects:

$$\langle r \rangle \sim \frac{1}{\Lambda_P} \qquad \text{but} \qquad m \ll \Lambda_P \tag{1}$$

where Λ_P is the appropriate scale associated with the preon theory. In the following section I shall discuss some of the possible dynamical ideas which may lead to Eq. (1).

2. DYNAMICAL CONSTRAINTS IN PREON MODELS

To my knowledge there have been only two suggestions made which lead naturally to Eq. (1). The first suggestion can be traced to the Cargèse Lectures of 't Hooft[5]. He pointed out there that if the preon theory contained a global symmetry group which preserved chirality, then the quarks and leptons would end up by having zero

---

*I am discounting here the possibility that for each family there is associated a distinct preon flavor.

mass as a result of this global chiral symmetry. Thus $\Lambda_P >> m = 0$. Associated with this suggestion, however, there arise three serious dynamical constraints:

(1) In the only case we know of a confining theory with a global chiral symmetry (QCD in the zero quark mass limit), chirality is spontaneously broken by the vacuum. Thus there are zero mass Goldstone bosons, to be identified with the pions, but no zero mass fermions.

(2) Even if chirality were not to be spontaneously broken, the solution $m = 0$ is not realistic. In real life, quarks and leptons do have small (with respect to Λ_P) masses. How then do these small masses arise?*

(3) Certain consistency conditions must be obeyed by the resulting composite quarks and leptons. Namely, the triangle anomalies[7] of the global chiral symmetry calculated at the preon theory level and at the composite level must agree with each other:

$$\mathrm{Tr}\ \lambda_a\{\lambda_b,\lambda_c\}_{\mathrm{preon}} = \mathrm{Tr}\ \lambda_a\{\lambda_b,\lambda_c\}_{\mathrm{composite}} \tag{2}$$

A second dynamical mechanism for achieving Eq. (1) has recently been suggested by Bardeen and Visnjic[8]. They suppose that the preon theory is supersymmetric and that the supersymmetry is spontaneously broken. In this case, as is well known,[9] one get instead of Goldstone bosons associated Goldstone fermions for each broken supersymmetric generator. Thus one has again $m = 0$ for the quarks and leptons, independent of what Λ_P is.

One may worry that having quarks and leptons as Goldstone fermions of a broken supersymmetry may lead to immediate contradictions with experiment. Indeed, some time ago, Bardeen and de Wit and Freedman[10] showed that neutrinos could not be Goldstone fermions because the weak interactions do not show any trace of the low energy theorems which would arise from having a spontaneously broken supersymmetry. The way that Bardeen and Visnjic[8] avoid this no-go theorem is to suppose that the supersymmetry at the preon level is not exact but is actually explicitly broken by the weak interactions. If the scale associated with the spontaneous breakdown of the supersymmetry is sufficiently large compared to M_W, then effectively the low energy theorems due to the spontaneous breakdown of the supersymmetry are no longer applicable.

Even though Bardeen and Visnjic[8] manage to avoid this first hurdle, there remain many open and difficult questions to be

---

* This can be called the Composite Model Hierarchy problem.

answered before their scheme can be implemented. Perhaps the two most crucial ones are: 1) How can one manage to break the supersymmetry? and; 2) it is sensible to envisage supersymmetries with so many broken generators? (For 3 SU(5) families of quarks and leptons one needs 45 broken supersymmetry generators!) It is clearly early days yet for this scheme, and I shall not consider it further here. Rather, for the rest of this section, I want to return to discuss the dynamical problems raised by imposing a global chiral symmetry on the preon theory. I shall particularly concentrate on the first two issues, discussed at the beginning of this section, because the satisfaction of the anomaly constraints has been already extensively discussed in the literature[11].

If the preon theory is based on some SU(N) group, it is difficult to imagine that any chiral symmetry of some global flavor group is not spontaneously broken, as happens in QCD - at least if all preons are fermions. Recently, Barbieri, Maiani and Petronzio[12] have suggested that if the preon theory is based on a group O(n) and if the preons are put in the n-dimensional vector representation, then there may be a dynamical reason why chirality could be unbroken. In this example, since the fermions are in the adjoint representation, it is possible to construct composite states which are singlets under O(n) by simply combining the preons χ_i, $i = 1, \ldots, n$, with the O(n) gluons, g_i:

$$\psi \sim \chi_i g_i \tag{3}$$

Spinless states in the theory, constructed from the combination of two preons:

$$\sigma \sim \chi_i \chi_i \tag{4}$$

are then, perhaps, unstable in that they can disassociate into a pair of ψ-states ($\sigma \to \psi\psi$). If this occurs, then it is unlikely that σ can develop a vacuum expectation value

$$\langle\sigma\rangle \sim \langle\chi_1 \chi_2\rangle = 0 \tag{5}$$

Thus, in these models, due to the simple construction of the bound states (hypercolor screening) it may be never possible to develop vacuum expectation values which violate chirality.

This idea of hypercolor screening may perhaps also be applicable in other cases, even if the fermions are not put into the adjoint representation of the group [13]. For instance, if one has scalar preons which transform in the same way under hypercolor as the fermions do, then one can again screen the hypercolor in a manner similar to (3). Banks and collaborators [14] have investigated whether the O(n) hypercolor screening idea works by doing strong coupling calculations on the lattice and appear to see no evidence for chiral symmetry breakdown. I shall report below on an analogous investigation[15], in two dimensions, which shows opposite behaviour.

Before discussing this two-dimensional counterexample, I want to mention a different suggestion, due to Harari and Seiberg[16], for preserving chirality in preon models. The idea put forth by Harari and Seiberg[16] arose out of considerations specific to their rishon model, based on an SU(3) hypercolor group. Their preons are of two types, a T and a V rishon, which transform under hypercolor and ordinary color as:

$$T \sim (3,3) \qquad V \sim (3,\bar{3}) \tag{6}$$

Harari and Seiberg argue that the presence of the confining color group inhibits the formation of hypercolor (and color) singlet condensates which break chirality. Thus, in contrast to what happens in pure QCD, they expect that

$$\langle \bar{T}T \rangle = \langle \bar{V}V \rangle = 0 \tag{7}$$

because of the presence of color besides hypercolor.

I should mention that in the rishon model it is necessary to suppose that the limit in which the color coupling constant $g_c \to 0$ is nontrivial. Obviously, in that limit the rishon model would just go into a 6-flavor version of QCD and thus (7) presumably ceases to hold. Furthermore, it is clear that in terms of this "flavor" SU(6) the anomaly constraints do not work,

$$\mathrm{Tr}\ \lambda_a\{\lambda_b,\lambda_c\}_{\text{rishons}} \neq \mathrm{Tr}\ \lambda_a\{\lambda_b,\lambda_c\}_{\substack{\text{quarks}\\ \text{leptons}}} \tag{8}$$

since quarks and leptons do not form a full SU(6) representation. This is not really inconsistent, if the limit $g_c \to 0$ is nontrivial, because for g_c strictly equal to zero then there is no reason why quarks and leptons should remain massless. For $g_c \neq 0$, quarks and leptons may be massless, but then one has no longer a "flavor" SU(6) to which to apply the 't Hooft[5] anomaly constraints.

The idea that the presence of another confining group somehow stabilizes a theory against chiral symmetry breakdown is attractive but rather speculative, since it requires non-smooth behaviour in the coupling constant of the additional confining group. It appeared interesting to test this idea and that of hypercolor screening in a solvable model and, in collaboration with W. Buchmüller and S. Love, I investigated these matters in various versions of QCD_2[15]. Of course, two dimensions are not the best place to investigate the breakdown of continuous symmetries since one knows, from Coleman's theorem[17], that in fact no such breakdown is allowed! Nevertheless, there are instances in two dimensions where one gets phenomena essentially analogous to real symmetry breakdown.* Correlation functions drop off only as powers of the distance, because of the presence of zero mass bosonic bound states. These states are not really Goldstone bosons, but their presence is necessary to establish this almost long range order. In QCD_2 one knows that in the chiral limit and in the large N limit, where the number of colors goes to infinity but g^2N is kept fixed, there exists precisely such a zero mass state[19]. This state can be interpreted as a signal of "quasi chiral breakdown". What we investigated was whether the presence of an extra confining group or of having the fermions in the adjoint rather than the fundamental representation changed this pattern. If the ideas of Harari and Seiberg[16] and/or of hypercolor screening were correct, and if these ideas still were applicable in this two-dimensional case, then one would expect to see the zero mass bosonic state disappear.

Considering QCD_2 in the large N limit when there are two confining groups or when the fermions are in the adjoint representation is not easy to do since, in either case, one has to worry about another parameter growing like N. Fortunately, we were able to answer the more restricted question related to the presence or not of the zero mass bosonic bound state without having to appeal to the large N limit. I summarize below our findings:[15]

(1) The zero mass bound state found by 't Hooft[19] is not an artifact of the large N limit. It persists for arbitrary N.

(2) This state still appears if the confining group is changed from SU(N) to SU(N) x SU(M), N and M arbitrary.

(3) If the fermions are put in the adjoint, rather than in the fundamental representation the zero mass bound state is still in the theory.

(4) By studying the zero mass discontinuity of the chiral U(1) anomaly one can show that in all the above cases the zero mass bosonic state saturates this discontinuity. Thus no zero mass fermionic states, which couple to the U(1) current, are present in the theory.

I shall not enter into all the details of how these results were proven, since this is done in Ref. 15. Basically, we looked

---

* The most famous example of this kind is the so-called Berezinski-Kosterlitz-Thouless phase. This shows up, for instance, in the SU(N) Thirring model for large N. For a discussion see Ref. 18.

at the various versions of QCD_2 in the light come gauge and studied the effect of the " Hamiltonian" P_+ on a trial bosonic state

$$|\varphi;P_-\rangle = \frac{1}{\sqrt{N}} \int_0^1 dx\, b_i^+[(1-x)P_-]\, d_i^+[xP_-]\, \varphi(x)|0\rangle \tag{9}$$

where b^+ and d^+ are creation operators for fermions and antifermions in the model. In general P_+ on this state does not leave one in the two-particle sector, except for SU(N) groups in the large N limit where one just reproduces 't Hooft's[19] equation for the wave function $\varphi(x)$. However, <u>if $\varphi(x)$ is a constant</u>, one can show rather readily that, in the chiral limit,

$$2P_+P_-|\varphi = \text{const};P_-\rangle \equiv M^2|\varphi = \text{const};P_-\rangle = 0 \tag{10}$$

This property is true in all cases 1)-3) above and it establishes our results.

Having a specific eigenstate for the zero mass bosonic state allows one to compute its contribution to correlation functions and one can show that this state does lead to power behaved correlation functions. Most importantly, one can compute the contribution of this state to the discontinuity of the U(1) current in the model. One knows that at $q^2 = 0$ this discontinuity just reflects the anomaly coefficient[20]. For instance, putting the fermions in the fundamental representation of an SU(N) version of QCD_2 one has[20]

$$\begin{aligned} &\text{Im}\, i \int d^2x\, e^{iqx} \langle 0|T(J^+(x), J^+(0))|0\rangle \\ &\quad = (g^2N)q_-^2\, \delta(q^2) + \ldots \end{aligned} \tag{11}$$

Calculating the contribution of the exact zero mass bosonic state to this discontinuity yields precisely the same value for the coefficient of $\delta(q^2)$. Since only zero mass intermediate states can give a contribution at $q^2 = 0$ one can conclude that no zero mass composite fermions exist in the theory (at least fermions such that the fermion-antifermion states can couple to the U(1) current). This is true for all the cases 1)-3) considered and establishes our result 4).

Our results show that, at least in two dimensions, confining theories in the chiral limit do not appear to have zero mass fermion bound states, even in theories where some of the suggestions discussed for not breaking chirality are implemented. It is of course difficult to extrapolate our results to four dimensions. Indeed, in some recent papers [21,22] it has been argued that the situation may not be even that clear in two dimensions. For not quite vanishing fermion masses in QCD_2 ($m \ll g$) one can show, using a method first suggested by Baluni[23], that there are some soliton solutions of mass m. If these solutions are interpreted as fermions, then it would appear that the theory, in the near chiral limit, has also low mass fermions, besides the light bosons we discussed. It is somewhat a mystery, at least to me, what happens to these fermions in the strict massless limit, since there is no room for them in the anomaly equations.

Before closing this section, I would like to make some remarks concerning what I called earlier the Composite Model Hierarchy problem. Let us suppose that one is able eventually to construct a preon model where chirality is in fact not spontaneously broken and zero mass quark and lepton bound states ensue. How can these states then acquire the needed "experimental" masses? There appear to be two options for this, which involve assuming either that the chiral symmetry at the preon level is not perfect, or that indeed there is some spontaneous breakdown of chirality, but that this breakdown is small and can be ignored as a first approximation. Having a small explicit breakdown at the preon level puts one, so to speak, back to square one. Again one would have little control on the origin of quark masses. Perhaps more interesting is the possibility of having a small amount of spontaneous chiral breakdown as the origin of the quark and lepton masses. This appears to be more convincing physically in that the dynamics may indeed force somehow chiral noninvariant vacuum expectation values to have a scale much less than Λ_P.

There is, of course, a troublesome issue associated with any spontaneous breakdown, and that is that one necessarily generates unwanted Goldstone bosons. It turns out, however, that if these Goldstone bosons are sufficiently weakly coupled then perhaps they do not have any untoward physical effects. This happens, for instance, in the case of Majorons[24] where the combination of weak coupling and their pseudoscalar nature makes them essentially undetectable. Similarly, hypoaxions [25] arising from spontaneous breakdown of chiral U(1) global symmetries, may also be tolerated, provided the scale Λ_P is sufficiently big and the magnitude of the vacuum expectation value causing the breakdown is sufficiently small, since their coupling to matter scales like $1/\Lambda_P$.

The rishon model of Harari and Seiberg[16] provides an example where one expects both hypoaxions and Majorons. At the preonic level one has both an axial and a vector U(1) global symmetries, which are not finally present at the quark and lepton level. Whence one expects these associated Goldstone bosons, coupling proportionally to Λ_P^{-1}. The presence of these Goldstone bosons in the theory can be used to set a minimum scale for Λ_P. Hypoaxions must be sufficiently weakly coupled so as not to cause stars to cool down too rapidly by their emission. This leads to an estimate, for the rishon model[16], of $\Lambda_P \gtrsim 10^8\text{-}10^9$ GeV.

It is worth remarking that the Bardeen and Visnjic[8] scheme, which has massless quarks and leptons as a result of their being Goldstone fermions, is more amenable to generating small masses for these states. The supersymmetry is supposed to be spontaneously broken by some vacuum expectation values $< F_a >$, thereby generating the Goldstone fermions, and it is also supposed to be explicitly broken by the weak interactions, so as to avoid trouble with low energy theorems. In analogy with what happens in QCD, where the pion mass squared is a product of the quark masses that break explicitly the chiral symmetry, times the vacuum expectation values which break the chiral symmetry spontaneously:

$$m_\pi^2 = -m\{\frac{<\bar{u}u>}{f_\pi^2}\} \tag{12}$$

one would expect for the quark and lepton masses a formula

$$m_f \sim h <F> \tag{13}$$

where h details the amount of explicit breaking of the supersymmetry. The particular dependence of the above formula on the flavor of the leptons and quarks, however, is far from clear at this stage.

3. PHYSICAL CONSTRAINTS IN PREON MODELS

Preon models are constrained physically by the fact the the proton is very stable ($\tau_P \gtrsim 10^{30}$ years). Because quarks and leptons are supposed to be made up of the same constituents, it is clear that processes like $P \to e^+ \pi^o$ cannot be a priori forbidden. In fact, by dimensional analysis, one can write down simple interaction Lagrangians which should arise in preon models and which can mediate proton decay. The most dangerous such Lagrangian will involve operators of dimension 6 (4 Fermi-interactions) scaled by Λ_P^{-2}:

$$\mathcal{L}_{eff} \sim \frac{1}{\Lambda_P^2} \ell qqq \tag{14}$$

The above Lagrangian would give too fast a rate for the proton to decay ($\tau_P \sim \Lambda_P^4$) unless Λ_P itself were in the range of GUT masses $\Lambda_P \gtrsim 10^{14}$-$10^{15}$ GeV.

One may well ask if the above conclusion is strictly true or whether it can be avoided somehow. Clearly, if one could associate baryon number with preon number, then one could forbid directly proton decay. This suggestion, however, is not really tenable if all preons are fermions, since then leptons would end up by having baryon number! Casalbuoni and Gatto[26] have studied this problem in detail from a somewhat more general (and more reasonable) point of view than just simply trying to identify baryon number with preon number. They looked at models containing some hypercolor group G_H and flavor group G_F and studied which representations of G_F allowed for a definition of baryon number at the preonic level, still yielding the usual quarks and leptons in the composite spectra. To help them further determine the G_H's and G_F's they made use of 't Hooft anomaly constraints.[5,11] Their results, although interesting, lead one unfortunately to rather "unrealistic" models having very large G_H's, G_F's and having many preons.

One can avoid the proton decay problem in a much simpler way if one allows for some scalar preons. I shall illustrate this point with a model due to Barbieri, Masiero and Mohapatra.[27] These authors consider a preon model based on $G = SU(4)_H \times SU(3)_c \times U(1)_{em}$ and have four kinds of preons (two spinor and two scalar):

$$\begin{aligned} &F^u_\alpha \sim (4,1,\tfrac{1}{2}) \qquad && F^d_\alpha \sim (4,1,-\tfrac{1}{2}) \\ &\phi^c_{\alpha i} \sim (\bar{4},3,\tfrac{1}{6}) \qquad && \phi^\ell_\alpha \sim (\bar{4},1,-\tfrac{1}{2}) \end{aligned} \tag{15}$$

Clearly one can make four distinct hypercolor singlets with quantum numbers appropriate to the ones of the quarks and leptons:

$$u_i \sim F^u_\alpha \phi^c_{\alpha i} \;;\quad d_i \sim F^d_\alpha \phi^c_{\alpha i} \;;\quad \nu \sim F^u_\alpha \phi^\ell_\alpha \;;\quad e \sim F^d \phi^\ell_\alpha \tag{16}$$

One can associate a U(1) connected with fermion number F and a separate U(1) attached to ϕ^c and ϕ^l. However since an interaction term $(\phi^c)^3\,\phi^l$ can be written down, the model only has an overall global $U(1)_F \times U(1)_{\phi^c - 3\phi^l}$. It is easy to check that these U(1)'s are just linear combinations of baryon and lepton number

$$\begin{aligned} B &= \frac{1}{4}F + \frac{1}{12}(\phi^c - 2\phi^\ell) \\ L &= \frac{1}{4}F - \frac{1}{4}(\phi^c - 3\phi^\ell) \end{aligned} \tag{17}$$

In this model then proton decay is forbidden at the preonic level. However, one may still obtain some proton decay, if condensates form which spontaneously break baryon number. (One must then worry about the associated Goldstone boson.) For instance, if a condensate

$$\langle FFFF \rangle = M^6 \tag{18}$$

formed, then proton decay could proceed through this breakdown yielding a rate

$$\tau_p \sim \frac{\Lambda_p^4}{m_{prot.}^5} \left(\frac{\Lambda_p}{M}\right)^6 \tag{19}$$

This could be long enough, for $\Lambda_p \ll M_{GUT}$, provided the condensate scale were much smaller than Λ_p. Whether this can happen is unclear to me.

A somewhat different suggestion for preventing proton decay, in the context of the rishon model, has been proposed by Harari, Mohapatra and Seiberg[28]. They remark that in the model the current

$$J^\mu_{A+} = \bar{T}\gamma^\mu\gamma_5 T + \bar{V}\gamma^\mu\gamma_5 V \tag{20}$$

has both color and hypercolor anomalies. Thus there is really not a $U(1)_{A_+}$ global symmetry in the model, because of the nontrivial contributions of the color and hypercolor instantons. What there is, however, is a remnant discrete Z_{12} subgroup. In terms of this subgroup one can associate a number X, which takes the values +1 or -1 depending on whether one is dealing with left- or right-handed rishons. Now since the neutrino and the electron are made up of two right-handed rishons and a left-handed rishon, or vice versa depending on the final helicity, while the up and down quarks are products of three left-handed (or three right-handed) rishons one has

$$X_\ell = \pm 1 \; ; \quad X_q = \pm 3 \tag{21}$$

Whence an effective interaction like that in Eq. (14) is forbidden by X-number conservation. The first allowed term, which could mediate proton decay has the schematic form[28],

$$\mathcal{L}_{eff} = \frac{1}{\Lambda_p^6} \partial\{qqq\ell\ell\ell\} \tag{22}$$

where the derivative is needed for helicity reasons.

An interaction like (22) gives a very long proton decay, even for moderate values of Λ_p. In fact, one can obtain in general faster decay rates by replacing some of the explicit fields in (22) by appropriate condensates. These condensates will break X-number, but since this number is due to a discrete symmetry no particular harm is done. It is worth remarking that in the rishon model the condensates which break B-L by two units should dominate over the condensates which conserve B-L:

$$\frac{\langle \ell\ell \rangle_{B-L=2}}{\langle \ell\ell \rangle_{B-L=0}} \gg 1 \tag{23}$$

Whence one expects to see preferentially in the model decays like $P \to \nu\pi^+$ or $P \to e^-\pi^+\pi^+$ instead of the standard SU(5) B-L conserving decay $P \to e^+\pi^o$.

The above two examples have shown that at least it is not impossible that the proton lifetime in composite models is longer than 10^{30} years. However, it is clear that it is not possible to give any precise prediction for the value of this lifetime. One point should, however, be made. Since one must necessarily suppress the dimension 6 terms in the effective Lagrangian for proton decay, it no longer follows[29] that in this decay B-L should be conserved. Thus, a possible signal for compositeness at scales less than M_{GUT} may indeed be the observation of B-L violating proton decay.

A problem which is potentially even more damaging to composite models with a scale $\Lambda_P \ll M_{GUT}$ is the matter of the universe's baryon asymmetry. In collaboration with A. Masiero and T. Yanagida, I investigated this point recently[30] and shall summarize our findings here. Before doing so it may be useful to recall the conditions necessary for establishing a non-zero asymmetry in the early universe. One must have a theory where:

(1) $\Delta B \neq 0$ processes are allowed.

(2) C and CP are violated.

(3) The processes involving baryon violating particles must be out of thermal equilibrium at some temperature and, at lower temperatures, all the baryon violating interactions must remain out of equilibrium.

The observation made with Masiero and Yanagida[30] is that it is essentially impossible to establish the asymmetry at temperatures $T \gg \Lambda_P$. This is because for these temperatures one is dealing directly with preons and not their constituents, quarks and leptons. In the preon theories now under consideration[5], in general, none or almost none of the necessary conditions for establishing an asymmetry obtain. For instance:

(1) Baryon number, as we have seen, is difficult to define in preon models. Once it is defined, however, it is difficult to violate.

(2) Preon theories are mostly gauge theories where C is conserved, unless the preons are put in particular C-violating representations. Further, there is no reason to violate CP at the preon level.

(3) Preons are generally taken to be massless and massless particles follow equilibrium distributions.

From the above consideration, it is fair to conclude[30] that one must generate the baryon asymmetry at temperatures below Λ_P, where the "effective" composite theory begins to play a role. This, however, is also a difficult thing to do, if $\Lambda_P << M_{GUT}$, because it is hard then to satisfy the out of equilibrium requirement. In terms of the rate (decay or interaction) of particles which have baryon violating channels one needs to have that

$$\Gamma_{\Delta B \neq 0} < H \simeq \frac{T^2}{M_P} \tag{24}$$

where H is the universe's expansion rate. Now in general one expects that $\Gamma_{\Delta B \neq 0}$ is proportional to T and one sees that (24) will not be satisfied for $T \lesssim \Lambda_P << M_{GUT} < M_P$, unless the constant of proportionality in $\Gamma_{\Delta B \neq 0}$ is very small. Writing

$$\Gamma_{\Delta B \neq 0} = h^2 T \tag{25}$$

we see that for $T \simeq \Lambda_P$ we require

$$h \lesssim \sqrt{\frac{\Lambda_P}{M_P}} \tag{26}$$

For Λ_P in the 10^5-10^6 GeV range this requires $h \lesssim 10^{-7}$! There is, in general, no good reason for having such suppressed rates. In fact, particles which are heavy (i.e., near Λ_P in mass), and which may have baryon violating channels, will most likely also have some fast decay channels.

There is a further point that must be worried about. If the rate (25) describes the decay of a particle S which has 2-body channels qq and ql, then in fact h must be even further limited, since S-exchange would give an unacceptably large proton decay. At any rate, because the baryon asymmetry in general necessitates the computation of an interference diagram involving a final state rescattering which is baryon number violating, one expects that

$$\frac{n_B}{n_\gamma} \sim h^2 \tag{27}$$

Given the bound (26), or the potentially stronger bound from proton decay, one sees that the amount of n_B/n_γ that can be generated is tiny. (There are of course also many other suppression factors, coming from the amount of CP violation needed and the available phase space, which make things even worse.)

If $\Lambda_P << M_{GUT}$, so that the effective coupling constant h is necessarily very small, it appears that to obtain a sufficient amount of n_B/n_γ it is necessary to avoid a quadratic dependence of this quantity on h. It is, in fact, possible to construct a model where n_B/n_γ is only proportional to h. This was done in Ref. 30 by supposing that there existed two massive bound states ($m \sim \Lambda_P$), of the preonic theory which had decay channels with distinct baryon number and which could communicate with each other. Then, the necessary interference diagrams for the baryon asymmetry, having one of these states as intermediary, needs only be of order h^3 and hence

$$\frac{n_B}{n_\gamma} \sim h \tag{28}$$

However, the model we constructed[30] is very artificial and it is far from being believable. For instance, there is no reason why CP should be violated in the decay of the massive states.

In conclusion, it appears very difficult to obtain a baryon asymmetry in preon models with $\Lambda_P << M_{GUT}$. For $\Lambda_P \simeq M_{GUT}$, however, some of the problems we encountered should be alleviated. For instance, the out of equilibrium condition is not then so difficult to be satisfied. It should be pointed out, nevertheless, that serious problems remain even in this case. For example, the calculation of n_B/n_γ will necessitate a careful interrelation between composite states of mass of order Λ_P with the, essentially massless, quarks and leptons of the theory.

4. CONCLUSIONS AND OUTLOOK

I have discussed some of the dynamical and physical issues confronting composite models of quarks and leptons. On the dynamical side, it is clear that we are still very far from understanding if it is possible to construct bound states which are light but almost point-like. Much more work is needed in this direction, both along 't Hooft's chiral suggestion[5] and the recent supersymmetric suggestion of Bardeen and Visnjic[8]. Indeed, it would be very nice if alternative mechanisms could be found to produce light but almost point-like bound states.

The physical constraints on preon theories imposed by proton decay and the universe's baryon asymmetry seem to make it very hard to conceive of a compositeness scale $\Lambda_p << M_{GUT}$. Indeed, some theorists are beginning to seriously consider compositeness at, or above, the GUT scale.[31] It is, however, very difficult for me to conceive that physics at scales of energy of the order 10^{15} GeV should ultimately determine such fine splittings as $m_\mu - m_e \simeq 10^{-1}$ GeV.

The present prospects of composite models of quark and leptons seem rather glum - we have so many more problems than answers! However, if one really wants to understand the family problem this appears to be the only way to proceed. Only time will tell if this is the right direction.

ACKNOWLEDGEMENTS

I have benefitted greatly from discussions I have had with W. Buchmüller, H. Harari, S. Love, A. Masiero and T. Yanagida.

REFERENCES

1. For a review of GUTs see, for example:
P. Langacker, Phys. Rep. 72:185 (1981);
J. Ellis, Grand Unified Theories in "Gauge Theories and Experiments at High Energy", K.C. Bowler and D.G. Sutherland, ed., Scottish Universities Summer School in Physics, Edinburgh (1981).
2. A.J. Buras, J. Ellis, M.K. Gaillard and D.V. Nanopoulos, Nucl. Phys. B135:66 (1978).
3. I. Bars and M. Günaydin, Phys. Rev. Lett. 45:859 (1980).
4. See, for example:
D.V. Nanopoulos, these Proceedings.
5. K. Matumoto, Prog. Theor. Phys. 52:1973 (1974);
J.C. Pati, A. Salam and J. Strathdee, Phys. Lett. 59B:265 (1975);
H. Terazawa, K. Akama and Y. Chikashige, Phys. Rev. D15:480 (1977);
H. Harari, Phys. Lett. 86B:83 (1978);
M. Shupe, Phys. Lett. 86B:86 (1978);
G. 't Hooft, Cargese Lectures (1979);
H.P. Dürr and H. Saller, Phys. Rev. D22:1176 (1980);
R. Casalbuoni and R. Gatto, Phys. Lett. 93B:47 (1980);
R. Barbieri, L. Maiani and R. Petronzio, Phys. Lett. 96B:63 (1980);
H. Harari and N. Seiberg, Phys. Lett. 98B:269 (1981), ibid

100B:41 (1981), ibid 102B:263 (1981), Weizmann preprint WIS-81/38 (1981);
O.W. Greenberg and J. Sucher, Phys. Lett. 99B:339 (1981);
R. Chanda and P. Roy, Phys. Lett. 99B:453 (1981);
L. Abbott and E. Fahri, Phys. Lett. 101B:69 (1981);
H. Fritzsch and G. Mandelbaum, Phys. Lett. 102B:319 (1981);
R. Barbieri, A. Masiero and R.N. Mohapatra, Phys. Lett. 105B:369 (1981).
H.P. Dürr, to appear in Proceedings of Heisenberg Symposium, Munich (1981).

6. For some interesting ideas on this problem see, however: G.B. Gelmini, Nucl. Phys. B189:241 (1981).
7. S. Adler, Phys. Rev. 177:2426 (1969); J. Bell and R. Jackiw, Nuov. Cim. 60A:47 (1969).
8. W.A. Bardeen and V. Visnjic, FNAL preprint 81/49 (1981).
9. L. O'Raifeartaigh, Nucl. Phys. B96:331 (1975).
10. W.A. Bardeen (unpublished); B. de Wit and D. Freedman, Phys. Rev. Lett. 35:827 (1975).
11. G. 't Hooft, Cargese Lectures (1979); S. Dimopoulos, S. Raby and L. Susskind, Nucl. Phys. B173:208 (1980); A. Zee, Phys. Lett. 95B:290 (1980); G. Farrar, Phys. Lett. 96B:273 (1980); Y. Frishman, A. Schwimmer, T. Banks and S. Yankielowicz, Nucl. Phys. B177:157 (1981); J. Preskill and S. Weinberg, Phys. Rev. D24:1059 (1981); I. Bars, Phys. Lett. 106B:105 (1981); C. Albright, Phys. Rev. D24:1969 (1981).
12. R. Barbieri, L. Maiani and R. Petronzio, Phys. Lett. 96B:63 (1980).
13. R. Barbieri, private communication.
14. T. Banks and V. Kaplunovsky, Tel-Aviv preprint, TAVP 930-81; (1981); T. Banks and A. Zaks, Tel-Aviv preprint, TAVP 944-81 (1981).
15. W. Buchmüller, S.T. Love and R. D. Peccei, MPI-PAE/PTh 70/81, (1981), to appear in Phys. Lett. B.
16. H. Harari and N. Seiberg, Weizmann preprint, WIS-81/38 (1981).
17. S. Coleman, Comm. Math. Phys. 31:259 (1973).
18. E. Witten, Nucl. Phys. B145:110 (1978).
19. G. 't Hooft, Nucl. Phys. B75:461 (1975).
20. Y. Frishman, A. Schwimmer, T. Banks and S. Yankielowicz, see Ref. 11.
21. D. Amati and E. Rabinovici, Phys. Lett. 101B:407 (1981).
22. S. Elitzur, Y. Frishman and E. Rabinovici, Phys. Lett. 106B:403 (1981).
23. V. Baluni, Phys. Lett. 90B:407 (1980).
24. Y. Chikashige, R.N. Mohapatra and R.D. Peccei, Phys. Lett. 98B:265 (1981); G. Gelmini and M. Roncadelli, Phys. Lett. 99B:411 (1981).
25. M. Dine, W. Fischler and M. Srednicki, Phys. Lett. 104B:199 (1981).

26. R. Casalbuoni and R. Gatto, Univ. of Geneva preprint 1981/06-279 (1981).
27. R. Barbieri, A. Masiero and R.N. Mohapatra, Phys. Lett. 105B:369 (1981).
28. H. Harari, R.N. Mohapatra and N. Seiberg, CERN preprint TH-3123 (1981).
29. S. Weinberg, Phys. Rev. Lett. 43:1566 (1979); F. Wilczek and A. Zee, Phys. Rev. Lett. 43:1571 (1979).
30. A. Masiero, R.D. Peccei and T. Yanagida, MPI-PAE/PTh 76/81 (1981).
31. J. Ellis, these Proceedings; Y. Achiman, Wuppertal preprint WUB 81-13 (1981). M. Yasue, INS preprint, INS-Rep.-425 (1981).

INSTANTONS IN SUPERSYMMETRIC QUANTUM MECHANICS

J.W. van Holten

Theoretical Physics Division
CERN
1211 Geneva 23, Switzerland

1. INTRODUCTION

The rôle of classical solutions in supersymmetric field theories is a very intriguing one. A few phenomena in supersymmetric theories which bear witness to this are the following.

- The generation of central charges in the supersymmetry algebra from topological charges connected with solitons and monopoles, as in the two-dimensional Wess-Zumino model and the four-dimensional $N = 4$ supersymmetric Yang-Mills theory[1)].
- The breaking of supersymmetry. Instanton effects can lead to supersymmetry breaking, because instantons may connect fermionic states in one topological sector of the theory to bosonic states in another one[2),3)].
- The quantum corrections to the mass of the particles corresponding to classical extended objects seem to vanish in various cases, indicating the almost unchanged persistence of these objects in the quantum theory. This takes place, e.g., for the solitons in the three-dimensional CP^{N-1} model[4)].
- Finally, a more speculative possibility is that the well-known vanishing of the β function of $N = 4$ Yang-Mills theory results from the theory describing a gas of non-interacting monopoles. Indications of a similar behaviour in the much simpler model of supersymmetric quantum mechanics are discussed below.

In order to study such phenomena related to classical solutions, it is convenient to study first a simple model for which I take supersymmetric quantum mechanics. This is equivalent to a (0+1) dimensional relativistic field theory. Depending on the choice of the potential, such a model may admit instanton solutions, i.e., finite action solutions to the imaginary time field equations.

In the following, I will first present the model, discuss its supersymmetry and the existence of instanton solutions. Subsequently I will exhibit the rôle of instantons in the breaking of supersymmetry. Finally, I will show that it is possible to reformulate the theory in terms of one bosonic field, corresponding to the instantons of the original theory, such that the path integral becomes a pure Gaussian. However, this field satisfies unusual boundary conditions.

2. N = 2 SUPERSYMMETRIC QUANTUM MECHANICS

Supersymmetric quantum mechanics of a particle with one bosonic and one fermionic degree of freedom, q and ψ, is defined by the Lagrangian[2),3)]:

$$\mathcal{L} = \frac{1}{2}\left(\frac{\partial}{\partial t} q\right)^2 - \frac{1}{2} V'(q)^2 + \frac{1}{2}\bar{\psi}\left(i\sigma_2 \frac{\partial}{\partial t} + V''(q)\right)\psi . \tag{2.1}$$

Here V' and V'' are the derivatives with respect to q of a function $V(q)$ called the superpotential. The two real components $\psi_{1,2}$ of the anticommuting fermion variable together form a one-component complex spinor:

$$\psi_\pm = \frac{1}{\sqrt{2}}(\psi_1 \pm i\psi_2) . \tag{2.2}$$

Equivalently, the theory may be described in terms of the Hamiltonian operator:

$$\hat{H} = \frac{1}{2}\hat{p}^2 + \frac{1}{2}V'(\hat{q})^2 - \frac{1}{2}V''(\hat{q})[\hat{\psi}_+, \hat{\psi}_-] \tag{2.3}$$

supplemented by the canonical (anti-) commutation relations:

$$[\hat{p},\hat{q}] = -i, \quad \{\hat{\psi}_+,\hat{\psi}_-\} = 1, \quad \hat{\psi}_\pm^2 = 0 . \tag{2.4}$$

This commutator algebra can, as is well known, be realized by the operator representations

$$\hat{p} = -i\frac{\partial}{\partial q}, \qquad \hat{\psi}_+ = \frac{\partial}{\partial \zeta},$$
$$\text{and} \tag{2.5}$$
$$\hat{q} = q, \qquad \hat{\psi}_- = \zeta,$$

where ζ is an anticommuting c number. A second and sometimes more convenient way to represent the fermionic operators, is by the 2x2 matrix representation

$$\hat{\psi}_+ = \begin{pmatrix} 0 & 1 \\ 0 & 0 \end{pmatrix}, \qquad \hat{\psi}_- = \begin{pmatrix} 0 & 0 \\ 1 & 0 \end{pmatrix} . \tag{2.6}$$

The theory defined here possesses an $N = 2$ supersymmetry generated by two charges $\hat{Q}_\pm$:

$$\hat{Q}_\pm = (\hat{p} \pm i\, V'(\hat{q}))\, \hat{\psi}_\pm, \tag{2.7}$$

which satisfy the algebraic relations

$$\{\hat{Q}_+,\hat{Q}_-\} = 2\hat{H}, \quad [\hat{H},\hat{Q}_\pm] = 0, \quad \hat{Q}_\pm^2 = 0 . \tag{2.8}$$

This is the Poincaré supersymmetry algebra in (0+1) dimensions. It is easy to verify that the $\hat{Q}_\pm$ transform the bosonic and fermionic variables into each other. Besides the supercharges, there is one other conserved quantity in the theory, which is the fermion number

$$\hat{n} = \hat{\psi}_+ \hat{\psi}_-, \quad [\hat{H},\hat{n}] = 0 . \tag{2.9}$$

This operator has eigenvalues 0 and 1, as is easy to see in the matrix representation (2.6). Of course, this is nothing but the expression of the Pauli principle in this model.

3. THE SUPERPOTENTIAL AND INSTANTONS

The dynamics of the discussed theories depends on one specific function of the bosonic variable q, the superpotential $V(q)$. The interesting properties of the model turn out to depend only on the general form of this potential: its number of minima and maxima and its behaviour at infinite q. The two most interesting cases for illustrative purposes are the supersymmetric harmonic oscillator and supersymmetric q^4 theory (Fig. 1). In the first case, the superpotential and its derivatives are:

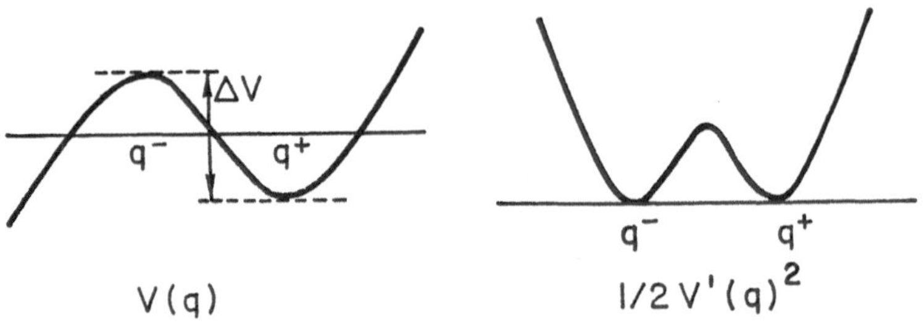

Fig. 1. The potential of q^4 theory

$$V(q) = \tfrac{1}{2}\,\omega\, q^2, \quad \tfrac{1}{2}\, V'(q)^2 = \tfrac{1}{2}\,\omega^2 q^2, \quad V''(q) = \omega \quad . \tag{3.1}$$

There is no coupling between the bosons and the fermions, supersymmetry is unbroken, and the spectrum consists of two disjoint sets of harmonic oscillator states, distinguished by their fermionic quantum number (0,1) and a relative shift ω only.

In the second case the superpotential is cubic:

$$V(q) = \frac{\lambda}{3}\, q^3 - \mu^2 q, \quad \frac{1}{2}\, V'(q)^2 = \frac{\lambda^2}{2}\left(q^2 - \frac{\mu^2}{\lambda}\right)^2, \quad V''(q) = 2\lambda q \ . \tag{3.2}$$

The bosonic potential is of the familiar double-well type, with two classical minima at $q_\pm = \pm(\mu/\sqrt{\lambda})$, while there is a Yukawa-type coupling of q to the fermions. As is well known, in theories of this type the spectrum of states is decisively influenced by the possibility of tunnelling of the particle between the minima of the potential. In the classical theory such a behaviour is signalled by the existence of instanton solutions to the imaginary time field equations[5),6)]. Integrating the field equations once, and looking for zero-fermion solutions, we find that the instantons satisfy

$$\begin{cases} \dfrac{\partial}{\partial\tau}\, q_c \pm V'(q_c) = 0, & \tau = it = \mp \displaystyle\int^{q(t)} \frac{dq}{V'(q)} \\ \psi_c = 0 \ . & \end{cases} \tag{3.3}$$

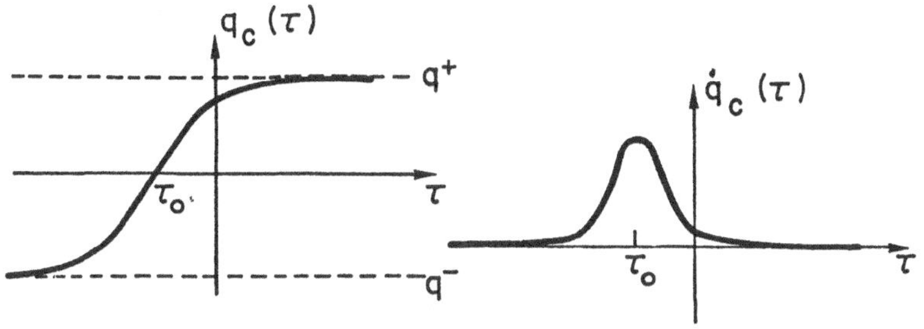

Fig. 2. Instanton solution of q^4 theory

As is clear from Fig. 2, this solution interpolates between the two classical minima of the bosonic potential, as expected.

4. SUPERSYMMETRY BREAKING

Before continuing, let me briefly discuss the connection between the matrix representation and the anticommuting c number representation of the fermions. It is given by the following:

- in the matrix representation, the wave function has two components:

$$\Psi(q) = \begin{pmatrix} \psi_1(q) \\ \psi_2(q) \end{pmatrix} .$$

  The upper- and lower-component describe one- and zero-fermion states, respectively;

- in the anticommuting c number representation, the wave function depends on the two variables (q,ζ); expanding in terms of the anticommuting variable ζ gives:

$$\Psi(q,\zeta) = a_1(q) + \zeta a_2(q) .$$

  In this representation it is a_1 (a_2) which corresponds to the one- (zero-) fermion component, and hence we may identify it with ψ_1 (ψ_2) above.

After having established this simple connection, I next write the Hamiltonian in matrix notation as:

$$\hat{H} = -\frac{1}{2}\frac{\partial}{\partial q^2}\,\mathbb{1} + U(q) \qquad (4.1)$$

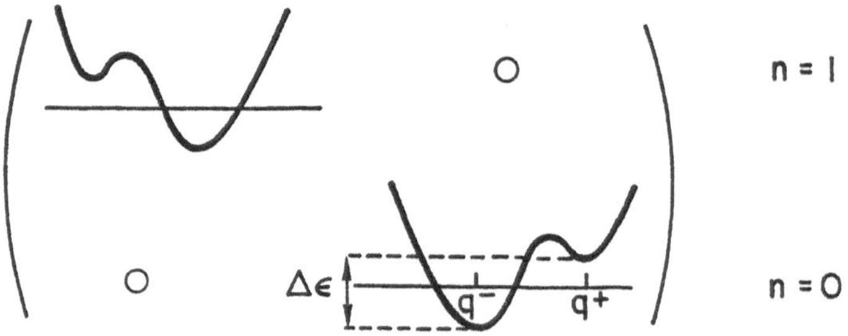

Fig. 3. Graphical representation of the potential U(q)

where the potential U(q) has the matrix form:

$$U(q) = \frac{1}{2} V'(q)^2 \mathbb{1} + V''(q)\, \sigma_3 \tag{4.2}$$

as indicated in Fig. 3. Note that the Hamiltonian is diagonal with respect to fermion number. The effect of the fermions is to lift the degeneracy between the two wells of the bosonic potential. The split in energy between the wells is $\Delta\varepsilon = V''(q_+)$. However, except for a reflection, no difference exists between the one- and zero-fermion potentials, hence the energy spectrum of the $n = 0,1$ states will be degenerate. Therefore one might expect that it would be possible to change the fermion number in the ground state of the system without adding any energy to it. This would correspond to the existence of a zero-mass fermion in a field theory.

There is only one objection to this: the ground states of the $n = 0,1$ spectra are localized in different wells of the potential. In field theory language: they correspond to different vacuum-expectation values of the field q. Hence, as long as the particle is in the lowest energy state, fermion number cannot be changed, unless the particle simultaneously tunnels to the other well of the potential. It is exactly at this point that the instantons become crucial: they mediate the tunnelling, and hence assist in the possibility of creating zero-mass fermions. At the same time they make the transformation of one vacuum state into the other possible one, whence the vacua are no longer invariant under supersymmetry. As a result, supersymmetry is broken, and the zero-mass fermions actually become the Goldstone fermions corresponding to this broken supersymmetry.

As is well known, in a broken supersymmetry phase the ground state energy of the system must be different from zero. Thus for the consistency of our explanation, the instantons must make a finite contribution to the ground state energy. On the other hand, this additional energy may not spoil the picture of states localized in the lowest well of the potential, which results from the energy splitting between the two originally degenerate minima. This instanton contribution to the energy has been calculated to be[3]:

$$\varepsilon_o = \frac{V''(q_+)}{\pi} e^{-2\Delta V} << \Delta\varepsilon \quad . \tag{4.3}$$

This fulfils all the requirements.

5. REFORMULATION OF THE THEORY

After this discussion of the rôle of instantons in supersymmetry breaking, I turn now to an entirely different aspect of the supersymmetric models under consideration. The object is to rewrite the path integral of these theories in a very simple way by integrating over the fermionic variables, and then making a transformation of the bosonic variables such that its Jacobian cancels the fermion determinant.

Take the path integral

$$K(q\zeta t|q'\zeta't') = \int_{(q',\zeta')}^{(q,\zeta)} DqD\psi \; e^{i\int_{t'}^{t} \mathcal{L}(q,\psi)dt} \tag{5.1}$$

where the functional integral extends over all trajectories between the points (q',ζ'), (q,ζ) of superspace. It was shown in Ref. 3) that the fermionic integral can be completely performed, leading to the result (in imaginary time):

$$K_E(q\zeta\tau|q'\zeta'\tau') = \int_{q'}^{q} Dq \; \{\zeta \; e^{\int \mathcal{L}_E^{(-)}d\tau} - \zeta' \; e^{\int \mathcal{L}_E^{(+)}d\tau}\} \tag{5.2}$$

where $\mathcal{L}_E^{(\pm)}$ is the zero/one-fermion part of the Euclidean Lagrangian:

$$\mathcal{L}_E^{(\pm)} = -\tfrac{1}{2}(\frac{\partial}{\partial\tau} q)^2 - \tfrac{1}{2}V'(q)^2 \pm \tfrac{1}{2}V''(q) \quad . \tag{5.3}$$

Now we define the new variable

$$x^{\pm}(\tau) = q(\tau) \mp \int_{\tau}^{\tau_o} V'(q)d\tau \tag{5.4}$$

where τ_o is a fixed endpoint of integration. The Jacobian of this transformation may be found by discretizing time, calculating the resulting finite dimensional determinant, and only then taking the continuum limit. This way, the following result is established:

$$\det\left(\frac{\partial X^{\pm}}{\partial q}\right) = e^{\pm\int_{\tau}^{\tau_o} V''(q)d\tau} \quad . \tag{5.5}$$

Moreover, the action in terms of this new variable, calculated from (5.3), is simply:

$$S_E^{(\pm)} = \pm\Delta V - \tfrac{1}{2}\int_{\tau}^{\tau_o}\left(\frac{\partial}{\partial\tau}X^{\pm}\right)^2 d\tau \tag{5.6}$$

whence we find for the path integral:

$$\begin{aligned} K_E(q\zeta\tau|q'\zeta'\tau') &= e^{-\Delta V_{\zeta}}\int DX^- \, e^{-\frac{1}{2}\int(\dot{X}^-)^2 d\tau} \\ &\quad - e^{\Delta V_{\zeta'}}\int DX^+ \, e^{-\frac{1}{2}\int(\dot{X}^+)^2 d\tau} \end{aligned} \tag{5.7}$$

The boundary conditions on the field $X^{(\pm)}$ are given by:

$$\begin{aligned} X^{\pm}(\tau_o) &= q(\tau_o) \\ X^{\pm}(\tau) &= q(\tau) \mp \int_{\tau}^{\tau_o} V'(q)\, d\tau \quad . \end{aligned} \tag{5.8}$$

The result (5.7) is deceptively simple: it looks like a free field theory. That it is not follows from the boundary conditions (5.8). Namely, the transformation (5.4) does not map points of q space into points of X space. Rather it maps <u>paths</u> in q space into points in X space, and is thus very non-<u>local</u>. As to the nature of these paths we observe that for a constant value of $X^{\pm}$ we have

$$\frac{\partial}{\partial\tau}X^{\pm} = 0 \quad \leftrightarrow \quad \frac{\partial}{\partial\tau}q \mp V'(q) = 0 \quad . \tag{5.9}$$

Hence constant values of $X^{\pm}$ correspond to the instanton paths in q space!

This remarkable result leads us to the following conclusion: if we interpret the path integral (5.7) with the boundary conditions that $X^{\pm}$ is fixed at τ, τ_0, then the theory is a theory of free instantons and anti-instantons. These boundary conditions translated into q space mean that one has to specify the instanton on which the particle moves at τ, τ_0. However, the actual boundary conditions from which we started fixed the values of q at τ, τ_0, not those of $X^{\pm}$.

At the same time, and equivalently, the wave function $\Psi(q,\zeta,\tau)$ is transformed into a non-local functional $\Psi(X(\tau))$ of the path $X^{\pm}(\tau)$. There does not exist a unitary transformation from the "q representation" to the "X representation" of the wave functions.

What is the importance of this result to other supersymmetric theories? It seems there is good reason to believe that it is of more general validity; namely, it was shown in Ref. 7) that, for any rigidly supersymmetric theory, one can find a transformation like (5.4) order by order in perturbation theory, such that the path integral becomes Gaussian. It would be interesting to know whether such a transformation can always be linked to classical solutions.

In this context, it can be remarked that there are strong indications for our result to be valid for the solitons in the (1+1) dimensional complex Wess-Zumino model. As was mentioned in the Introduction, one might attempt to relate the corresponding result for the monopoles in $N = 4$ Yang-Mills theory to the other properties of that model. However, at present, this remains mere speculation.

The work reported here was carried out in collaboration with Per Salomonson at CERN.

REFERENCES

1. D. Olive and E. Witten, Phys. Lett. 78B:97 (1978).
2. E. Witten, Nucl. Phys. B188:513 (1981).
3. P. Salomonson and J.W. van Holten, CERN preprint TH.3148 (1981).
4. S. Rouhani, Nucl. Phys. B182:462 (1981).
5. A.M. Polyakov, Nucl. Phys. B120:429 (1977).
6. G. 't Hooft, Phys. Rev. D14:343 (1976).
7. H. Nicolai, Phys. Lett. 89B:341 (1980); Nucl. Phys. B176:419 (1980).

GALILEAN APPROXIMATION OF MASSLESS SUPERSYMMETRIC THEORIES

S. Ferrara
CERN, Geneva, Switzerland

and

F. Palumbo
CERN, Geneva, Switzerland and
Laboratori Nazionali di Frascati dell'INFN, Italy

There is no systematic way of studying dynamical symmetry breaking. In supersymmetry in particular, while it has been shown that fermion condensation can provide dynamical breaking[1], no model has been shown to give rise to fermion condensation.

Recently arguments have been given that Galilean approximations to relativistic theories can provide reliable information[2] on spontaneous symmetry breaking. These arguments are supported by the analysis of the Goldstone model[3], the Higgs model[4] and the Schwinger mechanism[5], whose essential features are conserved[2] in the Galilean approximation.

We have therefore studied the Galilean limit in supersymmetry. Here the situation is somewhat different from that of the above-mentioned models. The key point in those cases was to obtain the Galilean approximation by contraction (velocity of light $c \to \infty$) of the Poincaré group, while conserving all the other symmetries of the Lagrangian (charge symmetry, chiral invariance, gauge invariance,...). But now the Poincaré algebra is a subalgebra of the supersymmetry algebra, and its contraction implies the contraction of the whole supersymmetry algebra. This contraction is presented here in the simplest case, in which all the particles are massless. In such a case the relativistic theory undergoes a dimensional reduction to 0 + 1 dimensions[2], i.e., to quantum mechanics. We have supersymmetric quantum mechanics with chiral invariance for the pure Wess-Zumino model, and with non-Abelian local gauge invariance for Yang-Mills theories. An additional internal O(3) invariance arises from dimensional reduction.

The 0 + 1 dimensional field theory that we have obtained is not trivial, but it can obviously be solved, possibly by numerical methods. We will confine ourselves here to showing that spontaneous mass generation in the Wess-Zumino model[6] and gauge or supersymmetry breaking in the Fayet model[7] are reproduced in the limit. This adds confidence to the possibility of obtaining reliable information on symmetry breaking by this method.

Finally it should be mentioned that a quantum mechanical model of supersymmetry, exhibiting dynamical breaking, has been constructed by Witten[8] and further investigated by Salomonson and Van Holten[9]. As we will see, however, this model cannot be obtained as the limit of a relativistic model, and therefore its mechanism need not be relevant to relativistic supersymmetry.

THE LIMIT FOR THE WESS-ZUMINO MODEL

The Wess-Zumino Lagrangian reads

$$\begin{aligned}\mathcal{L} = & -|\partial_\mu\phi|^2 - ic\bar{\psi}_L \not{\partial}\psi_L + |H|^2 + \frac{1}{\sqrt{2}}(H+H^*)X + \\ & + \frac{g}{c}(\phi^2 H + \phi^{*2}H^*) - ig(\phi\bar{\psi}_L\psi_R + \phi^*\bar{\psi}_R\psi_L),\end{aligned} \tag{1}$$

where

$$\begin{aligned}\psi_L &= \frac{1}{2}(1 + i\gamma_5)\psi \\ \psi_R &= \frac{1}{2}(1 - i\gamma_5)\psi\end{aligned} \tag{2}$$

and ψ is a Majorana spinor.

This Lagrangian is invariant under the supertransformations

$$\begin{aligned}\delta\phi &= \sqrt{2}\, i\, \bar{\varepsilon}\, \psi_R \\ \delta\psi_R &= \frac{1}{c}\not{\partial}\phi\frac{1}{\sqrt{2}}(1+i\gamma_5)\varepsilon + \frac{1}{c}H\frac{1}{\sqrt{2}}(1-i\gamma_5)\varepsilon \\ \delta H &= \sqrt{2}\, i\, \bar{\varepsilon}\not{\partial}\psi_R\end{aligned} \tag{3}$$

and, for X = 0, under the chiral transformations

$$\begin{aligned}\delta\phi &= -i\alpha\phi \\ \delta\psi_R &= -i\,\frac{\alpha}{2}\,\psi_R \\ \delta H &= 2i\alpha H\end{aligned} \tag{4}$$

If we perform the limit $c \to \infty$ leaving the fields finite we break supersymmetry. This is conserved if we rescale ϕ according to

$$\phi = c\varphi \tag{5}$$

This yields

$$\begin{aligned}\mathcal{L} \underset{c\to\infty}{\to} \mathcal{L}^{(o)} &= |\partial_t\varphi|^2 + \psi_L^* i\partial_t\psi_L + |H|^2 + \frac{1}{\sqrt{2}}(H+H^*)X + \\ &+ gc(\varphi^2 H + \varphi^{*2}H^*) - igc(\varphi\bar{\psi}_L\psi_R + \varphi^*\bar{\psi}_R\psi_L)\end{aligned} \tag{6}$$

with the constraints

$$\partial_K\varphi = \partial_K\psi_L = 0 \tag{7}$$

These constraints are Galilei-invariant and under these constraints so is $\mathcal{L}^{(o)}$. It is convenient to introduce Pauli spinors by the definition

$$\begin{aligned}\psi_L &= \frac{1}{\sqrt{2}}\begin{bmatrix} \chi \\ -\chi \end{bmatrix} \\ \psi_R &= \frac{1}{\sqrt{2}}\begin{bmatrix} \sigma_2\chi^* \\ \sigma_2\chi^* \end{bmatrix}\end{aligned} \tag{8}$$

In terms of χ $\mathcal{L}^{(o)}$ becomes

$$\begin{aligned}\mathcal{L}^{(o)} &= |\partial_t\varphi|^2 + \chi^* i\partial_t\chi + |H|^2 + \frac{1}{\sqrt{2}}(H+H^*)X + \\ &+ gc(\varphi^2 H + \varphi^{*2}H^*) + gc(\varphi\chi\sigma_2\chi + \varphi^*\chi^*\sigma_2\chi^*)\end{aligned} \tag{9}$$

We can now perform the limit on the supertransformations (3). Expressing the Majorana parameter ε by the Pauli spinor η according to

$$\varepsilon = \begin{bmatrix} \eta \\ \sigma_2\eta^* \end{bmatrix} \tag{10}$$

and using Eqs. (5), (7) and (8) we obtain

$$\begin{aligned} c\delta\varphi &= \eta\chi^* - \eta^*\sigma_2\chi^* \\ c\delta\chi &= i\partial_t\varphi(\eta - \sigma_2\eta^*) + H(\eta + \sigma_2\eta^*) \\ c\delta H &= -i\partial_t(\eta^*\sigma_2\chi^* + \eta\chi^*)\end{aligned} \tag{11}$$

It is easy to check that these transformations close, giving time translations on the fields, and that they leave $\mathcal{L}^{(o)}$ invariant.

The limit on the chiral transformations is obvious and gives

$$\delta\varphi = -i\alpha\varphi$$
$$\delta\chi = i\frac{\alpha}{2}\chi \tag{12}$$
$$\delta H = 2i\alpha H$$

The form of the limiting supertransformations (11) shows that supersymmetry breaking can be discussed in terms of the auxiliary fields or of fermion bilinears as in the relativistic case. As in the relativistic case it can also be shown that for X ≠ 0 the chiral multiplet acquires a mass. This can be seen in the following way. Since the fields in the $c \to \infty$ limit are space independent they do not carry momentum, so their energy (being at zero momentum) should be interpreted as a mass. Now, while for X = 0 the excitation energy starts from zero, for X ≠ 0 it acquires a gap equal for bosons and fermions.

The above can be more clearly seen by quantizing $\mathcal{L}^{(o)}$. The canonical procedure gives rise to the usual ambiguities due to the possibility of different ordering of non-commuting operators. These ambiguities can be avoided if the Hamiltonian is given as the anticommutator of the spinorial generators, whose relativistic expression is

$$J^o_\alpha = -\ell^{\frac{3}{2}}\{\partial_0\phi\psi_L + \partial_0\phi^*\psi_R - \gamma^0(H\psi_L + H^*\psi_R)\} \tag{13}$$

where ℓ is an arbitrary parameter with the dimension of a length that we introduce in order that J^o_α has the dimension of (energy density)$^{1/2}$.

For $c \to \infty$

$$J^o_\alpha \to Q_\alpha = -\frac{\ell^{\frac{3}{2}}}{\sqrt{2}}\left[\chi\partial_t\varphi + \sigma_2\chi^*\partial_t\varphi^* - iH\chi - iH^*\sigma_2\chi^*\right] , \tag{14}$$

and eliminating the auxiliary fields and introducing the canonical momenta

$$\pi = \partial_t\varphi^*$$
$$\pi^* = \partial_t\varphi , \tag{15}$$

$$Q_\alpha = - \frac{\ell^{\frac{3}{2}}}{\sqrt{2}} \left[\pi^* \chi + \pi\sigma_2 \chi^* + i\left(\frac{X}{\sqrt{2}} + gc\varphi^{*2}\right)\chi + \right.$$
$$\left. + i\left(\frac{X}{\sqrt{2}} + gc\varphi^2\right)\sigma_2 \chi^*\right] \tag{16}$$

It is now convenient to introduce dimensionless scalar and spinor fields φ_i and ξ

$$\varphi = \sqrt{\frac{Z}{\ell^3 \Lambda}} \frac{1}{\sqrt{2}}(\varphi_1 - i\varphi_2)$$
$$\chi = \frac{1}{\ell^{\frac{3}{2}}} \frac{1}{\sqrt{2}}(\xi + \sigma_2 \xi^*) \tag{17}$$

which satisfy canonical commutation relations

$$[\varphi_i, \pi_j] = i\delta_{ij} \; ; \; \{\xi^*, \xi\} = 1 \tag{18}$$

In Eqs. (17) Z is a dimensionless parameter while Λ has the dimension of an energy.

The Hamiltonian density is

$$\mathcal{H} = \{Q^*_\alpha, Q_\alpha\} = \frac{1}{\ell^3} \frac{\Lambda}{Z} \{\frac{1}{2}\pi_1^2 + \frac{1}{2}\pi_2^2 + \frac{1}{2}g_r^2 \left[(\varphi_1^2 - \varphi_2^2 + X_0)^2 + \right.$$
$$\left. + 4\varphi_1^2\varphi_2^2\right] + 2g_r\varphi_1 - 2g_r\varphi_1\xi^*\xi + g_r\varphi_2 i(\xi^*\sigma_2\xi^* - \xi\sigma_2\xi)\} \tag{19}$$

where

$$g_r = g_0 Z^{\frac{3}{2}} \; , \; g_0 = \frac{g}{\sqrt{c}} \tag{20}$$

$$X_0 = \frac{\ell^{\frac{3}{2}}}{\sqrt{\Lambda}} \frac{X}{g_r} \tag{21}$$

By performing the shift

$$\varphi_1 = \varphi_0 + \varphi_1^1 \tag{22}$$

with

$$\varphi_0^2 = -X_0 \tag{23}$$

(note that X_0 can always be taken negative)

we get

$$\mathcal{H} = \frac{1}{\ell^3}\frac{\Lambda}{Z}\{\frac{1}{2}\pi_1^2 + \frac{1}{2}\pi_2^2 + \frac{1}{2}m\varphi_1^{1\,2} + \frac{1}{2}m\varphi_2^2 + m\dot{\xi}^*\dot{\xi} +$$

$$+ \frac{1}{2}g_r(\varphi_1^{1\,2} + \varphi_2^2)^2 + 2g_r^2\sqrt{-X_0}\,\varphi_1^1(\varphi_1^{1\,2} + \varphi_2^2) + \tag{24}$$

$$+ 2g_r\varphi_0 + 2g_r\varphi_1^1 - 2g_r\varphi_1^1\dot{\xi}^*\dot{\xi} + 2g_r\varphi_2 i(\xi^*\sigma_2\dot{\xi}^* - \dot{\xi}\sigma_2\dot{\xi})\}$$

with

$$m = 2g_r\sqrt{-X_0} \tag{25}$$

We see that for $X_0 \neq 0$ bosons and fermions have an energy spectrum with a gap m. The chiral multiplet has become massive.

We finally comment on the Witten model[8)], whose Hamiltonian is

$$H = \frac{1}{2}\left(p^2 + W^2(x) + \sigma_3\frac{dW(x)}{dx}\right)$$

where σ_3 is the Pauli matrix and W(x) an arbitrary function.

Witten has shown that if W(x) has an even number of zeros dynamical symmetry breaking occurs.

In order to compare the above Hamiltonian with the Hamiltonian (19) let us observe that the two spin states can be put into one-to-one correspondence with states of 0 or two fermions. Witten's Hamiltonian however contains a single boson field rather than two, and requires a particular relation between the quartic bosonic term and the Yukawa coupling which is not realized in the Hamiltonian (19).

THE LIMIT FOR YANG-MILLS THEORIES

It is most convenient to use the formulation of supersymmetric Yang-Mills theories in the Wess-Zumino gauge[10]. In component fields the gauge degrees of freedom of the vector multiplet are eliminated from the start.

We will not report on the relativistic formulae which can be found for instance in the work by de Wit and Freedman[11], but we will give directly their Galilean limit. For the Lagrangian we have

$$\mathcal{L}_G^{(o)} = \frac{1}{2}(\mathcal{D}_t A_k)^a(\mathcal{D}_t A_k)^a + \lambda^* i\,\mathcal{D}_t\lambda + \frac{1}{2}D^a D^a +$$
$$- \frac{1}{4}(gc)^2(f^a_{bc}A^b_i A^c_i)^2 - \frac{i}{2}gcf^a_{bc}A^b_k(\lambda^{a*}\sigma_k\sigma_2\lambda^{c*} + \tag{26}$$
$$+ \lambda^a\sigma_2\sigma_k\lambda^c)$$

Here A^a_k are the magnetic potentials and λ^a is a Pauli spinor. The covariant derivative is

$$\mathcal{D}^a_{tb} = \delta^a_b\partial_t - gcf^a_{bc}V^c \tag{27}$$

with V^c the electric potentials. All the fields V^a, A^a_k, λ^a and D^a are of course in the adjoint representation of the internal symmetry group. The spatial index k has become the internal index of an additional O(3) invariance.

We must mention, however, one ambiguity in the derivation of Eq. (26). This is related to the possibility of choosing that the product gc should remain finite for $c \to \infty$, or that g should be finite. This second alternative is the correct one in the coupling of Yang-Mills fields with massive matter fields in non-supersymmetric theories where the fields are x dependent, while the first one appears to be "natural" in the present case, although the possibility of keeping an x dependence in this case should also be investigated.

The limit for the Lagrangian for the matter fields gives

$$\mathcal{L}_M^{(o)} = (\mathcal{D}_t\varphi)^+(\mathcal{D}_t\varphi) + \chi^* i\,\mathcal{D}_t\chi + H^{a*}H^a +$$
$$+ (gc)^2(L_b A^b_k\varphi)^+(L_a A^a_k\varphi) + gcA^b_k\chi^*\sigma_k L_b\chi +$$
$$+ gc\varphi^+ L_b\,\varphi D^b - igc(\lambda^{a*}\varphi^+ L_a\chi + \lambda^a\varphi^+\sigma_2 L_a\chi - \tag{28}$$
$$- \chi^* L_a\varphi\sigma_2\lambda^{a*} - \chi^* L_a\varphi\lambda^a)$$

In the above equation L_a are the generators of the internal symmetry group in the representation of the matter fields φ, χ and H, so that the covariant derivative is

$$\mathcal{D}_t = \partial_t + igcL_a V^a$$

The supertransformations are

$$\begin{aligned} c\delta V^a &= -i(\eta^* \lambda^a + \eta \lambda^{a*}) \\ c\delta A_k^a &= -i(\eta^* \sigma_k \sigma_2 \lambda^{a*} + \eta \sigma_2 \sigma_k \lambda^a) \\ c\delta\lambda^a &= (\mathcal{D}_t A_k)^a \sigma_k \sigma_2 \eta^* + gc\, f^a_{bc} A_i^b A_j^c \sigma_i \sigma_j \eta + iD^a \sigma_2 \eta^* \\ c\delta D^a &= -[\eta^* \sigma_2 (\mathcal{D}_t \lambda^*)^a + \eta \sigma_2 (\mathcal{D}_t \lambda)^a] - \\ &\quad - gc\, f^a_{bc} A_k^b (\eta^* \sigma_k \lambda^c + \lambda^{c*} \sigma_k \varepsilon) \end{aligned} \tag{30}$$

$$\begin{aligned} c\delta\varphi &= -\eta^* \chi - \eta \sigma_2 \chi \\ c\delta\chi &= i\mathcal{D}_t \varphi(\eta + \sigma_2 \eta^*) + H(\eta - \sigma_2 \eta^*) + igc\, L_b A_k^b \sigma_k (\eta + \sigma_2 \eta^*) \\ c\delta H &= -i(\eta^* - \eta\sigma_2) \mathcal{D}_t \chi + igc\, L_b A_k^b (\eta^* - \eta \sigma_2) \sigma_k \chi + \\ &\quad igc\, L_b \varphi (\eta^* \lambda^b - \eta \lambda^{b*} + \eta^* \sigma_2 \lambda^{b*} - \eta \sigma_2 \lambda^b) \end{aligned} \tag{31}$$

The Lagrangians $\mathcal{L}_G^{(o)}$ and $\mathcal{L}_M^{(o)}$ are invariant under the above supertransformations and under the gauge transformations

$$\begin{aligned} \delta_G(\theta)\lambda^a &= f^a_{bc}\, \lambda^b \theta^c \\ \delta_G(\theta) D^a &= f^a_{bc}\, D^b \theta^c \\ \delta_G(\theta) V^a &= -\frac{1}{gc} (\mathcal{D}_t \theta)^a \\ \delta_G(\theta) A_i^a &= f^a_{bc}\, A_i^b \theta^c \end{aligned} \tag{32}$$

and analogous for the matter fields.

The commutator of two supersymmetry transformations gives rise to a time translation of parameter $\xi_0 = -(\varepsilon_2^* \varepsilon_1 + \varepsilon_2 \varepsilon_1)$ plus a gauge transformation of parameter $\xi_k A_k = -(\varepsilon_2^* \sigma_k \sigma_2 \varepsilon_1^* + \varepsilon_2 \sigma_2 \sigma_k \varepsilon_1) A_k$. We give, as an illustration, the following equations

$$\begin{aligned} [c\delta_1, c\delta_2] V^a &= -2i \delta_G(\xi_k A_k) V^a \\ [c\delta_1, c\delta_2] \lambda^a &= -2i\xi_0 (\mathcal{D}_t \lambda)^a - 2i \frac{1}{gc}\, \delta_G(\xi_k A_k) \lambda^a \\ [c\delta_1, c\delta_2] \varphi &= -2i\xi_0 \mathcal{D}_t \varphi - 2i \frac{1}{gc}\, \delta_G(\xi_k A_k) \varphi \end{aligned} \tag{33}$$

Since the structure of the supertransformations and of the Lagrangian is the same as that of the relativistic models, Fayet's model[7] conserves in the limit its feature concerning the breaking of the gauge symmetry or of supersymmetry. The quantization of the gauge Lagrangian models can be performed by evaluating the anticommutator of two spinorial generators as in the case of the Wess-Zumino model. It happens however a new feature related to new constraints coming from gauge invariance. This requires a detailed discussion which will not be given here.

REFERENCES

1. S. Dimopoulos and S. Raby, Nucl. Phys. B192:353 (1981); M. Dine, W. Fischler and M. Srednicki, Inst. Adv. Study, Princeton preprint (1981).
2. F. Palumbo, Nucl. Phys. B197:334 (1982).
3. J. Goldstone, Nuovo Cimento 19:154 (1961).
4. P.W. Higgs, Phys. Lett. 12:132 (1964); Phys. Rev. Lett. 13:508 (1964); Phys. Rev. 145:1156 (1966); F. Englert and R. Brout, Phys. Rev. Lett. 13:321 (1964); G.S. Guralnik, C.R. Hagen and T.W.B. Kibble, Phys. Rev. Lett. 13:585 (1964); T.W.B. Kibble, Phys. Rev. 155:1554 (1967)
5. J. Schwinger, Phys. Rev. 125:397 (1962); ibid. 128:2425 (1962).
6. J. Wess and B. Zumino, Phys. Lett. 49B:52 (1974).
7. P. Fayet, Nuovo Cimento 31A:626 (1976).
8. E. Witten, Nucl. Phys. B 188:513 (1981).
9. P. Salomonson and J.W. van Holten, CERN preprint TH-3148 (1981).
10. J. Wess and B. Zumino, Nucl. Phys. B78:1 (1974); S. Ferrara and B. Zumino, Nucl. Phys. B79:413 (1974).
11. B. De Wit and D.Z. Freedman, Phys. Rev. D12:2286 (1975)

INDEX